"In this tour de force, LeDoux artfully guides the reader from the unconscious defensive system, through attention and memory, to the conscious experience of fear and anxiety. His traverse from the unconscious to the conscious experience of emotion is rich in scientific detail and yet exquisitely readable. LeDoux completes his masterpiece with provocative discussions of therapies for anxiety. This book is a fascinating revelation of the evolution in LeDoux's own scientific thinking and in the field at large and is a must read for any student of learning, memory or emotion."
—**Michelle G. Craske, Ph.D., Professor of Psychology and Director, Anxiety Disorders Research Center, UCLA**

"*Anxious* is a profound, exciting and immensely useful work about one of our most troubling—and puzzling—emotions. Joseph LeDoux takes us behind the scenes of our own minds to show us not only how anxiety is constructed in the brain but how it can be deconstructed. This is neuroscience at its very best: helpful and hopeful without a hint of hyperbole."
—**Mark Epstein, M.D., author of *Thoughts without a Thinker* and *The Trauma of Everyday Life***

"This marvellous book is science at its best ... an absolute must read for clinicians and basic scientists as well as for anyone else interested in anxiety and its disorders."
—**Eric R. Kandel, Senior Investigator, Howard Hughes Medical Institute; author of *In Search of Memory* and *The Age of Insight*; recipient of the 2000 Nobel Prize in Physiology or Medicine**

"An exquisite and unique attempt to truly relate how neural cells lead to felt conscious states in the human mind—the toughest problem in all of science. LeDoux has thrown down the gauntlet and set the standard."
—**Michael Gazzaniga, Professor of Psychology, University of California, and author of *Tales from Both Sides of the Brain* and *Who's in Charge? Free Will and the Science of the Brain***

"LeDoux is a true leader in the field of cutting-edge neuroscience and psychology, yet he also has an uncanny ability to write beautifully and clearly ... A must-read for anyone interested in the intersection of the mind and brain, and how an understanding of psychology and neuroscience can change ourselves and the world around us!"
—**Kerry J. Ressler, M.D., Ph.D., Professor of Psychiatry and Behavioral Sciences, Emory University; Scientific Council Chair, Anxiety and Depression Association of America**

ALSO BY JOSEPH LEDOUX

The Emotional Brain

Synaptic Self

ANXIOUS

THE MODERN MIND
IN THE AGE OF ANXIETY

Joseph LeDoux

ONEWORLD

A Oneworld Book

First published in Great Britain and Australia by Oneworld Publications, 2015

Copyright © Joseph LeDoux 2015

ISBN 978-1-78074-767-5
eISBN 978-1-78074-768-2

Printed and bound in Sweden by ScandBook AB

Oneworld Publications
10 Bloomsbury Street
London WC1B 3SR
England

Stay up to date with the latest books,
special offers, and exclusive content from
Oneworld with our monthly newsletter

Sign up on our website
www.oneworld-publications.com

*To all the researchers who have worked
with me over the years and to all my colleagues
who have helped move our understanding
of fear and anxiety forward*

PREFACE

When I completed my previous book, *Synaptic Self*, which was published in 2002, I wasn't certain I wanted to write another book for a general audience. I had gotten the idea that the way to really have an impact on the field was to write a textbook in my particular area, behavioral and cognitive neuroscience. My agents, John Brockman and Katinka Matson, urged me not to, as did my editor at Viking, Rick Kot, each of them warning me that I would regret it as a publishing experience. After struggling with the project for almost a decade, I had to admit that they were correct. I discovered that the textbook format was far too restrictive—it had to be fresh and innovative . . . so long as it was just like every other competing book. After each chapter was reviewed by a number of teachers from a mix of universities, colleges, and junior colleges around the country, I began to feel little connection to the edited text that was resulting and concluded that my role was more to be a name on the cover than to actually drive the content.

A few years ago I ran into Rick at a reading by our friend Rosanne Cash, who wrote *Composed* under his editorship, and he asked with a wry smile, "How's that textbook going? I've been waiting for you to bail out of that and do another book with me." I was thrilled that he was still interested in working with me, and I negotiated, with some help from Eric Rayman, my way out of the textbook and prepared a new proposal for Katinka. *Anxious* was the result. Rick loved the idea, and so here we are.

Anxious is different from my other books. While *The Emotional Brain* and *Synaptic Self* can be thought of as a series of connected essays that hang together around a single theme, in *Anxious* each successive chapter builds on the previous ones to argue for a new view of emotion, especially the emotions fear and anxiety. Although the book is called *Anxious*, fear and anxiety are complexly entwined, and must be understood both separately and together.

As an overview, here are the key points that *Anxious* addresses. First, the science of emotion, and especially the science of fear and anxiety, now finds itself at an impasse, dictated by the way we discuss emotions in relation to the brain. For example, researchers use words like "fear" to describe the brain mechanisms that cause rats to freeze when in danger, and also to describe the conscious feeling that humans experience if they think that they will be seriously harmed physically or psychologically. The general idea is that a fear circuit in the brain is responsible for the feeling of fear, and when it is activated, whether in a rat or a human, the feeling of fear occurs, along with responses characteristic of fear (such as freezing, facial expressions, changes in body physiology). The feeling of fear is often said to mediate between the threatening event and those responses. Because these circuits are conserved throughout mammals, including humans, we can study human fear by measuring freezing in rats. The key circuits crucially involve the amygdala, which is generally described as the seat of fear in the brain.

In fact, most of what I have just described is wrong. Because my work and writings are in part responsible for these misconceptions, I feel some responsibility to try to straighten out the story before it goes further off track. One of the main goals of this book is to provide a new view of fear and anxiety, one that more accurately distinguishes what we can learn from animals from what we can best learn from humans, and what fear itself really refers to in the context of the human brain.

Don't get me wrong: I am not arguing that we have to study brain mechanisms related to emotions exclusively in humans. There is much we have learned, and can continue to learn, and can, in fact, only learn, from animal research. But we do need a rigorous conceptual framework for understanding what the animal work does and does not mean for understanding the human brain. I offer my view of such a framework, which I think provides a new perspective on fear and anxiety, and the disorders related to these states.

The suggestions I make in this book concern in part the words we use to describe certain phenomena, but my argument is not simply about semantics. Words have extended meanings that imply a great deal. For example, some researchers who study fear in rats by measuring freezing behavior say that they are not studying what most people think of as fear, but rather some nonsubjective physiological state that they *call* fear. While this scientific redefinition of fear makes it more tractable as a research problem, it has three disadvantages. First, using fear in a nonconventional way to describe a physiological state that connects threats with responses often leads researchers to write and talk about this state as if it was referring to the conscious feeling of fear. Second, even when the researchers adhere to that definition, everyone thinks that they are actually studying the feeling of fear. And third, we in fact do need to understand the feeling of fear, and ignoring it is not the solution.

As scientists we have an obligation to be precise in how we describe our research. This is especially important when the work is being used to conceptualize human problems—in this case, fear and anxiety disorders—and develop treatments for them. But because conscious feelings of fear and anxiety arise from circuits in the brain that differ from the circuits that control the expression of defensive behaviors like freezing, and are likely vulnerable to different factors, they need to be understood separately. Certainly the circuits that control defense responses and give rise to feelings of fear interact, but this does not mean that they are the same.

Failure to make such distinctions accounts for poor outcomes of studies that have attempted to develop new pharmacological treatments for fear and anxiety in animals, as the studies assess the effects of drugs on behavioral responses but then expect the drugs to actually make people feel less fearful or anxious. We have long known that there is discordance in how treatments affect the way people feel when threatened as opposed to the behavioral and physiological responses they express in such situations.

One of the key issues to point out is that people can be shown pictures of threats in such a way that they are not conscious of the stimulus, and have no conscious feeling of fear. But their amygdala is activated by the threat, and bodily responses, such as changes in perspiration, heart rate, or pupil size occur, showing that the detection and response to threat is independent of conscious awareness. If we don't need conscious experience to control responses to threats in humans we should be cautious about concluding that conscious states cause rats to respond to threats. I am not saying that rats or other animals lack consciousness. All I am saying is that we should not simply assume that because they may respond the way we do when threatened they feel what we do. The problem is that scientific studies of animal consciousness are not easily performed.

Implied above is that fear and anxiety are conscious feelings. As such, we need to understand consciousness in order to understand fear and anxiety. Several chapters of *Anxious* are devoted to giving a progress report on the current state of our understanding of consciousness in neuroscience, psychology, and philosophy (at least from my perspective). Included is the controversial topic of animal consciousness, which, as I just mentioned, is extremely difficult to study scientifically. I suggest guidelines about how we might be more scientific in our approach to this subject.

My view of consciousness dates back to my graduate work on split-brain patients, which I conducted with my mentor, Michael Gazzaniga, at SUNY Stony Brook. We concluded that one important role of consciousness is to make sense of our complex brains. Much of what our brain does, it achieves nonconsciously. Our conscious minds then construct an explanation of what we experience. In this sense, consciousness is a self-narrative built from bits and pieces of information we have direct conscious access to (perceptions and memories) and also from the observable or "monitorable" consequences of nonconscious processes. Emotions are, as some now say, cognitive or psychological constructions.

Finally, I discuss issues related to therapy. One key argument I make is that contrary to popular opinion, the behavioral procedure called extinction is not the main process at work in exposure therapy. Extinction plays a role, but exposure therapy actually involves many more mechanisms, and it is possible they actually interfere with the ability to extinguish. Another principle I challenge is that avoidance is always a bad thing for people with anxiety, for I believe that a form of proactive avoidance can be very useful. These and a number of other ideas for improving psychotherapy come directly from animal research. The key is to know what we can and can't learn from animals, and to not conflate the two.

I have dedicated this book to the many graduate students, postdoctoral, and technical researchers in my lab over the years who have contributed to the work with which I have been credited, for they deserve as much credit as I, and in some cases more. In alphabetical order they are:

Prin Amorapanth, John Apergis-Schoute, Annemieke Apergis-Schoute, Jorge Armony, Elizabeth Bauer, Hugh Tad Blair, Fabio Bordi, Nesha Burghardt, David Bush, Christopher Cain, Vincent Campese, Fernando Canadas-Perez, Diana Cardona-Mena, William Chang, June-Seek Choi, Piera Cicchetti, M. Christine Clugnet, Keith Corodimas, Kiriana Cowansage, Catarina Cunha, Jacek Debiec, Lorenzo Diaz-Mataix, Neot Doron, Valerie Doyere, Sevil Durvaci, Jeffrey Erlich, Claudia Farb, Ann Fink, Rosemary Gonzaga, Yiran Gu, Nikita Gupta, Hiroki Hamanaka, Mian Hou, Koichi Isogawa, Jiro Iwata, Joshua Johansen, O. Luke Johnson, JoAnna Klein, Kevin LaBar, Raphael Lamprecht, Enrique Lanuza, Gabriel Lazaro-Munoz, Stephanie Lazzaro, XingFang Li, Tamas Madarasz, Raquel Martinez, Kate Melia, Marta Moita, Marie Monfils, Maria Morgan, Justin Moscarello, Jeff Muller, Karim Nader, Paco Olucha, Linnaea Ostroff, Elizabeth Phelps, Russell Philips, Joseph Pick, Gregory Quirk, Franchesa Ramirez, J. Christopher Repa, Sarina Rodrigues, Michael Rogan, Liz Romanski, Svetlana Rosis, Akira

Sakaguchi, Glenn Schafe, Hillary Schiff, Daniela Schiller, Robert Sears, Torfi Sig-
urdsson, Francisco Sotres-Bayon, Peter Sparks, Ruth Stornetta, G. Elizabeth
Stutzmann, Gregory Sullivan, Marc Weisskopf, Mattis Wigestrand, Ann Wilensky, Walter Woodson, Andrew Xagoraris. Also included are Elizabeth Phelps, my
long-standing collaborator, and her team at NYU, as they have done human versions of our rodent studies and verified that our findings apply to people.

For assistance with the ancient roots of the modern word "anxiety," I am
grateful to my son, Milo LeDoux, who was trained in classics at the University of
Oxford, and is now a student at the University of Virginia School of Law, and
Peter Meineck, clinical associate professor of classics at NYU and founder of the
Aquila Theatre. The cognitive therapist Stefan Hofmann of Boston University
helped me tremendously by providing key papers for me to read to help me better
understand cognitive therapy and its relation to extinction. Isaac Galatzer-Levy,
a colleague from the NYU Langone Medical Center Department of Psychiatry,
read several chapters and made helpful comments.

I am also grateful to my illustrator, Robert Lee, for his patience in working
through my various incomplete and sometimes incoherent rough drafts of the art.

Special thanks to William Chang, my longtime assistant, who has suffered
graciously through many writing projects, and without whom completion of this
project would have been far more onerous a task.

I have been continuously funded by the National Institute of Mental Health
since 1986, and much of the research discussed here was made possible by its support. Recently, I have also been supported by the National Institute of Drug Abuse.
In the past I have also received funding from the National Science Foundation. I
am grateful to Robert Kanter and Jennifer Brour for their support.

In 1989 I joined the faculty of Arts and Sciences at NYU, where I have been
a member of the Center for Neural Science and Department of Psychology. In
recent years I have received appointments in psychiatry and in child and adolescent psychiatry at NYU Langone Medical School. NYU has been a loyal and
generous friend to me and my research.

In 1997, through a collaboration between NYU and New York State, I was
appointed as director of the Emotional Brain Institute. This is a multisite program with laboratories at NYU and at the Nathan Kline Institute for Psychiatric
Research. Through support of this program by NYU and New York State we hope
to make new gains in understanding fear and anxiety. Some of the studies
described in this book have been conducted in this context.

John Brockman, Katinka Matson, and everyone at Brockman Inc. are incred-
ible agents. I am grateful for all they have done for me over the years, starting with
The Emotional Brain.

At Viking, I can't lavish enough praise on Rick Kot. He was the editor of *Synaptic Self* as well, and I hope of future books that may be lurking deep down in the synaptic recesses of my brain. Rick's assistant, Diego Núñez, has been terrific in helping navigate the end-of-book steps.

I want to express my love and thanks to my brilliant and beautiful wife, Nancy Princenthal. Nancy and I were engaged in major book projects at the same time, both headed toward publication in the spring/summer of 2015. In spite of special challenges that she faced in completing her biography of the late artist Agnes Martin, she was a friend, companion, critic, and editor when each role was needed.

How did the title *Anxious* finally come about? In 2009 my band, The Amygdaloids, released an album titled *Theory of My Mind* on the Knock Out Noise label, on which Rosanne Cash sang two songs with me. One of the pieces that we recorded that didn't make it onto the album was called "Anxious." I always liked the song, and had been thinking of releasing it separately. That's when the idea that the book should be called *Anxious* hit me. And it didn't take long to make the next mental leap: Why not release both *Anxious* the book and *Anxious* the CD simultaneously, since my songs are related to the themes in the book? Below you will see a barcode that can be scanned for a onetime free download of the songs on *Anxious*.

Use a "scanning app" with your smartphone and scan the code provided here to obtain a free download of songs from *Anxious* (the CD) by The Amygdaloids. The app should take you directly to The Amygdaloids' website, where you will see instructions on how to download the songs. This offer is expires on January 14, 2017. If any problems arise, email a copy of the copyright page (on the back of the title page) of the book to amygdaloids.anxiousdownload@gmail.com with the subject "Anxious CD Download."

Enjoy the book, and the music.

CONTENTS

CHAPTER 1

THE TANGLED WEB OF
ANXIETY AND FEAR

"He who fears he shall suffer already suffers what he fears."

—MICHEL DE MONTAIGNE[1]

"While I was fearing it, it came, but came with less of fear. . . .
'Tis harder knowing it is due, than knowing it is here."

—EMILY DICKINSON[2]

A nxiety is a normal part of life—there's always something to worry about, dread, fret over, or be stressed by. But we aren't all anxious to the same degree. Some people are "nervous Nellies" and worry about everything; others are "cool as a cucumber" and seem to just take it all in stride.

My mother was a worrier. Not in the extreme, but she could be preoccupied and fidgety and sometimes complained about sleepless nights. She had a good reason to be this way. My father was more or less carefree, the kind of guy who could put the day behind him and fall asleep within minutes of his head hitting the pillow. If she didn't worry, their business, a mom-and-pop store, could not have thrived. She kept everything together, both at work and at home. And while loving and kind, she also sometimes suffered under the pressures of keeping all the balls in the air on a daily basis. My own temperament lies somewhere between theirs, and when I feel the stress of daily life pulling me toward anxiety and worry, I try to channel a bit of my father's disposition to balance things out. But it's a temporary measure, as I tend to revert to who I am, to my own level of anxiety, quite quickly.

Not surprisingly, one's general level of anxiety is a fairly stable personality trait,[3] a significant component of temperament.[4] We vary from our personal hot spot from time to time, but we always return to our resting place. It's as if "conservation of anxiety" is a law of human nature.

What makes us each have his or her own individual anxiety level? In part, it is because we each experience and respond to the world differently. Anxiety is very subjective: What's really stressful to one person may hardly matter to another. It's not as simple a matter as just having the ability to let the small stuff slide. People who are dispositionally anxious see more things as stressful than less anxious people; for the more anxious, fewer experiences fall in the "small stuff" category.

But simply stating that we are each different only begs the question: What is it that makes us each psychologically distinct? The answer, of course, is that we each have a one-of-a-kind brain. As I explained in *Synaptic Self*,[5] while all human brains are similar in overall structure and function, they are wired differently in subtle, *micro*scopic ways that make us individuals. These differences come about both because of the unique combination of genes we get from our two parents and because of the experiences we have as we go through life. Nature and nurture are partners in shaping who we are, and that partnership is played out in each of our brains.

ANXIETY: ANCIENT YET NEW[6]

The English word "anxiety" and its European equivalents (e.g., *angoisse* in French, *angoscia* in Italian, *angustia* in Spanish, *Angst* in German, and *angst* in Danish) come from the Latin *anxietas*, which, in turn, has roots in the ancient Greek *angh*.[7] Although *angh* was sometimes used by the Greeks to mean burdened or troubled (i.e., *angh*uished), it was primarily employed in reference to physical sensations, such as tightness, constriction, or discomfort. For instance, the word "angina," a medical condition in which chest pains occur in relation to heart disease, comes from *angh*.[8]

Literary and religious writings and works of art over the ages reveal that people have always recognized the mental state we commonly refer to as anxiety today, even though they did not typically label it using *angh* or its linguistic descendants.[9] For example, the famous Greek sculpture *Laocoön and His Sons*, shown in Figure 1.1, illustrates anxiety (anguish, worry, and/or dread) in faces of

Laocoön and his offspring, who are entwined with and are being bitten by snakes as punishment by the gods for having attempted to expose the ruse of the Trojan horse.[10] Ares, the Greek god of war, had two sons, Phobos (the god of fear) and Deimos (the god of dread), who accompanied him into battle to spread their namesake emotions.[11] In the New Testament, the reader is told in Matthew 6:27, "You cannot add any time to your life by worrying about it." The philosopher and theologian Thomas Aquinas noted in the thirteenth century, "When a man dreads the punishment which confronts him for his sin and no longer loves the friendship of God which he has lost, his fear is born of pride, not of humility."[12] Indeed, in the Christian world, anxiety was often connected with sin and redemption.[13] In the 1800s, for example, Søren Kierkegaard, who at the time was a little-known Danish theologian and philosopher, conceived of anxiety as the key to human existence: a sense of dread over our freedom to choose. It began, Kierkegaard said, when Adam struggled between Eve's apple and God, and remains a factor in every choice that humans make.[14]

Figure 1.1: The Anguish of Laocoön and His Sons.

In spite of its long history, however, the word "anxiety" was not primarily thought of as a troubled, worried state of mind and a source of psychopathology until the early twentieth century. This transformation began when Sigmund Freud made anxiety the centerpiece of his psychoanalytic theory of mental disorders.[15] Earlier psychopathologists such as Emil Kraepelin[16] had formulated ideas about anxiety, but it was Freud who introduced the concept of pathological anxiety to the world at large.[17]

According to Freud, anxiety is the root of most if not all mental maladies[18] and central to any understanding of the human mind: "There is no question that the problem of anxiety is . . . a riddle whose solution would be bound to throw a flood of light on our whole mental existence."[19] He saw anxiety as a natural and useful state but also a common feature in mental problems that plague people in everyday life. Ever since, anxiety has been viewed as a state of mind characterized by worry, dread, anguish, and apprehension.

Anxiety was, for Freud, first and foremost "something felt," a special "character of unpleasure."[20] Like the Greeks, he made a point of distinguishing *Angst* (anxiety) from *Furcht* (fear). Anxiety, Freud said, relates to the state itself, and disregards the object that elicits it, whereas fear draws attention precisely to the object.[21] Specifically, Freud noted that anxiety describes a state of expecting danger or preparing for it, and of dreading it, even though the actual source of harm may be unknown; fear, however, requires a definite object of which to be afraid.[22] He also distinguished between *primary anxiety*, which has an immediate object (essentially fear), and *signal anxiety*, which is objectless and involves a more diffuse or uncertain feeling that harm may come in the future (essentially anxiety).

In Freud's view anxiety is born out of a need to keep impulses based on stressful thoughts and memories, mostly about childhood, out of consciousness. Through the defense mechanism of repression, these impulses are hidden in the unconscious mind. When repression fails, the troubling impulses reach consciousness, and neurotic anxiety results. The impulses then need to be repressed again or "satiated" through neurotic "enactments" to relieve anxiety. The goal of Freud's psychoanalytic method was to bring the cause of the neurotic anxiety, or what came to be called *anxiety neurosis*, into consciousness and eliminate its clandestine, subversive power.

Existential philosophers such as Martin Heidegger[23] and Jean-Paul Sartre[24] offered a different view of mental life, and especially anxiety, one centered on consciousness.[25] Sartre, for example, rejected Freud's emphasis on the pathological and unconscious aspects of mind. He famously said, "*l'existence précède l'essence*" (existence precedes essence), by which he meant that we create ourselves by the conscious choices we make.

The existentialists viewed anxiety as an integral part of human nature, rather than a disorder. In this they were greatly influenced by the writings of Kierkegaard. In *The Concept of Anxiety*, published in 1844, before Freud was born, Kierkegaard made a distinction between fear, which has a specific object (similar to Freud's *Furcht*, or primary anxiety), and anxiety, a kind of unfocused, object-less, future-oriented fear (comparable to Freud's *Angst* and signal anxiety but with much less emphasis on pathology and a greater focus on consciousness).[26] Because of its lack of an objective focus, Kierkegaard argued that anxiety (dread) was caused by "nothingness": the despair that comes from the realization that we are not grounded in the world and are defined only by the practices in which we engage. It is through choice that we prevent the return to nothingness.[27] Kierkegaard became well known only after the existentialists adopted him, and Freud apparently was not aware of his writings when developing psychoanalytic theory.[28]

Kierkegaard believed that experiencing anxiety was essential for a successful life, for without it one could not advance. As he noted, "Whoever is educated by anxiety is educated by possibility."[29] The well-adjusted person faces anxiety and moves ahead.[30] His emphasis on the importance of anxiety to success is borne out by research showing that there is an optimal relation between cognition and anxiety in performing life's tasks; with too little anxiety, one is not motivated, but with too much, impairments result.[31] As pointed out by leading anxiety researcher David Barlow, without anxiety, "[t]he performance of athletes, entertainers, exec-utives, artisans, and students would suffer; creativity would diminish; crops might not be planted. And we would all achieve that idyllic state long sought after in our fast-paced society of whiling away our lives under a shade tree. This would be as deadly for the species as nuclear war."[32]

Therapies arose from both the Freudian and existentialist camps but with different goals. Freud's psychoanalysis sought to rid the person of unconscious psychic conflict caused by past experiences; he viewed the analyst as an arche-ologist digging through layers to uncover the past. Existential therapy viewed anxiety and other sources of inner strife as a condition of human life that is best coped with by using our freedom to make choices about our actions as we go for-ward in life. Mainstream psychiatry today is biologically oriented and in this sense more aligned with Freud's view that anxiety can become a pathological condition for which treatment is required to heal the troubled brain. Yet, while contemporary biological psychiatry recognizes the importance of Freud's semi-nal contributions,[33] it is divorced from his psychoanalytic theory.[34]

The popularity of both Freud and Sartre helped make anxiety a cultural byword in the United States following World War II[35] (Figure 1.2). In 1947 the

Figure 1.2: Anxiety in Popular Culture of the Mid-Twentieth Century.
(Clockwise, from upper left) W. H. Auden's 1947 poem, *The Age of Anxiety*; Leonard Bernstein's 1947–1949 Symphony, *The Age of Anxiety*; the cover of *Mad* magazine from 1956, introducing Alfred E. Neuman's trademark expression, "What? Me worry?"; and advertisements for the 1967 film *Valley of the Dolls*, the 1977 film *High Anxiety*, and the 1958 film *Vertigo*. *(Center)* The Rolling Stones' 1966 45 rpm hit, "Mother's Little Helper."

poet W. H. Auden published a book-length poem called *The Age of Anxiety*.[36] Although the piece itself was complex and difficult and said to have actually seldom been read,[37] its title had a tremendous impact. The composer Leonard Bernstein almost immediately produced a symphony with the same name.[38] The phrase "age of anxiety" has since been used to characterize everything perilous about the modern world[39] and has appeared in the titles of many books, paired with subjects as diverse as science, motherhood, "the transforming vision of Saint Francis," and "mindblowing sex." In 1956 *Mad* magazine celebrated anxiety by putting a cartoon character, Alfred E. Neuman, and his motto, "What? Me worry?" on the cover. In film, Freud's view of anxiety was a popular theme for Alfred Hitchcock, figuring prominently in *Spellbound* (1945), *Stage Fright* (1950), and *Vertigo* (1958). In the 1960s Woody Allen made anxiety his signature foible, the centrifugal force of his cinematic humor. Mel Brooks capitalized on the cultural fascination with anxiety, spoofing Hitchcock's *Vertigo* and its Freudian themes in *High Anxiety* (1977). The Rolling Stones' 1966 hit "Mother's Little Helper" was about British housewives getting through the day on Valium (a highly prescribed antianxiety drug at the time). The use of drugs to control anxiety also played a key role in *Valley of the Dolls*, Jacqueline Susann's popular novel that was made into an equally popular film. ("Dolls" was Susann's nickname for the pills abused by the characters.) In Alan J. Pakula's film *Starting Over* (1979), when the main character has a panic attack in Bloomingdale's, his brother pleads with the other customers for a Valium; everyone in the area pulls out a pill bottle.[40] The psychoanalyst Rollo May, who earlier helped fuse Freud and Kierkegaard in psychiatry,[41] proclaimed in 1977 that "[a]nxiety has certainly come out of the dimness of the professional office into the bright light of the market place."[42] Simply entering the word "anxiety" into a Google search returns more than 42 million hits.

FROM FEAR TO ANXIETY AND BACK AGAIN

Scientists and mental health professionals today are greatly influenced in their views of fear and anxiety by both Freud and Kierkegaard, who each regarded fear and anxiety as perfectly normal, yet unpleasant, feelings. In fear, as we have seen, the focus is on a specific external threat, one that is present or imminent, whereas in anxiety the threat is typically less identifiable and its occurrence less predictable—it is more internal, and in the mind more of an expectation than a fact, and can also be an imagined possibility with a low likelihood of ever occurring.[43] Tables 1.1 and 1.2 summarize common similarities and differences between fear and anxiety.

A simple analysis of the English language suggests that the words "fear" and "anxiety" can describe a range of emotions (Figure 1.3). We've seen some of these above: fear, panic, terror, anxiety, anguish, dread, worry. There are actually more than three dozen English words that are either synonyms, variants, or aspects of "fear" and "anxiety."[44] Some of these are shown in Figure 1.4.

Words usually exist because they account for something important in the lives of the people who use them—Inuit, as is well known, have many words for snow. Fear and anxiety, it would seem, are significant to us. Indeed, each generation since Auden's time has claimed a special relation to anxiety, insisting that it is more anxious than the last.[45]

How are we to deal with the semantic complexity of these terms and the

Table 1.1: Similarities Between Fear and Anxiety

Presence or anticipation of danger or discomfort

Tense apprehensiveness and uneasiness

Elevated arousal

Negative affect

Accompanied by bodily sensations

Based on Table 1.1 in Rachman (2004)

Table 1.2: Differences Between Fear and Anxiety

	FEAR	ANXIETY
Threat is present and identifiable	yes	no
Evoked by specific cues	yes	no
Connection to threat is reasonable	yes	no
Usually episodic (specific onset and offset)	yes	no
Overall quality of an emergency	yes	no
Overall quality of sustained vigilance	no	yes

Based on Table 1.2 in Rachman (2004) and Table 1.2 in Zeidner and Matthews (2011)

implications of linguistic imprecision for our understanding of the underlying mechanisms of fear and anxiety? Some emotion researchers treat all (or at least many) of the terms as measures of intensity of varying degrees of fear: On the low end are *concerned, nervous, jittery, apprehensive,* and *worried,* with *threatened, scared,* and *frightened* in the middle, and *panicked* and *terrified* at the other end.[46] Another approach retains the centrality of fear and anxiety as categories of aversive experience and identifies specific members of these two families. *Frightened, panicked, scared,* and *terrified* are viewed as states that have an objective cause and an imminent consequence and thus are considered forms of fear, whereas *anguish, worry, dread, nervousness, concern, trepidation,* and *troubled* are viewed

Figure 1.3: The Lexicon of Fear and Anxiety.

Figure 1.4: Some Variations on the Themes of Fear and Anxiety.

(FROM MAKARI [2012].)

as variants of anxiety because the source, or cause, is more amorphous and the consequences less certain.

But even this simple solution hints at a source of confusion in the study of fear and anxiety. While these terms are sometimes used to define *categories* (families) of experience, they are also often used more specifically to refer to distinct *types* of experience: In this context "fear" is considered as only one particular form of fearful experience among many possible others, whereas "anxiety" is likewise one particular form in the range of anxious experiences. It is also unclear the extent to which the examples in each category are truly distinct states of fear or anxiety or just slight variations, or even synonyms, of the identical states. But in spite of such complications, we at least have guidelines to help us separate the two broad categories: Fear states occur when a threat is present or imminent; states of anxiety result when a threat is possible but its occurrence is uncertain.

DEFINING FEAR AND ANXIETY

While we can often separate fear from anxiety conceptually on the basis of the nature of the threat, in our daily lives fearful and anxious states are not completely independent. It is probably impossible to feel fear without also being anxious—as soon as you are afraid of something, you begin to worry about what the consequence of the danger at hand will be. For example, the sight of an agitated person near you waving a gun compels the feeling of fear, but worry (or anxiety) quickly takes over as you fret over what the person will do. As Montaigne notes in one of the epigraphs to this chapter: "He who fears he shall suffer already suffers what he fears."

Likewise, when you are anxious, the perceived threat potential of stimuli related to your anxiety can rise such that things you typically encounter that might not usually trigger fear now do so. For example, if you encounter a snake in the course of a hike, even if no harm comes anxiety is likely aroused, putting you on alert. If farther along the trail you notice a dark, slender, curved twig on the ground, an object you would normally ignore, you might now momentarily be prone to view it as a snake, triggering a feeling of fear. Similarly, if you live in a place where terror alerts are common, benign stimuli can become potential threats. In New York City, when the alert level rises, a parcel or paper bag left under an empty subway seat can trigger much concern.

Ultimately, the question we have to ask is: Can we really make a distinction between fear and anxiety given that both are anticipatory responses to danger and thus closely entwined? I think we can, and must. As I will describe in later chapters, somewhat different brain mechanisms are engaged when the state is triggered by an objective and present threat as opposed to an uncertain event that may or may not occur in the future. An immediately present stimulus that is itself dangerous, or that is a reliable indicator that danger is likely to soon follow, results in fear. Anxiety may well also be present, but if the initial state is triggered by a specific stimulus, it is a state of fear. However, when the state in question involves worry about something that is not present and may never occur, then the state is anxiety. Fear can, like anxiety, involve anticipation, but the nature of the anticipation in each is different: In fear the anticipation concerns if and when a present threat will cause harm, whereas in anxiety the anticipation involves uncertainty about the consequences of a threat that is not present and may not occur.

Fear and anxiety, as I will argue later, both involve the self. To experience fear is to know that YOU are in a dangerous situation, and to experience anxiety is to worry about whether future threats may harm YOU. This involvement of the self in fear and anxiety is a defining feature of these and other human emotions.

DISORDERED ANXIETY AND FEAR

While fear and anxiety are perfectly normal experiences, sometimes they become maladaptive—excessive in intensity, frequency or duration—causing the sufferer distress to the extent that his or her daily life is disrupted.[47] When this happens, an *anxiety disorder* exists.[48] For historical reasons, as I will soon explain, problems involving maladaptive fear and anxiety are typically grouped together under the label "anxiety disorders." Table 1.3 compares normal and pathological expressions of fear and anxiety.

What constitutes an anxiety disorder in the United States is dictated by the *Diagnostic and Statistical Manual* (DSM) of the American Psychiatric Association.[49] Although the World Health Organization has its own system, the two are for the most part compatible.[50] The DSM has recently gone into its fifth edition (DSM-5), but to understand its classification of anxiety disorders, it will be useful to examine prior versions first.[51]

The DSM classification system, which was introduced in the mid-twentieth century, was initially dominated by psychoanalytic ideas, which resulted in mental disorders being categorized into states of either psychosis or neurosis. Psychotic conditions were considered to involve thought disturbances, including delusions and/or hallucinations, a break with reality, and, in general, an inability to function in normal social situations. Neuroses involved several conditions in which one suffered from distress (sometimes debilitating distress) but without significant distortions of thought, or loss of touch with reality. The neurotic conditions most related to fear and anxiety included anxiety neurosis (excessive

---

Table 1.3: Everyday vs. Pathological Fear and Anxiety

EVERYDAY ANXIETY	ANXIETY DISORDER
Worry about paying bills, landing a job, or other important life events	Constant and unsubstantiated worry that causes significant distress and interferes with daily life
Embarrassment or self-consciousness in an uncomfortable or awkward social situation	Avoiding social situations for fear of being judged, embarrassed, or humiliated
A case of nerves or sweating before a big test, business presentation, stage performance, or other significant event	Seemingly out-of-the-blue panic attacks and the preoccupation with the fear of having another one
Worry about an actual dangerous object, place, or situation	Irrational worry about and avoidance of an object, place, or situation that poses little or no threat of harm
Making sure that you are healthy and living in a safe, hazard-free environment	Performing uncontrollable repetitive actions such as excessive cleaning or checking, or touching and arranging
Anxiety, sadness, or difficulty sleeping immediately after a traumatic event	Recurring nightmares, flashbacks, or emotional numbing related to a traumatic event that occurred several months or years before

Based on http://www.adaa.org/understanding-anxiety

---

worry, dread), phobic neurosis (irrational fears), obsessive neurosis (repetitive thoughts), and war neurosis (mental problems in soldiers that stemmed from stress, exhaustion, and specific battlefield experiences).

With the arrival of DSM-III in 1980, anxiety neurosis was divided into two separate states, a partition based on research findings by the psychiatrist Donald Klein.[52] Klein had been studying a new experimental drug, imipramine, to treat hospitalized schizophrenic patients in the hope of reducing their high levels of anxiety. The patients claimed that their anxiety levels were unchanged, but the staff noted a dramatic decrease in the frequency with which these patients would show up at the nurses' station complaining of physiological symptoms (inability to breathe, racing heart, dizziness) and psychological distress (feeling terrified that they were about to die). These brief bouts of intense fear (or, as they came to be called, *panic attacks*) were lessened after several weeks of treatment. Benzodiazepines, drugs like Valium, by contrast, reduced chronic anxiety but did not help with panic attacks. Findings such as these led Klein to distinguish between two broad kinds of anxiety disorders: *generalized anxiety disorder* (GAD) and *panic disorder*. While Freud anticipated this distinction, because he discussed anxiety as a general condition that sometimes had physiological symptoms similar to those in a panic attack, he did not distinguish these as different subcategories of anxiety neurosis.

Let's look at these two conditions in a bit more detail. Generalized anxiety (worry, nervousness, apprehension, dread) is what most laypeople have in mind when they use the term "anxiety." People with GAD have prolonged, uncontrollable, and excessive worry and tension about life situations (including family, work, finance, health, romance, and other circumstances) to the point of interfering with normal routines.[53] Panic disorder, by contrast, is typified by brief, intense attacks during which a person is overcome with the feeling that he or she is suffocating or experiencing a heart attack—remember that *angh*, the Greek root of the English word "anxiety," referred to physical sensations more than the mental states of worry and dread that occur in GAD.[54]

DSM-IV, which was published in 1994, integrated conditions related to some of the other forms of neurosis (specifically phobic, obsessive, and war neuroses) with GAD and panic disorder. Two broad categories of phobic conditions were included: *specific phobias* (in which one experiences anxiety about encountering certain objects, such as snakes or spiders, or physical situations, such as high elevations or tight, closed spaces) and *social phobias* (anxiety about attending social events such as parties or situations in which one has to speak publicly). Also added was *obsessive-compulsive disorder* (OCD), which involves recurrent thoughts (e.g., concerns about germs) that intrude into consciousness and are accompanied by repetitive actions (e.g., excessive hand washing) that are aimed at reducing distressing feelings. Finally, *posttraumatic stress disorder* (PTSD) was also categorized as an anxiety disorder. In this condition, thoughts and memories about a past, often life-threatening event led to feelings of detachment, sleep problems, and hypersensitivity to trigger cues. Although soldiers throughout history have suffered psychologically from experiences in battle, PTSD came to be recognized as a specific condition after the Vietnam War. This term replaced earlier designations, such as *war neurosis, nostalgia,*[55] *shell shock, battle fatigue,* and *combat stress reaction.* But the PTSD diagnosis was not limited to battlefield conditions and instead also included a response to any kind of severe traumatic experience, such as an automobile or other accident, or rape, torture, or other forms of physical abuse.

With the arrival of DSM-5 in 2014, some retractions and reorganization occurred. In addition to using an Arabic rather than a Roman numeral to identify the edition, two of the DSM-IV disorders in the anxiety disorder domain were removed from DSM-5 and placed into separate categories: PTSD became part of *trauma and stressor-related disorders* and OCD part of *obsessive-compulsive and related disorders*.

The term "anxiety disorders" thus originally arose to subsume two states of anxiety (GAD and panic) and was retained when additional conditions were added. But the label "anxiety disorders" gives short shrift to the fact that most of the conditions included also involve fear (e.g., fear of specific objects or situations in specific and social phobic disorders; fear elicited by somatic sensations, such as heart palpitations or shortness of breath, in panic disorder). Therefore, I prefer to describe these states as *fear and anxiety disorders*, conditions in which maladaptive fear and/or anxiety plays a central role. With this in mind, I break with the DSM-5 categorization and include PTSD in my discussion of fear and anxiety because it involves maladaptive fear (fear of trauma-related cues).[56] Some of the accepted characteristics typically associated with these fear and anxiety disorders are shown in Figure 1.5.

Together, fear and anxiety disorders are the most prevalent of all psychiatric problems in the United States, affecting about 20 percent of the population, more than twice the number who suffer from mood disorders such as depression and bipolar disorder, and twenty times the number with schizophrenia.[57] The economic cost of fear and anxiety disorders is estimated to exceed $40 billion annually.[58] These conditions have a significant impact on the workforce. For example, a study from Australia found that anxiety and affective disorders resulted in 20 million work impairment days annually, mostly involving absences.[59]

But the problem is actually more pervasive than the 20 percent anxiety prevalence statistic implies. Problems with threat processing and maladaptive fear and anxiety are factors in many other psychiatric conditions. GAD and depression often occur together, and fear and anxiety can play a role in schizophrenia, borderline personality disorder, autism, and eating and addictive disorders. Moreover, many individuals are impaired by uncontrollable fear or anxiety without having received an official psychiatric diagnosis of their condition. These issues can also trouble those whose health is compromised by illnesses such as cancer, heart disease, and other chronic physical ailments. Even many people who are considered otherwise healthy in mind and body can from time to time suffer from bouts of excessive fear and worry. A better understanding of the nature of these conditions, and the brain mechanisms involved, would be extremely helpful to just about everyone.

What determines who will be likely to suffer from a fear or anxiety disorder? For example, why does only a relatively small proportion of people exposed to a

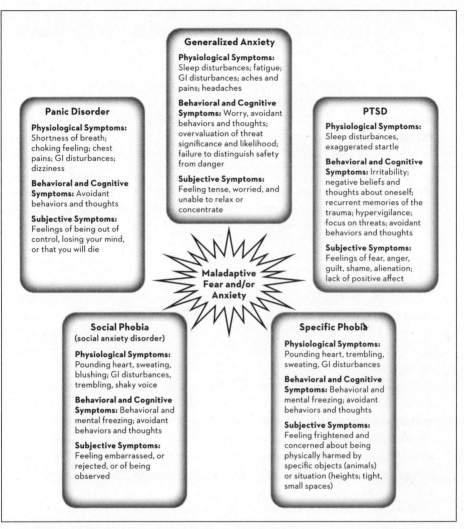

Figure 1.5: Major Symptoms of Fear / Anxiety Disorders.

trauma develop PTSD?[60] David Barlow has proposed that three factors make people vulnerable to these disorders[61] (Figure 1.6). One is genetics or other *biological factors* in the brain. The heritability of anxiety is estimated to be between 30 percent and 40 percent, which is considerably less than that of some other conditions.[62] But the rates rise when one examines particular anxious traits, such as the tendency to be inhibited and withdrawn in situations involving uncertainty. Genetic influences on anxiety and other mental disorders are complex and

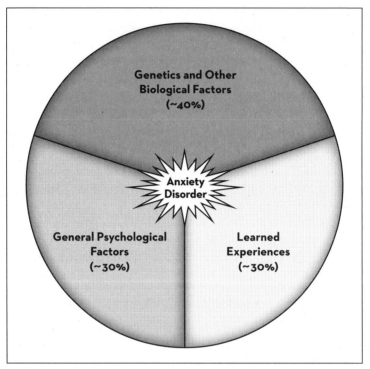

Figure 1.6: Vulnerability to Pathological Anxiety.
(Based on Barlow [2003].)

involve interactions between multiple genes. Individual differences in brain organization that arise from environmental influences, and interactions between genetic and environmental factors, are also important. Another source of vulnerability involves *general psychological processes,* such as an individual's tendency to perceive situations as unpredictable and uncontrollable. The third factor listed by Barlow is *specific learning experiences.* If a child is given excessive attention when ill, he may continue to use "sick behaviors" as a way to attract attention and sympathy. Similarly, if a child observes a parent or other adults using such strategies, he might adopt them as well. Early life situations in which one experiences negative consequences that cannot be controlled following situations involving uncertainty may well predispose one to feel less in control in later life. It should be noted that psychological processes and learning experiences are, in the end, also biological in nature, because they are products of the brain and as such are also subject to genetic influences and the influence of gene-environment interactions, or what is called epigenetics.

The social scientists Allan Horwitz and Jerome Wakefield have called for prudence in the use of the term "disorder" when talking about mental problems. In their books, *All We Have to Fear*[63] and *The Loss of Sadness*,[64] they point out that "disorder" implies that something physical is not working as it should. In people who are having trouble with anxiety, they argue, the brain is often doing what it is supposed to do—it's just doing it in the wrong context. Fear and anxiety triggered by strangers, snakes, heights, and the like served our ancestors well by helping them avoid dangers but can cause distress in the modern world. Horwitz and Wakefield are particularly concerned about the rising rates of diagnosis of fear and anxiety disorders and the alarming increase in the use of medications to treat brains that are working fundamentally correctly, from an evolutionary point of view. They do accept that disordered fear and anxiety can exist and describe criteria to distinguish disordered from normal states. Regardless of what one thinks of their gauges for defining disorders, their book is important because it raises important social issues about psychiatric diagnosis and treatment.

THE CENTRALITY OF THREATS

This book is called *Anxious,* and it discusses how feelings of anxiety and related states (including worry, concern, dread, disquiet, apprehension, nervousness) come about. But the close connection between fear and anxiety demands that these two emotions be understood together. A key factor that links them is that they both depend on mechanisms in the brain that detect and respond to threats to well-being.

Threats, whether present or anticipated, real or imagined, demand action. As many others have noted, threat detection provides preparation for fight or flight.[65] We are all familiar with the *fight-flight response*, the defensive emergency reaction that is triggered when we encounter present or anticipated threats, and that moves into overdrive when we are under stress (see Chapter 3). This whole-body reaction is mobilized to help us survive an encounter with danger. When it is in play, our conscious mind is consumed with fear or anxiety, and often with both. Threat processing is at the heart of fear and anxiety.

Particularly important is the fact that threat processing is altered in each of the fear and anxiety disorders (Figure 1.7).[66] In this book I will be less concerned with the disorders themselves than with the question of how threat processing contributes to maladaptive feelings of fear and anxiety in these disorders.[67] People who suffer from them are hypersensitive to threats, which seize and hold their attention, a condition sometimes called hypervigilance. They are also impaired in distinguishing things that are dangerous from those that are safe and overestimate the significance of perceived threats. Even when threats are not present,

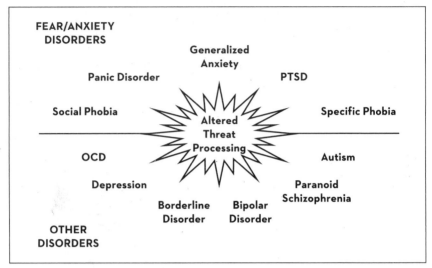

Figure 1.7: Alterations of Threat Processing Occur in Many Psychiatric Disorders.

they worry excessively that threats will occur and constantly scan the environment to try to understand why they feel distressed. They go to extremes to escape from or avoid threats, so much so that these avoidance strategies interfere with daily life.

SHOWING MY HAND

Any understanding of fear and anxiety presumes an understanding of emotion. So before we go further, I want to make clear what my own view of the subject is, using the emotion fear for illustrative purposes. In many ways my core view of emotion has not changed since the 1980s.[68] But recently I have begun to discuss it somewhat differently in an effort to sharpen the conceptualization of this complex psychological function and its relation to brain mechanisms.[69]

Traditionally, emotion theories have focused on conscious feelings.[70] For example, in the late nineteenth century, William James, the father of American psychology, proposed that fear is a conscious feeling that occurs when we find ourselves responding to danger; the feeling of fear, for him, was the perception of body signals that are unique to defending against danger.[71] Not all theorists have agreed with James about how conscious feelings come about, but many have concurred that the feeling *is* the emotion. Freud, as mentioned above, said that anxiety is "something felt" and also noted, "It is surely the essence of an emotion that we should feel it."[72] More recently, the Dutch psychologist Nico Frijda claimed that emotions are primarily "hedonic experiences." Lisa Barrett, James Russell, Andrew Ortony, and Gerald Clore, and others, emphasize that emotions are psychologically constructed conscious experiences.[73] Clore notes that "emotions are never unconscious."[74]

Other theorists, though, have found conscious experience to be unnecessary, or even a detriment, to understanding emotion. For example, in the early twentieth century, behaviorists argued strongly that consciousness, being unobservable, had no place in psychology; they insisted that behavior alone should be the focus of inquiry.[75] This led to the idea that fear was a relation between stimuli and responses rather than a specific feeling.[76] When behavioral psychologists later turned to physiology in an effort to understand how stimuli and responses are connected in the brain, fear became a central motivational state—a physiological state of the brain that organized responses to dangerous stimuli. But like the behaviorists, these physiological theorists, for the most part, also shunned conscious experiences— central states were physiological intermediaries between stimuli and responses, not subjectively felt states.[77] While this approach provided a way of studying the emotions like fear similarly in animals and humans, it achieved this goal by ignoring the feeling of fear, which is what most people think fear is.

But even those who argue that emotions *are* conscious experiences sometimes claim that such experiences are just one aspect or component of emotion. For example, the Swiss psychologist Klaus Scherer views emotion as a process consisting of cognitive appraisals, expressive responses, physiological changes, and conscious feelings.[78] In this view, fear is what happens when we cognitively appraise a situation as dangerous, express certain behaviors in response to it, are physiologically aroused, and feel fear. This approach seems logically cumbersome to me because it regards fear as both the overall process and the specific feeling of being afraid; fear (the experience) is thus a component of fear (the process).

Yet another theory is that emotions are hardwired in the brain and unleashed in the presence of trigger stimuli.[79] In this view, championed by those who adhere to basic emotions theory, innate behavioral reactions, physiological responses, and conscious feelings all flow from a fear center or network. As I will argue later, although threats do indeed release innate behavioral and physiological patterns, the feeling of fear is not itself innate, a view consistent with the ideas of psychological construction theories of emotion.

Fear, anxiety, and other emotions are, in my view, just what people have always thought they were—conscious feelings. We often feel afraid while we freeze or flee in the presence of danger. But these are different consequences of threat detection—one is a conscious experience and the other involves more fundamental processes that operate nonconsciously. The failure to distinguish the conscious experience of fear and anxiety from more basic unconscious processes, I argue, has led to much confusion. The more basic processes contribute to emotional feelings, but they evolved, not to make conscious feelings, but instead to help organisms survive and thrive. For the sake of avoiding confusion, the more basic nonconscious processes should *not* be labeled as "emotional."

Feelings of fear, in my view, result when we become consciously aware that our brain has nonconsciously detected danger.[80] How does this happen? It all starts when an external stimulus, processed by sensory systems in the brain, is nonconsciously determined to be a threat. Outputs of threat detection circuits then trigger a general increase in brain arousal and the expression of behavioral responses and supporting physiological changes in the body. Signals from the behavioral and physiological responses of the body are sent back to the brain, where they become part of the nonconscious response to danger (sensory components of these responses can be "sensed" just like sights or sounds). Brain activity then comes to be monopolized by the threat and by efforts to cope with the harm it portends. Threat vigilance increases—the environment is scanned to figure out why we are aroused in this particular way. Brain activity related to all other goals (eating, drinking, sex, money, self-fulfillment, etc.) are suppressed. If, via memory, environmental monitoring reveals that "known" threats are present, attention becomes focused on these stimuli, which are consciously "blamed" for the aroused state. Memory also allows us to know that "fear" is the name we give to experiences of this type (starting in childhood, we build up templates of what it is like to be in states we label with emotion words). When the various factors or ingredients are integrated in consciousness, an emotion, specifically the conscious feeling of fear, is thus compelled. But this can only happen if the brain in question has the cognitive wherewithal to create conscious experiences and interpret the contents of these experiences in terms of implications for one's well-being. Otherwise, the brain and body responses are a motivational force that guide behavior in the quest to stay alive, but the feeling of fear is not a part of the process. This does not mean that the feeling of fear is a mere by-product. For once it exists, it opens up the resources of the conscious brain to the quest to survive and thrive.

I am not alone in arguing for this view of emotions as cognitively assembled conscious feelings.[81] The recent notion that emotions are "psychological constructions"[82] is perhaps the cognitive-based theory of emotions that is closest to my own view.

One important implication of the ideas to be developed in this book is that they reveal a disconnect between the troubling feelings of fear and anxiety that motivate people to seek help and the way research on those feelings—including research on how to find new treatments to allay them—is conducted and interpreted. Consciousness is no longer a taboo subject in science, and much progress has been made in this area in recent years. Yet research on fear and anxiety disorders in animals, and also in humans, that I and others have conducted is often focused on how the brain detects and responds to threats, processes that operate nonconsciously. Although this work is very relevant to understanding conscious fear and anxiety, it has to be understood in the proper context. Responses to threats, in spite of common practice, are not foolproof markers of conscious feelings, even in humans, and likewise should not be assumed to be so in animals.

A major aim in this book is to provide a framework that will allow a better understanding of the connection between research, therapy, and conscious feelings. But to do that we have to be careful about when we should call upon consciousness and when we should not. We can't understand fear and anxiety if we ignore consciousness, but neither should we overemphasize its role.

LOOKING AHEAD

Having established the tangled web of fear and anxiety from the point of view of threat processing by the brain in the present chapter, I will summarize in the following one how my current views evolved in the course of my three-decade struggle to scientifically understand the emotional brain. Subsequent chapters cover defense in the animal kingdom and the brain mechanisms that enable animals, including humans, to detect and respond defensively to threats. I then address the question of what we have inherited from animals. Contrary to the popular opinion of laypeople and many scientists, I argue that we have not inherited feelings like fear or anxiety from animals; we have instead inherited mechanisms that detect and respond to threats. When these threat-processing mechanisms are present in a brain that can be conscious of its own activities, conscious feelings of fear or anxiety are possible; otherwise threat processing mechanisms motivate behavior but do not necessarily result in or involve feelings of fear or anxiety. Organisms that have the capacity to be conscious can feel fear; otherwise they cannot have such experiences. Thus, if we want to understand feelings of fear and anxiety, we have to understand consciousness, and several chapters on this topic follow. One examines the physical basis of consciousness, another examines the role of memory in consciousness, and a third explains how conscious feelings of fear and anxiety emerge when nonconscious consequences of threat processing are consciously experienced. The final three chapters turn to brain mechanisms related to fear and anxiety and their disorders, and offer a reconception of these disorders. The final chapters also suggest how research on brain mechanisms can offer new ways to help people better cope with such troubling feelings.

Anxiety and its partner, fear, are, as Freud said, riddles, and seeking their solution will take us through many aspects of how the brain and its mind work. Many topics in psychology and neuroscience, ranging from basic mechanisms of defensive behavior in animals to decision making by humans, from automatic nonconscious processing to conscious experience, from perception and memory to feelings, will be covered. Some of these involve complex brain mechanisms, but rest assured that the principles, which will be my focus, for the most part are easily grasped.

CHAPTER 2

RETHINKING THE EMOTIONAL BRAIN

"Neuroscientists use 'fear' to explain the empirical relation between two events: for example, rats freeze when they see a light previously associated with electric shock. Psychiatrists, psychologists, and most citizens, on the other hand, use . . . 'fear' to name a conscious experience of those who dislike driving over high bridges or encountering large spiders. These two uses suggest . . . several fear states, each with its own genetics, incentives, physiological patterns, and behavioral profiles."

—JEROME KAGAN[1]

After working on what I referred to as the "emotional brain" for more than thirty years, I concluded that this terminology needed rethinking.[2] Jerome Kagan's quote in the epigraph above hints at my reasoning but doesn't go far enough. Kagan implies that there are two different kinds of fear states, with different underlying brain systems, that are elicited by fear-arousing stimuli, one involving conscious feelings and the other involving behavioral and physiological responses. In contrast, I believe we should restrict the use of emotion words like "fear" to *conscious* feelings, such as the feeling of being afraid. Brain systems that detect threatening stimuli and control behavioral and physiological responses elicited by these stimuli should not be described in terms of fear. The latter systems operate nonconsciously in humans, and although they contribute to feelings of fear, they are not fear mechanisms per se. This chapter explains why I think the distinction between mechanisms that detect and respond to threats outside the realm of consciousness, as opposed to mechanisms that create conscious feelings of fear, is so important for our conception of fear and its partner, anxiety.

IN THE BEGINNING

When I started my research on the emotional brain in the early 1980s, ideas like the limbic system theory were popular.[3] It proposed that our reptilian ancestors were dominated by reflexes and instincts. Then, with the emergence of mammals, a new brain system (the limbic system) evolved to make feelings, enhancing the adaptive potential of the newest vertebrates. Later in mammalian evolution the neocortex appeared and provided a thinking brain that made reasoning and emotional control possible. Although the limbic system concept inspired much research, the goal was often as much about validating the concept as about understanding the emotional brain. If some region outside the limbic system happened to be implicated, rather than calling into question the limbic system theory, the criteria for inclusion of brain areas were simply changed. As a consequence, the limbic system theory lost its connection to the theory of brain evolution upon which it was based.[4] (Unfortunately, the limbic system theory continues to be prominent in both lay and scientific discussions of the emotional brain in spite of its evolutionary basis having been discredited.)

I thought that a different approach was needed, one that made minimal starting assumptions about how emotions are organized in the brain. The approach I took was to follow the flow of information in the brain from the sensory system that processes a stimulus to the muscles that control the responses to it. Somewhere along that pathway would be mechanisms that detect the significance of the stimulus and trigger the appropriate responses. Limbic areas might well be implicated, but the point was to have an objective approach to the circuitry rather than one that presumed to know the answer before the research was done.

I was led to this "information flow" conception by prior experiences I had had at SUNY Stony Brook while doing doctoral research with Michael Gazzaniga.[5] We studied patients in whom the two sides of the brain were separated in an effort to control epilepsy. In most people, information in one hemisphere is automatically and instantly transferred to the other hemisphere, a process that enables the two sides of the brain to work together seamlessly in daily life.[6] In our *split-brain patients*, however, a stimulus presented to one hemisphere remained in that hemisphere (Figure 2.1). For example, if you show a split-brain patient's right hemisphere a picture—say, of an apple—he will be unable to name the stimulus because the ability to speak is located in the left hemisphere. He can reach into a bag containing several items with his left hand (which is connected to the right hemisphere) and identify the apple by touch and retrieve it with ease. The right hand (which is connected to the left hemisphere) can't do this with any degree of accuracy because the left hemisphere did not see the stimulus. When studying

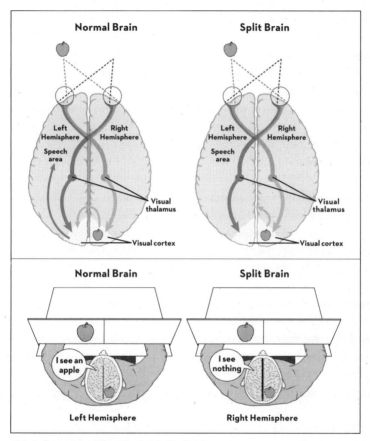

Figure 2.1: Information Flow in the Split-Brain.

In the human brain, visual stimuli that appear to the left side of center are transmitted to the right hemisphere, and stimuli to the right side of space are transmitted to the left hemisphere. Connections between the two hemispheres (not shown) allow stimuli seen by one to be seen by the other. Through these connections, each hemisphere can construct a complete view of perceptual space, and visual stimuli presented in the left field of view and thus directed to the right hemisphere can be talked about by the left hemisphere, where speech is typically controlled. In the split-brain patient, because the connections between the two hemispheres are surgically sectioned as a treatment for epilepsy, if an apple is presented to the left side of space and thus directed to the right hemisphere, it remains in the right hemisphere and the patient cannot verbally describe the stimulus, as the left hemisphere does not have access to the sensory information.

split-brain patients, you can't help but picture signals flowing from point to point in the brain to construct what we are seeing, remembering, thinking, and feeling, and in controlling behavior.

Conceiving of brain functions in terms of information flow had become inevitable with changes in the psychological zeitgeist. For decades behaviorists had dominated psychology, shunning all talk of mind, consciousness, and other unobservable inner factors (whether in the mind or brain) in explaining behavior.[7] A science of psychology, they said, must be based on observable events—stimuli and responses. But by the 1970s behaviorism had been supplanted by the cognitive approach, which treated the mind not as a place where conscious experiences occur so much as an information-processing system that connects stimuli with responses, and that does not necessarily involve consciousness.[8] Split-brain studies fit perfectly within this intellectual framework.

After I received my PhD in 1978, Gazzaniga and I both moved to Manhattan to work at Cornell Medical School. Initially, I was exploring the consequences of brain damage on language and attention,[9] but my real interest, which had begun with a study we had conducted in a split-brain patient, was in the brain mechanisms of emotion.[10] When we showed the patient's right hemisphere an emotional stimulus, the left hemisphere could not name it but could rate its emotional valence. This suggested that the cognitive processes involved in perceiving what a stimulus is are separable in the brain from the processes that evaluate its emotional significance. I wanted to figure out how emotional significance was added to a stimulus as it flowed, in the form of information, through the brain. Because there was no way to pursue detailed brain mechanisms in humans, I turned to studies of rats.

The field of neuroscience was officially born as a discipline when the Society for Neuroscience was formed in 1969[11] and by the end of its first decade was coming of age. In 1979, with Gazzaniga's help, I got a position in the Neurobiology Laboratory at Cornell, run by Don Reis, which had all the latest and greatest tools for studying the brain. At the time, everyone in our young field was aware of Eric Kandel's pioneering research on learning and memory.[12] Kandel had begun his work in the 1960s with studies of how memory is made in the brains of rats, though he soon concluded that brain research was not at a sufficiently advanced state to be able to tackle a complex issue like memory in complex animals. He accordingly recast his research in a way that would enable him to make progress with the tools available at the time. Specifically, he picked a simple organism (a sea slug, the invertebrate *Aplysia californica*) and focused on several simple forms of learning in which this organism engages. He then determined the complete neural circuit by which sensory information flows to motor output during the behaviors, and he isolated the cells and synapses that change during learning and the molecular mechanisms in the cells and synapses that made the cellular changes possible. With this "what, where, and how" strategy, Kandel revolutionized research on learning and memory, earning him a Nobel Prize in 2000.

A word or two is in order here about the behavioral backbone of the Kandel strategy. There are two basic approaches to studying learning in the laboratory: classical and instrumental conditioning. As is well known, in the early twentieth century Ivan Pavlov discovered that after pairing a sound with food, the sound alone would elicit the salivary reflex in dogs.[13] Thus, classical (or Pavlovian) conditioning brings an inborn reflex under the control of a novel stimulus. By contrast, in instrumental conditioning, pioneered by Edward Thorndike in the late nineteenth century, a new response is learned because of the success of that response in obtaining a positive, or avoiding a negative, outcome.[14] A typical example is a rat that learns to press a lever to receive food. This was called instrumental conditioning because the response is instrumental in achieving the outcome (it is also called operant conditioning).[15] When Kandel started his research, most work on learning and memory in the brain was being done using instrumental conditioning because it was thought to be more relevant to complex human behavior than simple Pavlovian conditioned reflexes.[16] But Kandel recognized that research on the neural basis of instrumental conditioning in mammals was going nowhere fast and that more progress might be made by using Pavlovian conditioning and other simple learning procedures in an organism with a less complex nervous system.[17] That insight made possible his pioneering work.

Table 2.1: Pavlovian and Instrumental Conditioning Compared

PAVLOVIAN (CLASSICAL) CONDITIONING	INSTRUMENTAL (OPERANT) CONDITIONING
Developed by Pavlov	Developed by Thorndike and Skinner
A reinforcing unconditioned stimulus (US) is presented when a certain stimulus (the conditioned stimulus, CS) occurs	A reinforcing unconditioned stimulus (US) is presented when a certain response (the conditioned response, CR) is performed
An association forms between the CS and US	An association forms between the CR and US
Later, the CS *elicits* an innate conditioned response that is motivationally related to the US	Later, when similar motivational conditions occur, the *learned* CR is *emitted* because the CR has resulted in the reinforcing US in the past

Based on Gluck et al, 2007.

By the time I began my research on emotion in Reis's laboratory, well over a decade had passed since Kandel had started his project, and the field of neuroscience had advanced considerably. It was now possible to map connections between neurons in different brain areas, record cellular responses from neurons, disrupt neural activity, and measure molecules related to learning, not just in invertebrates but also in the brains of mammals and other vertebrates. Inspired by Kandel's success, researchers were thus beginning to explore memory in the brains of complex animals using Pavlovian conditioning.[18]

Some researchers had also begun to use Pavlovian conditioning to study emotional behavior, especially defensive or fear behavior, in mammals.[19] I was particularly taken with the behavioral studies of Pavlovian fear conditioning in rats being done by Robert and Caroline Blanchard and by Robert Bolles and his students Mark Bouton and Michael Fanselow.[20] These researchers showed that when an innocuous stimulus, such as a tone, was paired with mild electric shocks, it elicited freezing behavior, whether tested after a few minutes, or days or weeks later (Figure 2.2). Freezing is an innate defensive response that is just as important to animals as its more familiar partners, fight and flight.[21] (Indeed, as discussed in the next chapter, the fight-flight reaction is now often described as freeze-fight-flight.) The tone also produces increases in blood pressure, heart rate, and respiration and releases hormones like adrenalin and cortisol,[22] providing physiological support for energy-demanding defensive behaviors.

The meaning and even the value of the term "innate" in relation to behavior has been debated over the years.[23] It is now widely recognized that individual experience affects the way genetic programming is expressed, leading to the lack of a clear boundary between what is innate and what is learned. Although some shun the use of the term "innate," others think that it is useful, as some behaviors are decidedly more dependent than others on characteristics that are expressed so consistently within a species that the opportunity to learn the behavior seems limited. Freezing is one such example.

Fear conditioning is an example of associative learning, a process by which the brain forms memories about the relation between events. In the language of psychological learning theory, the tone in the example above is a conditioned stimulus (CS), the shock an unconditioned stimulus (US), and the responses elicited by the CS after conditioning are conditioned responses (CRs). During fear conditioning, the brain thus learns the relation between the CS and the US. After conditioning, the CS tone becomes a warning signal that danger is imminent. When the CS appears, it thus elicits the conditioned fear responses because it activates the CS-US association, which controls freezing and other fear CRs.

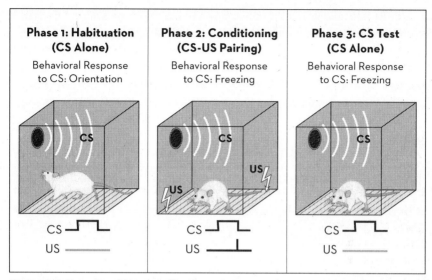

Figure 2.2: Fear Conditioning: The Procedure.

So-called fear conditioning is a variant on Pavlovian conditioning in which an innocuous conditioned stimulus (CS), often a tone, is paired with an aversive unconditioned stimulus (US), typically footshock. My laboratory has used this procedure extensively in rats, but it can be used similarly in a wide variety of animals, including humans. On the first day of typical study the rat is exposed to the CS alone (habituation). The next day one or more CS-US pairings occur (conditioning). One or more days later, the conditioned response is tested by presenting the CS alone (CS test). As described later, in order to better separate processes that give rise to feelings of fear from those that underlie the detection and response to danger, we have begun to use the expression "threat conditioning" in place of "fear conditioning."

Although freezing is said to be a conditioned response, the response is not learned. What gets conditioned is the ability of the CS to elicit the response.

In spite of much elegant work on Pavlovian fear conditioning at the behavioral level, this procedure was not being used in any systematic way to study how fear mechanisms operate in the brain.[24] Most research in this area was still employing complex instrumental conditioning tasks (especially tasks in which animals learned somewhat arbitrary responses to avoid shocks).[25] I thought that with all the tools available in the Reis laboratory, I could use the Kandel strategy, in conjunction with Pavlovian conditioning procedures, to trace the flow of information that enabled a meaningless stimulus to elicit fear responses in mammals (rats) following fear conditioning. This should be possible, I reasoned, because the responses, which are expressed the same in all rats, are driven by a specific stimulus that was completely under my control as an experimenter. As a result, I might be able to trace the flow of the stimulus processing from the CS sensory system to the CR motor system. So that's the approach I took initially at Cornell and then at NYU after I moved there to set up my own laboratory in 1989.[26] And work by my laboratory and the laboratories of colleagues using Pavlovian fear conditioning was very successful in achieving, in a few short years, what instrumental avoidance conditioning had failed to do—identification of the brain areas and connections between them that constituted what came to be known as the brain's fear system.

THE FEAR SYSTEM

I started my work on the neural basis of fear conditioning by determining which areas of the auditory system were required for the auditory CS to elicit freezing and blood pressure responses. Then, using anatomical connection-tracing techniques, I pinpointed possible output targets of the key auditory processing areas. One of the targets suggested by the tracing studies was the amygdala. When we lesioned this area, or disconnected it from the auditory system, the fear conditioned responses were eliminated. Within the amygdala, we also found an area that receives the auditory CS input (the lateral amygdala, LA) and connects with an area (the central amygdala, CeA) that sends outputs to downstream targets that separately control freezing and blood pressure conditioned responses. Further, we were able to locate cells in the LA input region that received both the auditory CS and the shock US. This was an especially important discovery because the integration of the CS and US at the cellular level was thought to be required for fear conditioning to occur. After the circuit and cellular changes involved in the process were identified, we turned to the molecular mechanisms in the LA that underlie the learning and expression of conditioned fear, many of which were the same as those discovered by Kandel and others in invertebrates.[27] In doing this work I have been fortunate to have had a fantastic group of people with me over the years, and I have dedicated this book to them.[28] They have contributed not just in terms of their lab skills and work ethic, but also intellectually.

The laboratories of several close colleagues also made many important contributions to this area of research. Initially, Bruce Kapp, Michael Davis, and I were the main players in this game.[29] But soon Michael Fanselow, who had started out studying fear conditioning at the level of behavior,[30] turned to questions about brain mechanisms as well.[31] Each had students who eventually also started laboratories,[32] and others joined in this exciting new area of work.[33] Fear conditioning became one of the most popular areas of research in neuroscience, and one known for having made great strides in relating brain to behavior.

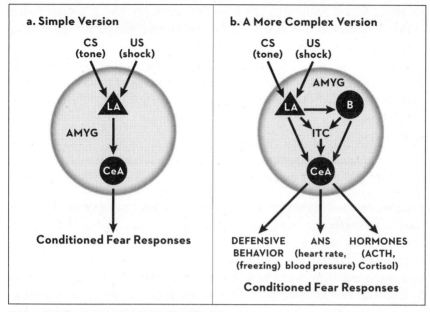

Figure 2.3: Fear Conditioning: The Circuit.

a. The Simple Version. The basic circuitry underlying the acquisition and expression of fear (threat) conditioning involves conditioned stimulus (CS) and unconditioned stimulus (US) sensory transmission to the lateral nucleus of the amygdala (LA), where a CS-US association is learned and stored. LA communicates with the central nucleus of the amygdala (CeA), which then connects with areas that control conditioned fear responses. **b. A Slightly More Complex Version.** The LA connects with the CeA directly and by way of other amygdala areas, such as the basal nucleus (BA) and intercalated cells (ITC). CeA then connects with downstream targets that separately control freezing, autonomic nervous system (ANS), and hormonal conditioned responses. Additional details are described in Chapters 4 and 11.

A simplified and more complex version of the amygdala-centered fear conditioning circuit uncovered by this collective body of work is illustrated in Figure 2.3.[34] A still more elaborate version of the circuitry will appear in later chapters. Through this research, the amygdala came to be viewed as a key component of the brain's fear system.[35]

FEAR AS A STATE THAT INTERVENES BETWEEN THREAT STIMULI AND FEAR RESPONSES

That amygdala-based circuits described above were part of a fear system was widely accepted. But the question of what exactly a fear system does has turned out to be a tricky issue.

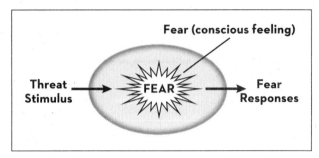

Figure 2.4: The Darwinian (Commonsense) View of Fear.
Darwin adopted the everyday or commonsense view, treating fear
and other emotions in people as "states of mind" that control emotional responses, and that have been inherited from our animal
ancestors.

The most obvious answer to this question is that a fear system makes fear: A
threat activates a fear system in the brain, and the result is a feeling of fear. This
feeling then drives the expression of behavioral defense responses and physiological accompaniments (Figure 2.4). Consequently, the defensive response is often
used as a sign that a person or animal is feeling fear.

William James called this the commonsense view of fear: We run from a bear
because we are afraid of it.[36] Although he rejected it, the commonsense view has
thrived. Charles Darwin was a proponent of this idea, calling fear a "state of
mind" that accounts for the expression of fear behavior.[37] This is also the view
most laypeople have of fear, and when journalists write about the brain's fear
system, they typically adopt this perspective as well. Some scientists also argue
that innate circuits involving the amygdala are responsible for the feeling of fear.[38]
But this is not the only perspective.

Behavioral research on fear conditioning took off in the 1940s and 1950s due
to the rise of O. Herbert Mowrer's influential theory that fear conditioning plays a
key role in maladaptive fear and anxiety in people.[39] Fear conditioning researchers
treated fear as a state that intervenes between threats and defense responses,[40] but
not in the way conceived by Darwin's and other commonsense theories. Coming
from behaviorism, most researchers avoided reference to conscious state and feelings.[41] Instead, fear was a central state, specifically a *defensive motivational state*,[42]
a physiological response in a hypothetical brain circuit[43] (Figure 2.5). Most of these
early central state researchers were not physiologists, and the physiological states
were conceptual placeholders, what are sometimes called *intervening variables* or
hypothetical constructs,[44] rather than actual brain mechanisms. However, when
researchers from this tradition started studying the brain, they applied the central

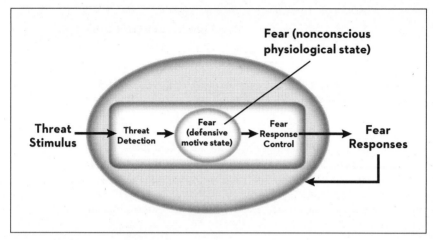

Figure 2.5: The Central State View of Fear.
Physiological psychologists trained in the behaviorist tradition treat fear as a physiological state (rather than a subjective feeling—"state of mind") that controls fear responses. However, calling the physiological state "fear" often leads to confusion about what fear really is, as even those who treat it in physiological terms often talk and write in a way that implies that the physiological state is the neural embodiment of the feeling of fear.

state term to the circuits uncovered. Thus, amygdala activity came to be viewed as the neural embodiment of a central state called fear.[45]

Central states, as a link between threats and fear responses, served a function similar to the Darwinian notion of feelings. But unlike Darwin, proponents of central states did not necessarily call upon a conscious feeling as the link between stimuli and responses. The entire question of conscious emotion was simply not considered relevant. For example, Michael Fanselow, a leading figure in the study of the central state of fear in animals, stated: "Our job is to redefine the concept of motivation in a scientific manner, and those new definitions should replace the layman's informal view. I don't see how subjective experience helps us do that."[46] According to Fanselow's mentor, Robert Bolles, "Human experience . . . cannot be invoked to lend ancillary validity to a construct that is otherwise anchored to behavioral phenomena. . . . Its surplus meaning must always remain surplus."[47] And Robert Rescorla, another prominent behavioral fear conditioning researcher, noted, "I do not think that reference to subjective experiences (by which I mean private experiences not subject to independent inter-observer ver-ification) is especially useful."[48] Because many central state proponents removed subjective experience (consciousness) from the causal chain, central states were *de facto* nonconscious states for them. But not all researchers had this view.

Mowrer, for example, was more like Darwin, treating fear central states as fearful subjective experiences that drive defensive behaviors. And even those who did eschew subjective states often wrote and spoke as if fear, the state, meant fear the feeling, describing threatened rats as "afraid," "frightened," "frozen in fear," "anxious," and the like. How would a naïve reader or listener know that their fear did not mean "fear."

FEAR AS A COGNITIVE CONSEQUENCE OF THREAT PROCESSING

I took a different approach from the ones described above. I thought that the Darwinian commonsense idea was flawed because it attributed too much to conscious fear, and the central state view was flawed because it ignored conscious fear. I believed conscious and nonconscious states both played roles, but the roles needed to be kept separate.

I was compelled toward my view by the split-brain studies I had done earlier. Gazzaniga and I noticed that the left hemisphere of a split-brain patient often commented on behaviors that were produced by the right hemisphere. One might expect that in people with this condition the left hemisphere would be surprised when it saw its body doing something for a reason that it (the left hemisphere) was not aware of. Instead, the left hemisphere took these unexpected behaviors in stride and wove them into its stream of thought. This was a fascinating phenomenon, so we designed some studies to explore it.[49] Basically, we coaxed the right hemisphere to respond behaviorally and then simply asked the left hemisphere, "Why did you do that?" Without hesitation, time and time again, the left hemisphere came up with an explanation. For example, when the patient was urged to stand up via stimuli presented to his right hemisphere, in response to our query as to why he did that, his left hemisphere said he needed to stretch; when he waved, it was because he thought he saw a friend out the window; he scratched his hand because it itched. These were fabrications on the part of the conscious brain, explanations for why the body responses were being generated. Gazzaniga and I proposed that the human brain does this all the time.[50] While we are not always privy to the motivations underlying the responses controlled by our brains, consciousness ties the loose ends together by coming up with an interpretation that unifies the mind and behavior by filling in the blanks of an otherwise incomplete mental pattern. Gazzaniga called this the *interpreter theory* of consciousness.[51] And I used this idea to explain how nonconscious processes that underlie emotional responses contribute to the conscious feelings that we experience.

In the mid-1980s, around the time I was starting work on fear conditioning in rats, I developed a model, based on our split-brain conclusions, of unconscious emotional processing in the brain. Specifically, in a 1984 book chapter, I proposed that emotional stimuli, transmitted to the brain via sensory systems, are processed nonconsciously to initiate emotional responses.[52] In 1996 I published the *Emotional Brain,* in which I framed the model in terms of brain mechanisms of fear, proposing that threat stimuli activate the amygdala in the process of eliciting fear responses; the amygdala processing, I argued, is automatic and requires neither conscious awareness of the stimulus nor conscious control of the responses.[53] This conclusion was supported by a handful of findings at the time but has since been bolstered by many studies showing that the amygdala can process threats and trigger conditioned responses without a person's being aware of the actual stimulus[54] and without having any feeling of fear.[55] This conclusion is consistent with our common experience of inadvertently responding to something and only afterward realizing that danger was present—as when one jumps back from a speeding bus and then, on the basis of this reaction, consciously realizes that he or she was in danger. The neural processes that enable us to consciously know that danger is present are slow compared with those that unconsciously control certain built-in protective (defense) responses. It is now recognized that the contrast between fast and automatic and slower, more deliberate processes is a fundamental organizational principle of the human mind and brain.[56]

In the 1984 chapter, I had also hypothesized how conscious feelings come about. I suggested that sensory processing in the brain splits into two channels, one that detects the emotional significance of stimuli and controls emotional responses and another that engages cognitive processing and leads to conscious feelings (see Figure 2.6). In the *Emotional Brain,* continuing the fear theme, I argued that the conscious feeling of fear is due to the representation in consciousness, via attention and other neocortical cognitive processes, of unconscious ingredients that are consequences of activation of the amygdala threat-processing circuit. I proposed that we could study the nonconscious aspects of fear in animals and humans alike, but that the conscious feelings of fear are best studied in humans. (More about that later.)

The amygdala circuit, I proposed, thus contributes to fear in two ways. It has a direct role in detecting threats nonconsciously and controlling subsequent behavioral and physiological fear responses, but it also has an indirect role, via cognitive systems, in the emergence of a conscious feeling of fear. Specifically, I proposed that the nonconsciously controlled brain and body consequences are raw materials that, when cognitively interpreted, contribute to conscious feelings of fear. When I used the term "fear system," then, I was referring to this entire

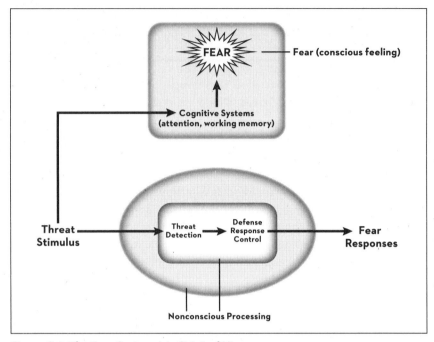

Figure 2.6: The Fear System: My Original View.

In 1984, I described emotional stimuli as being processed through two channels that diverge from the sensory pathway in the brain (LeDoux, 1984). One channel directs the stimulus to circuits that nonconsciously detect and respond to the stimulus, and the other channel directs the stimulus to cognitive systems that give rise to conscious emotional feelings. This idea is shown here in terms of threat stimuli and how they separately trigger so-called fear responses and feelings of fear. Over the next decade, the amygdala emerged as being responsible for the unconscious processing of threats, as described in the *Emotional Brain* in 1996. While I still adhere to the same basic idea about the two channels of threat processing, I no longer refer to this group of circuits as a "fear system." Saying that the amygdala is part of a fear system led to the view that the feeling of fear is a product of the amygdala, in spite of the fact that I argued that the amygdala is a nonconscious processor and that fear is a product of cognitive systems in the neocortex. I therefore now use more descriptive terminology to help make the distinction clear (see Figure 2.7 and Table 2.2).

process, including the amygdala's role in both controlling fear responses and in providing ingredients that indirectly contribute to conscious feelings of fear.

In retrospect, I now believe that it was a mistake to use the expression "fear system" to describe the role of the amygdala in detecting and responding to threats, and also erroneous to talk about fear stimuli and fear responses in this context. To refer to the circuit that detects and responds nonconsciously to threats as part of a fear system, as I did, and as central state proponents have done, unnecessarily complicated things. Because the most commonly accepted meaning of fear is the conscious feeling of being afraid, those outside the field who came across research on the fear system naturally thought that it was a system that generated feelings of fear. Thus, even though research on the so-called fear system was really mostly about systems that operate nonconsciously in detecting and responding to threats, and thus at odds with the commonsense view, our findings were interpreted as supporting the commonsense contention that feelings of fear are unleashed from the amygdala by threats. It didn't help that fear system researchers often wrote in a way that strongly implied that the neural activity in the amygdala is the underpinning of the feeling of fear; for example, describing rats as "scared" or "frozen in fear" or "anxious." When this is done, the difference between the central state and the commonsense view of fear vanishes.

Don't get me wrong. I'm not innocent in all this. Although I called upon separate brain circuits to account for nonconscious threat detection versus conscious feelings, I also talked about the amygdala as being involved in fear. I was basically thinking in terms of conscious versus unconscious fear. But I slowly began to realize that this was confusing, and finally, in 2012, I wrote a long piece called "Rethinking the Emotional Brain" in the journal *Neuron*. In it, I introduced the ideas of survival circuits and global organismic states. These, I suggested, provide nonconscious ingredients that are cognitively interpreted in the assemblage of conscious feelings, so-called emotions. In 2013 I was elected to the National Academy of Sciences and asked to write an inaugural article in the *Proceedings of the National Academy of Sciences*. I chose to write "Coming to Terms with Fear," which delved into issues about the language of fear and how fear arises as a cognitive interpretation of nonconscious ingredients generated by survival circuits that the global organismic states engender. These ideas are discussed in the following sections.

COMING TO TERMS WITH FEAR AND ANXIETY

Although research on fear conditioning has been very successful, we are now at a crossroads. We could easily continue along this same path and produce more findings, perhaps even important findings, but I believe we should take another approach, one based on a sharper conceptualization of what is actually being studied.

The language of fear, anxiety, and other emotions comes by necessity from our *folk psychology:* commonsense intuitions derived from introspections about how the mind works that have been handed down through the ages.[57] Scientists often start from such words and the intuitions behind them. But as Francis Bacon argued hundreds of years ago, scientists should be vigilant when using such common language terms and especially should guard against tacitly granting reality to things simply because we have words for them.[58] Everyone knows what the words "leprechaun," "unicorn," and "vampire" mean, but few believe they refer to actual beings.

The use of the vernacular language of fear to describe a system that detects and responds to threats is an example of what Bacon had in mind. It reifies fear and makes it *a natural kind*, something that is wired into the brain by evolution.[59] This belief justifies a search in the brain for a specific, innate locus of the phenomenon called fear. Findings about how the brain detects and responds to threats are used to conclude where fear lives in the brain because it is assumed that the same system that controls such responses gives rise to the feelings associated with them. (This is the essence of the Darwinian commonsense approach.[60]) We are told that this primitive emotion, fear, is inherited from animals and that every human in the world, no matter where he or she lives, and even if isolated from the rest of humanity, has the same or a similar fundamental (primal) experience when in danger and expresses a fear response to it in the same way.[61] The home of this universal fear is frequently assumed to be the amygdala, which until fairly recently was an obscure region of the brain but has now widely come to be known as the brain's "fear center."

Problems created by the ambiguous use of the word "fear" in scientific research can be illustrated by considering a 2012 finding that led magazines such as *Nature, Science, Wired, Scientific American,* and *Discover* to run dramatic headlines such as these: "Humans Can Feel Terror Even if They Lack the Brain's Fear Center," "Scaring the Fearless," "Evoking Fear in the Fearless," "Researchers Scare Fearless Patient," and "What Scared the Fearless Woman?" The fuss concerned the surprising finding that a woman with bilateral amygdala damage could still experience "feelings of fear."[62] But the only reason this would be considered surprising was if one believed that the amygdala is the primary wellspring of fearful feelings and that amygdala-controlled responses are reliable markers of these feelings. As I've said, and will explain in detail later, amygdala-controlled responses are *not* unequivocal signatures of fearful feelings. When we scientists use the term "fear" to refer to the neural mechanisms underlying both conscious feelings and nonconsciously elicited responses, we are inviting confusion.

The problem is not limited to fear. Jeffrey Gray's *behavioral inhibition theory* is a prominent animal model of human anxiety.[63] According to Gray and Neil McNaughton, the behavioral inhibition system of the brain is activated when goals are in conflict—for example, the need for food versus the risk of being exposed to predators. This conflict causes one's brain to attribute more risk, more harm potential, to stimuli and situations than we otherwise would, thus leading to a central state of behavioral inhibition that promotes risk avoidance rather than food seeking. Gray and McNaughton equated this brain state with anxiety because rats took more risk in conflict situations when treated with drugs, such as benzodiazepines, that relieve anxiety in people. But were they referring to a conscious feeling of anxiety that involves dread, foreboding, and worry? Or were they scientifically defining anxiety to mean a nonconscious brain state of behavioral inhibition that leads to motivational conflict and behavioral arrest? Gray and McNaughton sometimes claimed the latter (the central state version) but also often wrote in ways that could be interpreted the other way (in terms of conscious feelings). Certainly many followers of this approach, and there are legions, believe that the anxious feelings are direct products of the behavioral inhibition system.

A recent study showed that benzodiazepines relieved a so-called behavioral inhibition response in crayfish[64] (as a Cajun, I am always momentarily surprised when this is not spelled as "crawfish"). After receiving electric shock in a certain location, the crayfish remained immobile for an extended period (a behavior that was viewed as risk assessment) and then avoided the shock area, whereas the drugged crayfish were less inhibited (more exploratory). The authors claim that their results may lead to a new view of the emotional status of invertebrates. The study was published in *Science*, whose website led with the headline "Anxious Crayfish Can Be Treated Like Humans." The *New York Times* announced, "Even Crayfish Get Anxious," while the BBC, slightly more tempered, noted, "Crayfish May Experience a Form of Anxiety."

The theory of behavioral inhibition could, in fact, easily account for motivational conflict, behavioral arrest, and risk assessment in animals (including crayfish, rats, and people) without requiring the conscious experience of anxiety. Unfortunately, just as the meaning of defensive motivation gets tied up with subjective feelings when the motivational state and its brain system are labeled with the term "fear," the meaning of behavioral inhibition becomes entangled with subjective states when it or its brain system is labeled with the term "anxiety." Defensive motivation and behavioral inhibition are not the same as the conscious experience of fear and anxiety. That is not to say that the defensive motivation and behavioral inhibition states are unrelated to fear and anxiety, as they do make important contributions, but more is required to feel afraid or anxious.

In June 2014, a psychology website's headline read: "Fear Center in Brain Larger Among Anxious Kids."[65] The story that followed described a study that measured the level of anxiety in a large group of children based on a questionnaire answered by their parents.[66] The brains of these children were then imaged and the findings related to the parents' assessments. The results showed that the larger the amygdala of the child, the higher the level of anxiety rated by the parents. Let's consider what this actually means. In this study the parents did what animal researchers often do: They based a conclusion about anxiety, an inner feeling, on observations of behavior—their child seemed nervous, edgy, or had trouble concentrating or sleeping. Thus, although the size of the amygdala might well correlate with certain behaviors, whether it was related to feelings of anxiety was not tested. The website's headline was inaccurate in three respects: (1) What was being measured was behavioral activity, not the feeling of anxiety; (2) the kids were not anxious in the clinical sense, in spite of some being described as "anxious" in the story; and (3) the amygdala is not the fear center (and certainly not the anxiety center) if by fear or anxiety we mean a conscious feeling.

Fear and anxiety are hardly the only emotions that are viewed in these inaccurate and confusing ways. As we saw above, a number of emotions, including anger, sadness, joy, and disgust, are often considered to be wired into brain circuits.[67] The same problem arises in these cases—the conflation of innate systems that detect and respond in predictable ways to significant stimuli with systems that give rise to conscious feelings.

The science of the conscious mind is different from other kinds of science.[68] Physicists, astronomers, and chemists don't need to take seriously commonsense ideas about nature because people's beliefs and attitudes about the stars, matter and energy, and chemical elements don't affect the subject under investigation.[69] The fact that we commonly say (and some may actually believe) that "the sun rises in the east" does not have any scientific bearing on the fact that sunrise is an illusion. But psychologists do have to pay attention to folk psychology because people's common beliefs about the mind influence their thoughts and actions in daily life and are thus an important part of what psychology is all about.[70] Folk psychology is a window into the things that interest people and affect their lives.[71]

Typically, when a science matures, whatever vernacular terms it may have used are replaced with scientific ones.[72] Some argue that neuroscience will eventually do this for descriptions of mental states.[73] Terms such as "fear," "joy," and "sadness" will, in this view, be replaced with proper scientific language that does not have familiar, everyday connotations.

But the psychologist Garth Fletcher made a helpful distinction between using folk psychological ideas as explanations about how the mind works as opposed to using folk psychology as a way to identify things about the mind we want to understand and pursue scientifically.[74] He agrees that commonsense explanations about how the mind works will be replaced as psychological science progresses, but he thinks that the other side of folk psychology will continue to have a legitimate role because people's subjective experiences, their beliefs, fears, desires, and so on, affect how they approach their lives.

I side with Fletcher on this. If we want to understand conscious feelings, there's no getting around the use of words like "fear," "anxiety," "joy," "jealousy," "pride," and so forth. We run into problems when we label nonconscious processes with the words about conscious feelings. The conscious state takes on characteristics of the nonconscious processes: We assign the *feeling* of fear the responsibility for the defense responses elicited by threats. At the same time, nonconscious processes take on properties of the conscious feeling: the process of detecting and responding to threats comes to be the function of fear. The result is that it becomes very difficult to disentangle the concepts. We need a solution to get us out of this terminological quagmire.

A PROPOSAL

When scientifically discussing fear and anxiety, we should let the words "fear" and "anxiety" have their everyday meaning—namely, as descriptions of conscious experiences that people have when threatened by present or anticipated events. The scientific meaning will obviously go deeper and be more complex than the lay meaning, but both will refer to the same fundamental concept. In addition, we should avoid using these words that refer to conscious feeling when discussing systems that nonconsciously detect threats and control defense responses to them.

Thus, rather than saying that fear stimuli activate a fear system to produce fear responses, we should state that *threat stimuli* elicit *defense responses* via activation of a *defensive system*.[75] Because "threat" and "defense" are not terms derived specifically from human subjective experiences, using them would go a long way toward making it easier to distinguish brain mechanisms underlying the conscious feeling of being afraid or anxious from mechanisms that detect and respond to actual or perceived danger. Similarly, what we now call fear conditioning can simply be called what it is: threat conditioning. So, in place of "fear CSs" and "fear CRs," we can refer instead to "threat CSs" and "defensive CRs" (Table 2.2).

Table 2.2: Coming to Terms with Fear Conditioning

OLD TERMS	NEW TERMS
Conditioned Fear Stimulus	Conditioned Threat Stimulus
Conditioned Fear Response	Conditioned Defense Response

Some think we should stay the course—that the value of our work will be diminished if we separate the feeling of fear or anxiety from the mechanisms that detect and respond to threats. But separating the processes does nothing to diminish research on their individual contributions and instead paves the way to a richer understanding of how fear and anxiety emerge from neural circuits. For example, if, as I suggest, anxious feelings arise from mechanisms that go well beyond those that control the behavioral and physiological symptoms that also trouble anxious people, more effective therapies are more likely to emerge if we acknowledge the separate mechanisms involved than if we disregard the differences. Recall from Chapter 1 that modern understanding of panic disorder began when Donald Klein found a difference in the ability of a drug treatment to affect the conscious feeling of being terrified of dying (a cognitive interpretation) without changing physiological symptoms (which are direct consequences of survival circuit activation).

DEEP SURVIVAL

Problems with the way fear has been conceived become clear when we consider the widespread capacity to detect and respond to danger in the animal kingdom. This ability is necessary to survive and is present in every animal, whether it's a worm, slug, crayfish, bug, fish, frog, snake, bird, rat, ape, or human. Should we argue that crayfish, worms, and cockroaches escape from threats because they are driven by feelings of fear or anxiety? Or should we simply state that they possess mechanisms that enable them to detect and react to danger? Many are happy to agree to the latter characterization with respect to invertebrates, and even for fish and frogs; fewer are willing to do so where mammals are concerned. But if conscious feelings of fear are not required for a human to respond to danger, why should we resist the idea that defensive responses in other mammals reflect non-conscious processes as opposed to conscious feelings?

What we should be concluding, it seems to me, is not that humans have inherited fear per se from our animal ancestors, but rather that, through a long line of evolutionary history, we have inherited from them the capacity to detect and respond to danger. Problems result when it is assumed that this capacity depends on a feeling of fear intervening between threat stimuli and defense responses in human or nonhuman animals. This assumption compels the search for things that cannot be readily measured in nonhuman animals and forces researchers to bend the rules of evidence in order to conclude that such states exist. If, however, we accept that this capacity to detect and respond to danger does not require consciousness, then we will not be driven to search for elusive processes. We can, then, just study the specific topics that interest us without having to endlessly debate whether animals experience what we experience.

It is not my intention here to deny conscious feelings in animals. My aim, rather, is to highlight problems that hinder measurement of feelings in animals scientifically, to suggest a path forward that allows us to study those aspects of brain function that we know from objective evidence are shared between humans and other animals, and to focus on studies of humans for those functions that can be verified only in our species.

The lines of descent underlying threat detection may go much deeper than already discussed—as deep, in fact, as single-cell organisms, which also have to determine what's harmful and beneficial in their world. Viewed in this light, the capacity to detect and respond to threats is a deep survival mechanism, one that is as crucial to the life of an individual bacterial cell as it is for the numerous cells in a complex organism, whether in a fly, rat, or human. In animals, detecting and responding to danger is not only something that each cell in the body does on its own, it is also a function of a defense system in the brain, which enables the organism as a whole to defend itself. The evolutionary function of this ancient capacity is not to generate emotions like fear or anxiety, but simply to help ensure that the organism's life continues beyond the present.

We have, in short, been looking at the brain from a very human-centered point of view—as if our conscious introspections can tell us how ancient survival mechanisms that operate nonconsciously are organized in the brain. As I've noted, the conscious mind is compelled to explain what the brain does, even when it does not know.[76] We think we react to danger because we feel fear, and it is this belief that leads scientists to search for the fear in the brains of animals by looking for circuits that control defense responses. But instead of trying to locate fear in the brains of animals, we should be trying to understand how processes that are similar in animals and humans—namely, nonconscious processes that detect and respond to threats—contribute to feelings of fear we experience.

SURVIVAL CIRCUITS AND GLOBAL ORGANISMIC STATES

The innate view of emotion is said to apply to states that are wired into ancient subcortical circuits that have been inherited from our animal ancestors.[77] We certainly do have circuits that control innate responses that are commonly associated with emotions in people. But these are not emotion circuits; they are not feeling circuits; they are *survival circuits*.[78]

I recently introduced the expression "defensive survival circuit" as a way to discuss brain mechanisms that are often labeled as fear circuits.[79] To me, this term is preferable over "fear circuit" or "fear system" because it doesn't imply that defensive behaviors are propelled by conscious feelings of fear. Thus, the amygdala circuitry that has been a subject of my research does not make fearful feelings; it detects threats and orchestrates defensive responses to help keep the organism alive and well.

Defensive survival circuits are one of several classes of survival circuits that are common to most animals. Others include circuits for acquiring nutrients and energy sources, balancing fluids, thermoregulation, and reproduction.[80] The circuits involved in these functions are conserved within and between mammalian species, and to some this is true across vertebrates. The nervous systems of invertebrates are organized differently—for example, although they don't have an amygdala or any other brain areas that are present in vertebrates, they do have circuits that perform survival functions that are similar to, and likely precursors of, comparable functions in vertebrates.[81] Because similar functions are also present even in single-cell organisms lacking nervous systems, these functions predate, evolutionarily speaking, neurons, synapses, and circuits[82] and as such are primitive precursors of survival functions in more complex organisms with nervous systems.[83] Survival circuits do not exist to make emotions (feelings). They instead manage interactions with the environment as part of the daily quest to survive.

Survival circuits are activated in situations in which well-being is potentially challenged or enhanced. The overall response of the brain and body that results is a *global organismic state.*[84] For example, activation of a defensive survival circuit results in a *defensive motivational state.*[85] Such states involve the whole organism (that is, body as well as brain) as part of the task of managing resources and maximizing chances of survival in situations where challenges or opportunities exist.[86] Global organismic states in mammals and other vertebrates,[87] like the survival circuits that initiate them, are elaborations of similar states in invertebrates.[88]

When a defensive survival circuit detects a threat, as I pointed out in Chapter 1, it not only triggers defensive reactions; it also activates brain areas that control the widespread release of chemical signals, including neuromodulators and hormones.[89] As a result, the organism becomes highly aroused and vigilant—attuned to the sensory environment, focusing on the clear and present danger, but also being on the alert for other potential sources of harm. The threshold for the expression of additional defensive responses is lowered, whereas other motivated behaviors, such as eating, drinking, sex, or sleep, are suppressed. This global defensive motivational state reflects the wholesale mobilization of brain and body resources for the purpose of staying alive and helps ensure that the subsequent actions that are performed in an effort to cope with danger in more complex ways guided by past instrumental learning are suited to the external circumstances—escape or avoidance when in danger. In other motivational circumstances, global organismic states function similarly: for example, helping to guide the approach to food or drink when energy supplies or fluids are low, etc.

The idea of global organismic states is closely related to that of central states (see earlier discussion). And the idea of defensive motivational states has also

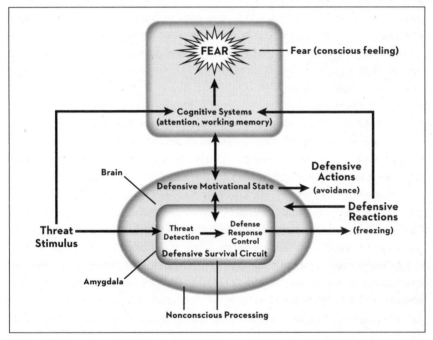

Figure 2.7: The Survival Circuit View of Fear and Defensive Motivation.

Because my traditional view of the fear system (Figure 2.6) was often misconstrued as implying that the amygdala is the seat of fear in the brain, I have revised my terminology. In the current model the term *fear* is no longer used to describe functions of the amygdala. I now describe the amygdala circuit that detects and responds to threats as a *defensive survival circuit*. One consequence of survival circuit activation is the establishment of a *defensive motivational state* throughout the brain. This state is not the neural instantiation of a feeling of fear. The state (or neural components of it) provides neural ingredients that when cognitively interpreted give rise to a feeling of fear. This view differs from the commonsense approach in that fear does not cause defense responses. It differs from the central state view in that both defense responses and the central state are consequences of activation of the survival circuit. While the defensive motivational state does not cause innate defensive reactions and changes in body physiology that accompany them, it does contribute to motivation of instrumental behaviors that allow the organism to act, rather than simply react, in the face of danger.

been around for some time.[90] But earlier views treated defensive motivational states as the cause of defensive responses. My view, in contrast, is that the defensive motivational state is a global consequence of activating a defensive survival circuit (Figure 2.7). Defensive responses in the present conception are thus not caused by, but instead actually contribute to, defensive motivational states. But as just noted, once a global motivational state is present, it helps guide instrumental behaviors in the effort to survive and thrive.

Although defensive motivational states can occur in both simple and complex organisms, only animals that have the ability to be consciously aware of their own brain's activities can experience the state we commonly refer to as fear. I propose that defensive motivational states, or at least components of such states, are ingredients that, along with other factors such as perceptions and memories, contribute to conscious feelings. Thus, when a defensive survival circuit has been activated in your brain and its consequences linked to the present stimulus and to your memories regarding it and similar stimuli, all in relation to your awareness that the event is happening to YOU, a feeling of fear arises.

Ultimately, feelings like fear require that we somehow have the *concept* of fear, based on words and their extended meanings, in our minds.[91] We learn such concepts because they are important for our well-being and account for significant experiences in our lives. We also learn to associate these concepts and relevant words with the consequences of defensive survival circuit activity. Every culture ends up having these concepts and corresponding words because every human brain has built-in defensive survival circuits that produce similar kinds of innate reactions and supporting changes in brain and body physiology. But the feeling of fear is not a direct product of a survival circuit. It is a cognitive interpretation, which is, in canonical instances, based on the consequences of survival circuit activation. And because survival circuits have an innate foundation within a species, they provide at least some universal signals that are the basis for cognitive interpretation, and thus help make fear feel like a familiar experience when in danger, and also make self-reports of fear across individuals have similar content as well. Animals obviously cannot label and interpret survival circuit activity in the ways made possible by human language. They may experience something, but it is incorrect, in my opinion, to assume that their experience is the same as, or even similar to, what a human often experiences when a defensive survival circuit is active in his or her brain.

In sum, fear is not something that is unleashed from an innate circuit. Instead, it is, as I and some others argue, a conscious state that emerges when certain kinds of nonconscious ingredients coalesce and are cognitively interpreted.[92] If so, then the search for innate circuits that unleash fearful feelings is

the wrong approach for understanding fearful feelings. Innate circuits are important to survival but are not feeling circuits.

THE MENTAL LIFE OF ANIMALS

Many people, and some scientists, believe that we can use animal behavior as a
means of revealing what is on the animal's mind. If by "mind" we mean what most
contemporary scientists do—cognitive functions that involve information processing and that largely operate nonconsciously—we can learn quite a lot from
studies of animals. As the esteemed psychologist Karl Lashley noted in 1950, we
are never consciously aware of information processing; we are only aware of its
outcome when it generates conscious content.[93] These nonconscious mental
functions are sometimes referred to as the *cognitive unconscious*.[94]

The ability to use complex cognitive (mental) functions that process information to control behavior is thus different from the ability to have conscious experiences.[95] Both are "mental" in the sense that they involve the use of internal
representations of the world. But animals can satisfy nutritional and fluid needs by
consuming food and drink, have sexual intercourse, writhe when injured, and
freeze or flee when threatened, all without the necessity of conscious awareness
that they are doing so. Specifically, when animals engage with their environments
in daily life, they rely not only on innate responses and conditioned reactions, but
also on goals, values, and decisions, each of which involves complex cognitive processes but which do not necessarily require conscious awareness of those processes.

When it is assumed that conscious states in animals underlie behaviors that
seem to be similar to the ways humans would act in a certain situation, this is
"argument from analogy" rather than from scientific evidence.[96] Thomas Chamberlin noted at the end of the nineteenth century that scientists must consider
multiple hypotheses in scientific research in order to prevent biases in data interpretation.[97] Consciousness should therefore only be attributed to an organism if
there is both compelling evidence that the behavior expressed by the organism
depends on consciousness and compelling evidence that the behavior cannot be
explained in terms of nonconscious processes.[98] Seldom do both considerations
go into discussions of animal consciousness.

Signs of cognition and intelligent behavior are often simply interpreted as evidence that animals are conscious and have feelings similar to our own. For example,
in the introduction to *The Inner World of Farm Animals*[99] by Amy Hatkoff, the
esteemed primatologist Jane Goodall writes: "Farm animals feel pleasure and sadness, excitement and resentment, depression, fear, and pain. They are far more aware
and intelligent than we ever imagined . . . they are individuals in their own right."[100]

And in a 2013 interview, Goodall said she had seen many examples of compassion, altruism, calculation, communication, and even some form of conscious thought; she concluded that animals "show emotions similar to those we describe in ourselves."[101] She speaks with authority when she generalizes from behavior to conscious thought and feelings, but how does she really know what animals experience?

Animals certainly act in ways that show they can solve complex problems inside their heads (they behave intelligently), and it goes without saying that each animal is, as Goodall noted, an individual. But scientific practice, at least in my view, cautions us, in the absence of rigorous and compelling evidence, to avoid the attribution of conscious feelings to animals on the basis of our intuition, no matter how strong, that animals *should* have such feelings.[102] It also goes without saying that the lack of compelling scientific evidence that animals have the kinds of states that we experience as fear, love, joy, sadness, and so forth does not in any way justify their callous or abusive treatment, whether for research, recreational, cosmetic, or nutritional purposes. The treatment of animals in research and in society at large is today held to a high standard by the laws of the United States and many other countries. The assumption underlying these laws, that animals do have feelings, is based on a particular ethical position adopted by our society rather than on scientific data. So long as philosophers, scientists, and the public recognize the difference (that different considerations underlie ethical and scientific conclusions), the integrity of both societal values and science can be preserved.[103]

Whether animals are conscious ultimately depends on how consciousness is defined. I will discuss this in more detail in later chapters, but animals are obviously conscious in the sense that, when awake, they are alert and behaviorally responsive to meaningful stimuli. The key issue is whether they are conscious in a mental state way. One problem is that in animals it is very difficult to distinguish a behavioral response controlled by a nonconscious cognitive (mental) process from one that depends upon conscious awareness. There is also a distinction to be made between being aware that some stimulus is present and being aware of one's self as the entity that is experiencing that stimulus. We can make these distinctions in humans through self-report, but in animals, the lack of language is a considerable barrier, as discussed in Chapters 6 and 7.

It's natural to think that a rat or a cat fleeing from danger feels fear. But, as I've noted, there is considerable evidence that the feeling of fear is not necessarily what causes humans to respond in protective ways in the face of danger; responses to danger can even occur without any feeling at all. The opposite is also true. One may feel extremely fearful while looking completely fearless on the outside: For example, the fear recounted after heroic actions by soldiers in battle or of parents protecting their children from harm. If we cannot with certainty conclude what a human is experiencing on the basis of how they react to danger, how can we claim that we know what animals are feeling on the basis of their reactions?

The primatologist Frans de Waal supports the idea that animals have conscious feelings, but he also admits that "we cannot know what they feel."[104] He even admits that too much is made of animal consciousness in his specialty, empathy, noting that in humans much of what is labeled as empathy occurs automatically—that is, nonconsciously. But he bolsters his case that animals have feelings by noting that it's not so easy to be certain about what people are feeling, either, because the only way to know about human feelings is to rely on verbal self-report, which he distrusts. In light of this, he concludes that "postulating feelings in animals is not as big a leap as it may seem."[105]

Unlike de Waal, I don't believe we have the same problems when generalizing from our private experiences to the minds of other people as we do when generalizing to other animals. In the absence of brain damage, severe psychological disturbances, or genetic mutations, all humans are endowed with the same general capacities. If my brain can be conscious, so can yours. Our experiences may differ, but we all have capacity for the same kinds of experiences. The same cannot be said of other species. The human brain differs in significant ways from even our closest primate cousins, not so much in terms of the areas that are present but in their patterns of connectivity[106] and cellular organization.[107] As we go further and further back in our branching evolutionary history, the differences become more profound; for example, certain areas of the prefrontal cortex that have been implicated in complex cognitive functions in primates are lacking in other mammals.[108] Furthermore, because our species is naturally endowed with language, we can share information about common topics that are comprehensible to everyone, whether they involve memories of a trip or the amazing sight of the sun setting over the ocean. Self-reports are not perfect (we may sometimes be mistaken in our recollections or may occasionally intentionally mislead others), but, as many highly regarded cognitive and brain scientists and philosophers have noted,[109] verbal self-reports are the best way to verify and compare conscious experiences between two organisms.

A major reason why verbal reports are so important is that they provide a way to distinguish brain states that involve consciousness from brain states that do not. Because other animals can't speak, we can't directly assess their conscious mental states separately from their nonconscious ones. This is not a statement about whether animals have conscious experiences but rather about the difficulty, in the absence of verbal report or other acceptable, rigorously based forms of reporting/commentary,[110] of knowing whether they do, and if they do, what such experiences might be like.

Humans have a particular thing in mind when they say they feel fear or anger or joy. And many, probably most, people find it useful to assume that their pets have similar kinds of experiences. But when such assumptions lead scientists to look for fear or other emotions in animals, they end up pursuing something that cannot be readily verified. In some sense, we can't help attributing thoughts and feelings to animals,[111] as anthropomorphism may be an innate feature of the human brain.[112] It is certainly a part of everyday folk notions about how the mind works.[113] We are even prone to anthropomorphizing when observing inanimate objects. For example, if human subjects watch a video screen on which a large triangle is chasing a small circle, and the triangle keeps bumping into the circle even when the circle changes course, the subjects interpret the triangle as aggressive and the circle as fearful.[114]

However, the fact that a belief or attitude is natural, or even innate, does not necessarily mean that it is scientifically correct.[115] For millennia common sense has told humans that the earth is flat, and still does. We are perfectly fine continuing to make that assumption while driving a car, even over long distances, because in our conscious experience at any given moment, it *is* effectively flat. Not all practical activities of daily life need to be based on scientific facts. We can treat our pets as if they have feelings (I certainly do), even as we acknowledge that their brains work differently, and possibly in ways that are not capable of making conscious experiences such as those we have.

The problem arises when we transfer such everyday assumptions into our understanding of the brain. Earlier, I noted that the part of psychology that is concerned with people's conscious feelings, beliefs, and desires may always depend on the use of language from everyday folk psychology (sometimes scientific terms in fact become part of folk psychology). But because we cannot make the same assumptions about the brains (and thus minds) of other animals that we can about humans, we cannot simply transfer the language of human conscious experience to them. Animals cannot label and interpret survival circuit activity in the ways made possible by human language. They may experience something, but it is incorrect to assume that their experience is the same as, or even similar to, what a human experiences when his or her defensive survival circuit is active. Too often, the language of human conscious experience is used to describe processes in animals that work nonconsciously in humans. We need a sharper conception of these processes in humans so that we can know how to talk about both human and animal behavior.

As I noted in Chapter 1, to experience fear is to know that YOU are in danger. This involvement of the self in fear, and the rapid and inevitable morphing of fear into anxiety, makes human fear and anxiety unique. Even if consciousness is present in other animals in some form, it can't exist in the way made possible by the human brain.

The key issue, for me, is: How far can we go in explaining behavior without having to call upon conscious states in animals? I think pretty far, since like human behavior much of animal behavior is controlled nonconsciously. Remember, nonconscious does not mean nonmental. It just means the organism was not explicitly aware that the process was happening in its brain. As I noted above, attribution of consciousness to explain a particular class of behavior, or a specific exemplar within such a class, should, whether in humans or other animals, only be done when the behavior cannot be explained by processes that operate nonconsciously.

I have no problem with speculation about the existence of consciousness in animals. My problem is when those speculations come to be taken as a given, based on assumptions rather than data. Good science can be driven by wild speculation. But when speculation about what seems to be true is treated as truth, we run into problems. Scientists have an obligation to draw a line in the sand between speculation and data, and to help others outside their field recognize where the divide occurs, even if it is fuzzy at the edges. This is especially important in areas of research that are of interest to the general public and that are being used to help understand and treat problems that plague people. Miscommunication can result in inaccurate translation of scientific findings into clinical applications.

RETHINKING THE EMOTIONAL BRAIN

The view of emotion that I've described in this chapter can be summarized succinctly. Often, when scientists (including me) have used the label "emotion," we have been talking about the consequences of survival circuit activation. These circuits do not exist to make us or any other animal feel a certain way. Their function is to keep the organism alive. Emotion is the feeling an organism has when it consciously experiences these consequences. Keeping separate the processes that detect and respond to significant events from the processes that generate feelings is thus key to making progress in understanding what emotions actually are and how they work. Although these processes are related, conflating them only impedes a genuine understanding of the emotional brain.

In recent times scientists have either attributed too much to conscious feelings (the commonsense view) or have not granted feelings enough of a role (the typical central state view). My aim here is to develop a view that strikes a balance, giving feelings a central role in the science of fear and anxiety while not giving them more credit than they deserve.

CHAPTER 3

LIFE IS DANGEROUS

"As soon as there is life there is danger."

—RALPH WALDO EMERSON[1]

A number of years ago I had an Australian colleague who freely shared bits of wisdom from down under. He often used an expression that stuck in my mind: "Time for a kangaroo's breakfast—a quick pee and a look around." That colloquialism appealed to the template of quaint yet crude folksy sayings stored in my brain from a rural Louisiana childhood. But as a scientist, I was troubled by the order of the sequence. I didn't know much about kangaroos, but imagined that they lie down when sleeping and stood up to pee. I also assumed that these animals live among predators, in which case they should first have taken a quick look around while close to the ground before standing up and announcing their presence on the landscape. Why risk life for a few seconds' head start on bladder relief?

For most animals, life is a constant struggle to survive from moment to moment, day to day, and year to year. But not only do they have to be on the lookout for bloodthirsty brutes seeking a meal, they also have to find food, drink, and shelter, and if their species is to survive, they have to reproduce. Each of these survival activities comes with some risk of being eaten by hungry predators or beaten by territorial enemies. If you've ever watched nature shows on television, you know that animals don't usually lounge around after a meal or a coital encounter—it's more a matter of "eat and run" and "slam-bam, thank you, ma'am." Behavioral choices in the wild reflect the fact that life is a dangerous undertaking.

Fortunately, although humans have to a great degree developed ways to get through the day without having to be concerned about being devoured, other threats do plague the "anxious animal," which our species certainly is. We may face fewer physical threats on a daily basis, but we more than make up for their absence with our brain's capacity to anticipate threatening events, including some that may never happen. Yet the animal ways are still in our brains and are called upon whenever we encounter a barking dog, are challenged by an aggressive colleague or stranger, or face any kind of situation that has the potential to cause us physical or psychological harm.

In this chapter we continue our journey toward understanding how fear and anxiety work in the brain by exploring the ways in which animals, including humans, respond to present or future threats. This discussion will set the stage for considering in later chapters how conscious feelings of fear and anxiety come about. The focus of this chapter on laboratory studies of conditioning procedures in animals is also the basis for many of the ideas about ways to enhance the effectiveness of psychotherapy, described in later chapters.

THREATS GALORE

Predation is the ultimate threat in the wild, where the food chain rules. In the ocean little fish are eaten by bigger fish, which are eaten by even bigger ones. On land small mammals such as mice eat insects and seeds and are, in turn, eaten by cats, foxes, and carnivorous birds. We humans can often choose what to eat because we have created technological ways to overcome the larger size, greater strength, and superior predatory skills of many of our prey. You might say we are the ultimate predators.

But being subject to predation in the name of nutrition is not the only source of danger in life. Other members of one's own species can do a lot of damage as well, fighting over food, territory, and mates, and sometimes for no particular reason at all. Scientists distinguish *predatory aggression* from *conspecific aggression*, in which the former is directed toward another species and the latter toward members of one's own.[2]

Bad things also happen in nature that don't involve being eaten or attacked by another animal. Consumption of rotten food can cause harm. So can dehydration— every cell in the body depends on fluids being kept in balance. And let's not forget about temperature extremes, in which protection from the elements through shelter is also essential to life. When inner body temperature changes significantly, cells don't do well; and when our cells suffer, so do we.

These examples reflect the key survival needs that all organisms must satisfy in order to persist as individuals: defense against external harm, maintenance of energy and nutrition supplies, fluid balance, and thermoregulation.[3] Each is controlled by innate circuits in the brain—the survival circuits discussed in the previous chapter. Although reproduction is not necessary for the survival of the individual, it is, of course, the basis of species continuity and has its own survival circuits.

Survival functions are not independent of one another.[4] Searching for food and drink, for example, often leaves an animal susceptible to predation, which puts defense in conflict with foraging. When a predator is detected, foraging and other survival activities are suppressed. Foraging, for its part, burns energy and can lead to heat and fluid loss, further increasing the need for food and drink. When energy supplies are low, activity levels go down to conserve resources for foraging. Shelter is used to regulate body temperature and also to hide from predators. When any of these survival needs are not met, or the activities associated with them are compromised, things go south. Life is indeed dangerous.

FACTORY-INSTALLED DEFENSES

All species have inborn ways of dealing with perennial threats such as those described above. Although there are many sources of harm in life, the defense mechanisms in the brain that evolved to deal with predation are a key foundation upon which fear and anxiety are built.

The classic antipredation options are typified by the expression "the *fight-flight response.*" Coined by Walter Cannon in the early twentieth century, this well-worn nugget describes behaviors that occur in emergency situations where life or well-being is on the line.[5]

"Frozen in fear" is another common expression that describes a vital defensive behavior. As noted by Darwin, "The frightened man at first stands like a statue motionless and breathless or crouches down as if instinctively to escape observation."[6] Indeed, freezing is a typical defensive response that is expressed in many species when threatened.[7] But doesn't freezing set you up for a quick end? In fact, the contrary is the case: Freezing is actually a quite effective antipredation response.[8] For one thing, it helps reduce detection. Motion is an important cue used by predators, as movement can be seen at greater distances than other visual details. Also, if predator and prey are in proximity, movement is an innate trigger for attack.

For many animals the key defensive strategy involves selection from a three-part menu: freeze first; flee if you can; and fight if you must[9] (Figure 3.1).

Freezing, fleeing, and fighting are defensive *reactions* automatically triggered

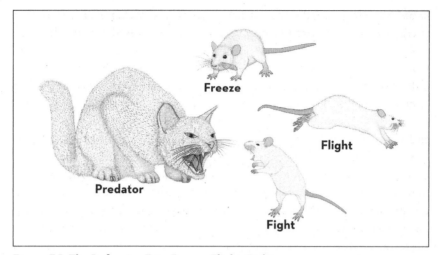

Figure 3.1: The Defensive Trio: Freeze, Flight, Fight.
In the presence of danger, many animals call upon a trio of defensive strategies, including freezing, fleeing, fighting.

by external stimuli and are expressed the same way (or very similarly) in all members of a species. Later we will contrast such reactions with defensive *actions*, responses that are learned by the individual because of their success in preventing harm.

The trio of freeze, flight, and fight are fairly universal behavioral defensive reactions in mammals and other vertebrate species. But some species have other options available,[10] such as "playing dead," which is also called *tonic immobility*. Like freezing, this behavior can help prevent attack, but whereas in freezing muscles are contracted and poised to be used in fight or flight, in tonic immobility the body is flaccid. Another such response is *defensive burying*: Rodents will use their paws and head to shovel bedding materials (in the laboratory) or dirt (in the wild) toward an aversive stimulus. Other behavioral options include making loud noises; retreating into a shell; rolling into a tight, impenetrable ball; choosing to live in a predator-free or predator-inaccessible area, such as underground; or relying on safety in numbers by living and foraging in a group.

In addition to these behavioral defense reactions, there are a variety of other options, mostly involving either permanent or inducible body characteristics.[11] Some animals have armor, sharp spikes, or poisonous spines. Others use crypsis, a kind of natural camouflage that helps avoid being detected and attacked by changing skin or feather coloration to blend in with certain features of the environment, or deimatic behavior, which involves making one's body look bigger, stronger, meaner, venomous, or assuming some other threatening trait. Darwin pointed out that the goose bumps that appear on our arms and legs in response to a threat are a vestige of our hairier ancestors, who made their fur stand on end to make themselves appear larger in stature.

Predator and prey are opponents in a continuous game of hide and seek. But in spite of the way I've described this dynamic so far, the seeking is not always done by vision. Many mammalian predators rely on odors (especially pheromones present in urine, feces, and fur)[12] for tracking, whereas prey use vocalizations as warning signals.[13] For example, rodents have evolved the ability to issue alerts by emitting ultrasonic vocalizations that are not detectable by the auditory system of feline predators.[14]

Life is not static—evolution is an ongoing process, not an end state. Prey are an element of the adaptive environment of the predator, just as predators are for prey. For that reason it makes perfect sense that over the course of time, predator and prey evolve to adapt more effectively to each other.[15] If, for example, a prey species has some trait that is useful in avoiding a predator, those predators that have their own traits that put them at an advantage relative to that defensive feature in prey will increase in number. Pressure then returns to the prey to adapt to this newly evolved population of predators. This process has been called the evolutionary arms race.[16]

We tend to think of these antipredator strategies as ways of protecting one's own hide. However, as evolutionary biologists point out, inborn defenses are not necessarily all directed toward self-survival. Some defenses are useful in protecting the survival of one's mates, offspring, or other members of the social group or species.[17]

THE PHYSIOLOGICAL SUPPORT OF DEFENSE

In the late 1890s Walter Cannon, who would later introduce the fight-flight concept, was researching the digestive system of animals.[18] He noticed that when an animal was stressed, its digestion was disrupted—specifically, the peristaltic contractions of stomach muscles ceased. This led him to pursue the role of the nervous system in emotionally arousing situations. He went on to explore how the autonomic nervous system (ANS) controls the physiology of the body in challenging circumstances, such as those involving threats or other stressors. Cannon called this the *emergency response,* a term he used interchangeably with fight-flight response.

The ANS has two components, the sympathetic and parasympathetic divisions (Figure 3.2). These each send nerve fibers to the various tissues and organs

Sympathetic	Target Organ	Parasympathetic
Dilates Pupil	 Eye	Constricts Pupil
Inhibits Tearing	Lacrimal Glands 	Stimulates Tearing
Inhibits Salivation	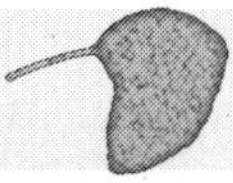 Salivary Glands	Stimulates Salivation
Increases Heartbeat	Heart 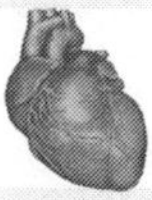	Decreases Heartbeat
Constricts Arteries	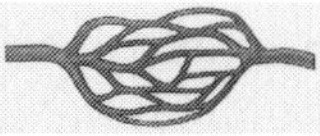 Blood Vessels	No Effect
Bronchial Dilation	Lungs 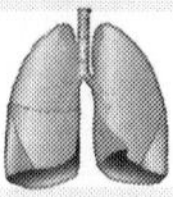	Bronchial Constriction
Increased Release of Epinephrine and Norepinephrine	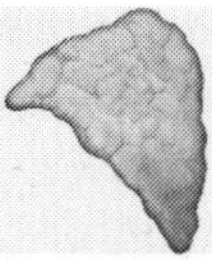 Adrenal Gland	Decreased Release of Epinephrine and Norepinephrine
Stimulates Glucose Release	Liver 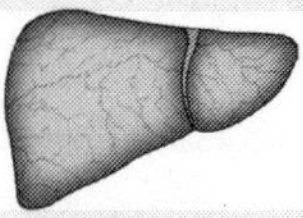	Inhibits Glucose Release
Inhibits Digestion	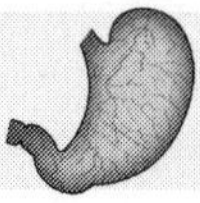 Stomach	Stimulates Digestion
Stimulates Intestines	Intestines 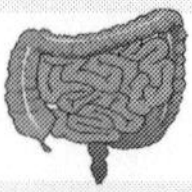	Inhibits Intestines
Contraction of Rectum	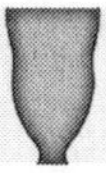 Rectum	Relaxation of Rectum
Relaxes Urinary Tract	Bladder 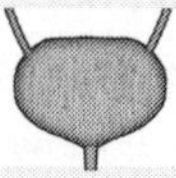	Contracts Urinary Tract
Stimulates Erection	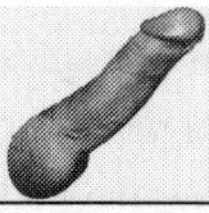 Reproductive Organs	Stimulates Ejaculation and Vaginal Contraction

Figure 3.2: Some Functions of the Sympathetic and Parasympathetic Divisions of the Autonomic Nervous System.

The two divisions of the autonomic nervous system (ANS) often work in opposite ways, such that if the sympathetic division stimulates an organ, the parasympathetic will inhibit it. In this way, the ANS can arouse the body to meet demands and restore balance when circumstances change.

of the body and regulate their function. The classic view is that the *sympathetic division* takes control in situations that require energy mobilization, such as when life or well-being is on the line, and the *parasympathetic division* counters the sympathetic response and restores balance (or *homeostasis*) in the body once the threat is over.[19] While this view is still generally accepted, it is now recognized that the two divisions interact in more complex ways than originally thought.[20]

Cannon noted that defensive behaviors are energy demanding and that activation of the sympathetic nerves is essential for managing energy resources. Thus, sympathetic nerve activity increases respiration, which helps convert lactic acid into glucose and provide muscles with their main source of energy. In addition, it raises heart rate, increasing the flow of blood through the circulatory system to aid in the delivery of energy to the muscles. Sympathetic nerve activity also commands the adrenal medulla to release the hormone adrenaline (epinephrine), which, Cannon proposed, stimulates the conversion of glycogen in the liver into glucose, adding more energy supplies. The sympathetic nerves also participate in the redistribution of blood in the body in order to direct energy to the muscles needed for fight or flight. To do this, blood flow has to be reduced in areas such as the gut and skin and increased in the limbs. This is achieved by constricting and relaxing blood vessels in the relevant body tissues. The decreased flow in the skin has the added benefit of reducing blood loss in the event of a wound. Cannon used the term *"sympathoadrenal system"* to describe this combination of sympathetic nerves and adrenal medullary hormones underlying the fight-flight reaction.

It is often not appreciated that there are two distinct classes of physiological adjustments controlled by the ANS. The first is an innate physiological response that anticipates a certain innate behavior.[21] Thus, when the defense system detects danger, it initiates both behavioral and physiological responses that have been "wired in" by their usefulness. This is what Cannon's emergency reaction was all about. But in addition, when any behavior is performed, whether the behavior is innate or learned or just a random occurrence, metabolic support is needed to carry out the response to its completion. Such homeostatic adjustments occur on the fly rather than by way of innate programming and are regulated by specific momentary needs of the body. This helps explain why physiological responses correlate better with simple innate reactions than with complex learned emotional behaviors,[22] as the latter can be quite variable from person to person and thus do not show a reliable pattern across individuals the way responses associated with innate behaviors do.

Hans Selye, working around the same time as Cannon, extended the emergency reaction system to include the adrenal cortex and its hormone, corti-sol.[23] Cortisol, a steroid hormone, also contributes to energy regulation. It is released from the adrenal cortex by *adrenocorticotrophic hormone* (ACTH), which itself is released by the pituitary gland.

As a result of Cannon and Selye's work, the emergency reaction (or *alarm response*, as Selye called it) came to be viewed as being controlled by two com-plementary physiological axes: the *sympathoadrenal axis,* involving the sympa-thetic nervous system and adrenaline released from the adrenal medulla, and the *pituitary-adrenal axis,* involving the release of cortisol from the adrenal cortex (Figure 3.3). The response of the sympathoadrenal axis is rapid, occurring within seconds of encountering a threat. The pituitary-adrenal axis, by contrast, is slower and not fully expressed for minutes or even hours.[24]

It is common to discuss the sympathoadrenal and pituitary-adrenal systems as responsible for making us feel "stressed out." This flows naturally from Can-non's idea of an emergency reaction and Selye's of an alarm response, as well as Selye's notion that stress leads to three phases of response: alarm, resistance, and exhaustion. In modern times, the work of Bruce McEwen, Robert Sapolsky, Gustav Schelling, Benno Roozendaal, and James McGaugh has shown how the negative consequences of stress, especially those mediated by cortisol, not only affect memory and other cognitive functions but also compromise immune function and lead to illness.[25] But as these investigators all emphasize, the pur-pose of the so-called stress response is to help the organism adapt, not to wear us out or make us feel bad. It is only when the stressful event is prolonged and espe-cially intense that the negative consequences result, and resistance gives way to exhaustion.

The psychiatrist Donald Klein has proposed another physiological response, the *suffocation alarm response,*[26] which is triggered by internal physiological threat signals, such as an excess of carbon dioxide (hypercapnia), leading to "air hunger" (shortness of breath). Whereas the sympathoadrenal and pituitary-adrenal responses are relevant to all forms of fear and anxiety, the suffocation alarm system is particularly relevant to a subgroup of patients with panic disorder. These people, Klein suggests, have a hypersensitive suffocation alarm system, which falsely detects a dangerous level of CO_2 and leads to hyperventilation, which in turn produces an actual rise in CO_2 (due to short, fast inspiration). The resulting dizziness and light-headedness lead the person to misinterpret the physiological changes and worry and dread follow in the panic-stricken person. Klein's hypothesis is supported by data[27] but is viewed as controversial by some researchers.[28]

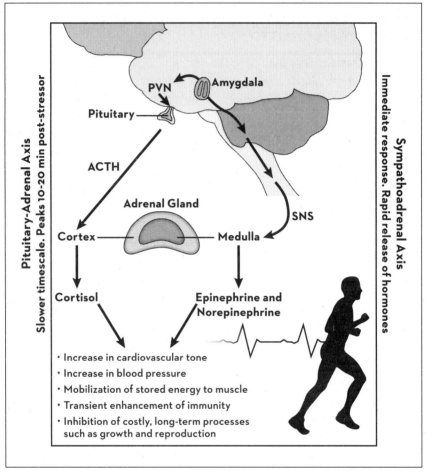

Figure 3.3: Endocrine Support of Defense: Sympathoadrenal and Pituitary-Adrenal Systems.

The sympathoadrenal system (also known as the fight-flight system) and the pituitary-adrenal system are both responsive to threat processing by the amygdala. The sympathoadrenal system involves nerves from the sympathetic nervous system (SNS) that terminate on various target organs and tissues, including the adrenal medulla. SNS activation of the adrenal medulla releases epinephrine and norepinephrine into the bloodstream, allowing these hormones to influence many of the same organs and tissues affected by the nerves of the SNS (see Figure 3.2). Hormones of the adrenal medulla do not cross the blood-brain barrier and have to affect the brain indirectly. The pituitary-adrenal system involves the paraventricular hypothalamus (PVN), connections to the pituitary gland, and the release of adrenocorticotropin hormone (ACTH) into the bloodstream. ACTH then binds to receptors in the adrenal cortex to release cortisol, which, in turn, is distributed to many sites within the body and within the brain as well.

(BASED ON RODRIGUES ET AL [2009].)

SELECTION PROCESSES CONTROLLING INNATE DEFENSIVE REACTIONS

Defending themselves is not just something that animals do occasionally—for many, it is a way of life. In the wild an encounter with a predator or other forms of danger is always possible. Animals, or rather their brains, have to adjust their behavior in accordance with this momentary threat potential while still engaging in routine daily activities. When a threat does suddenly appear, the brain has to quickly resolve what action to take, as delays or mistakes can be costly. So how *does* the brain determine what to do? We'll look at ideas that have been proposed to account for the choice of defense responses in the framework of the defensive behavioral trio described above: freezing, fleeing, and fighting.[29]

The classic view is that distance from the predator is the key factor determining whether to freeze, flee, or fight—with freezing being the optimal response at moderate distances, fleeing at closer ones, and either fighting or fleeing when the predator is about to strike or has made contact.[30] But Robert and Caroline Blanchard, pioneering researchers in this area, proposed a more subtle rule: Although distance is important, other factors also contribute, such as the environmental support stimuli.[31] They argued that whether you flee, freeze, or fight when in proximity to danger depends on the situation: Escape occurs if there is a way out; otherwise freezing results. Fighting only occurs if the predator is directly front and center and is about to attack or has already done so.[32]

Research by Michael Fanselow suggested that this theory might need modification.[33] He modeled predator-prey interactions by giving rats a training session in which a light predicted an electric shock and reasoned that if the Blanchards' environmental conditions theory was correct, the rats should freeze if they were trapped but should flee if they were provided an escape route. His findings showed that freezing resulted regardless of the environmental options, which led him to propose a very influential idea called *predatory imminence theory*.

According to this theory, the defensive behavior of prey has to be understood in relation to the momentary imminence of the predator. In an effort to thwart being eaten, the prey's behavior changes systematically as the imminence of the predator changes. The prey's goal is to remove itself from the predatory sequence, and the behaviors that are relevant in doing so differ depending on where the predator and prey are in the sequence, which, from the point of view of the prey, can be broken down into three major stages.

The first is the baseline condition, in which a predator has not been detected. This is called the *preencounter* stage. Once the prey detects the predator, the *encounter* stage begins,[34] in which freezing is the dominant or default response. If this enables the prey to avoid being detected by the predator, the animal can then flee to safety, provided that there is an escape route. (Escape thus has a role but is secondary to freezing in priority.) If and when the predator also detects the prey, and manages to close in, then the next mode in the sequence occurs. *Circa-strike* refers to the time immediately before or after the predator makes physical contact with the prey. Here the prey's options become fighting or fleeing (or, in some animals, playing dead). The predatory imminence theory is depicted in Figure 3.4.

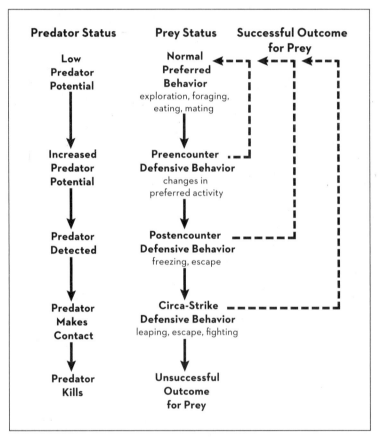

Figure 3.4: Fanselow's Predatory Imminence Theory.

According to this theory, the defensive behavior of a prey organism can be understood in terms of the relation of the prey to the predator at various points in the predatory sequence. The prey's goal is to exit from the sequence as soon as possible to prevent an unsuccessful defensive effort, which will likely result in harm, or even death.

(BASED ON FANSELOW AND LESTER [1988].)

According to Fanselow and Robert Bolles, threats activate a *defensive motivational state* in the brain, which restricts the animal's behavioral options to its species-specific defensive repertoire.[35] Given that freezing, fleeing, and fighting are innate response programs that are hardwired into brain circuits, the response selection problem can be reduced to a question of circuit activation. Threats activate defensive survival circuits, and this lowers the threshold for the expression of each of the defensive responses. Freezing has the lowest threshold and so is activated first. But then the prey's changing position in the imminence sequence triggers the activation of a new response and the inhibition of other options. As the sequence unfolds, the particular state of activation and inhibition of each response can rapidly change: Freezing gives way to fleeing or fighting, either of which may, in turn, give way to the other.

As in other central motive state theories, the defensive motivational state proposed by Fanselow and Bolles is presumed to dictate the responses that occur. But as I noted in Chapter 2, I differ on this point. The defensive motivational state, in my view, is a *consequence*, not a cause, of the responses that result when a survival circuit is activated by a threat: the survival circuits cause brain arousal and the expression of defensive behaviors and supporting physiological changes, which produce signals that feed back to the brain; the defensive motivational state is the result, not the cause, of all this. That said, once a defensive motivational state exists, it may contribute to the selection of additional responses to help cope with the threat. In particular, avoidance and other learned instrumental responses that help cope with potential danger are greatly influenced by the defensive motivational state.

Fanselow and Bolles, as discussed in Chapter 2, did not regard the defensive motivational state as a subjective experience (conscious feeling) of being afraid.[36] They and other central state proponents viewed subjective states as unnecessary (and counterproductive) in understanding how environmental conditions are translated into behavioral outcomes via processes in the nervous system of animals or people. By default, then, they assume that defensive motivational states are nonconscious states. I obviously agree that defensive motivational states are nonsubjective (nonconscious) states elicited by threats. But in contrast to these theorists, I think the subjective experience—the conscious feeling of fear and anxiety—can and must be accounted for if we are to truly understand fear and anxiety. In research on humans, this can be done.

In summary, the prey's present relation to the predator (is there a predator in the area, has it detected you, and how close is it?) and environmental conditions (do they support escape?) are significant aspects in determining what the prey will do defensively. But some other factors are also important.[37] One is the nature of the threat: Not all predators are equally dangerous. Another is the group dynamic: Do others have to be protected (mate, offspring, other members of the group)? If so, fighting may have to trump fleeing and freezing. Still another is whether physical defenses are available (armor, camouflage). Another key factor is learning and memory—past experience about similar situations and about the success of responses in those situations that the organism can draw upon.

DEFENSIVE EXTRAS IN THE SERVICE OF SURVIVAL: THE ROLE OF LEARNING AND MEMORY

Given that inborn defenses are automatically elicited by circumstances, how do prey ever manage to perform novel adaptive responses to threatening conditions? It's often useful for evolution to do the "thinking" initially in an encounter with danger, but we know that there's more to behavioral control than innate reactions to innately programmed or learned stimuli.

Learning is a particularly important supplement to evolution's defensive gifts. Learning gives you a leg up in the quest to survive and thrive; rather than having to start from scratch each time, memory allows the past learning to enhance survival now.

We will consider several ways that learning helps cope with danger. In Chapter 2, I discussed how learning is studied scientifically using Pavlovian and instrumental conditioning. Here, I will elaborate. Through Pavlovian threat conditioning, stimuli associated with danger in the past come to elicit innate defense *reactions* in anticipation of the actual danger in the present. Through instrumental conditioning, novel *actions* are acquired via their consequences with respect to their particular outcomes that allow escape from or avoidance of harm. *Habits* are instrumental actions that have become so ingrained that they lose their relation to the outcomes that established them and are routinely repeated in the relevant context. Let's look at these forms of behavioral learning in more detail.

ONCE BITTEN, TWICE SHY: PAVLOVIAN CONDITIONING OF DEFENSE REACTIONS

A rabbit is having a cool drink from a pond on a hot summer afternoon. Suddenly it is attacked and wounded by a bobcat but manages to escape. The rabbit is likely to store information about that experience, such as cues associated with the bobcat itself (such as its odor and any sounds that it made as it was about to strike) as well as cues about the location where the event occurred. This is Pavlovian conditioning in the real world.

Pavlovian conditioning is not just part of the daily experiences of animals in the wild but also the fundamental way that the human brain learns about threats. As noted in Chapter 2, it is generally considered to be an example of associative learning, in which relations are formed between stimuli (between the CS and US). The US changes the meaning of the CS in such a way that the CS can elicit innate defenses and physiological responses. It is thus stimulus-stimulus association learning—learning about the predictive value of the CS in warning that the US is likely to follow. It is not response learning: The response is innate and simply comes to be elicited by the CS. Pavlovian conditioning thus makes it possible for a novel stimulus that occurs in connection with danger to initiate defensive responses in anticipation of the predicted danger.

Conditioning occurs not only to the specific CS that predicts the US, but also to the context or situation in which the event occurs. In our rabbit example, the rabbit came to be conditioned not only to the cues directly related to the bobcat, but also to the place where it encountered the predator. In the laboratory, animals will freeze when placed back in the chamber where conditioning occurred. For this reason, conditioned responses elicited by the CS itself are usually tested in a novel context; otherwise it is hard to separate the effects of the cue from those of the context. Pavlovian conditioning to a specific CS as opposed to the background stimuli present are often distinguished with the terms *"cued conditioning"* (Figure 3.5) and *"contextual conditioning"* (Figure 3.6).

Some scientists use the odors of predators instead of neutral sounds or sights in laboratory experiments in an effort to create a more naturalistic version of Pavlovian conditioning. Although predatory odors are innately threatening and will themselves elicit freezing and other defense responses,[38] they can also serve as CSs. Thus, pairing a predatory odor with a shock US can produce conditioned responses that are stronger than those to the odor itself.[39]

Although the usual view is that Pavlovian conditioning links a weak biologically neutral stimulus with a strong biologically significant stimulus,[40] "weak" and "strong" here are relative terms. They depend on the inner state of the organism, the environmental conditions that exist at the time, and the organism's stored history of past internal and external states of these types.

The effects of conditioning can be undone, or, more accurately, suppressed, by *extinction:* repeated exposure to the CS without the US following[41] (Figure 3.7). If the rabbit visits the watering hole a few times and nothing happens, the cues there will, through extinction, lose their potency as threatening stimuli. Extinction is not memory erasure but instead a form of new learning in which the original memory that indicated that the CS is dangerous is inhibited by new information indicating that the CS is safe. Just as the initial learning of threat

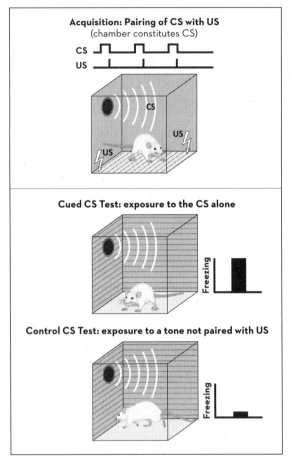

Figure 3.5: Pavlovian Threat Conditioning: Cued Conditioning.

In cued conditioning, a specific stimulus, such as a tone, is the conditioned stimulus (CS) paired with an unconditioned stimulus (US), such as footshock. The conditioned response elicited by the CS later is typically measured in a novel chamber so as to separate responses elicited by the tone CS from responses conditioned to the chamber where the shock US occurred (see Figure 3.6). Freezing behavior is commonly measured, but a variety of other responses, such as autonomic nervous system changes, can be measured as well. A tone that has not been paired with the US will typically elicit much less freezing than the tone paired with the CS.

conditioning is said to involve a CS-US association, the learning of extinction is said to depend on a "CS–no US" association. However, the original memory, which is still present, is susceptible to being revived in various ways, such as by the passage of time, returning to the place (context) where the conditioning experience occurred, or by pain and stress.[42] As we will see later in the book,

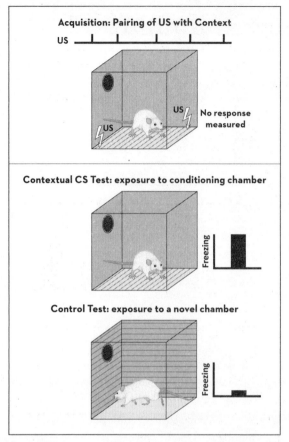

Figure 3.6: Pavlovian Threat Conditioning: Contextual Conditioning.

In context conditioning, an unconditioned stimulus (US), such as footshock, occurs in a certain chamber but its occurrence is not signaled by a phasic cue conditioned stimulus (CS). The context itself is a continuously present CS. Conditioned responses are then elicited when the subject is returned to the conditioning context, and are considerably weaker in a different context.

extinction plays a key role in *exposure therapy*, which is a mainstay in the treatment of anxiety, and the fragility of extinction is a problem for treatment.[43]

Another important variant of threat conditioning is safety learning[44] (Figure 3.8). People with anxiety disorders often are impaired in detecting the difference between threat and safety.[45] Laboratory studies of safety conditioning typically involve two CSs, one that is paired with shock and one that is not.[46] The unpaired stimulus is the safety signal. Obviously, it is very useful to learn to discriminate safety from danger. However, sometimes people become overly reliant on safety

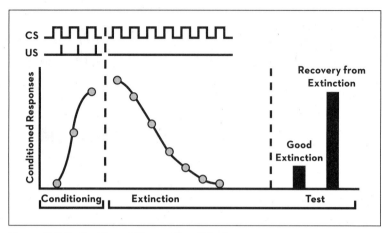

Figure 3.7: Extinction of Threat Conditioning.

Extinction is a process by which repeated presentation of the conditioned stimulus (CS) without the unconditioned stimulus (US) weakens the ability of the CS to elicit conditioned responses. When extinction is successful, the conditioned responses tested some time after extinction training are weaker. However, a variety of conditions can result in the return of previously extinguished responses.

cues. For example, if one can only feel safe in a social situation when accompanied by a friend, this can become a problem since it's not always possible to call upon these kinds of support. One goal of therapy is to help "wean" the anxious person from the use of safety signals.[47]

A key advantage of Pavlovian threat conditioning as a research tool is that it can be used very similarly in humans and animals.[48] But two variants of Pavlovian conditioning are particularly relevant to humans. These are observational and instructed conditioning (Figure 3.9).

Through *observational learning*[49] one can develop a conditioned response to a CS by simply watching someone else be conditioned with that CS paired with shock.[50] People often learn about danger from observing its effects on others, such as seeing someone being harmed, either in real life or on television or in films. Although some animals also exhibit instances of social transmission of threat information,[51] this is a capacity in which our species specializes.

The other particularly human variant is called *instructed Pavlovian conditioning*, in which information about potential threats is conveyed by verbal instruction.[52] Children, for example, learn about danger from the teachings of parents and caretakers. Companies instruct employees how to stay safe at work. In laboratory experiments, simply telling a subject that a CS is likely to be

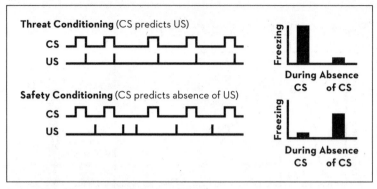

Figure 3.8: Threat vs. Safety Conditioning.
Just as the brain can learn through conditioning that a stimulus predicts harm, it can also learn that stimuli predict safety (the absence of harm). In safety conditioning, the conditioned stimulus (CS) becomes a predictor that the unconditioned stimulus (US) will not occur. Thus, in safety conditioning, and in contrast to threat conditioning, the absence of the CS results in conditioned freezing responses.

followed by a shock is sufficient for the subject to develop a conditioned response to the CS, even if the shock never occurs.[53]

STAYING OUT OF HARM'S WAY: INSTRUMENTAL AND HABITUAL AVOIDANCE

The expression of innate defense responses like freezing and related physiological reactions in the presence of innate or learned threats is of course very useful. But organisms can also learn completely new behaviors—novel responses that are acquired as a result of their success in escaping or avoiding harm. For example, if the rabbit in the scenario above manages to escape from a bobcat at the watering hole by squeezing into a small hole in a nearby tree, the success of this maneuver will be stored and may well be called upon if the rabbit detects a bobcat or other predator in the future, and a small hole in which to escape is available. It may even be called upon as a way to avoid being noticed. Although there may be limits as to what one can do to escape or avoid harm, and differences in the ease with which different kinds of actions can be learned, a range of actions can serve as ways to escape or avoid harm.

Learning to perform an action to escape and/or avoid harm requires that the default response, freezing, be suppressed: One can't take action if frozen in place.[54] Unlike freezing, escape and avoidance are not species-specific defense responses.

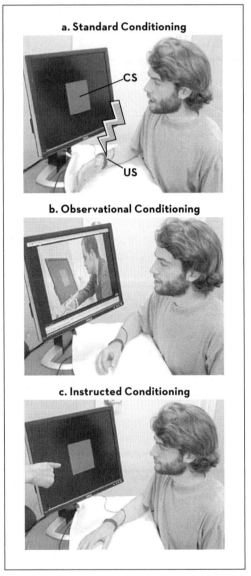

Figure 3.9: Observational and Instructed Conditioning in Humans.

Humans are especially adept at learning by observation and instruction. In observational threat conditioning, the participant watches someone else receive an unconditioned stimulus (US) in relation to a conditioned stimulus (CS). When the participant is then exposed to the CS, conditioned responses are expressed, even though the US was never directly experienced with the CS. Similarly, people can be instructed that when a certain CS appears they are likely to receive a US. Even though the US never occurs, the CS acquires the ability to elicit conditioned responses. (Image provided by Elizabeth Phelps.)

Animals can use many different kinds of behaviors to escape and avoid (e.g., running away, jumping, rearing, climbing, swimming, pulling a chain, pressing a lever, and on and on), depending on what sorts of conditions they are in. These are not inherently or exclusively escape or avoidance responses; they are just motor actions that can, through learning, be used to escape or avoid, as instructed by past learning.

As we've seen, behaviors that are learned as a result of the success of their outcome are said to be *instrumental responses* (responses that are instrumental in obtaining a goal or outcome). The ability to acquire new instrumental behaviors provides the organism with a wider range of options in dealing with danger. Instrumental, goal-directed learning is often described as *response-outcome (R-O) learning*.[55] In the laboratory, instrumental learning of actions to cope with danger is studied by using active avoidance conditioning tasks (Figure 3.10). In a typical experiment, a rat is placed in a box with two compartments.[56] A sound is played, and at the end of it a shock is delivered. The rat, of course, freezes the next time it hears the sound. So far, this is standard Pavlovian threat conditioning with a sound CS and a shock US. But if the CS and US are repeated, the shock US will begin to elicit random movements, and at some point the animal will end up in the other compartment, where there is no shock. It then learns that it can escape from the shock by running to the other compartment. It eventually also learns that escaping to the other compartment when the tone comes on will avoid the shock altogether. Once a rat has learned the avoidance response, the presence of the CS, by virtue of its relation to the US, then becomes an *incentive,* a stimulus that motivates behavior: the CS not only tells the brain when to perform the learned avoidance response, it also regulates the vigor of the avoidance behavior.

Some have argued that conditioned avoidance responses may look like they are instrumentally learned but are really just species-specific defenses.[57] However, our findings, described in the next chapter, show that the neural circuits that underlie innate reactions like freezing and learned actions like avoidance are distinct, making these unique kinds of behavior and not simply variants of species-specific defense responses.

Many of the criteria for evaluating instrumental responses come from appetitive conditioning studies with food or addictive drug reinforcers, and, for technical reasons, it has been difficult to conduct studies of this type in research using aversive stimuli (especially shocks) as reinforcers. I am less concerned with whether avoidance responses are in some abstract sense strictly instrumental than with whether these are an interesting category of responses that deserve to be investigated. I have little doubt that this is the case. Studies described below are consistent with this view, and my laboratory is now vigorously pursuing these issues.

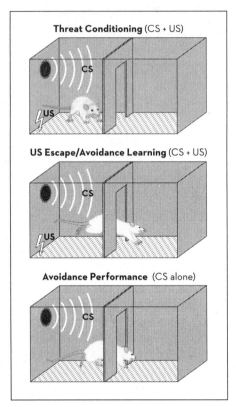

Figure 3.10: Active Avoidance.

Active avoidance conditioning involves a tone conditioned stimulus (CS) and a shock unconditioned stimulus (US). At first the subject freezes to the CS. Over time, though, it learns that
if it crosses to the other side of the chamber when the tone appears, then the shock US can be
escaped from or even avoided altogether. Responses such as these that are learned by their
consequences are thought to be goal-directed or instrumental responses. In contrast to the
reactions elicited by a Pavlovian CS, instrumental responses are *actions* that are emitted in
the presence of the CS.

The outcome achieved in successful avoidance conditioning likely depends
on the fact that the response both prevents the shock US from occurring and
also terminates and/or prevents exposure to the threatening CS. That CS termination can, on its own, produce the learning of a new response has been shown
through studies using a task called *escape from threat*[58] (often less appropriately
called *escape from fear*[59]). In this procedure, rats undergo Pavlovian conditioning in one chamber, and then some time later are placed in a new chamber where
the CS is presented. The rats freeze, but if they make any movement, the CS is
terminated. Over time they learn to shuttle or perform other responses that

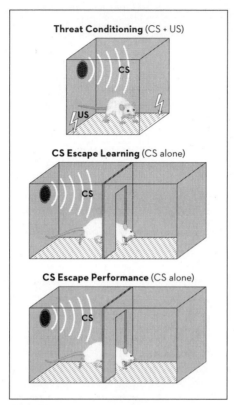

Figure 3.11: Escape from Threat.

Escape from threat is an active avoidance task in which the Pavlovian and instrumental phases are separated. First, Pavlovian conditioning occurs with a tone conditioned stimulus (CS) and a shock unconditioned stimulus (US). Then the subject is placed in a novel chamber and learns to shuttle to the other side when the CS comes on. This allows escape from the CS. Then, over time, the subject learns to shuttle back and forth continuously to avoid the CS altogether. This task is thus motivated and reinforced by CS termination rather than by the US, since the latter never occurs in this chamber.

turn the CS off. The only reinforcement in this scenario is escape from the CS—there is no shock involved in the learning of the new response.[60] Essentially, this separates the Pavlovian and instrumental components of avoidance learning into two separate procedures and allows the reinforcing effects of the CS to be assessed independently of reinforcement by the shock US. Studies performed by Chris Cain in my laboratory with this task support the idea that escape from the CS, rather than just avoidance of the US, contributes to avoidance learning (Figure 3.11).[61]

In avoidance and escape from threat, termination or prevention of the CS is the reinforcement that strengthens the response. Because it involves the removal or prevention of a stimulus, it is called *negative reinforcement;* an example of positive reinforcement would be the use of food to reinforce a response in an animal that has not eaten for a while. Thus, positive and negative in this context do not imply valence (goodness or badness) but instead presence or absence. And because the stimulus is reinforcing because of prior Pavlovian conditioning (association of the CS with a US), it is a *conditioned reinforcer.* Avoidance and escape from threat are thus dependent on *conditioned negative reinforcement.*[62]

The most common view of the nature of the negative reinforcement signal resulting from CS escape/avoidance is that it results from the relief of fear.[63] This idea is central to the theory of avoidance proposed in the 1940s by O. Hobart Mowrer and his colleague Neal Miller.[64] They argued that avoidance is a two-factor learning process. First, the warning sound, which predicts shock, becomes a Pavlovian CS. Then actions that enable escape from the shock, and eventually from the CS, are learned instrumentally by their outcome. Mowrer and Miller proposed that the Pavlovian CS elicits a state of fear, and during the instrumental phase, responses that allow escape from the shock reduce the fear. These responses are learned because fear is an unpleasant experience, and its reduction is reinforcing.

The idea that the CS elicits "fear" and that escape from the CS results in "relief" builds on hedonistic theories that assume reinforcement depends on the subjective experience of the pleasure of reward or the displeasure of punishment or pain.[65] Therefore, when the CS is terminated, the reinforcement comes from the dissipation of fear. I question the value of viewing brain states elicited by threats as subjective feelings. Although some fear-reduction theorists treat fear as a nonsubjective central state, that still begs the question of how the reduction of a state of fear, whether subjective or nonsubjective, increases the likelihood of a behavior. As we will discuss in the next chapter, in neuroscience, reinforcement is thought of as a cellular and molecular process that occurs in specific circuits. Calling upon relief from a feeling of fear to explain learning raises more questions than it answers in terms of how this might occur in the brain. Cellular processes such as those underlying reinforcement and motivation are better thought of as contributing ingredients that help construct feelings, rather than as feelings themselves. Conscious feelings, I argue, are cognitive elaborations of the more basic nonconscious processes.

The two-factor Mowrer-Miller theory, in spite of being criticized by some,[66] remains an important part of the conceptual foundation underlying the use of exposure therapy to control fear or anxiety.[67] I believe that this two-factor theory can be salvaged by a reconceptualization of the nature of the reinforcement signal. The reinforcement, in my view, is not due to fear reduction but to a reduction of components of the nonconscious defensive motivational state that the CS triggers. That is, behaviors that eliminate the CS are reinforced because the CS is no longer activating the defensive survival circuit, and this, among other things, changes the level of neuromodulators, which are known to be important as reinforcement signals in survival circuits and instrumental action control areas.[68] This will be explained in terms of specific circuits and chemical modulators in the next chapter.

Avoidance can become very persistent: an animal or person who learns how to successfully avoid the actual danger may never again experience it. The response is self-perpetuating because, as noted, there is never an opportunity to test whether the CS is still a reliable predictor of the US. As a result, the underlying association between the response and the negative reinforcing effects of the CS and US can never be extinguished; because the negative outcome fails to occur, avoidance is further reinforced.[69] The anxious person also develops false beliefs that the avoidant actions prevented the negative outcomes,[70] and these provide a layer of cognitive support of pathological fear and anxiety that has to be treated as well (see Chapters 10 and 11).

When an avoidance response becomes self-sustaining in this way, it is no longer goal directed and has become an automatic *stimulus-response habit*.[71] The CS automatically triggers the avoidance response, even if the CS is no longer associated with the US. Just as freezing is an innate response that is automatically elicited by a CS acquired during Pavlovian conditioning, a habit is a learned response that was instrumental (goal directed) but that loses its relation to the goal and then comes to be automatically elicited by stimuli that had been connected with the goal.

Habitual avoidance prevents the brain from entering into a defensive state— there is nothing to defend against if you know how to avoid danger.[72] Habitual avoidance learning can simplify life by making it less stressful[73] but can also have an adverse side—it can become so mechanized that it is performed when not needed or even when it is maladaptive. Many people with anxiety disorders, for example, will go to great lengths to avoid situations that elicit anxiety, even when such behavior is detrimental to other life goals.[74] We will consider the two faces of avoidance in more detail later, when we discuss pathological anxiety toward the end of the book.

Humans don't always have to go through prolonged training to avoid harm. We are able to use observation and instruction and can create avoidance concepts or schemas upon which to draw as stored action plans.[75] When we encounter threats, they can trigger avoidance and motivate its performance. Given the hypersensitivity of anxious people to threats, learned or schematized avoidance can be easily activated and drive behavior in pathological ways.

The CS thus plays at least four different roles in relation to avoidance. Initially, it is a Pavlovian CS associated with a shock and elicits freezing. Then, if freezing can be overcome, the CS functions as a reinforcer that makes escape, and ultimately avoidance, learning possible. Once the avoidance response is learned, the CS becomes an incentive that motivates the performance of the avoidance response in anticipation of the threat, or escape if the threat is present. And if avoidance becomes habitual through extensive repetition, the CS becomes a trigger of the habit.

Table 3.1: Four Roles of a CS in Avoidance

1. Pavlovian Conditioned Stimulus (CS): Elicits innate defense responses (freezing and support physiological changes) after being associated with an aversive unconditioned stimulus (US)

---

2. Conditioned Negative Reinforcer: Promotes learning of responses that terminate exposure to the CS and US (escape) and ultimately prevent exposure to the CS and US (avoidance)

---

3. Conditioned Incentive: Motivates performance of the learned avoidance response

---

4. Habit Trigger: If avoidance becomes habitual, the CS will trigger the response even if it is no longer associated with prevention of the CS and/or US

THREATS ALSO FUNCTION AS INCENTIVES THAT GUIDE LEARNED ACTIONS

When we undergo Pavlovian and instrumental learning, we not only learn to react or act, but also pick up information about the stimuli and responses themselves. In particular, we learn the incentive value of Pavlovian conditioned stimuli, the value of responses, and the value of outcomes (reinforcers) in relation to responses. These various values are useful when deciding what to do in novel situations, whether to approach or avoid certain stimuli, and estimating what kind of outcome might be expected for certain possible courses of action.[76]

Stimuli that have acquired incentive value through association with positive or negative outcomes can have profound effects on behavior. An animal foraging for food can use food-related Pavlovian cues as incentives to help locate suitable sources of nutrition while using predator-associated cues as incentives to help stay safe in the process. The contribution of Pavlovian incentives to decision making is often studied by examining the effects of a CS on some instrumental behavior.[77] For example, if a rat has learned an instrumental response such as pressing a bar to get food, a Pavlovian CS previously associated with the same or even a different food will facilitate performance of the food-motivated instrumental response; a CS associated with water will either produce less or no facilitation because the motivation underlying the response is different from the motivation underlying the incentive value of the CS; and a CS associated with shock will suppress food-motivated behavior. The opposite occurs for aversive instrumental responses; a Pavlovian CS associated with shock facilitates performance of shock-motivated avoidance.[78]

We humans use learned incentives to select which products to buy and which people to befriend and trust. But incentives can also steer us in maladaptive directions. Cues associated with food can give rise to cravings in the absence of hunger and stimulate overeating, just as drug-related cues can give rise to cravings and induce relapse in addicts.[79] In social situations, use of the wrong cues to judge trustworthiness can get people into trouble—for example, trusting someone because he is attractive and funny rather than because he is reliable. And, as noted above, incentives can also motivate maladaptive avoidance responses in people with problems related to fear and anxiety.[80]

Incentives are the flip side of drives.[81] Drives such as hunger are said to motivate from within: They *push* us toward goals that can satisfy biological needs. Incentives, in contrast, *pull* us toward goals. Although both are important aspects of motivation, incentive motivation plays an especially important role in everyday decision making, even about how to satisfy biological needs. Nutritional needs, for example, can be satisfied in many ways, and the incentive value of different options often determines how we decide what to eat, and can even lead to eating when we don't need to biologically. Similarly, when in danger, we may initially freeze but then have to make a decision about our next action. This involves an assessment of the risks implied by the aversive incentives present.

RISKY BUSINESS

So far we've looked at defense in relation to specific, detectable, and immediately present threats. But not all threats are of this type. Sometimes organisms find themselves in unfamiliar situations, are exposed to unexpected stimuli (such as a sudden noise), or are in conditions where danger is possible or even likely, all of which raise the alert level. In each of these circumstances, risks have to be assessed about threats that are not actually present and whose occurrence and likelihood are uncertain. Because there is no actual threat, such behaviors are typically said to be more relevant to anxiety than fear. Uncertainty is present when there are conflicting goals (approach versus avoid) or a mismatch between what we are expecting and what actually happens. Uncertainty about the future and how to prepare for various possible outcomes is a significant factor in fear and anxiety disorders.[82]

Risk is defined in terms of both external and internal factors. Proximity to a threat is an external factor, though some threats are inherently more dangerous than others (a snake at your feet versus a snake behind a glass window at the zoo). Internal factors include other motivational conditions that are at play at a particular moment (the need to eat versus the risk of exposure to harm) as well as individual traits (one's inherent tolerance/aversion to risk) based on genetic background and past experiences.[83] The amount of risk in a given situation can also vary as the situation unfolds over time. (Recall the predatory imminence hierarchy—risk is low in the preencounter phase, rises rapidly when the predator is detected, and changes again if the predator is close enough to strike.) Risk also rises when one needs to approach a dangerous situation.

For example, consider the behavior of a rat in a situation in which it has not eaten for a while and has entered a potential danger zone in search of food.[84] The rat will actively avoid bright, unprotected areas, remaining frozen next to the closest structure that provides some protection. It uses small movements of its head, whiskers, and nostrils to scan the visual, acoustic, and olfactory cues present. Larger movements, when they occur, are very slow, often with the body stretched out and close to the ground. Such risk assessment behaviors allow active evaluation without attracting attention. If danger is not detected, foraging can proceed, but still in small, cautious increments. These are highly significant experiences for prey; once completed, even if danger was not encountered, hours or even days may pass before routine daily activities, such as eating, drinking, or sex, fully resume. Better safe than sorry.

Because the uncertain nature of future events is a particularly important factor in fear/anxiety disorders,[85] many laboratory tests that are designed to be animal models of human anxiety create situations in which the outcome is *not* predictable by the stimuli that are present.[86] This is achieved in various ways, such as changing the reliability with which a Pavlovian threat CS occurs as a predictor of the US,[87] using a prolonged CS that creates uncertainty about when the perceived threat will end,[88] or placing the animal in an open space where it is unprotected or in situations where it faces some sort of conflict.[89]

As we saw in Chapter 2, Jeffrey Gray and Neil McNaughton argue that risk assessment behaviors in situations of uncertainty, especially situations where there is a conflict between the need to approach and avoid, are the result of a central state of behavioral inhibition.[90] According to their theory of anxiety, when an animal or human is in this state, the negatively valenced cues increase in significance, which counters the tendency to approach a risky goal in spite of the need to do so; the result is the avoidance of harm by remaining immobile. This kind of avoidance strategy is called *passive avoidance* and is quite different from the active avoidance behaviors described above (Figure 3.12). In passive avoidance, harm is avoided or postponed, not by taking action, but by risk assessment during inaction.

Inaction in the course of passive avoidance is very difficult to distinguish from freezing if one is simply observing behavior. However, there is evidence that the two do differ. For example, some drugs, especially benzodiazepines, that reduce passive avoidance do not affect freezing to a specific stimulus.[91] It is generally

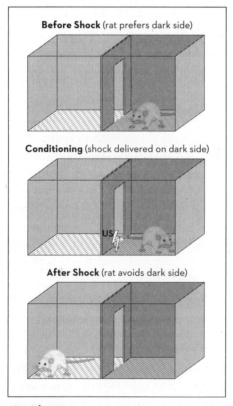

Figure 3.12: Passive Avoidance.

In contrast to active avoidance, in passive avoidance harm is prevented by withholding a response rather than performing one. One way to structure the passive avoidance task is to take advantage of rats' innate preference for dark over brightly lit areas. If a rat is placed in a chamber with both dark and light areas, it will quickly move to the dark area. It then receives a shock in the dark area and is removed from the apparatus. The next day, the rat, when put in the bright area, will avoid going to the dark area—avoiding shock by withholding its natural response.

accepted that passive avoidance is not simply a freezing reaction but is instead, at least in part, an instrumental action learned by its consequences.

Some people who suffer from severe anxiety avoid exposure to stressful situations by staying at home in spite of the potential adverse consequences of missing work and being socially isolated. This passive avoidance behavior, like the active forms of avoidance described above, can also become a learned habit that successfully avoids harm. Because the threat is prevented, the passive avoidance response is reinforced and becomes stronger and stronger.

It is important to emphasize the complexity and modularity of risk assessment. Different brain systems use different criteria for deciding what to do in risky situations.[92] For example, people who smoke know consciously that it is bad for their long-term health but do it anyway because this activity is controlled by nonconscious systems that operate by different rules and that win out over conscious control systems.

WHO'S IN CHARGE?[93]

A key question is: who (or what) is actually doing the deciding when we make decisions in daily life, whether about threatening or more mundane circumstances. The term "decision making" may seem to imply that the conscious mind is doing the heavy lifting, but in human decision-making research, there is much discussion about the role of nonconscious factors as well.[94] Following Daniel Kahneman's pioneering work,[95] current models adopt a dual-processing approach involving two decision systems (in which "system" has a psychological rather than a neural meaning).

System 1 is a fast, implicit system that operates automatically and without the need for conscious intervention. Most if not all of the effects of Pavlovian incentives on actions involve this automatic process. Advertisers who place products in the context of sexual symbolism, for example, are hoping to create Pavlovian incentives (by pairing the product with sexual arousal) that influence behavior nonconsciously. System 1 also uses mental shortcuts, so-called *heuristics*.[96] If you find yourself in danger—say, faced with a bear on a country road—you may decide to run based on the generalization that large, fat animals on all fours are slower than light two-legged ones. A strategy like this allows decisions to be made rapidly on the basis of limited information, economizing on mental workload, hopefully without sacrificing too much in terms of accuracy. Although heuristic decision making is natural and often useful, it can also steer you wrong. Bears, in spite of their size, are swift on their feet; and many cases of medical misdiagnosis result from decisions based on heuristics rather than a more extensive evaluation.[97]

System 2 is slower and deliberate, and is often said to involve careful reasoning and consciousness. But the degree of rationality and the involvement of consciousness in System 2–type decisions are both debated. The idea that we are rational decision makers is part of our folk psychology and leads people to believe that they are in charge of their behavior.[98] In actuality, much research suggests that we lack direct knowledge of the processes and motivations underlying our decisions and behaviors[99] and instead we often confabulate explanations for them after the fact, making them seem more rational than they are.[100] Our sense that our conscious minds are running the show is part fact and part fiction.[101] Thus, even slow, System 2 decisions are not necessarily based on well-reasoned, conscious decision-making processes.[102] Furthermore, just because you are consciously aware of some decision you made does not mean that you are aware of the motivational causes that went into the decision-making process. We have to distinguish the outcome of a decision-making process from the process itself. And it is always hard for the conscious mind to presently know which was in play when we made a decision in the past, even if the decision occurred just a moment earlier.

We become conscious of brain activities on a need-to-know basis. Our initial response to danger often needs to be fast as we are better off responding first nonconsciously with what has worked in the past. Once we are consciously aware that we are in danger, though, the resources of consciousness can also be used to help solve the problem at hand. Through consciousness we can use factual information and personal experiences stored as memory to help evaluate the implication of possible actions for our present and imagined future self.

Still, the default assumption to prove wrong about any decision is that it was made nonconsciously. Although consciousness plays an important role in human decision making, if we give it more credit than it deserves, we run the risk of masking its real contributions. The trick is figuring out when we really consciously decide as opposed to when, after the fact, we consciously explain decisions that were made unconsciously. This is an especially thorny issue for our legal system.[103]

NEXT, THE BRAIN

Research on the role of the brain in psychological processes depends on the availability of behavioral tests to measure the psychological process. Fortunately, in the area of threat and defense, as we've seen in this chapter, very detailed behavioral approaches are available. We will build on these in the next chapter, where the brain mechanisms of threat processing and defensive behavior are discussed.

CHAPTER 4

THE DEFENSIVE BRAIN

"It's a dangerous business, Frodo, going out your door."

—J.R.R. TOLKIEN[1]

All living things are composed of cells. Some organisms consist entirely of a single cell—bacteria, for example. That one cell has to take care of everything required for life, which, although efficient, also limits what the organism can do. Complex organisms (i.e., animals) have many cells organized into systems that perform specialized functions, an arrangement that provides more flexibility in the daily quest to stay alive and well. For example, mammals like us have digestive, respiratory, circulatory, reproductive, and musculoskeletal systems, among others. The unique function of each is made possible by the kinds of specialized cells it has and how they interact—cells of the digestive system process food and convert it into sources of energy and nutrition; cells in the respiratory system take in air and extract oxygen for use in metabolism; cells of the endocrine system release hormones to regulate metabolism and other functions; cells in the cardiovascular system move blood around in the body in such a way as to distribute energy, nutrients, oxygen, and hormones to tissues in need of these resources; and cells of the musculoskeletal system make motility (behavior) possible. The nervous system, which includes the brain and spinal cord and nerve pathways from these to various body organs, glands, and tissues, coordinates all of the other systems so that the body can work as an integrated unit.

This chapter explores the role of the nervous system, especially the brain, in defense, one of the most important behavioral activities in which animals engage. Eating, drinking, sex, and other survival behaviors can be postponed for long periods without life-threatening consequences. But in dangerous or potentially dangerous situations, a delayed response to threats can be deadly. The brain thus has to rapidly select musculoskeletal patterns that constitute the most appropriate behavioral response. It also has to manage cardiovascular, endocrine, respiratory, and many other responses that provide the physiological support needed for energy-demanding defensive behaviors. Before turning to the details of how defensive activities are controlled by the brain, it will be useful to first briefly summarize some basic principles of brain structure and function.

A FEW KEY POINTS ABOUT BRAIN ORGANIZATION

The brain has two kinds of cells: *neurons* and *glial cells* (Figure 4.1a). Neurons are involved in communication of information. Glia play complex roles in the brain,[2] one of which is to help neurons do their job. Although glial research is on the rise, our main concern here is with neurons themselves.

Most cells in the body communicate with one another by releasing chemicals that affect other cells in close proximity. Neurons, though, can communicate across vast distances as well as in their local vicinity. This is due to the fact that neurons are unique in having fibrous appendages (*dendrites* and *axons*) that extend out of the cell body (also called the soma). These fibers send and receive inputs from other neurons. Neurons have many dendrites, which are especially important in receiving inputs from other neurons; typically, though, neurons have only one axon, the main structure that they use to send out messages. Although there is only one, the axon branches, allowing a single neuron in one area to communicate with many neurons in another or several other areas.

Two messaging systems underlie neuronal communication. One, within the neuron, involves transmission from the soma to the ends of the axon, and the other involves communication from the axon to other neurons, often by way of their dendrites. The communication process starts when an electrical firestorm, called an *action potential*, is generated in the soma (Figure 4.1b). This electrical response then travels down the axon. When it reaches the ends of the axon branches, the second part of neuronal communication begins; the action potential results in the release of a chemical transmitter from the axon endings (Figure 4.1c). The neurotransmitter then binds to receptors on other neurons, primarily on their dendrites (fibrous extensions that receive inputs) but also on the soma or axon.

The junction between two neurons is the *synaptic cleft*, or *synapse* for short. It is via synaptic transmission that neurons communicate with one another.

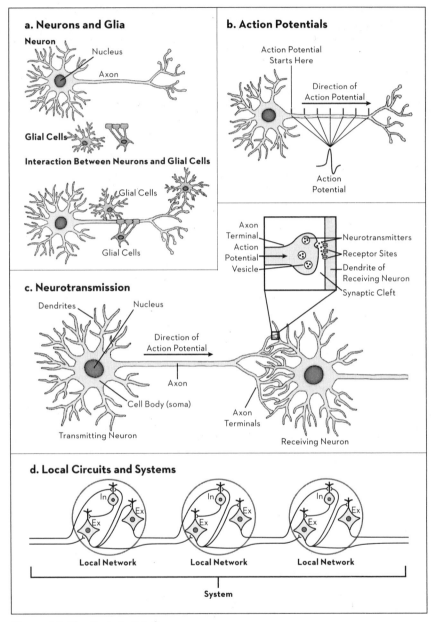

Figure 4.1: The Brain in Brief.

See text for explanation. Neuron and glial drawings in **a, b,** and **c,** based on http://www.ninds .nih.gov/disorders/brain_basics/ninds_neuron.htm. **e.** Drawing of the vertebrate brain based on Figure 2.4 in Bownds (1999). **i.** The drawings of lateral cortex in different mammals are based on Figure 2.4 in Bownds (1999).

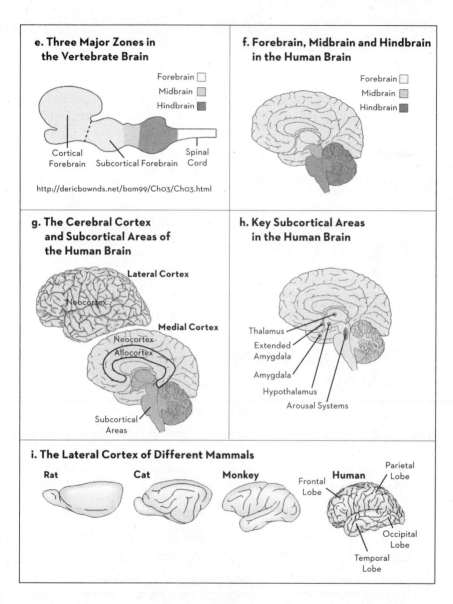

e. Three Major Zones in the Vertebrate Brain

Forebrain ☐
Midbrain ▨
Hindbrain ■

Cortical Forebrain   Subcortical Forebrain   Spinal Cord

http://dericbownds.net/bom99/Ch03/Ch03.html

f. Forebrain, Midbrain and Hindbrain in the Human Brain

Forebrain ☐
Midbrain ▨
Hindbrain ■

g. The Cerebral Cortex and Subcortical Areas of the Human Brain

Lateral Cortex

Neocortex

Medial Cortex

Neocortex
Allocortex

Subcortical Areas

h. Key Subcortical Areas in the Human Brain

Thalamus
Extended Amygdala
Amygdala
Hypothalamus
Arousal Systems

i. The Lateral Cortex of Different Mammals

Rat   Cat   Monkey   Human

Frontal Lobe   Parietal Lobe   Occipital Lobe   Temporal Lobe

Some neurons are excitatory and help trigger activity in other neurons, whereas others are inhibitory and suppress activity in other neurons. Synaptic connections between neurons within a particular brain area or subarea form *local circuits* or *networks* (Figure 4.1d). Connections between circuits in different areas form *systems* that have specific functions.[3]

The brains of all vertebrates have three major zones (Figure 4.1e and 4.1f). The *hindbrain*, necessary for the primary functions of life (such as breathing and heartbeat), is the most similar across vertebrates. Damage there is often lethal. The *midbrain*, which is responsible for normal patterns of sleep and wakefulness, is also fairly similar across species, though not to the same degree as the hindbrain. The greatest differences across vertebrates are in the *forebrain*, which has several components.

The forebrain of mammals and other vertebrates consists of the cerebral cortex and underlying subcortical areas. Cortical areas account for much of the volume of the human brain (Figure 4.1g). The cerebral cortex of various mammals is depicted in Figure 4.1i.

Cortical areas belong to either the neocortex or allocortex. The *neocortex* is the wrinkled outer layer, the most prominent and visible part of the human brain (Figure 4.1g). It was so named because it was thought to be a new addition to the brain that appeared with evolution of mammals, a view that has since been challenged.[4] It has six identifiable layers, or laminae, of neurons. The *allocortex* has fewer laminae (typically only three or four) and is not visible unless the two halves of the brain are pulled apart, as it is located in the medial wall of the hemispheres.

Table 4.1: Abbreviations of Key Brain Areas Discussed in This Book

NEOCORTEX	SUBCORTICAL FOREBRAIN
Prefrontal Cortex (PFC)	**Amygdala (Amyg)**
PFCL, lateral prefrontal cortex	BA, basal nucleus of the amygdala
PFCDL, dorsal lateral PFC	CeA, central nucleus of the amygdala
PFCM, medial prefrontal cortex	LA, lateral nucleus of the amygdala
PFCDM, dorsal medial PFC	**Extended Amygdala**
PFCVM, ventral medial PFC	BNST, bed nucleus of the stria terminalis
Parietal Cortex (PAR)	**Basal Ganglia**
	CPu, caudate-putamen (dorsal striatum)
	NAcc, nucleus accumbens (ventral striatum)
	Subcortical Midbrain
	PAG, periaqueductal gray region

The allocortex is thus often called the *medial cortex*,[5] and the neocortex is called the *lateral cortex*.[6] Sometimes medial cortical areas are referred to as the *limbic cortex*. I avoid this term because of its relation to the controversial idea known as the limbic system theory of emotion.[7]

Subcortical areas lie below the cortex (see Figure 4.1g). Although there are many such regions, several will be examined repeatedly in this book, most of which are part of the forebrain: areas of the amygdala, extended amygdala (a group of areas related to but somewhat different from the amygdala), basal ganglia, thalamus, and hypothalamus. Areas of the subcortical midbrain that will be discussed include the periaqueductal gray region and a group of areas collectively known as arousal systems (Figure 4.1h).

For ease of discussion, I will use only a few abbreviations when speaking about brain areas. These are noted in Table 4.1.

STIMULATING TIMES

By the end of the nineteenth century, it was known that the expression of rage behaviors (defensive attack or fighting) depends on subcortical brain areas because damage to the neocortex did not disrupt such reactions.[8] Later, Walter Cannon used the term "sham rage" to describe these responses because he believed that, without the involvement of the cortex, the feeling of rage could not be experienced.[9] As we saw in the previous chapter, Cannon was particularly interested in the autonomic nervous system, and he showed that diffuse activation of its sympathetic division occurs during sham rage, as indicated by increases in blood pressure and heart rate, piloerection (goose bumps), sweating, and release of epinephrine from the adrenal medulla.

The same pattern of sympathetic activation had been found by researchers earlier in the twentieth century using the technique of electrical stimulation to explore the brain mechanisms controlling the autonomic nervous system in anesthetized animals.[10] The rationale for this approach was that because neurons generate electrical responses when they are activated, artificially stimulating them should mimic the response that occurs when they are activated naturally. With this method, the hypothalamus, a subcortical region at the base of the forebrain, was identified as playing a key role in controlling body functions via the sympathetic nervous system. On the basis of these findings, Cannon hypothesized that the hypothalamus is the subcortical area responsible for integrating defensive (rage) behavioral and physiological responses in emergency situations.[11]

Cannon's student Philip Bard pursued the hypothesis.[12] Because it was not technically possible at that time to electrically stimulate the hypothalamus in awake, behaving animals, Bard used the lesion method. When he disconnected the hypothalamus from the cortex and other upper areas of the forebrain, animals could still exhibit ragelike behaviors and supporting physiological responses when provoked. But when he disconnected the hypothalamus from lower areas in the midbrain and hindbrain (which connect the brain with the spinal cord in ultimately executing behavioral and autonomic nervous system responses), the rage response and physiological changes no longer occurred (Figure 4.2).

By the 1940s the technique of electrical stimulation had been refined as a tool for studying brain function, and findings in anesthetized animals further confirmed the role of the hypothalamus in the control of the sympathetic nervous system.[13] But particularly important was the development of the ability to stimulate the brain of awake animals while they engaged in normal daily activities.[14] With this approach, stimulation of the hypothalamus and other subcortical areas was found to elicit a variety of behaviors that are critical for the survival of the individual or species, including defensive, eating, drinking, and sexual behaviors. These effects were thought to reflect the fact that survival behaviors are innately programmed in ancient subcortical circuits. Just as Cannon and Bard hypothesized, behavioral defense responses and supporting physiological changes are both controlled by the hypothalamus; moreover, the behavioral and physiological

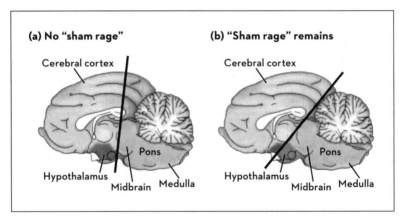

Figure 4.2: How Cannon and Brad Produced Sham Rage.
When brain areas below the hypothalamus were separated from the hypothalamus and the rest of the forebrain, provocation of the animal resulted in little or no rage behavior. But if the hypothalamus was left connected to the brainstem, provocation resulted in rage. The behavior was called sham rage since Cannon and Brad felt that without the cortex being able to participate in the response control, the true experience of rage could not occur.

(BASED ON LEDOUX [1987], AS MODIFIED BY PURVES ET AL [2001].)

responses in rage could both be elicited from the exact same region of the hypothalamus by electrical stimulation.[15]

The upstream and downstream areas that enable the hypothalamus to play
its role in defense were also identified.[16] These were the amygdala and periaqueductal gray matter (PAG), respectively. Thus, the defense response elicited by
amygdala stimulation could be disrupted by damaging the hypothalamus, and
the effects of hypothalamic stimulation could in turn be disrupted by damaging
the PAG. The amygdala, hypothalamus, and PAG thus appeared to be linked in a
serial defense circuit (Figure 4.3). These findings added weight to the idea, which
had emerged from other research, that the amygdala and related areas of the
so-called limbic system were key areas in the processing of "emotional" stimuli
and, via outputs to lower areas of the brain, in controlling "emotional" responses.[17]
But, as I mentioned in previous chapters, the limbic system theory, though still
very popular, is a scientifically dubious notion.

Recall from Chapter 3 that the sympathetic response that occurs in relation
to defensive behaviors is not simply a homeostatic adjustment to the behavior.
Both defensive behavior and the preparatory physiological response are innately
programmed reaction patterns.[18] Homeostatic adjustments occur to match the
metabolic needs of the specific behavior as it unfolds, but the initial physiological

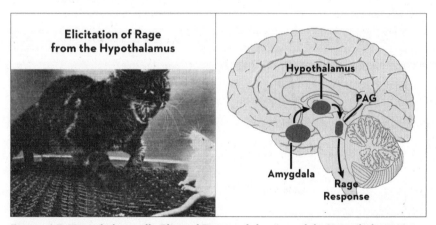

**Figure 4.3: Hypothalamically Elicited Rage and the Amygdala-Hypothalamic-
PAG Rage Pathway.**

(Left) Electrical stimulation of the hypothalamus elicited rage and attack responses. (From
Flynn [1967].) *(Right)* Rage could also be elicited from the amygdala and periaqueductal gray
area (PAG). Lesions of the PAG prevented elicitation of rage from hypothalamic and amygdala
stimulation and lesions of the hypothalamus prevented responses from amygdala stimulation.
The amygdala, hypothalamus, and PAG thus seemed to be serially connected in the elicitation
of rage.

response is a prewired reaction that unfolds much the same in all animals within a species.[19]

In situations like defense, physiological support is needed not only by the body. The brain, too, needs to be mobilized and energized, a process known as *arousal.*[20] Brain stimulation studies in the 1940s provided the first clues about how brain arousal is regulated.[21] Regions were found in the central core of the brain (especially in the midbrain and some areas of hypothalamus and thalamus) that, when stimulated in the absence of anesthesia, aroused the organism (Figure 4.4). This *arousal system* was shown to control sleep-wake cycles and alertness and vigilance while awake.[22] Initially thought to be a product of a diffuse, undifferentiated, but interconnected, collection of neurons known as the *reticular formation*, arousal is now known to be regulated by populations of neurons that make and release specific chemicals called *neuromodulators,*[23] including norepinephrine, dopamine, serotonin, acetylcholine, and others, which will be discussed later. The amygdala, as we will see, is involved in triggering brain arousal, as well as arousal of the body, in the presence of threats.

Although the electrical stimulation method has continued to be used as a research tool to study innate defensive behavioral and physiological responses,[24] it provides only a crude picture of the brain circuits involved in behavior and is problematic for other reasons.[25] For example, it is typically assumed that neurons in the area stimulated are part of the response control circuit, but this is not necessarily the case. Electrical stimulation activates not only the neurons in a given location, but also the axons that pass through it. It is therefore possible that the actual neurons that control the behavior are located elsewhere and are being activated remotely via their input or output axons. Over the years efforts have been made to overcome the limits of the electrical stimulation method,[26] and researchers today have the benefit of new approaches based on genetic tools that allow targeting of specific kinds of neural elements (neurons versus fibers) with unique physiological roles (excitatory versus inhibitory cells) and/or neurochemical signatures (cells containing a particular gene product—a protein such as an enzyme that synthesizes a particular neurotransmitter or hormone, for example).[27] Such novel techniques have led to a reevaluation of some of the classic conclusions based on electrical stimulation. For example, the regions of the hypo-. thalamus classically implicated in several innate responses, including attack[28] and eating,[29] are no longer thought to be involved.

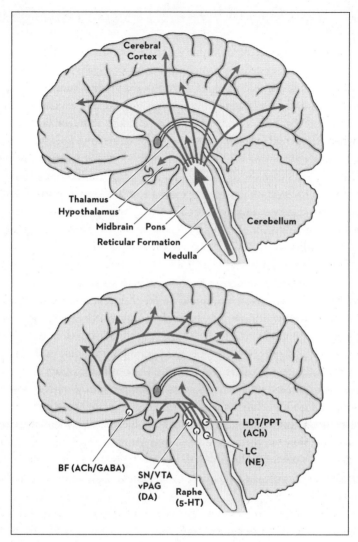

Figure 4.4: The Arousal System Then and Now.

The original view of arousal (top) postulated a diffuse network in the brainstem called the reticular formation that controlled sleep and wakefulness and alertness while awake (modified from Starzl et al [1951]. The new view (bottom) is that arousal functions are mediated by specific populations of neurons each of which manufactures a different neuromodulator, and these modulators are responsible for sleep, wakefulness, arousal, vigilance, etc. (based on España et al [2011]). Abbreviations: BF, basal forebrain; SN/VTA, substantia nigra/ventral tegmental area; LC, locus coeruleus; LDT/PPT, lateral dorsal tegmentum/pedunculopontine area; ACh, acetylcholine; DA, dopamine; 5-HT, serotonin; NE, norepinephrine.

GET REAL

Much of our current understanding of the brain's control over defensive behavior and its physiological support is based on studies that have used real threats, sensory stimuli that activate defense circuits in a natural way as opposed to inducing defense by direct stimulation of the brain. The direct stimulation approach, whether based on electrical stimulation or current advanced technologies, is mainly useful for looking at the response control circuits (in other words, the output circuits). But by using threatening stimuli, the complete circuit can be identified, starting with the sensory system that processes the threat and continuing all the way through the motor system that controls the musculoskeletal, autonomic, and endocrine responses.

Because this sensory-based approach examines the neural control of innate defense responses from the point of view of the eliciting stimulus, I will focus here on the kinds of stimuli that are most relevant to human threat processing—specifically, auditory and visual stimuli. Further, I will consider primarily auditory and visual stimuli that have acquired threat significance through Pavlovian threat conditioning, because this approach has revealed much of what we know about the threat processing circuits.[30]

The basic circuit through which conditioned threats are processed by the amygdala was discussed in Chapter 2, where we learned that the two key areas of the amygdala involved are the lateral (LA) and central (CeA) nuclei. In this chapter we will discuss this topic in greater detail, looking specifically at what goes on within these amygdala areas and how they interconnect, as well as how other areas, such as the prefrontal cortex and hippocampus, contribute to threat processing by the amygdala (Figure 4.5).

A stimulus becomes threatening by way of its association with something harmful. If you are bitten by a dog, the sight of that dog (or even a different dog) puts you into defensive mode and enables you to begin to protect yourself in anticipation of being bitten again ("once bitten, twice shy"). In order for this to happen, information about the sight and bite has to converge onto the same neurons in the amygdala. This convergence leads to an increase in the strength of the relation between the two stimuli. In 1949, the Canadian psychologist Donald Hebb proposed that when weak and strong stimuli activate the same neurons, the strong stimulus changes the chemistry of the neurons in such a way as to enable the weak stimulus to activate the neurons more strongly in the future.[31] In our dog-bite scenario, the strong sensation from being bitten modifies the chemistry of the neurons such that the sight of a dog alone comes to strongly activate the neurons in the future (Figure 4.6).

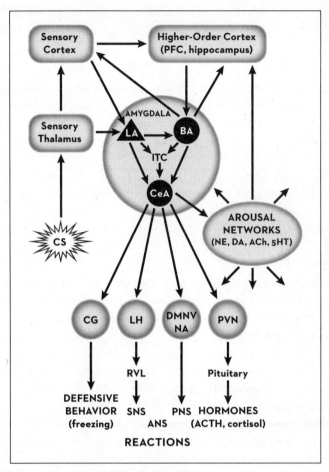

Figure 4.5: Brain Processing of Conditioned Threats and Control of Conditioned Defense Reactions.

The basic circuitry underlying the acquisition and expression of fear (threat) conditioning was depicted in Figure 2.3. Additional details about the processing of previously conditioned threats are shown here. The threat stimulus is transmitted to sensory processing areas of the thalamus and the cortex, both of which transmit to the lateral amygdala (LA). The LA in turn connects with the CeA directly and by way of other amygdala areas, such as the basal nucleus (BA) and intercalated cells (ITC). CeA then connects with downstream targets that separately control freezing behavior, sympathetic and parasympathetic responses of the autonomic nervous system (ANS), and hormonal secretions. CeA also activates the arousal system in the brain that releases neuromodulators such as norepinephrine (NE), dopamine (DA), acetylcholine (ACh), and serotonin (5HT). Processing in this circuit is regulated by higher-order areas of the cortex, such as medial temporal lobe areas (including the hippocampus and surrounding cortical areas) that contextualize threats and the various regions of lateral and medial prefrontal cortex (PFC) that modulate the intensity and persistence of the conditioned responses. Additional details are described in Figure 4.10 and in Chapter 11.

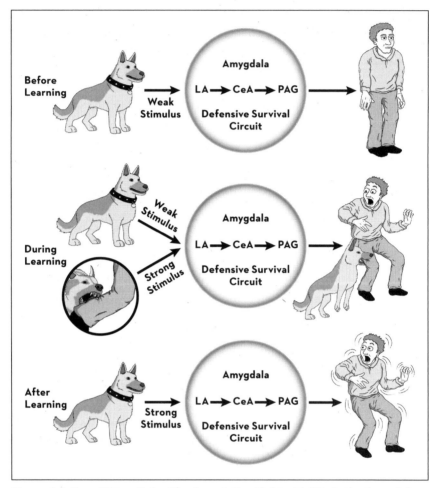

Figure 4.6: Once Bitten, Twice Shy: Associating Sight with Bite in the Amygdala.
Before being bitten by the dog, his sight is a weak stimulus (in terms of its ability to activate the defensive survival circuit involving the lateral [LA] and central [CeA] amygdala and periaqueductal gray area [PAG]). While being bitten, the co-occurrence of sight of the dog (weak stimulus) is paired with the bite (strong stimulus). Later, the sight of the same dog, or even a different dog, activates the sight-bite association in the amygdala and thereby elicits defensive behavior, such as freezing, via the PAG.

In an experimental setting, a tone or light that precedes an electric shock becomes a conditioned threat. This occurs by way of convergence of information about the tone or light (the CS) onto neurons in the LA that also receive information about the shock (the US).[32] The strong US then changes the ability of the weak CS to activate the neurons.[33] Numerous studies by my laboratory and others have confirmed that when the CS is paired with an aversive US, LA neurons do respond more strongly to the CS.[34] Further, we and others have identified many molecules that contribute to the induction of these changes during learning and the stabilization of these changes in the storage of memory.[35] Once the associative memory has been formed, the CS can, on its own, strongly activate LA neurons (Figure 4.7).

The LA has several subareas.[36] Evidence suggests that the learning and long-term storage of the association occurs in its dorsal region.[37] When the CS occurs later, it activates the association there; via connections to the medial LA, the information is then distributed to several other amygdala areas. Ultimately, the CeA is activated to control the expression of the conditioned responses, including behavioral defense responses (especially freezing) and supporting physiological changes (see Figure 4.5).

There are several routes by which the LA and CeA communicate (see Figure 4.5). First, there are direct connections between them. Second, there are connections from the LA to other areas, such as the *basal amygdala* (BA), which then connect to the CeA. Third, both the LA and CeA connect with a group of neurons called the *intercalated cells*, which function as an interface between the LA/BA and the CeA.[38]

Information flow with the CeA involves complex interactions between two separate divisions that are reciprocally connected[39] (see Figure 4.5). Neurons in the lateral division of the CeA receive inputs about the CS-US association stored in the LA (via the connections just described) and connect to the medial division, which, in turn, connects back with the lateral division. The medial CeA also sends outputs to the PAG[40] to control freezing behavior.[41]

Unlike the electrical stimulation results, studies using conditioned threat stimuli do not require the involvement of the hypothalamus in order for the stimulus to elicit defensive behavior via the PAG—direct connections from CeA to PAG are responsible here. However, innate defensive behaviors elicited by unconditioned olfactory threat stimuli do seem to require connections from the amygdala to the hypothalamus, and from there to the PAG, with somewhat different circuits involved when defending against a predator or an aggressive conspecific.[42]

The CeA controls not only behavioral (freezing) responses but also changes in the body mediated by the autonomic nervous system and endocrine system[43] (see Figure 4.5). Connections from the CeA to the PAG, although involved in controlling defensive behavior, are not involved in autonomic nervous system and hormonal responses elicited by a CS.[44] Instead, connections from CeA to areas of the lateral hypothalamus,[45] and from there to motor neurons in the hindbrain (ventral medulla), control sympathetic nervous system responses, such as increases in heart rate and blood pressure,[46] all bypassing the PAG.[47] Connections from CeA

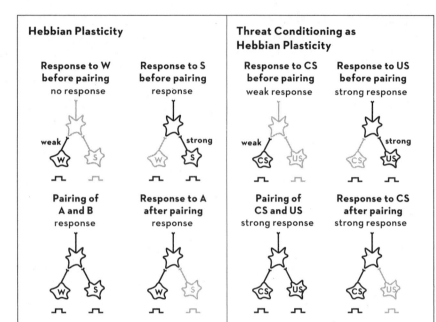

Molecular Mechanisms of Threat Conditioning

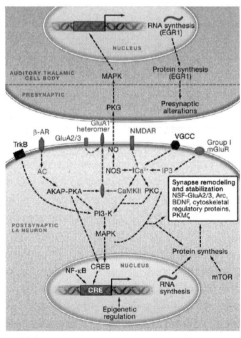

Figure 4.7 (caption on next page)

Figure 4.7: Hebbian Mechanisms Underlie Threat Conditioning. *(Left)*

Hebbian Plasticity occurs when the ability of a weak stimulus to activate a neuron is enhanced by co-occurrence with a strong stimulus that activates the same neuron. Pavlovian threat condition-ing is an example of Hebbian learning since the ability of the conditioned stimulus to activate a neuron is enhanced by co-occurrence with activity in response to the unconditioned stimulus. A variety of molecular changes in the pre- and postsynaptic neuron contribute to memory forma-tion during threat conditioning. Abbreviations can be found in the source article.

(BASED ON LEDOUX [2002] AND JOHANSEN ET AL [2011].)

to other hindbrain areas (the dorsal motor nucleus of the vagus, the nucleus ambiguus) control parasympathetic responses that help restore balance when the threat subsides.[48] And CeA connections to the paraventricular hypothalamus activate the pituitary-adrenal axis, releasing ACTH from the pituitary and cortisol from the adrenal cortex.[49]

As noted earlier, threats change not only the physiology of the body but also the physiology of the brain, arousing its level of alertness and vigilance, and increas-ing sensitivity to relevant (threatening) inputs.[50] Threat-induced brain arousal is controlled by still other outputs of the CeA, in this case outputs to the neurons that release norepinephrine, serotonin, dopamine, acetylcholine, orexins, and other neuromodulatory chemicals throughout the brain.[51] Arousal increases attention and vigilance in the face of threat or other significant environmental stimuli.[52]

Although the LA is a key site of plasticity in threat conditioning, plasticity also occurs in the sensory areas that send inputs to the LA[53] as well as in other areas of the amygdala, including the BA[54] and CeA.[55] However, plasticity in these other areas appears to depend on plasticity first occurring in the LA.[56] For this and other reasons, the LA is thought to be a key site of plasticity in learning about danger.

As noted above, the amygdala does not act alone to control defensive behav-ior. The medial prefrontal cortex (PFCм), especially its ventromedial region (PFCvм), connects with the amygdala (see Figure 4.5), and studies in rats show that these pathways play a key role in regulating information processing within the amygdala during the expression of defense responses.[57] The prelimbic region of the PFCvм regulates the expression of responses on a trial-by-trial basis, deter-mining the intensity of the responses for that occurrence of the CS. But particu-larly important for later chapters is the role of another region of PFCvм, the infralimbic region, in regulating changes that take place over repeated repetitions of the CS in the absence of the US. Thus, damage to the infralimbic region dis-rupts the ability to weaken the threat potential of the CS through extinction, which is a key process involved in exposure therapy for the treatment of maladap-tive fear and anxiety in humans. It is generally thought that the PFCvм regulates the amygdala,[58] and that this regulation is disrupted in people with pathological anxiety.[59] The amygdala can be thought of as the accelerator of defensive reactions, and the PFCvm as the brake to them.[60] The accelerator and brake analogy is discussed further and depicted in Chapter 11 (Figure 11.1).

The amygdala is also connected to the hippocampus (see Figure 4.5), which has a role in the contextual control of defense.[61] Thus, rats with damage to the hippocampus no longer freeze when they are in a conditioning chamber where shocks occurred, but still freeze if a tone paired with shock is presented in that chamber. Although threat conditioning generalizes across situations, we can also learn, through experience, to discriminate contexts that are dangerous from those that are safe.[62] For example, being exposed to dangerous animals in a zoo is usually not a cause for alarm.

Studies in humans cannot, so far, reveal in detail the contribution of subareas of the amygdala or cellular mechanisms in its functions. However, my NYU colleague Elizabeth Phelps and other researchers (Kevin LaBar, Ray Dolan, Arne Öhman, Mohamed Milad, Andreas Olsson, Daniela Schiller, Jorge Armony, Patrik Vuilleumier, Mauricio Delgado, and Fred Helmstetter, to name several) have confirmed the fundamental role of the amygdala in Pavlovian threat conditioning and extinction in people. Damage to the human amygdala prevents conditioning from occurring,[63] and functional imaging studies show that neural activity increases in the amygdala during conditioning, as well as when the subject is later exposed to the CS.[64] Further, these responses are present whether the subjects are consciously aware of the stimulus or not.[65] As in rats, the hippocampus is involved in the contextual control of defense responses,[66] and the PFCvm in extinction and other processes that regulate amygdala outputs.[67] In the previous chapter, we noted that humans learn about danger from observing others or via instruction. Imaging reveals that amygdala activity increases in connection with both observational and instructed conditioning.[68]

The neural control of freezing by a Pavlovian CS after threat conditioning in rodents is one of the best characterized neural systems responsible for behavior. Much is known not only about the neural system but also the cellular and molecular mechanisms involved in the learning and expression of the responses. The correspondence of findings in rats and humans shows that the rodent studies can be used to research mechanisms relevant to humans.

Before leaving the threat-processing circuitry, I want to emphasize two recent sets of studies by my laboratory that have shed new light on some of the details about how associative learning occurs within the LA during threat learning. In the first, led by Josh Johansen, we used a new technique called optogenetics to demonstrate that the Hebbian hypothesis is indeed a correct account of

a. Optogenetic Manipulation of a Brain Area

1. Obtain a genetic construct for a light-sensitive protein (channel rhodopsin, ChR2) that has been inserted into a virus.

2. Surgically inject the virus into the brain area and wait several weeks for the virus to transport the ChR2 into the neurons. ChR2 when activated will allow sodium ions to enter the neurons, and this will produce action potentials.

3. Surgically implant a fiber optic cable into the brain area and allow one day to recover.

4. Shine blue light through the cable to activate ChR2 and produce action potentials in neurons.

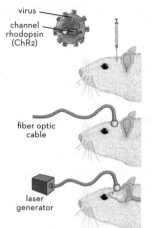

b. Use of Optogenetics to Explore the Hebbian Hypothesis in the Lateral Amygdala

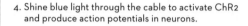

Inject virus

| Conditioning (CS+US) |
| CS: tone |
| US: laser stim of LA at |
| end of CS |

21+ days → 1 day → Test

Three Groups (each receives CS and laser stim of LA)

- ChR2 paired (virus injected then given paired CS + US)
- ChR2 unpaired (virus injected but given unpaired CS and US)
- GFP paired (injection of control substance and give paired CS+US)

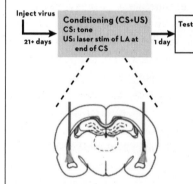

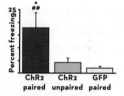

Figure 4.8: Optogenetic Demonstration of Hebbian Learning in the Lateral Amygdala.

The use of the optogenetic method to excite and inhibit neurons in specific brain areas was pioneered by Karl Deisseroth and Ed Boyden (Boyden et al, 2005). **a.** The steps involved in using optogenetics. **b.** The use of optogenetics to test the Hebbian learning hypothesis that during threat conditioning an association is formed when weak and strong inputs converge onto the same neurons in the lateral amygdala (LA). In this study, the channel rhodopsin viral construct (ChR2) or a control agent (GFP) was injected into the LA. After an incubation period, a conditioning session occurred in which a weak conditioned stimulus (CS: tone) was paired with a strong unconditioned stimulus (US: direct optogenetic depolarization of LA cells). This was sufficient to produce freezing to the CS in the paired animals that had the ChR2 but not the GFP injections during the test the next day. Animals that received the ChR2 injection but were conditioned using unpaired presentations of CS and the optogenetic stimulation did not freeze to the CS during the test. Part **a.** based on Buchen (2010), adapted by permission from Macmillan Publishers Ltd.: *Nature News* (vol. 465, pp. 26–28), © 2010. Part **b.** based on Johansen et al (2010).

threat conditioning—that strong activation of LA neurons by the US is sufficient to induce plasticity that changes the meaning of a CS in the LA and enables it to flow through the amygdala to activate downstream targets and trigger freezing. Of note, in this project, no aversive US was administered: the strong stimulus was simply neural, because we artificially activated LA cells as if they were being driven by a shock US. According to the Hebbian hypothesis, if LA cells are being strongly activated while a weak CS (tone) input arrives, the tone should, in principle, then be able to activate the circuitry and elicit defense responses. As depicted in Figure 4.8, this is exactly what we found.

The second study, by Linnaea Ostroff, used a conventional but very powerful technique, electron microscopy, to reveal how threat conditioning results in

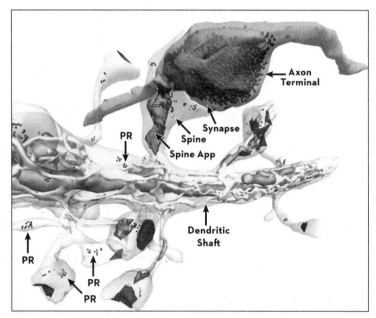

Figure 4.9: Visualizing Structural Changes in Lateral Amygdala Neurons After Learning.

Linnaea Ostroff's electron microscopic reconstruction of synapses in the lateral amygdala (LA) after threat conditioning. Note the neurotransmitter-containing vesicles (the small round structures) in the presynaptic axon terminal forming a synaptic connection with a dendritic spine. Two key findings are that learning results in the movement of the spine apparatus into the spine and in an increase in polyribosomes (PR), which are the basis for protein synthesis, a key step in memory formation. This is a static picture of structural changes that occur when the brain learns.

[IMAGE COURTESY OF LINNAEA OSTROFF.]

changes in the structure of synapses in the LA of rats, showing that learning physically alters the brain in animals when memories about danger are formed.[69] This finding is illustrated in Figure 4.9. The results, along with other evidence for so-called structural plasticity associated with various forms of learning,[70] suggest that physical changes in brain architecture enable learning to persist over time in the brain. This is likely true for all forms of learning, including the learning that underlies acquisition of pathological fear and anxiety, as well as the learning that occurs when these conditions are successfully treated with psychotherapy.[71]

BEYOND REACTIONS

Although Pavlovian conditioning results in the CS being able to automatically elicit innate defense reactions and physiological arousal, organisms can, in the presence of a threatening CS, also learn actions via their consequence in avoiding harm.[72] Recall from Chapter 3 that this is studied in the laboratory using active avoidance conditioning procedures.[73]

The neural mechanisms of avoidance[74] are not understood as well as those responsible for Pavlovian conditioning. However, given that Pavlovian conditioning is the first phase of avoidance (see the discussion of the two-factor theory of avoidance in the previous chapter), in *Synaptic Self* I argued that it should be possible to build on the great progress made in studying Pavlovian reactions to try to understand avoidance actions.[75] My laboratory thus began to work on aversive instrumental tasks with a tone as a warning signal and a shock as the US because this would, in principle, engage some of the same circuitry as Pavlovian conditioning with tone and shock.[76] We've made significant progress in the last few years thanks to the work of Chris Cain, Gabriel Lázaro-Muñoz, June-Seek Choi, Justin Moscarello, Rob Sears, Vin Campese, Franckie Ramirez, and Raquel Martinez.[77]

As described earlier, Pavlovian conditioning depends on the LA and CeA. We discovered that avoidance, in contrast, involved the LA and the basal amygdala (BA).[78] To understand why this difference is important, let's break down what goes on in the brain as the avoidance response is learned.

We've seen how, early in the avoidance learning process, stimuli present when the US occurs become Pavlovian CSs that elicit freezing, and how this dominant tendency of the CS to elicit freezing has to be inhibited before avoidance can be learned—you can't act if you are frozen in place.[79] The inhibition of freezing is achieved by a redirection of information flow within the amygdala, preventing the LA from activating the freezing circuits of the CeA and instead allowing signals to travel from the LA to the BA and control avoidance behavior.[80] If the CeA is damaged, avoidance is learned faster, because the conflicting

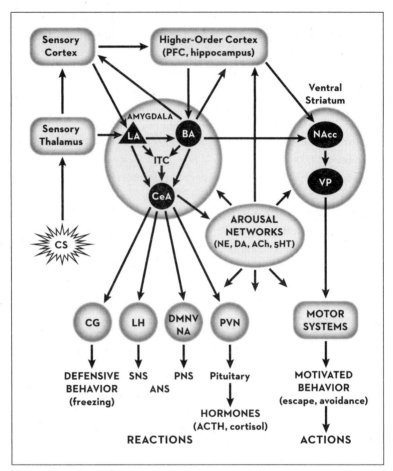

Figure 4.10: Defensive Action Circuitry Builds upon Pavlovian Reaction Circuitry.
The action circuitry is an extension of the reaction circuitry shown in Figure 4.5. The main difference is the connection from the basal amygdala (BA) to the NAcc of the ventral striatum, which allows the emission of actions under motivational influences signaled by information from the BA. For other abbreviations, see Figure 4.5.

freezing response is eliminated.[81] Thus, the CeA, although not necessary for avoidance, has a regulatory role in avoidance learning.

The redirection of information flow that allows avoidance to proceed is controlled by interactions between the amygdala and PFCvm.[82] A key output target of the BA is the *ventral striatum*, especially the *nucleus accumbens* (NAcc) and specifically its shell subdivision; damage to or functional inactivation of this region disrupts avoidance.[83] The active avoidance circuitry is shown in Figure 4.10 in relation to the circuitry of Pavlovian reactions.

Imaging studies in humans are consistent with the animal findings: The amygdala, NAcc, and frontal cortical areas have all been implicated in active avoidance behavior.[84]

The escape-from-threat variant of avoidance, which separates the Pavlovian and instrumental learning processes, is especially useful in isolating the role of the CS as a negative reinforcer during learning[85] (see Chapter 3). (Recall that in the instrumental phase of the escape task, the only reinforcement that occurs is CS termination; the shock US is not administered.) Karim Nader and Prin Amora-panth pioneered this work in our laboratory, showing that damage to the LA and the BA, but not the CeA, disrupts learning of CS-terminating behaviors, and suggesting that the connection from the LA to the CeA has to be suppressed as the connection from the LA to the BA is brought online.[86] The same amygdala circuits were later implicated in active avoidance conditioning. Thus, because the escape-from-threat task separates the Pavlovian and instrumental phases of learning, it is well suited for pursuing the nature of conditioned negative rein-forcement in the brain.

The most common hypothesis about the reinforcement signal in avoidance and escape from threat is that it involves the relief that occurs when CS termina-tion reduces a hypothetical central state of fear that the CS originally elicits (see Chapter 3). If we are to understand how CS termination can strengthen a behav-ioral response, however, psychological notions like fear relief are not very useful. Instead, we need to pursue the source of reinforcement at the synaptic and cellular level. Specifically, we would not only need to know how the avoidance behavior reduces CS-driven neural activity in particular circuits, but would also have to establish that such changes are required for the learning to take place. Our specific hypothesis, which I hinted at in Chapter 3, is that the key neural changes produced by conditioned negative reinforcement occur at synapses in the pathway from the BA to the NAcc, and regulation of these synapses by neuromodulators, such as dopamine, is a key molecular event. This hypothesis is inspired by earlier work in appetitive conditioning (involving reinforcing stimuli such as food, sex, or addic-tive drugs) by Barry Everitt, Trevor Robbins, and their colleagues,[87] and the work of Anthony Grace[88] and Kent Berridge[89] on goal-directed behavior circuits involv-ing the NAcc. While most of my research funding has over the years come from the National Institute of Mental Health (NIMH), the National Institute for Drug Abuse (NIDA) has funded our work on the role of negative reinforcement in avoid-ance because of the importance of negative reinforcement in addiction.

Once the avoidance response is well established and the response is per-formed habitually (see Chapter 3), the amygdala is no longer required for the process to be carried out.[90] This is indicated by studies showing that damage to the amygdala has no effect on entrenched avoidance. The exact circuits that are responsible for habitual avoidance once the amygdala relinquishes control are not known. In conditioning that uses food or addictive drugs, the amygdala is also initially involved and cedes control when habits set in:[91] At that point, the dorsal striatum, which sits above the nucleus accumbens (ventral striatum), is engaged.[92] Although it is tempting to suggest that the dorsal striatum may be part of the aversive habit circuitry as well, existing data do not support this.[93] Further elucidation of this circuit is especially important because of the role of habitual avoidance in anxiety disorders.[94] But to repeat a point made in the previous chapter, avoidance can be useful or maladaptive for people troubled by debilitating fear and anxiety;[95] this idea will be elaborated on in Chapter 11.

ACTIONS GUIDED BY PAVLOVIAN INCENTIVE VALUES

As noted in the last chapter, in situations in which challenges to well-being are present, you may not always have, from past learning, the exact kind of Pavlovian defensive reaction or instrumental avoidance response needed to save your skin. Nevertheless, the incentive values of stimuli acquired through Pavlovian conditioning can be used to guide decision making and choose a course of action in this novel context.

The role of Pavlovian incentives in guiding decision making and behavior is often studied by examining the effects of a Pavlovian CS on instrumental responses.[96] Research using appetitive incentives (food, sexually related stimuli, or addictive drugs) has consistently implicated several brain areas in these effects of Pavlovian incentive instrumental responses, including regions of the amygdala (lateral, basal, central), ventral striatum (NAcc), dorsal striatum, PFCvm, anterior cingulate cortex, and orbitofrontal cortex, as well as dopamine inputs to some of these regions.[97] Human studies have also implicated the amygdala, ventral striatum, and frontal cortex, suggesting some similarity in neural organization across mammalian species.[98] Although very little work has examined how aversive CSs affect aversively motivated responses like avoidance, my laboratory has begun to address this topic in rats and has shown that the LA and CeA play essential roles.[99]

The fact that the amygdala has been implicated in processing both appetitive and aversive incentive stimuli suggests to some that it may be a general value-processing area in the brain.[100] Others propose that subregions of the ventromedial, cingulate, and/or orbitofrontal cortex play this role.[101] Because all of these regions have been implicated in value processing and are strongly interconnected, it seems that they may interact in value coding during decision making.

UNCERTAINTY, RISK, AND THE BRAIN

In the examples discussed so far, immediately present stimuli are used to guide behavior in situations where a threat exists. As we've discussed, however, an important feature of anxiety, as opposed to fear, is uncertainty about whether and when an impending danger will occur, how long it will last, and what actions should be taken in response to it. Although the effects of risk and uncertainty have been explored most extensively in other research contexts,[102] I will focus here on the relation of uncertainty to risk assessment in situations of danger.

Although the amygdala had long been regarded a central neural component in fear and anxiety circuits, damage to the amygdala has not reliably been found to interfere with performance in tasks used in animal studies to screen drugs for their ability to reduce anxiety in people. These "anxiety" tests, in contrast to tests involving a CS that predicts when danger will occur, mostly involve placing animals in situations in which there is uncertainty about whether danger exists, or in situations in which a CS has an unpredictable start and finish. Another brain area, the *bed nucleus of the stria terminalis* (BNST), which is part of what is called the *extended amygdala*,[103] has often been found to play a role in such tests.[104] Studies in healthy humans have confirmed the role of the BNST in processing

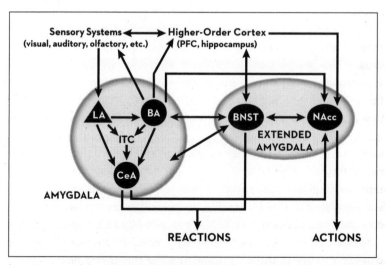

Figure 4.11: Amygdala and Extended Amygdala Connectivity: Reactions, Actions, and Threat Certainty.

The amygdala controls defensive reactions based on threats that are present or highly likely to occur, whereas the bed nucleus of the stria terminalis is proposed to control reactions and actions based on uncertain threats.

uncertainty.[105] The BNST thus seems to do for uncertainty what the amygdala does when there is a specific and certain threat stimulus.[106] This important distinction between the roles of the amygdala and BNST in situations involving certain versus uncertain threats was first noted by Michael Davis.[107]

The connectivity of the amygdala and BNST[108] helps illustrate their respective contributions to situations of certain versus uncertain harm (Figure 4.11). The BNST has many of the same output connections as the amygdala. Like the CeA, it links to circuits that control defensive behaviors such as freezing, as well as to circuits that control the autonomic nervous system, endocrine function, and brain arousal. Like the BA, it also connects with the hippocampus and PFCvm. This likely explains why the BNST can take over some defensive functions normally controlled by the amygdala if the latter is damaged.[109]

The inputs to the BNST, by contrast, are somewhat different from those that go to the amygdala, which may be a key reason why these structures have different roles in behavior.[110] The amygdala, by way of the LA, has extensive inputs from specific sensory systems. This enables particular threat cues to be evaluated and to trigger defensive responses by way of outputs of the central nucleus. The BNST, however, is more extensively connected with cortical areas involved in various aspects of cognitive processing, including the hippocampus and various regions of the prefrontal cortex (e.g., PFCvm, insula cortex, and orbitofrontal cortex). Although the hippocampus is best known for its role in memory, it is also associated with relational processing, including spatial relations, in which it is responsible for creating a map of the environment.[111] This accounts for the involvement of the hippocampus in the contextual regulation of threat conditioning mentioned above. Environmental mapping is also a key component of risk assessment in situations of conflict and uncertainty. In assessing risk, the hippocampus obviously has to draw upon memory, though attention and other executive functions of prefrontal areas are likely to make their own contributions to the estimation of the value of possible behaviors and their outcomes. The contribution of the hippocampus is especially interesting in light of Gray and McNaughton's behavioral inhibition theory, which proposes that the hippocampus, together with the septal region, is a major player in anxiety.[112] The so-called septohippocampal system has received renewed interest in light of new studies showing that anxiety-like behaviors can be increased or decreased by genetic manipulations of neural activity in the hippocampus.[113] The septum has also been implicated in Pavlovian threat conditioning and other defensive bahaviors.[114]

Another important set of inputs to the BNST comes from the amygdala: Connections from the BA and CeA give the BNST access to the amygdala's processing of specific threat cues. The BNST also connects back to the amygdala.

Recent studies exploring the function of different components of the BNST add further insight into its role in aversively motivated behavior.[115] For example, subregions of the BNST have cells that respond differently in situations of uncertainty and risk, with cells in one region triggering (and in others suppressing) risk assessment.[116]

Earlier I described the role of connections from the basal amygdala (BA) to the nucleus accumbens in defensive actions such as avoidance. Like the BNST, the NAcc is part of the extended amygdala (as well as being part of the striatum— brain terminology is not always consistent) and has connections with both the amygdala and BNST. Connections of the NAcc with the BNST may contribute to action control in situations where threat is uncertain.

The emergence of the BNST in various aversive behavioral tests that involve uncertainty suggests a way to integrate Gray and McNaughton's behavioral inhibition system with the defensive (freeze-flight-fight) system. The BNST sits at the crossroads between defensive circuits involving the amygdala and accumbens and risk-assessment circuitry involving the septohippocampal circuitry and prefrontal cortex. It thus may coordinate the two systems, balancing which dominates behavioral control, depending on the degree of uncertainty. At the same time, it should be noted that the amygdala, accumbens, prefrontal cortex, and septohippocampal system are all interconnected,[117] and the preferential role in one behavior or another is more a matter of functional segregation and selective recruitment of circuits by behavioral demands.

Uncertainty is the breeding ground of anxiety. But it is important to keep separate the brain state of uncertainty, which is by and large an unconscious factor that shapes brain processing and behavior, from the conscious feeling of anxiety, which can result from unconscious processing of uncertainty in ambiguous situations. They are related but not the same.

ALTERATIONS IN THREAT PROCESSING AND DEFENSIVE RESPONDING IN PEOPLE WITH FEAR AND ANXIETY DISORDERS

The brain mechanisms that are altered in fear and anxiety disorders can be understood in terms of the basic mechanisms underlying the processing of threats and the control of defensive responses in animals and healthy humans. The circuits that are responsible for these functions, as we have seen, are within the amygdala, ventral striatum (NAcc), extended amygdala (BNST), hippocampus, PAG, and various areas of prefrontal cortex (lateral and medial prefrontal cortex, orbitofrontal cortex, anterior cingulate cortex, and insular cortex).[118] To illustrate how these brain areas, and connections between them, contribute to anxiety, I will draw upon a summary by Dan Grupe and Jack Nitschke, who synthesized much of the literature and proposed an *uncertainty and anticipation model of anxiety*.[119]

Grupe and Nitschke noted that threat processing is altered in several ways in fear and anxiety disorders.[120] Anxious people exhibit: (1) increased attention to

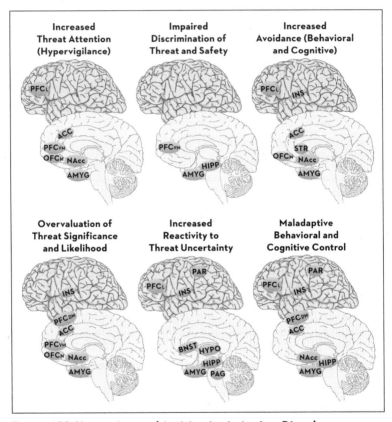

Figure 4.12: Uncertainty and Anticipation in Anxiety Disorders.

Grupe and Nitschke (2013) argue that anxiety involves anticipatory responses that occur in situations involving uncertainty. They propose that six fundamental processes are altered in the brain in anxiety disorders. Some key brain areas implicated in each process are shown. The illustration is a modification of their Figure 1. Abbreviations: ACC, anterior cingulate cortex; AMYG, amygdala; BNST, bed nucleus of the stria terminalis; INS, insular cortex; HIPP, hippocampus; NAcc, Nucleus Accumbens; PAG, periaqueductal gray; PAR, parietal cortex; PFCDM, dorsal medial prefrontal cortex; PFCL, lateral prefrontal cortex; PFCVM, ventral medial prefrontal cortex; OFCM, orbital frontal cortex; STR, dorsal striatum.

FROM GRUPE AND NITSCHKE (2013), ADAPTED WITH PERMISSION FROM MACMILLAN PUBLISHERS LTD.: *NATURE REVIEWS NEUROSCIENCE* (VOL. 14, PP. 448–501), © 2013.

threats; (2) deficient discrimination of threat and safety; (3) increased avoidance of possible threats; (4) inflated estimates of threat likelihood and consequences; (5) heightened reactivity to threat uncertainty; and (6) disrupted cognitive and behavioral control in the presence of threats. The role of the various brain areas in these six processes is illustrated in Figure 4.12. Below, I summarize some of the major conclusions drawn by Grupe and Nitschke about these processes and the brain mechanisms involved,[121] though I have added some information that I thought they overlooked. Still, this is a cursory summary.

1. Increased Attention to Threats (Hypervigilance)

A heightened sense of threat detection occurs in people who have generalized anxiety, which includes mostly everyone with a fear/anxiety disorder.[122] In extreme cases nearly anything can be threatening and trigger defensive behavior (freezing, avoidance), increase brain arousal (through the release of norepinephrine and dopamine), and initiate stress responses (through activation of the ANS and the release of stress hormones, especially epinephrine, norepinephrine, and cortisol). The tendency to view benign stimuli as threatening is referred to as *interpretation bias* and occurs in both generalized anxiety and specific disorders.[123] People who suffer from the latter have particular biases: Spider phobics can be especially sensitive to spider cues and not at all responsive to cues related to snakes or social situations; people with panic disorder can be unusually attuned to body sensations that might signal an attack; people suffering from combat PTSD can be highly sensitive to the sound of backfiring cars or the sight of blood or weapons.

Most often an overactive amygdala is implicated in these heightened responses to threats.[124] This focus on threats can prevent attention being paid to other factors that under normal circumstances might ameliorate the biased response. Amygdala activation engages the PAG, initiating defensive responses. Arousal systems in the basal forebrain and brain stem are also activated, facilitating processing in the amygdala, and in sensory areas of the cortex areas that are actively processing the threat.[125] Cognitive processing systems in the cortex, such as the prefrontal cortex and the anterior cingulate cortex, which are involved in working memory, attention, and other executive functions, are engaged.[126] The ventromedial and orbitofrontal cortex interact with the amygdala in processing the incentive value of threat stimuli, which may contribute to further focusing of attention on the threat.

2. Impaired Ability to Discriminate Threat and Safety

In healthy people, circuits involving the amygdala, PAG, PFCvm, and hippocampus are implicated in distinguishing threat from safety.[127] This ability is impaired in people with pathological fear/anxiety, again including GAD and more specific disorders.[128] For example, a failure of context discrimination may occur in panic disorder due to impaired hippocampal function.[129] One consequence of the brain's failing to make this distinction is that extinction is impaired, meaning that the normal processes that would weaken the threat value of stimuli fail to do so. In the healthy brain the PFCVM regulates the amygdala and allows the meaning of threats to change when experience shows they are no longer harmful[130] (e.g., after extinction). Emotional disorders, including problems with fear and anxiety, have long been thought to involve a failure of the prefrontal cortex to properly regulate the amygdala.[131]

3. Increased Avoidance

Excessive behavioral and/or cognitive avoidance is a leading hallmark of anxiety disorders.[132] It occurs in GAD, PTSD, panic disorder, and in various phobias. Avoidance is a way to prevent exposure to threat, and in anxiety disorders it becomes so habitual that the brain never has the opportunity to determine whether conditions have changed, and that what was once harmful is no longer. Avoidance thus leads to both constant threat expectancy and a failure of learning to identify safety. Because the events that are being fretted over never actually occur, cognitive avoidance is reinforced and strengthened, leading to the false belief that having made the choice to avoid is what prevented harm. Getting people to shed their behavioral and cognitive patterns of avoidance is an important part of many therapeutic approaches.[133]

As noted above, consistent with animal studies, avoidance imaging studies in humans reveal activity in the amygdala, NAcc, and dorsal striatum, and also implicate the insular, orbitofrontal, and cingulate cortex.[134] Successful treatment of excessive avoidance has been reported to reduce activity in the PFCVM, cingulate cortex, orbitofrontal cortex, and insular cortex, and to increase activity in the dorsal prefrontal region, which is involved in executive control.[135]

4. Heightened Reactivity to Threat Uncertainty

People with anxiety have trouble tolerating uncertainty, especially about threats.[136] Those with GAD, panic disorders, PTSD, and phobias have exaggerated responses to threats, especially when faced with uncertainty about whether the threat will occur or when it will end.

Areas implicated in exaggerated responses to uncertainty in people with anxiety disorders include the amygdala, BNST, hypothalamus, hippocampus, insula, and frontoparietal executive attention circuits.[137]

5. Overvaluation of Threat Significance and Likelihood

People with anxiety disorders, including GAD, phobias, and PTSD, view negative events as much more likely to occur, and expect more severe consequences as a result, compared with healthy controls.[138] This is called *judgment bias* and leads to anticipatory stress when any negative outcome is envisioned, however unlikely it may be. As we've seen, processing of values that enable learned threats to modulate action involve the amygdala; NAcc; orbitofrontal, insular, and cingulate cortices; and the PFCvm.

6. Maladaptive Behavioral and Cognitive Control in the Presence of Threats

Several ideas have been proposed about how control over behavior and cognition are altered in anxiety disorders. We explored this in the context of Gray and McNaughton's behavior inhibition system (focused on the hippocampus), my ideas about maladaptive avoidance systems (focused on the amygdala and nucleus accumbens), and Davis's notion of threat uncertainty (focused on the BNST). We also saw how these three views might be integrated. Grupe and Nitschke emphasized an additional point of view called the adaptive control hypothesis, originally proposed by Alexander Shackman and Richard Davidson.[139] This idea gives the anterior cingulate cortex a major role in behavioral control. This region, which is altered in anxiety disorders,[140] connects with the amygdala, BNST, hippocampus, and other areas discussed above, and thus can also be integrated into an overarching view of how uncertainty contributes to anxiety.

FROM THREAT TO CONSCIOUS FEAR AND ANXIETY

The circuits altered in anxiety disorders are, for the most part, those that have been implicated in normal aspects of threat processing and stress regulation in animals and healthy humans. Although the human data are at a less refined level of analysis, they nevertheless confirm that mechanisms discovered in the brains of animals are relevant to human fear and anxiety disorders.

But we have to be careful not to confuse malfunctioning cognitive processes with the feeling of fear or anxiety. Fear or anxiety is not an increased attention to threats, or failed safety versus threat discrimination, or increased avoidance, or heightened response to threat uncertainty, or overvaluation of the significance of perceived threats. Nor is it simply the combination of these. Fear and anxiety are unpleasant feelings, and fearful or anxious people want to eliminate them.

Although processes that Grupe and Nitschke have identified clearly help us better understand what issues need to be addressed in order to help people feel better, I think we need a subtler understanding of what fear and anxiety are if we are to take our conception to the next level.

Specifically, we need to understand how fearful and anxious feelings arise and persist in the stream of consciousness. At least two separate processes are involved: One includes the cognitive processes that underlie any kind of conscious experience, whereas the other includes all the factors that make emotional conscious experiences different from nonemotional ones.

Here's a snapshot of what is to come in the next several chapters. The goal is to arrive at an understanding of how threat detection and threat anticipation give rise to conscious feelings. Consciousness is personal; it's private; it's in each of our heads. It's mental, but it's also physical. There was a time when mental meant nonphysical, but we are past that belief. Mental processes and states are physical products of the brain. Because you are reading this book, you probably believe this. That's good, because in the coming chapters we get physical with the most mental of all mental functions—consciousness.

CHAPTER 5

HAVE WE INHERITED EMOTIONAL STATES OF MIND FROM OUR ANIMAL ANCESTORS?

"The conclusion that the animal hunts because it is hungry . . . will not satisfy the scientist who wants to know what is happening inside the animal when it is in this state. . . . Hunger, like anger, fear, and so forth, is a phenomenon that can be known only by introspection. When applied to another subject, especially one belonging to another species, it is merely a guess about the possible nature of the animal's subjective state."

—NIKO TINBERGEN[1]

It is common to think of emotions like fear, anger, and joy as primitive feeling states inherited from our animal ancestors. This view, which is deeply ingrained in our folk psychology, dates back at least to Plato, who viewed base emotions as wild animal impulses that had to be reined in by reason, much like a charioteer had to control his horses. Many scientific discussions also start from the assumption that emotional feelings are part of our evolutionary heritage. But what exactly does this mean? How are feelings like fear passed from species to species? The obvious answer is that they are encoded in neural circuits, and by virtue of inheriting these circuits from animals, we are endowed with the feelings, the emotions, they encode. Fear, for example, is often said to be a product of an innate neural circuit that controls not only defensive behaviors like freezing, flight, and fight but also the actual feeling of fear. Moreover, the feeling is often viewed as being responsible for the behaviors. The innate view of emotion, as we shall see, has an impressive scientific legacy, but as I said in Chapter 2, I believe it is wrong.

DARWIN'S THEORY OF EMOTIONS

The modern version of the idea that primitive human emotions are inherited from our animal ancestors was born on November 26, 1872, the date that Charles Darwin's book *The Expression of the Emotions in Man and Animals* was published.[2] Darwin had earlier advanced his revolutionary theory that species evolve through a process of natural selection,[3] and in this new work he argued that emotional "states of mind" evolve the same way.

Darwin's theory was inspired by his observations of what he called expressive actions: behavioral and physiological responses that occur in connection with emotions. He noted that "the chief expressive actions, exhibited by man and by the lower animals, are innate or inherited—that is, have not been learnt by the individual." As evidence for innate emotional responses in humans, Darwin noted that certain expressions of emotion, especially in the face, are similar in people around the world, regardless of their racial origins or cultural heritage and regardless of their isolation from other races or cultures. He also noted that the same emotional expressions occur in individuals who are born blind and had no opportunity to learn what these look like.

He drew extensively from the work of Guillaume-Benjamin-Amand Duchenne (de Boulogne) (1806–1875),[4] who took photographs of human facial expressions (which he induced by electrical stimulation of face muscles) and compared them to emotions as depicted in Ancient Greek sculptures[5] (including *Laocoön and His Sons*, as seen in Figure 1.1). Darwin seems not to have been aware of an earlier body of work by the German-Austrian artist Franz Messerschmidt (1736–1783), who also depicted emotional expressions. Figure 5.1 features facial expressions from Duchenne and Messerschmidt.

Darwin also noted that a number of emotional expressions are similar across species: "Some of the expressive actions of monkeys are . . . closely analogous to man." He cited expressions of pleasure, grief, anger, and fear, among others, and also cited the commonness of freezing and flight as a response to danger in many animals.

Because Darwin emphasized the outward expression of emotions, he is sometimes said to have not been concerned with their subjective aspects.[6] Although he certainly devoted more ink to behavioral characteristics than feelings, it is not the case that he ignored the latter. For him, as for many people then and now, emotional behaviors are fundamental signs (expressions of) emotional feelings. He explained his position this way: "Certain actions expressive of certain states of mind are the direct results of the constitution of the nervous system. . . . Trembling under the influence of fear . . . is an example." The key phrase in this

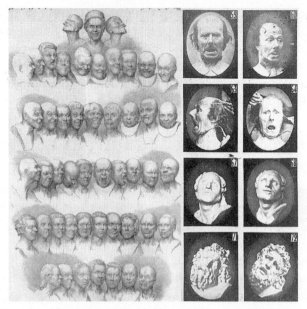

Figure 5.1: Emotional Expressions: Duchenne and Messerschmidt.
Darwin used the research and photographs of the French physiologist and physiognomist Guillaume-Benjamin Duchenne de Boulogne (1806–1875) in developing his ideas about the innateness of emotional expressions (right). Earlier, the German-Austrian sculptor Franz Xavier Messerschmidt (1736–1783) had constructed sculptures of facial expressions, including emotional expressions, but these seem not to have influenced Darwin (left).

quote is *actions expressive of certain states of mind.*" What he is implying here is that these mental states are the basis for innate behaviors: Threats elicit innate feelings of fear, and fear in turn elicits freezing, trembling, and flight. Darwin argued that because these emotional mental states give rise to behaviors that help organisms adapt and survive, the mental states were preserved in the nervous system via natural selection, passed on within species, and conserved as new species evolved. In Darwin's view, we feel fear when in danger because some prototype of fear present in our animal ancestor helped them survive, and it continues in us because it has been helpful to the survival of our species as well.

In contemporary psychology and neuroscience, as we've seen, terms like "mental" or "mind" do not necessarily refer to conscious processes. Perceiving, remembering, attending, thinking, planning, and deciding involve nonconscious processes that do much of the mental work and actually make conscious awareness possible. But in Darwin's time, "mental" was synonymous with "conscious." Darwin was clearly implying that conscious emotional states of mind (feelings) underlie emotional expressions.

Darwin's theory of evolution by natural selection was one of the greatest intellectual achievements in history. But his belief that our emotional feelings themselves are inherited from animal ancestors and are represented in essentially prepackaged form in the brains of all humans, although consistent with everyday folk wisdom and useful in daily life, takes us down the wrong path, in my opinion, in terms of understanding emotions and their underlying brain mechanisms.

DARWIN'S EMOTIONAL LEGACY IN EARLY PSYCHOLOGY

Darwin's interest in emotional behavior reflected a deeper interest in the evolution of the human mind. He believed that "there is no fundamental difference between man and the higher mammals in their mental faculties." But as one psychological historian pointed out, "Darwin bestowed a mental life upon man's cousins with a very open hand, without the self-critical zeal that marked his biological endeavors."[7] Darwin went so far as to proclaim that worms "deserve to be called intelligent, for they . . . act in nearly the same manner as a man under similar circumstances."[8] He often argued for such human qualities in animals, commonly characterizing their expressive behaviors with terms like "affectionate," "cheerful," "savage," "pleased by being caressed," "jealous," and so on. He likewise made generous use of anthropomorphically based anecdotes: "What a strong feeling of inward satisfaction must impel a bird, so full of activity, to brood day after day over her eggs."[9]

The zeal of the master on this topic was shared by his disciples. George Romanes, a close friend of Darwin's, wrote a book called *Mental Evolution in Animals,*[10] in which he described behavioral responses as "ambassadors of the mind" in humans and other animals.[11] According to Romanes, just as we use our own mental states to conceive of the mind of God, we use a similar anthropomorphism to understand the mind of animals by looking for behaviors we have in common with them.[12]

Like Darwin, Romanes is often criticized for treating anecdotes about animal behavior as scientific data.[13] (On the basis of largely innate behaviors triggered by innate stimuli, for example, he described earwigs as affectionate to their offspring and fish as jealous and angry.[14]) Such arguments based on analogy with human behavior are now viewed as on par with commonsense intuitions and should not, on their own, be taken as scientific evidence for mental state consciousness in other animals.[15] (Common sense is often a starting point in scientific research, but scientific conclusions require more.)

But Romanes was not alone in such theorizing. The tendency to attribute mental states—typically humanlike mental states—to animals on the basis of behavioral responses was so rampant in the waning years of the nineteenth century that one researcher, Lloyd Morgan, warned that scientists should resist the temptation to "humanize the brute." He argued that just because scientists necessarily start their exploration of animal behavior from their own subjective experiences does not justify the attribution of similar experiences to other animals.[16] This kind of attribution is desirable, he argued, when we interact socially with other humans, but is questionable when trying to understand animal behavior.[17] Morgan famously wrote that we should not call upon human mental states to account for animal behavior if a simpler, nonmental state explanation is available. This position is now known as Morgan's Canon. It is so difficult to resist the pull of folk wisdom, though, that even Morgan himself transgressed, using the phrase "coalescence of mental problems in a conscious situation" in a description of his dog's ability to open the garden gate.[18] Still, he acknowledged that although animals have intelligence, they lack reason—they think but "do not think the therefore."[19]

Perhaps animal researchers of the late nineteenth century should not be criticized too severely, as consciousness was in the air at the time. Psychology was just then beginning to break away from philosophy and emerge as a scientific endeavor.[20] It achieved this by applying experimental methods borrowed from physiology and physics to questions that had been framed by philosophers since ancient times about the nature of the mind, especially about consciousness. For example, early German psychologists pioneered an experimental approach to the mind that involved a form of personal introspection, but following strict procedures. With this, they sought to analyze the content of consciousness—for example, the elementary components or elements that make up the experience of a complex sensation (such as the flavor of a soup) or an emotion (such as an intense feeling of fear).[21]

In America psychological research was started by William James, who also focused on consciousness, but on its functions more than its contents.[22] As an admirer of Darwin, James sought to discover what made consciousness adaptive and thus subject to natural selection. But in another respect, James broke with Darwin and his commonsense approach, challenging the idea that feelings are the cause of emotional expressions and behaviors. James argued that we do not run from a bear because we are afraid, but instead we are afraid because we run.[23] He was correct on the first point (conscious feelings are not necessarily the cause of emotional behavior) and on the right track with the second (that feedback from nonconsciously controlled body responses plays a role in feelings). But in my opinion, he overstressed the role of feedback. Although feedback from the body does contribute to feelings, it is not the sole determinant of what we feel, as we'll discuss later.

Another important early American psychologist, E. L. Thorndike, was also influenced by Darwin. He argued that animals learn behaviors through trial and error, and that responses that lead to pleasure or that avoid pain (displeasure) are "stamped in."[24] This learning rule was called the "law of effect," a Darwinian principle applied to the individual; pleasure and pain help the organism survive, and behaviors that are connected to these hedonic states are acquired for future use.[25] Though he was generally opposed to mental explanations, in adopting a learning rule based on pleasure and pain, Thorndike followed a long tradition of British thinkers (Locke, Hume, Hobbes, Bentham, Mill, Bain, and Spencer) who also emphasized the role of such hedonistic feelings in motivation and learning; Bain and Spencer, in fact, both proposed learning rules similar to Thorndike's law of effect.[26]

By the 1920s the behaviorist revolt was under way in America against the mentalistic foundation of psychology. John Watson argued that psychology should not concern itself with private inner states in people or animals;[27] to be a legitimate science, it needed to focus on observable events—stimuli and responses. And during the behaviorist era, the subjective notions of pleasure and pain as the basis of learning were replaced by the concept of reinforcement. According to B. F. Skinner, a reinforcer is a stimulus that increases or decreases the likelihood that a behavior will be repeated.[28] Theoretical notions about inner feelings were supplanted by descriptions in terms of observable factors—especially the organism's history of reinforcement with certain stimuli in particular situations.

Emotions were also reinterpreted in such a way as to eliminate their subjective element. For example, for Watson fear became a Pavlovian conditioned reflex,[29] whereas for Skinner it was a behavioral disposition based on reinforcement history.[30] Others seeking objective "inner mediators" proposed that fear was an intervening variable,[31] a drive,[32] or a motivational state,[33] anything *but* a conscious feeling. Interestingly, though, they did not do away with the standard mental state terminology for conscious experience: They continued to use words like "fear," "anxiety," "hope," and "joy," but as descriptions of tendencies to act in a certain way rather than as terms for inner feelings.

Behaviorists didn't care much for brain states, either, as these, too, were internal and thus unobservable to the psychologist.[34] But brain research was advancing in parallel with behaviorism in the 1940s and 1950s, and the idea that emotions might be represented as central (brain) states was rising in popularity, even in behaviorist circles.

THE SEARCH FOR EMOTIONS IN THE BRAIN

Research on the brain mechanisms underlying behavior was, as described in Chapter 4, greatly facilitated by the development of electrical brain stimulation in the early twentieth century. Defensive, aggressive, feeding, sexual, and other seemingly innate behaviors could be readily elicited by brain stimulation.

Initially, this field was populated not by psychologists so much as by researchers trained in biology or neurology. These scientists were less concerned with and constrained by the rules of behaviorism and often argued that the innate behaviors elicited by brain stimulation were controlled by emotional states. For example, Walter Hess, one of the pioneers in the use of brain stimulation in animals, spoke of "mental motivations" and "experiences that have an emotional component" intervening between the electrical pulses and the behaviors.[35] It is also common to find mentions of fear, anger, rage, and pleasure in this literature. Although the only variable measured in stimulation studies was behavior, the assumption was that the circuit that gave rise to the behavior also gave rise to mental states, to feelings. And because the circuit was presumed to be conserved between man and animals, studies of emotional behavior in animals would be able to reveal the source of human feelings. In other words, the data were being interpreted in a way consistent with the Darwinian and the commonsense views of emotions— emotional responses reflect emotional states of mind (feelings).

Science is not only a process of collecting data but also of interpreting it, and there are obviously different ways of doing the latter. It's one thing to say that the brain areas from which innate behaviors can be elicited by electrical stimulation play a role in the control of those behaviors, and quite another to say that the circuit that controls defensive or aggressive response in animals is also responsible for feelings of fear or anger in them. The former interpretation stays close to the data, whereas the latter goes far beyond the data in ways that are not easily tested. Even in humans, as we've seen, behavior is not a foolproof way of knowing that a conscious state, like a feeling of fear, occurs in tandem with defensive behavior or physiological responses. Behaviors and associated feelings often do occur together, but not always. And when they do occur together, two key questions arise: Does the same brain system that controls the behavior also give rise to the feeling? Is the feeling the cause of the behavior?

As Lloyd Morgan observed, the fact that scientists must start from their own subjective experiences of emotional feelings does not justify the attribution of similar experiences to other animals. Niko Tinbergen, the father of ethology, made a similar point in the epigraph that opens this chapter: Attribution of hunger, fear, and other mental states to animals is mere guesswork.

Let's dig deeper into this point. That a food-deprived animal searches for food because its energy supplies are low is testable by measuring and manipulating energy-related chemicals (such as glucose) in relation to food-seeking behavior. That the animal is experiencing a mental state of "hunger" when it searches for food is an interpretation. The peril of this kind of speculation is borne out by the fact that people often seek out and eat when they are not food deprived, and animals too eat for reasons other than replenishing energy supplies—rats, for example, will work (press a bar) to get sweets, even nonnutritive sweets like saccharin, even if they are not "hungry" (food deprived).[36] Therefore, if we use eating as an indicator of a mental state called hunger, we will often be wrong. The neuroscientist Kent Berridge, who studies "emotional" (hedonic) reactions to tastes, cautions against assuming that these behaviors reflect conscious experiences of pleasure or aversion.[37]

By the middle of the twentieth century, many so-called physiological psychologists had turned to the brain to understand the motivational basis of behavior. Because most of these psychologists were trained in the behaviorist tradition, they were reluctant to call upon subjective states of consciousness, like fear or hunger, to explain the elicitation of defensive or eating behaviors by brain stimulation. Instead they introduced the idea of central motive states,[38] which we discussed in Chapter 2. In line with the behaviorist convention, mental state labels ("fear," "hunger") were retained in spite of the fact that the states were viewed as physiological rather than conscious states.

The idea that emotions could be explained as physiological states that detect stimuli and control behavior was a way to study the role of inner states and still be behaviorist. But this led to much confusion. For one thing, not all researchers adopted the nonsubjective approach—hunger and fear, for example, were used by some to mean nonsubjective physiological states, but others viewed them as states of consciousness. A second problem is that even those who claimed to be adherents of the physiological approach often spoke and wrote in a way that blurred the difference between subjective and nonsubjective interpretations of behavioral mediators.[39] It is therefore not surprising that scientists who were not in the field, and most laypeople, typically assumed that mental state terms referred to mental states, not to nonsubjective physiological states.[40] It didn't help that the very popular limbic system theory proposed that feelings of fear and other emotions emerge in limbic regions to intervene between eliciting stimuli and emotional responses in animals and people. The line between central state and commonsense approaches has always been blurry.

BASIC EMOTIONS THEORY: A MODERN DARWINIAN APPROACH TO EMOTION

Darwin was correct in stating that certain behaviors are innately wired in the brain.[41] But his view that emotional feelings cause behavior because conscious feelings and emotional responses are innately coupled in the brain demands further consideration.

Darwin's views are alive and well today in the theory of *basic emotions*, which originated in the writings of Silvan Tomkins in the 1960s.[42] Building on Darwin, Tomkins proposed that several *primary* (or *basic*) *emotions* are genetically built into the human brain by natural selection and expressed identically in everyone, regardless of racial or cultural background. Each of these innate emotions was said to be wired into an *affect program,* a hypothetical subcortical neural structure assumed to somehow involve the limbic and arousal systems. In the presence of a trigger stimulus for a given emotion, the affect program would be activated, and the bodily responses characteristic of that emotion would be expressed (Figure 5.2). The primary emotions that Tomkins identified were surprise, interest, joy, rage, fear, disgust, shame, and anguish. These primary emotions were contrasted with *secondary emotions,* such as guilt, shame, embarrassment, empathy, and so on, all of which are culturally determined. Like Darwin, Tomkins focused on universal expressions but used mental state (emotion) words to name the expressions and their underlying affect program.

Following Tomkins's lead, research by Caroll Izard,[43] Paul Ekman,[44] and their colleagues lent support to the notion of universal facial expressions, gathering data that people around the world indeed both expressed and recognized the emotions presumed to underlie particular looks of the face. The basic emotions that Ekman and Izard identified bear a close relation to those defined by

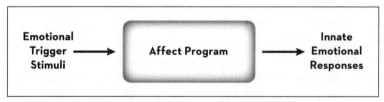

Figure 5.2: Affect Program.

Affect programs are hypothetical processes that are proposed, by basic emotions theorists, to mediate between emotional stimuli and emotional responses. Most theorists assume that affect programs are neural circuits, but are not committed to particular neural mechanism. When brain areas are discussed, often the limbic system is mentioned.

Tomkins, but other researchers have proposed lists of basic emotions that are less aligned with these.[45]

Basic emotions theory has had a wide impact in psychological and neuroscience research. Ekman's work has been particularly influential. He created sets of photographs of specific facial expressions of basic emotions that have been used in countless studies around the world to examine emotion cross-culturally.[46] These photos have also become a standard tool in assessing brain processing of emotions in healthy subjects and in patients with psychiatric disorders.[47] Examples of "Ekman faces" are shown in Figure 5.3.

Ekman has also had a huge impact on society and popular culture (Figure 5.4).[48] He became a consultant for the Central Intelligence Agency in training agents to recognize true emotions and to detect lying.[49] The popular television show *Lie to Me* concerns a psychologist, modeled on Ekman, who can identify, in real time, lying from facial expressions.[50] His methods were also used to analyze Alex

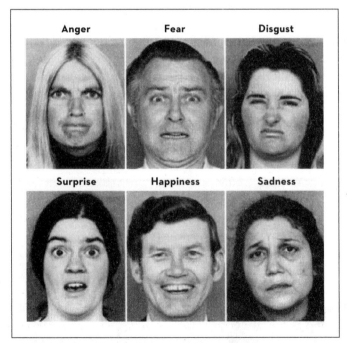

Figure 5.3: Ekman's Basic Emotions Expressed in the Face.
Ekman's original theory postulated six basic emotions (anger, fear, disgust, surprise, happiness, and sadness) each expressed in a characteristic, universal facial posture.

Figure 5.4: To Tell the Truth (or Not).
Paul Ekman's facial analysis scheme (called facial action coding, FAC) was used to assess whether the baseball star Alex Rodriguez was lying when he denied using performance-enhancing steroids on CBS's *60 Minutes*.

Rodriguez's facial expressions during a CBS *60 Minutes* interview in which he was questioned about whether he used performance-enhancing steroids (Figure 5.4).

In spite of its tremendous influence and support by many psychologists[51] and philosophers,[52] basic emotions theory is not universally accepted. Challenges to it are premised on logic (because different theorists have identified different basic emotions, they cannot actually be very basic[53]), philosophical objections (emotions are partly cognitive and involve intentions and beliefs, not just reactions[54]), methodological concerns (people are less accurate in matching emotion labels to faces if they have to generate the labels themselves than if they can choose from several options[55]), and research findings (facial expressions are not expressed in a singular, unitary fashion that unfolds automatically once elicited,[56] and the ability to judge feelings and other inner states from expressions is far less precise than previously thought, as it is often dependent on factors besides the facial muscles, such as vocal expression and pupil size[57]).

The psychologists Lisa Barrett and James Russell have been especially strong critics of basic emotions theory, questioning one of its implicit assumptions—namely, that emotions are "natural kinds," or biologically prepackaged psychological states.[58] They and others argue that emotions such as fear, which are assumed to be basic emotions, are in fact not singular entities with a biological existence established through natural selection and inherited from other animals.[59] Instead, they propose that the states of mind called basic emotions are psychologically constructed concepts that are labeled using culturally learned words. Words are indeed powerful dictators of beliefs, sometimes giving an existence to things that do not actually exist.[60] Although I don't accept all of Barrett and Russell's arguments,[61] I agree with their overall conclusions that the conscious feelings labeled with basic emotion terms are *not* prepackaged innate states that are unleashed by external stimuli but instead are cognitively assembled in consciousness.

Part of the problem in these debates is that different sides sometimes mean different things when they refer to basic emotions; for example, when basic emotions theorists talk about the emotion fear, they typically mean the entire brain and body response to some danger signal, and they use facial expressions as an indicator that the person is in this state called fear. But critics are often concerned with the conscious experience of fear, and question whether this is an innately programmed state. The rest of this chapter explicates this difference by analyzing the functions of affect programs.

WHAT DO AFFECT PROGRAMS DO?

Most basic emotions theorists are psychologists, not brain researchers, and tend to view affect programs as placeholders for brain mechanisms.[62] In general, they adhere to the idea that there are entities located in subcortical areas of the limbic system that constitute the affect program for each basic emotion, but are not strongly committed to hypotheses about which particular brain areas or circuits are responsible.

It's perfectly acceptable to use an expression like "affect program" to label hypothetical or even real hardwired circuits that control innate responses triggered by biologically significant stimuli. (Other labels that have been used for such circuits are "emotion command systems,"[63] "action programs,"[64] "innate emotion modules,"[65] and "neurocomputational adaptations."[66]) What is at issue is the question of what else these innate programs do besides detecting significant stimuli and controlling innate responses.

The problem starts with the use of terms derived from human introspective experiences (fear, anger, joy) to name the affect programs and their functions. One interpretation of these labels is that they are simply ways to connect scientific research on behavioral expressions to the psychological context in which they occur in everyday life. A researcher may be studying facial expressions that often occur when people say they feel afraid, and the affect program and behavior are both labeled with the term "fear." But "fear" is simply a convenient label as opposed to a literal reference to a subjective consciously experienced feeling of fear controlled by the affect program.

The second way to interpret the use of words like "fear," "anger," and "joy" in relation to affect programs is that the affect program is responsible for the mental state. This interpretation, which is at the core of the Darwinian (commonsense) view of what the fear system does (see Chapter 2), places the burden of explaining the mental state itself upon the affect program. As far as I can determine, this is

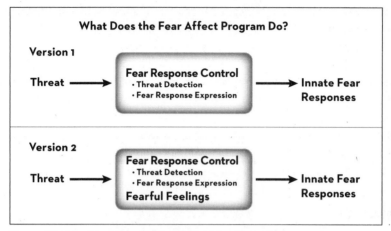

Figure 5.5: Two Versions of What Affect Programs Do.
All basic emotions theorists assume that affect programs mediate between emotional stimuli (such as threats) and emotional responses (such as fearful facial expressions). Some theorists also assume that affect programs also give rise to emotional feelings, and that the feelings are involved in connecting the stimuli and the response.

the standard interpretation of many basic emotions theorists. Some take a stronger stance on this matter than others, but most seem to assume that the fear affect program controls both fear responses and feelings of fear.[67] This, then, justifies the conclusion that expressive responses can be used to indicate when a feeling, an emotional state of mind, occurs.

To put it concretely, two versions of what a fear affect program does are illustrated in Figure 5.5. In one case, the affect program simply detects threats and controls responses. In the other, the affect program also gives rise to feelings of fear.

THE EMOTION COMMAND SYSTEM HYPOTHESIS

Jaak Panksepp's *emotion command system hypothesis* is a comprehensive and well-developed conception of how an innate affect program might actually work in the brain.[68] A key feature of his view is that "the mechanisms of affective experience and emotional behavior are intimately intertwined in comparatively ancient areas of the mammalian brain."[69] The ancient regions in question are part of the limbic system and are said to be conserved within mammals, including humans; functions mediated by the command systems are therefore also said to be conserved. It is possible, according to Panksepp, to determine how feelings

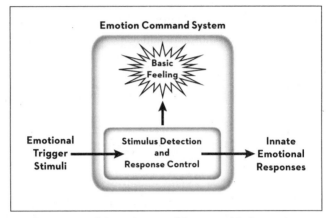

Figure 5.6: Panksepp's Model: Feelings and Emotional Behavior Intertwined in Emotion Command Systems.

Panksepp proposes that each basic emotion has a dedicated command system that detects specific emotional trigger stimuli, generates basic feelings, and controls specific innate emotional responses. These circuits are said to be located in subcortical areas, mostly involving the limbic system. Because the feeling and response associated with a given emotion are, in the theory, controlled by the same circuit, identification of the circuits that control responses reveals the circuits that control feelings. Because the circuits are conserved across mammalian species, studies of emotional response circuits in nonhuman animals reveals the neural basis of basic feelings in humans. Panksepp also proposes that the basic feelings are elaborated by cognitive processes in cortical areas. His view of the basic emotions is similar to the Darwinian theory except that Panksepp does not explicitly promote the idea that feelings are part of the cause chain that leads to responses. Feelings, for him, are more important in reinforcing behaviors that are successful in avoiding aversive and procuring desirable outcomes than in controlling the innate responses.

such as fear are represented in the human brain by studying the circuits that control innate behaviors in animals. This is so because the same circuit that controls fear behavior in animals gives rise to the feelings of fear in animals and humans[70] (Figure 5.6).

Panksepp distinguishes two kinds of conscious emotional feelings.[71] *Primary process affective states* are primitive conscious feelings (basic feelings) that are present in all mammals and that are encoded in emotion command systems; these include fear, rage, panic, and lust, among others. Then, through memory, attention, and language, humans can create *cognitive conscious feelings,* which are more elaborate versions of these emotions. His argument for emotional conservation across species is focused on the more basic kind.

Panksepp proposes that because the emotions humans feel are elaborations in cognitive consciousness of primary process affective states, we seldom experience the pure primary process versions of them, which makes these ancient emotions difficult to observe (and scientifically measure).[72] As a result, he notes, "one can never capture innate emotional dynamics in their pure form, except perhaps when they are aroused artificially by direct stimulation of brain areas where those operating systems are most concentrated."[73] In making his case, Panksepp thus relies heavily on the results of electrical stimulation studies in animals and humans.

In rats, Panksepp used electrical stimulation to map the areas from which behaviors related to each of several emotions could be elicited, and these areas constitute the command system. The fear command system, for example, involves the amygdala, anterior and medial hypothalamus, and the periaqueductal gray region. According to Panksepp, "fear—the subjective experience of dread, along with the characteristic bodily changes—emerges from the aforementioned circuit."[74] Because the rage circuit is interdigitated with the fear circuit, the full range of freeze-fight-flight behaviors is accounted for. In addition, a separate panic circuit is said to exist that underlies other aspects of fear and anxiety.

One criticism of Panksepp's approach is that brain stimulation only reveals behavioral output pathways. Panksepp admits that "we cannot directly measure subjective experience," but he believes that "behavioral evidence from all mammals that have been studied suggests that a powerful internal state of dread is elaborated by the fear system."[75] How fear[76] might emerge in primary process and cognitive consciousness from Panksepp's fear circuit is depicted in Figure 5.7.

As I mentioned in the previous chapter, electrical stimulation techniques are now regarded as imprecise and have in some cases led to a false understanding of circuitry.[77] Firm conclusions about emotion command systems that are based on electrical stimulation in animals should be put on hold until they are evaluated with the newer methods. It is possible that some, and maybe most, of the electrical stimulation results will hold up (some have been confirmed with chemical stimulations studies that do not suffer from the same problems as electrical stimulation[78]). My concern here, though, is not with the validity of the behavioral effects of electrical stimulation in animals so much as with the conclusion that these behaviors can be used as signs of feelings in animals and humans.

Panksepp makes a perfectly reasonable assumption—namely, that subcortical circuits shared by humans and other mammals have similar functions. I completely agree with this. We know, for example, that the amygdala has a very similar role in detecting and responding to threats in humans and rodents (see Chapters 2 and 8). But this leaves open the question of whether these circuits are responsible for feelings, in addition to behavioral and physiological response control. To help make the case for feelings, Panksepp turns to studies of electrical stimulation of the human brain.

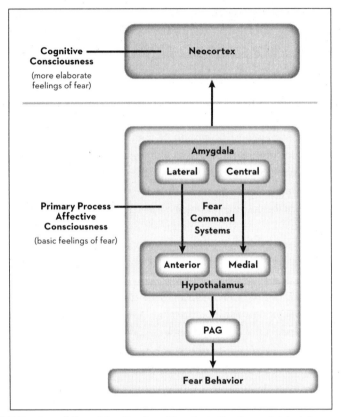

Figure 5.7: Basic and Cognitive Feeling Circuits in Panksepp's Model.
Brain areas involved in basic feelings of fear (primary process affective consciousness) and cognitively based feelings of fear are depicted. Basic feelings of fear depend on subcortical areas of the amygdala, hypothalamus, and periaqueductal gray (PAG), while cognitive feelings of fear depend on neocortical areas.

When areas of the human brain are stimulated electrically, first-person verbal reports of inner experiences can be obtained and are potentially very useful in relating brain circuits to experienced feelings. This is important because the lack of verbal reporting makes it difficult to verify conscious experiences in animals (see Chapters 2, 6, and 7). Panksepp relies heavily on the classic and much publicized work of Robert Heath done in the 1950s and 1960s.[79] Heath claimed that he had found specific sights from which a range of emotions (fear, anger, pleasure, etc.) could be elicited in humans, as revealed by the verbal reports of the patients about what they experienced. But Heath's conclusions have been called into question by other scientists who argue that contrary to the way the data have been portrayed, the findings do not in fact provide convincing support for the claim that stimulation of specific sites in the human brain elicit specific feelings. Both methodological and data interpretation issues have been identified. These issues are discussed further in the text box "Do Human Brain Stimulation Studies Reveal Specific Brain Sites Where Feelings Are Programmed?"

In sum, Panksepp is a thoughtful researcher who has argued that powerful emotional feelings result when subcortical emotion command circuits are activated in animals and people.[80] I agree with some but not all of his conclusions. In contrast to him I do not believe that it is possible to distinguish between conscious and nonconscious states elicited by electrical stimulation of subcortical areas, especially in animals. Panksepp recognizes the difficulties. He and Marie Vandekerckhove note that basic, subcortical, innate feelings are "implicit," "perhaps truly unconscious," and occur "without explicit reflective awareness or understanding of what is happening."[81] But "truly unconscious" states are not, by my definition, feelings. Feelings, even primitive ones, have to be felt (consciously experienced). Electrical stimulation most likely induces nonconscious central motivational states, such as those I described earlier, that result naturally when systems that control innate survival behaviors are activated (e.g., nonconscious defensive motivational states that are induced when a defensive survival circuit detects and responds to threats). I believe we should not assume conscious feelings in animals if nonconscious processes can account for the behavioral effects. With regard to humans, the conclusion that basic feelings are encoded in a prepackaged form waiting to be released from dedicated subcortical emotion command circuits that also control emotional behavior for each category of basic emotions is not compelling to me. In addition to the problems discussed in the text box, there are others. The electrical stimulation findings regarding feelings in humans should be much more robust and consistent if indeed these subcortical circuits encode innately programmed feelings. Moreover, because the evidence for the presence of conscious feelings in the stimulation studies, as in other kinds of research, is by way of verbal report, electrically elicited "subcortical feelings" do not reveal raw, primal emotions that are uncontaminated by cognitive consciousness.[82] A verbal report of a feeling by definition involves cognitive filtering of the information being described. As such it would seem impossible to separate the measurement of subcortical primary process feelings from cognitively constructed feelings. Panksepp seems to recognize this difficulty, noting this about the fear command system: "Whether the subjective experience of fear is mediated directly by this circuit or in conjunction with other brain areas will have to be addressed in future research."[83]

As I argue in this book, subcortical circuits provide nonconscious ingredients that contribute to feelings of fear and anxiety, but are themselves not the source of such feelings. The main difference between my view and Panksepp's is, therefore, whether subcortical systems are directly responsible for primitive emotional feelings or instead are responsible for nonconscious factors that are integrated with other information in cortical areas to give rise to conscious feelings. What Panksepp calls cognitive feelings are, I maintain, what feelings are. The subcortical states are, as he also says at times, "truly unconscious" and thus not feelings at all. They are, in my view, nonconscious motivational states. In the next several chapters, I describe how the cortical integration of information is required to account for what we humans experience as fear or other emotions.

Do Human Brain Stimulation Studies Reveal Specific Brain Sites Where Feelings Are Programmed?

I was first introduced to neuroscience by Robert Thompson, a charismatic psychology professor at Louisiana State University who studied learning and motivation in the brains of rats. I took an elective course with Thompson while working on a master's degree in marketing, and it was under his influence that I fell in love with brain research. Thompson's recommendation made it possible for me to be accepted into the graduate program for PhD work at SUNY Stony Brook.

Thompson told me about the work of Robert Heath, a researcher down the road in New Orleans who had placed electrodes in the brains of psychiatric and neurological patients.[84] Heath claimed to have found centers in the brain that, when stimulated electrically, would give rise to feelings of pleasure, rage, fear, and the like. (His work later inspired novels such as Michael Crichton's *The Terminal Man*[85] and Walker Percy's *Love in the Ruins*.[86]) Heath's studies were controversial because many were done on mentally ill patients, and questions arose about their consent to participate.[87] A number of additional stimulation studies followed at a variety of centers around the world but were mostly performed in the context of evaluating and treating severe epilepsy.[88]

A fundamental problem with Heath's stimulation studies was that they were not specifically designed to test whether feelings associated with basic emotions are wired into specific sites. The goal, instead, was to attempt to obtain a better understanding of the schizophrenic brain.[89] It is unclear whether a specific protocol was used for obtaining reports of subjective feelings and for translating what the patient said into data that could be tabulated. Thus, although Heath's studies are often discussed as having identified pleasure centers in the human brain, Kent Berridge and Morten Kringelbach examined the transcripts from the sessions for evidence that the patients described feelings of pleasure when stimulated but found little indication of this.[90] The patients were more likely to talk about vague sensations, or describe the urge to have sex or eat, rather than say that they felt pleasure. Similarly, the self-reports they provided when they said they felt "fear" are often metaphoric and involve situations in which one might feel fear: "entering into a long, dark tunnel" or "trying to escape."[91] Thus, researchers who were expecting to find specific feelings in these patients may have counted such examples as being indicative of fear or pleasure, even if the patient did not explicitly state that he was having these feelings.

Eric Halgren, a leading expert in human electrical stimulation, evaluated the field in the late 1970s and early 1980s.[92] He accepted that brain stimulation can elicit mental phenomena but concluded that once the general tendency for mental phenomena (thoughts, images, or specific emotional feelings such as fear, anger, pleasure, etc.) to be elicited by brain stimulation is taken into account, "there is no particular tendency for any category of mental phenomena to be evoked from any particular site."[93] In other words, the particular states were not consistently localizable to brain areas. He also noted that the kind of experience elicited was often more related to preexisting conditions, such as the patient's personality or demeanor, than to the site stimulated. (Anxious people, for example, were more likely to experience fear and anxiety when stimulated.) If the feeling of fear is hardwired into a fear command system, it should be experienced by everyone in a similar way when the fear command system is activated by stimulation.

In evaluating these data it is also useful to consider the nature of the process by which subjective feelings are assessed in humans. In a very interesting commentary on this topic, Berrios and Markova[94] detail the difference between measuring and grading. Measuring is an objective procedure that involves physical objects and their features. Grading, in contrast, "is carried out by means of categories that are external and reside not in the object itself (i.e., they are not internal to it) but in the eye of the evaluator." When categorizing the subjects' verbal responses using emotion labels ("fear," "pleasure," etc.), researchers are grading, not measuring. Thus, in Heath's studies, descriptions about sex are categorized as feelings of pleasure, whereas those about entering dark passages become instances of fearful or anxious feelings. Such "data" thus can reflect biases of the researcher.

The vague and variable nature of Heath's patients' descriptions of their subjective experiences elicited by brain stimulation suggests an alternative to the idea that feelings are genetically wired into subcortical emotion-operating systems. It seems equally possible that electrical stimulation of the brain creates a state of ambiguity or confusion. Artificial delivery of electric current (especially the relatively high levels used in the older studies) nonspecifically activates many neurons and causes them to fire action potentials that in turn activate the various areas to which the stimulated neurons are connected. The stimulations, for example, often produced increases in physiological arousal in the brain. Increased arousal enhances information processing widely, increasing vigilance and attention to the environment[95] (this is discussed further in Chapter 8). In situations in which people experience something unusual or unexpected, such as a sudden state of heightened arousal and vigilance, they often try to make sense of it.[96] This is a well-known phenomenon in psychology: Unexplained experiences create states of dissonance that motivate the conscious mind to explain, as best as it can, what might be happening. People gather as much information as possible and, in an effort to attribute a cause to the experience, verbally label the experience using common terms.[97] For example, in a famous study Stanley Schachter and Jerome Singer found that when arousal was artificially induced by giving experimental participants a shot of adrenalin, the participants looked to their social environment for cues to explain their state of arousal in order to label it—if they were in a room full of happy people, they felt happy; with sad people, they felt sad.[98]

That this may well have taken place in Heath's stimulation studies is suggested by his own observations. After delivering an electrical stimulus to the brain of one of his patients, she smiled. When Heath asked her why she was doing so, she replied: "I don't know. . . . Are you doing something to me? [Giggles.] I don't usually sit around and laugh at nothing. I must be laughing at something."[99] Her conclusion about what she was consciously experiencing built up over time and was more like a rationalization that she slowly constructed than a report of a feeling that was directly and immediately unleashed by stimulation of a brain site.

These comments from Heath's patient remind me of the findings described earlier that Mike Gazzaniga and I obtained in split-brain patients.[100] When we induced the right hemisphere to wave, stand, or laugh, and asked the left hemisphere why it had done so, the left hemisphere confabulated answers to make the actions seem reasonable ("I thought I saw a friend out the window, so I waved"). Heath's patient adopted the same tactic when confronted with the fact that she was smiling and giggling. The stimulation obviously elicited behavioral responses (smiling and giggling) but not the specific feeling typically associated with them.

Perhaps the most telling way to summarize the implications of Heath's work is to turn to his own words. He eventually concluded that the verbal reports from his schizophrenic patients might be "grossly unreliable and probably should not be accepted as valid. . . ."[101]

BODY FEEDBACK THEORIES OF BASIC EMOTIONS

The idea that subcortical affect programs are both innate-response control systems and repositories of feelings is not shared by all basic emotions theorists. Some argue that although affect programs do control responses, feelings result from alternative innate circuits—in particular, circuits that process feedback signals from the body during the expression of emotional responses.

As I discussed earlier, William James proposed a feedback theory of feelings in the late nineteenth century, arguing that we do not run from a bear because we are afraid but instead that we are afraid because we run.[102] The explanation for this is that feedback from the body during this behavior is felt by the brain as the emotion fear. Different emotions are experienced differently because they involve different body signatures that produce different patterns of feedback and thus different feelings.

Though popular for some time, James's theory was challenged in the 1920s by Walter Cannon, who argued that body feedback was too slow and too imprecise to signal the particular distinctions between fear, anger, joy, and sadness.[103] Cannon's critique focused on feedback signals resulting from visceral responses (responses of the autonomic nervous system and endocrine system) that occur during expressive behaviors. This led to a great deal of research attempting to find visceral (autonomic and endocrine) signatures of different emotions, research that continues today.[104] The results have clearly demonstrated some specificity to visceral feedback but have had only limited success in showing that such feedback plays a major causal role in determining feelings.

Early basic emotions theorists like Tomkins and Izard, however, pointed out that James emphasized feedback from the entire organism, not just the visceral organs controlled by the autonomic nervous system, and proposed that feedback from the facial muscles during innate facial expressions of basic emotions should have the requisite speed and specificity to determine what emotion one is feeling.[105] This launched a wave of research on the contribution of facial feedback to felt emotions.[106] Although some support for this theory has emerged, the findings have not convinced the field that facial feedback alone explains feelings.

Antonio Damasio's 1994 book *Descartes' Error* sparked renewed attention to body feedback as a potential source of feelings.[107] Like James, Damasio emphasizes the importance of feedback from the entire body, including signals arising from the inner organs and tissues and from the muscles and joints of the skeleton and face. By grounding his idea in modern neuroscience, however, he has been able to take the feedback theory to a new level.[108]

Damasio made a distinction between emotions and feelings.[109] For him, emotions are *action programs* that control innate behavioral and physiological responses. He also viewed drives as a second kind of action program that serves physiological needs (hunger, thirst, reproduction). Action programs are similar to Panksepp's emotion command systems; in contrast to command systems, however, action programs are not viewed as giving rise to feelings. Instead they are conceived as operating nonconsciously. Feelings, for Damasio, are conscious experiences that follow from the representation of action program–triggered emotional responses in body-sensing areas of the brain (Figure 5.8).

The conceptual space occupied by my own notion of survival circuits and global organismic states overlaps considerably with that of Damasio's action programs for emotions and drives; he and I also both emphasize feelings as conscious manifestations of these nonconscious processes. A key difference is that, for Damasio, feelings are primarily determined by body signals, whereas I think body signals are only one of many ingredients that contribute to feelings. We also differ on the value of postulating feelings in animals as the basis for certain human feelings; on this subject, he is more aligned with Panksepp.

Damasio calls the signals resulting from the expression of bodily responses during an emotional episode *somatic markers*.[110] These are said to be "read" by body-sensing areas of the cortex (somatosensory and insular cortex) and in subcortical areas in the hypothalamus and tegmentum (including the periaqueductal gray region). Each of these areas receives information from the body in the form of somatosensory and proprioceptive signals from the skin, muscles, and joints, and visceral sensory signals and hormones from the inner organs and tissues.[111] The feedback signals take several forms. Some originate from sensory nerves in the muscles or visceral organs of the body that send messages to sensory processing regions of the brain. Others involve hormones like cortisol that can enter the brain directly from the bloodstream and bind to receptors in areas such as the amygdala, hippocampus, neocortex, or in brain stem arousal systems.[112] Some hormones cannot enter the brain; epinephrine and norepinephrine released from the adrenal medulla are too large to diffuse from the bloodstream across the blood-brain barrier, a filter that keeps large molecules, such as toxins, out. These relatively large molecule hormones can nevertheless affect the brain

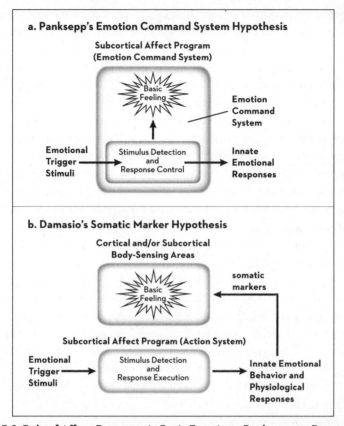

Figure 5.8: Role of Affect Programs in Basic Emotions: Panksepp vs. Damasio.

In Panksepp's theory, basic emotional feelings are products of the system (affect program or command system) that detects and responds to threats (also see Figure 5.7). In Damasio's theory, basic feelings arise when body-sensing areas of the brain receive feedback from behavioral and/or physiological responses elicited by emotional stimuli.

indirectly. For example, norepinephrine binds to receptors on the vagus nerve in the abdominal cavity. The ascending component of this nerve enters the brain and connects with arousal circuits.[113] Damasio's view of how body signals are processed by the brain is illustrated in Figure 5.9.

The body-sensing areas in the brain thus create a neural representation of the *body state*. Primitive emotional feelings are said to be the result of the aggregate of these somatic markers and the states they create. More elaborate feelings—full-blown emotions—result from the elaboration of these states by cognitive processes.

A key feature of Damasio's theory is the "as if" loop.[114] With this mechanism, he proposed that actual feedback from the body isn't necessary: By way of processing within the brain, body states can be re-created from memory, thus giving rise to feelings. In the case of both body feedback and "as if" re-creations of it, the neural representation of the body state can thus contribute to emotional feelings.

Damasio had long emphasized body maps in neocortical areas, especially areas of the somatosensory cortex and insular cortex, as the key factor in creating emotional feelings from body signals. Subcortical body maps were said to be too "coarse" to account for differences between feelings, whereas cortical body maps are more refined and could do the job.[115] Work by Damasio[116] and Bud Craig[117] drew particular attention to the insular cortex as a key body-sensing area for feelings, and Craig even proposed that the insular cortex was responsible for all

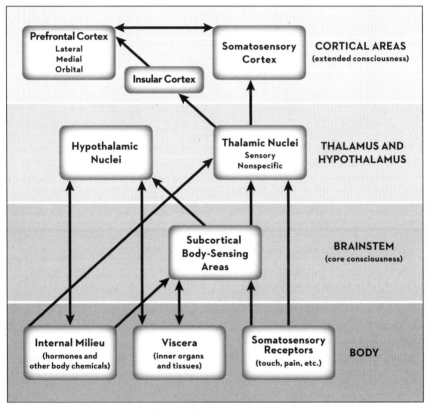

Figure 5.9: Processing of Body Signals in the Brain.

Brain areas proposed by Damasio to be involved in processing feedback from body signals. These are the foundation for feelings in Damasio's theory.

aspects of human consciousness.[118] The insular cortex now often comes up in discussions about how consciousness emerges from brain circuits.[119]

More recently, though, Damasio has shifted his emphasis from cortical to subcortical body-sensing areas in the brain stem (e.g., the hypothalamus, midbrain, and hindbrain) as the primary source of basic emotional feelings; in this scenario, the role of the insular and other body-sensing areas of the cortex is to cognitively represent and elaborate subcortically experienced conscious feelings.[120] This new focus on subcortical areas is based on the finding that damage to insular cortex does not eliminate feelings.[121] In arguing that basic feelings of fear, anger, disgust, sadness, and joy are consciously experienced subcortically, Damasio is in agreement with Panksepp, though they differ on how the feelings come about in subcortical areas (see Figure 5.8).

There is, however, relatively little direct evidence that subcortical body-sensing areas are responsible for feelings. Complicating the design of appropriate tests of the subcortical theory is the fact that damage to brain stem areas often results in coma,[122] a state in which all forms of consciousness and sentience are lacking, making it impossible to assess the role of such areas in any kind of psychological process. Also complicating such tests is the fact that body-sensing areas of the cortex are not localized in a particular region, and thus only a very extensive area of damage could completely eliminate the relevant cortical areas. Correlative studies show that subcortical areas are activated during both remembered emotion[123] and immediate emotional arousal,[124] but these findings do not demonstrate that these neural responses are actually responsible for the experienced feeling.

For more direct causal evidence, Damasio cites animal and human electrical stimulation studies.[125] But as we've seen, these are problematic as a way of accounting for feelings as well, so that the behavioral responses of animals, even though described by Damasio as "imbued with positive and negative valence,"[126] cannot simply be used as an indication that feelings are being consciously experienced by animals; again, as Tinbergen noted in the epigraph of this chapter, conclusions about mental states underlying animal behavior are mere guesses. The human work on brain stimulation, meanwhile, does not convincingly demonstrate precise mapping of conscious experiences (feelings or otherwise) to specific brain areas. Further, as described above, because these subjective experiences are assessed by verbal reports in human studies, the feelings reported cannot be said to be raw, primitive feelings. To be reported they must reflect representation of the information in cortical processing systems. How one would distinguish a conscious emotional experience originating solely in a subcortical circuit from unconscious subcortical processing that is cognitively represented, consciously experienced, and reported via cortical circuits is unclear.

Damasio is not very precise about the psychological nature of basic subcortical feelings. He and Gil Carvalho say that the insula is *not* necessary for consciously experiencing feelings, but they also describe information in subcortical areas as existing in an "implicit form" (i.e., it exists nonconsciously) that is then "explicitly represented" (i.e., consciously experienced) in the insula.[127] So are the subcortical states nonconscious or conscious? Or do they require cortical areas to be consciously experienced?

Damasio's work has helped illuminate the role of cortical and subcortical areas in mapping body states in the brain. However, the work does not demonstrate that feelings are directly experienced in subcortical body-sensing areas. In other words, the existence of a subcortical map of body-state information does not mean that the brain state related to that map is consciously experienced.

The question of whether sensory processing areas of the brain are, on their own, sufficient to create conscious experiences of sensory information has been debated extensively in the case of visual stimuli. As we will see in the following chapter, most researchers believe sensory processing alone is not adequate—nonconscious processing has to be re-represented via some higher cognitive process in order for a conscious experience to occur. Both Damasio and Panksepp acknowledge that this re-representation gives rise to a cognitive consciousness. But both also argue, less successfully in my opinion, that subcortical states are consciously experienced.

FEELINGS AS THE INTERFACE BETWEEN REACTION AND COGNITIVE SYSTEMS

Anthony Dickinson and Bernard Balleine have proposed another view of feelings across species, one that rejects the positions of both Panksepp and Damasio.[128] On the basis of an experience that Dickinson had with food poisoning, he and Balleine designed studies to test whether conscious feelings might function as a way to integrate two kinds of functional systems in the brain.[129]

One system is reactive: It embodies innate responses that can be controlled by innate or conditioned stimuli, and learned habits controlled by stimulus-response associations. The other system uses cognitive information to achieve goals. Both of these systems, they argue, operate nonconsciously. With these two systems, an animal can react appropriately to external stimuli in ways that reflect its current needs and values (reaction system), and can learn adaptive responses that achieve goals (cognitive system) without any conscious experience. This view challenges Panksepp's notion of an emotion command system that both produces responses to significant stimuli and creates feelings, because Dickinson and

Balleine contend that the reactive and cognitive systems function unconsciously and can even learn without consciously experiencing reinforcing stimuli. (For Panksepp, conscious feelings [e.g., pleasure and pain] are key to reinforcement and learning.) Further, Dickinson and Balleine also disagree with Damasio's theory that body feedback creates conscious feelings that control choices; their data suggest that behavioral choices do not depend on body feedback.

Dickinson and Balleine's theory about reactive and cognitive systems that operate to control different kinds of responses nonconsciously, and to even learn via reinforcement nonconsciously, are completely compatible with the view I developed in earlier chapters of this book. Their idea regarding nonconscious reactive systems is consistent with my own regarding survival circuits, and their notion that motivational states contribute to problem solving nonconsciously is consistent with my proposal that nonconscious global organismic motivational states occur in situations where challenges and opportunities exist. We also agree that these motivational states play a role in integrating and coordinating the different response systems to enable an organism to adapt to a given situation and benefit from an opportunity or cope with a challenge.

Where we diverge is on the nature of consciousness. For Dickinson and Balleine, consciousness evolved in mammals (and possibly birds) as a way to bridge reactive and cognitive systems. Dickinson admits that this hypothesis is a just-so story,[130] a speculation that cannot be easily proved.[131] My view about consciousness will be described in detail in the next several chapters.

PAIN AND PLEASURE

Pain and pleasure are often thought of as emotions. Although they are related to emotions, there is an important difference. Pain and pleasure are *hedonic states* that arise directly from sensory processing; they result when certain receptors detect specific kinds of stimuli, and axons connected to those receptors deliver the sensory information to the brain (when we talk about the pleasure of a friend's company or of doing crossword puzzles, we are going beyond the pure sensory hedonistic processes that are of interest in this discussion of innate feelings). For example, when receptors in the skin called nociceptors detect tissue irritation and damage, they pass the relevant information on to the brain, where pain is experienced. Other receptors in the skin send signals to the brain that are experienced as pleasure (a light touch on the back, arms, neck, or genitals; activation of certain taste receptors in the tongue and mouth).

It is important to recognize that the conscious feeling of pain or pleasure that humans associate with these sensory signals is but one consequence of what those signals do in the brain. In addition to giving rise to conscious hedonic experiences (feelings of pain and pleasure), they also elicit reflexes or other innate reactions, motivate complex actions, increase brain arousal, and reinforce learning. Each of these consequences, including the conscious feelings, has separate neural underpinnings, and we should not assume that observation of one of the nonconscious consequences (elicitation of body reflexes, motivation of more complex behaviors, or reinforcement of learning) means that a conscious feeling of pain or pleasure has occurred. And neither should we assume that all of these consequences involve the same brain mechanism.

For example, one common way that scientists have tried to get some traction on the problem is to determine whether an animal can learn to perform behaviors that allow it to escape from electric shock or that will provide it with sweet-tasting food or an addictive drug. Such instances of learned behavioral flexibility are examples of instrumental conditioning in which responses are reinforced and learned by their consequences. When an animal acquires such behaviors, the usual conclusion is that it must be motivated by the conscious feeling of pleasure or pain.

The philosophy of hedonism, which had been around since ancient times, was central to British philosophers (Locke, Hume, Hobbes, Bentham, Mill, Bain, and Spencer) that influenced Darwin and his followers, who in turn influenced Thorndike—the latter's law of effect, which emphasized pleasure and pain, was close in spirit to ideas proposed by Bain and Spencer.[132] More recently, Panksepp argued: "At some point in brain evolution, behavioral flexibility was achieved by the evolution of conscious dwelling on events and their meaning, as guided by internally experienced emotional feelings." These feelings are, he says, "a fundamental property of emotional command systems" and include "the various forms of affective consciousness that all mammals can experience."[133] In contemporary psychology and neuroscience, it is common for researchers to describe reinforcers (a nonsubjective behaviorist concept) as rewards (a hedonistic term implying a pleasurable feeling). But in spite of the long tradition in philosophy, psychology, and neuroscience of explaining learning and motivation in terms of consciously experienced hedonic states of pleasure and pain, the scientific basis for this intuitively appealing idea is not as solid as it may seem.

I earlier noted that Dickinson and Balleine argued that animals are able to respond to stimuli and learn instrumental responses from consequences without feeling anything. Indeed, as I will describe below, researchers who study addictive drugs have also been led to the conclusion that subjective pleasure is not the source of reinforcement.[134] If these researchers are correct, and I think they are, we need to distinguish between the feelings of pleasure and pain that may occur in conjunction with learning, from the nonconscious reinforcement mechanism that is the actual basis of learning.

Reinforcement, as we've seen, can be understood as a neural process involving cells, synapses, and molecules (see Chapter 4). Though instrumental learning was once conceived as a unique mammalian achievement, and thus might depend on a uniquely mammalian capacity to feel emotions, we know that other vertebrates and invertebrates (snails, flies, bees, crayfish) learn new behaviors by reinforcement.[135] Just because our species consciously experiences some feeling (say pleasure) when reinforcement occurs does not mean that the feeling is the source of the reinforcement. The feeling, once it exists in one's conscious experience, can influence subsequent behaviors and decisions but is not required for the reinforcement of learning itself.

Recall that in our optogenetic studies we were able to create threat conditioning to a tone by directly optically activating amygdala cells when the tone was played (Chapter 4); there was no "painful" unconditioned stimulus (no footshock) involved. We simply activated the neurons the way the sensory pathways that process the shock normally would. One might object that in real life, emotional learning creates stronger memories than nonemotional learning because of feelings: Pleasure or pain, or emotional feelings in general, may add some special spark to the more mechanical processes of stimulus convergence. But the strengthening that occurs in so-called emotional learning can also be accounted for in neural terms. We found that we could enhance the behavioral memory produced by the optical US if we also simulated the brain arousal system that releases the neuromodulator norepinephrine.[136] During real threat learning in life, activation of arousal systems is a natural consequence of the activation of the defensive survival circuit. So although it is true that in an emotionally arousing situation learning is more effective in creating memories, it is not necessarily because of the emotional feelings that may also be aroused. The stronger learning and the conscious feelings that result are both consequences of nonconscious survival circuit activity, such as the release of neuromodulators that separately affect survival circuits and circuits that give rise to conscious feelings. Another objection to the implications of our optogenetic studies might be that our experiments used simple Pavlovian conditioning rather than flexible instrumental learning. However, other studies demonstrate that instrumental responses can be learned by optogenetic stimulation of modulatory circuits involved in instrumental reinforcement.[137]

A pair of influential distinctions made by Kent Berridge is also relevant.[138] Berridge notes that what is often called "pleasure" in animal studies is better termed "liking." In contrast to pleasure, which implies subjective experience, liking is defined behaviorally and does not necessarily involve subjective experience. Neural activity related to liking is the neural basis of reinforcement. The second distinction is between liking and wanting. "Wanting," in behavioral terms, refers to the motivation to obtain reinforcers. Wanting is often based on a lack of something needed, such as food. But similar to liking, wanting is not a subjective state but rather a motivational drive to obtain the needed substance. (In a sense, liking behavior is similar to defensive reactions and wanting behavior to actions in my model.) In support of his ideas, Berridge and colleagues have shown that people can be prompted to pour and consume a drink by subliminal presentation of stimuli related to "pleasure," but only if they are water deprived. In spite of this increase in "wanting," they had no subjective experience of the stimulus or of any conscious feeling of pleasure in response to it.[139] Berridge and others have concluded that as important as conscious pleasure is to our lives, it was not the primary reason for evolution of brain liking systems.[140]

What about dopamine, a neuromodulator with a well-established role in learning via instrumental reinforcement?[141] Isn't it the chemical basis of pleasure? As such, might the release of dopamine cause pleasure, and pleasure in turn cause learning when a reward is involved in a real-life learning situation? This is certainly how it is described in the lay media.[142] For example, research on the effects of dopamine in behavioral reinforcement are described with dramatic headlines: "Sweet Treat Releases Feel Good Chemical," and "Dopamine Is the Chemical That Mediates Pleasure in the Brain."[143] But there are problems with this idea.[144]

First, dopamine and other modulators reinforce synaptic plasticity and otherwise contribute to behavior not only in mammals, or even vertebrates, but also in invertebrates.[145] Does this mean that invertebrates feel pleasure when they learn behaviors that are reinforced by food on the outside and dopamine on the inside? The effects of dopamine on neurons and learning are faithfully reproduced in studies of slices taken from a rat brain and kept alive in a dish of nurturing chemicals. Should we conclude, then, that this reinforcing effect on neural activity leads to pleasure in brain slices?[146] What dopamine actually does during learning is alter neural activity in a way that affects the likelihood that neural response will recur when the cellular and synaptic conditions recur later. When these changes occur in the brain of a living, awake animal, they help ensure that behavioral responses will be repeated in similar situations in the future.

Dopamine levels are highest when animals are food deprived and seeking food rather than when they are enjoying the fruits of their labors.[147] Optogenetic studies have also been used to precisely activate neurons and change behavior by releasing dopamine.[148] Typically, the results are quickly taken up by the media and characterized as the brain's "pleasure highway" and the "ability to feel pleasure."[149] But this is not what the data are about. Pleasure is the interpretation (or rather misinterpretation) of such data.

The distinction between the reinforcing and subjective consequences of pleasurable stimulation is especially apparent in studies of addiction. Although it is common to believe that drugs are abused because they give the user consciously experienced pleasure, research on addiction emphasizes the importance of nonconscious factors in drug-related behavior.[150] For example, in one study morphine addicts were connected with tubing through which either morphine or a placebo was infused when they pressed a button.[151] Button pressing was unaltered by the placebo but increased when morphine was administered, confirming that the presence of the drug was detected by the brain. But the key finding was that at low doses the morphine could affect behavior in spite of the fact that the subjects were unable to say whether they were receiving morphine or placebo on the basis of any subjective feeling of pleasure. The behavioral effects of morphine are thus dissociable from the subjective state of pleasure that may also result. Other studies similarly have confirmed that drug-taking behavior can occur in the absence of the subjective conscious pleasure produced by the drugs.[152] Moreover, much of drug taking ultimately comes to be driven by an effort to ward off the negative consequences of *not* taking rather than to obtain the high itself.[153] This is why NIDA is funding our research on negative reinforcement (see Chapter 4). Subjective feelings of pleasure are not the only factor that explains how addiction occurs and why it is so persistent.[154]

Certainly the sensory hardware that supports feelings of pleasure and pain in humans is present in animals. However, it is possible that more is required than the sensory process alone in order to consciously feel pain or pleasure. For example, when people are experiencing pain from an injury, but are distracted momentarily, the pain temporarily disappears. Likewise, one's attention can be directed away from painful sensations or other distressing events through hypnosis.[155] Brain areas involved in working memory and attention have been implicated in hypnosis.[156] Neural messages still reach the brain in hypnotic states, but in the absence of attention and the cognitive processes that it feeds, the sensations are not consciously experienced as pain.

In the field of pain research it is common to distinguish the sensory from the emotional or affective properties of pain,[157] but this is different from the distinction I am making. The so-called affective properties are more akin to nonconscious processes that are subsumed within the brain systems that underlie nonconscious defensive motivational states than to conscious feelings of pain.

I argue that stimuli that produce conscious feelings of pain and pleasure in humans can involve three separate neural states: sensory, motivational and conscious. In animals we can study the first two, which operate nonconsciously, without making assumptions about consciousness that are difficult to verify. We can observe behavioral responses consistent with painful feelings, but it is not easy to know what, if anything, is being felt. The sensory component of these systems can, like threats, trigger complex motivational states that organize behavioral responses that maximize survival potential. These complex responses occur in people when they consciously feel pain, but it does not mean that the feeling of pain is the cause of the response or even a necessary accompaniment to it.[158]

I am, like most people, drawn to the idea that animals feel pleasure and pain when they act as if they do have such feelings. But as a scientist, I am compelled to ask: How can we distinguish behavior due to presumed consciously felt hedonic states in animals from behavior due to nonconscious processes? This is especially problematic since we know that much of human behavior does not depend on consciousness. As we have seen, to make a case for animal consciousness it is not sufficient to provide evidence (even converging lines of evidence) that the behavior in question is consistent with the existence of a conscious experience. One also has to show that the behavior cannot be accounted for by processes that work nonconsciously.

DE-DARWINIZING HUMAN EMOTIONS

Our view of life on earth was radically changed by Darwin's theory of evolution by natural selection. Many of his ideas have been shown, through rigorous scientific tests, to accurately portray the continuity of organisms. I believe that one thing that Darwin got wrong, however, was the idea that humans have inherited conscious feelings from animal ancestors. In order for this to have occurred, we would have to have inherited brain circuits that encode these feelings—not just the responses, but also some version of the conscious feelings, as well.

One argument against the innate view of emotions like fear is that there are so many ways for humans to experience it: We can be afraid of a snake at our feet, a mugger, elevators, heights, tests, public speaking, tainted food, dehydration, hypothermia, reproductive failure, losing a friend, alien abduction, financial insecurity, failing an exam, not leading a meaningful or a moral life, the eventuality of death. This would seem to rule out a single brain circuit for fear (and/or anxiety) that could have been inherited from animal ancestors that would account for all these states.[159]

One could contend that variants of fear and anxiety are simply cognitive elaborations of the activity of a common underlying brain circuit, or of a small number of such circuits. As I have argued, however, although innate circuits that are relevant to feelings of fear and anxiety do exist, they are not feeling circuits (circuits that encode conscious feeling of fear or anxiety)[160] but rather survival circuits (circuits that control behaviors that help organisms survive and thrive in the face of challenges and opportunities in life). An important category of survival circuits are those that control defensive behaviors. Others are involved in energy and nutritional needs, fluid balance, temperature regulation, and reproduction. And one way that feelings arise is by the individual's becoming consciously aware of the consequences of survival circuit activity occurring in his brain and happening to him. These survival circuits generate nonconscious motivational states that contribute to, but are not the same as, the feeling of fear or anxiety. The implicit states influence conscious emotional feelings but remain outside of the view of our mind's eye, except when we notice their consequences in our behavior or in our body.

Though it is most often discussed in relation to predatory defense circuit activation, the feeling of fear does not have an exclusive contract with any particular subcortical circuit. We have to protect ourselves against many different kinds of challenges on a daily basis, not just predators. For example, if you are without food for an extended time, signals associated with low energy supplies can trigger the feeling of fear that you might starve. Or if you are trapped on a mountaintop and sense the temperature dropping, you might begin to be fearful or anxious that you will freeze to death. Fear/anxiety is the cognitive awareness that you are in danger, regardless of whether that danger triggers a defensive, energy regulation, fluid balance, or other survival circuit, or whether the danger is imagined, or whether it results from the contemplation of the meaning (or meaningless) of existence.

Emotions, in short, are states of consciousness pieced together by complex cognitive mechanisms. To understand how these feelings come about, we have to delve into the mechanisms of consciousness, which is what the next three chapters address.

CHAPTER 6

LET'S GET PHYSICAL: THE CONSCIOUSNESS PROBLEM

"The highest activities of consciousness have their origins in physical occurrences of the brain just as the loveliest melodies are not too sublime to be expressed by notes."

—W. SOMERSET MAUGHAM[1]

F ear and anxiety are conscious experiences, feelings that take over our conscious minds. But what *is* consciousness? Most everyone agrees that we humans possess it. But exactly what it is, how it works in the brain, and what other animals might have it remain contentious topics. Because we can't really begin to understand fear and anxiety without some understanding of consciousness, let's dive right into it.

SOME DEFINITIONS AND RESTRICTIONS

The word "conscious" is used in two different ways in everyday discourse. Sometimes it means being awake and alert and capable of interacting with one's surroundings as opposed to being asleep, anesthetized, or in a coma. This kind of consciousness is sometimes called creature consciousness, and contrasts with state consciousness, or as we'll call it here, mental state consciousness, which is the ability to be aware that one is experiencing something.[2] Mental state consciousness (awareness) depends on creature consciousness (wakefulness), but having creature consciousness does not guarantee mental state consciousness.

All animals have creature consciousness, but only animals that can be aware that they exist can have mental state consciousness.[3]

Table 6.1: Creature vs. Mental State Consciousness

FEATURE	CREATURE CONSCIOUSNESS	MENTAL STATE CONSCIOUSNESS
Awake and alert	yes	yes
Responsive to sensory stimuli	yes	yes
Execute complex behaviors	yes	yes
Solve problems and learn from experience	yes	yes
Awareness that a stimulus is present	no	yes
Awareness that one's self exists	no	yes
Awareness that it is one's self that is sensing, behaving, and solving problems	no	yes

When I use the word "conscious" (or "consciousness"), unless I say otherwise, I will be referring to mental state consciousness. As defined here, mental state consciousness will mean the ability to be aware that some state is occurring and to have knowledge of what that state is about. States that occur without one's explicit awareness of the content of the state and its occurrence are treated as nonconscious events. To the extent that nonconscious states involve cognitive processing and thus are part of the cognitive unconscious (see Chapter 2), they are mental states, though not conscious mental states.

This discussion will necessarily again raise the question of whether other animals have what we humans call consciousness, or something equivalent.[4] Some argue that other primates, especially apes, do have conscious experiences comparable to those of humans.[5] Others make the same case for mammals[6] or other vertebrates,[7] because their brains are also similar to ours in many ways. Another view is that there might be primitive kinds of consciousness that are shared by humans and other animals, in addition to more elaborate kinds that are more typical of humans[8] (e.g., the subcortical theories of feelings by Panksepp and Damasio, discussed in the previous chapter). But should such primitive versions be thought of as examples of mental state consciousness, or are they implicit, nonexperiential states that contribute to consciousness but are not themselves consciously experienced? Still others argue that some invertebrates are conscious.[9] Others go further, arguing that consciousness is simply biological information processing, which is pervasive in all forms of life, including plants and unicellular organisms.[10] An even more extreme view is that consciousness (or some form of it) is a feature of any physical entity that integrates information, such as devices based on electronic chips (a calculator or a smartphone).[11] Then there is the proposal that we aren't really conscious at all; we just think we are because of the way cognitive systems work in our brains—once we come to fully understand the brain, we will be able to do away with a concept like consciousness altogether.[12]

CONSCIOUSNESS AND THE UNCONSCIOUS CHALLENGE

René Descartes effectively formulated the modern consciousness problem when he tried to establish criteria for what can be known with certainty.[13] He concluded that the only thing a person can know directly and with confidence is his own inner experience, his own conscious states; what goes on in someone else's mind can only be inferred. For Descartes, mind, consciousness, and soul effectively meant the same thing: a nonphysical realm of existence that has no location in space or time, which humans have but other animals do not. Descartes described animals as mere "beast-machines," by which he meant that, lacking mental state consciousness, animals are simply physical entities that respond reflexively to the world.[14] Consciousness, he said, endows humans with rational thought, inner awareness, and free will, all of which beasts lack.

As we saw in Chapter 2, when experimental psychology emerged as a scientific alternative to philosophical speculation about the mind in the late nineteenth century, the inner realm of conscious experience was the primary topic of interest, and the new field hoped to use experimentation to solve Descartes' problem.[15] But by the 1920s consciousness had begun to lose its esteemed position due to challenges posed by behaviorists like John Watson and by Freud's theory of psychoanalysis. Behaviorists held that consciousness, being private and unmeasurable, is not a legitimate topic for an experimental science;[16] observable behavior is the only acceptable source of psychological data, whether studying animals or humans. For his part Freud argued that consciousness, although playing an important role, is just the tip of the mental iceberg—most of the mind is unconscious.[17]

By the middle of the century, cognitive science began to supplant both behaviorism and psychoanalysis, and with it a different version of the unconscious mind emerged.[18] The cognitive mind was an information-processing system that detects and responds to stimuli, learns and forms memories, and uses these to control behavior, mostly unconsciously.[19] We are consciously aware of the outcome of information processing (the content it creates in our conscious mind—the conscious experience itself) but never of the underlying processing.[20] Some of these implicit processes lead to conscious experience (conscious content), but in others the result of the processing is unconscious, or nonconscious. Many of the things we learn in life and use as we interact with our physical and social environment (such as syntactic parsing of sentences, depth perception, instrumentally reinforced actions, and habits) involve implicit, nonconscious processes and content to which we have no direct conscious access.[21] Although the Freudian unconscious is a repository for previously conscious content, a place where anxious thoughts and memories are shipped and kept under wraps,[22] the so-called cognitive unconscious refers to processes that perform functions that may or may not produce conscious content. When referring to processes like the latter, I prefer the term "nonconscious" to avoid confusion with the Freudian unconscious.

Information that is not present in consciousness at the moment, but that is readily accessible to be made conscious, is sometimes said to be in a preconscious state and is often distinguished from information that is inaccessible. Freud used this term, as have contemporary cognitive scientists.[23] For example, you are probably not at this moment thinking about what you had for dinner last night, but now that I've mentioned it, you very likely have retrieved that information.

MENTAL STATE CONSCIOUSNESS AND COGNITION

In Descartes' formulation, humans consist of two kinds of substances that give rise to two kinds of states: physical (beast-machine) states and mental (conscious) states. For him, mind or cognition was one and the same with consciousness. But Freud—and, later, cognitive science—offered a more subtle view, one in which the mind has both conscious and nonconscious aspects. This distinction comes into play when we are trying to determine which aspects of human behavior depend on mental state consciousness, and also when trying to address the question of whether animals have mental state consciousness. So how do we make the actual distinction?

The most straightforward way to distinguish mental state consciousness from nonconscious processes that control behavior is via language—by verbal self-report.[24] According to Descartes, man supplies evidence that he possesses a rational soul (consciousness) through speech.[25] The philosopher Daniel Dennett similarly notes that the hallmark of conscious states in humans is that they can be reported—we can talk about them.[26] People can also show that they are aware of something by nonverbal means, but this usually takes the form of a response to a verbal request, and most if not all of the time (in the absence of brain pathology), when a nonverbal report of consciousness can be given, a verbal report can also be offered. Although we may not be able to verbally describe with complete accuracy all of our conscious experiences, we can usually say whether we are having one. And just as the presence of a verbal report is the best evidence that someone was conscious of something, the absence of a verbal report is the best evidence that he was not (barring forgetfulness, deception, or mental dysfunction). Some go so far as to say that conscious experience is entwined with (depends upon) the ability to report on the experience.[27]

When dealing with the question of consciousness in human subjects, then, a key question is which kinds of brain events and resulting states are verbally reportable, and which are not. Studies of conscious versus nonconscious processing in humans often use a degraded presentation of visual stimuli to eliminate the ability to consciously see a stimulus. The presence of nonverbal responses (behavioral or physiological) in situations in which a subject fails to identify a stimulus verbally is viewed as evidence of nonconscious processing. To obtain these nonverbal responses to stimuli that were not consciously seen, the subjects typically have to be coaxed to respond by instructing them to choose from several the one that matches the stimulus presented but not consciously seen, even if they have to guess.

The classic way to prevent stimulus information from reaching conscious awareness in people involves subliminal stimulation, a procedure in which a visual stimulus is flashed very briefly—for only a few milliseconds, which is too short a time for it to reach awareness.[28] Another approach, which is more stringent and used more often today, is called masking (Figure 6.1). In this procedure one stimulus, the target, is followed a few milliseconds later by a second stimulus, the mask.[29] Masking has the same general effect on conscious perception as a brief flash, but because it helps prevent the stimulus itself from lingering, it blocks awareness more effectively. Under either condition, the typical result is that subjects deny having seen anything, yet through nonverbal measures the stimulus can be demonstrated to have been registered.[30] And there are other approaches as well.[31] The ability to measure a nonverbal response under conditions where verbal report does not occur is the reason that conscious and nonconscious states can be distinguished in people.

Studies of consciousness in other animals are a different matter, because evidence related to both nonconscious processes and mental state consciousness can only be obtained using nonverbal responses. How, then, can one determine when a nonverbal response is revealing nonconscious cognitive processes as opposed to consciously experienced mental states?

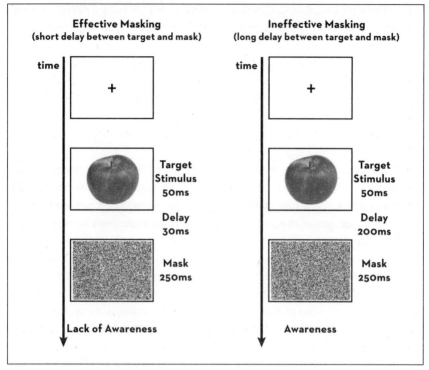

Figure 6.1: Preventing Conscious Awareness of Visual Stimuli Using Masking.

Masking is a psychological procedure in which the ability to be aware of a stimulus and report its presence is disrupted. If the target stimulus (the stimulus to be identified) is followed by a second stimulus (the mask) after a brief delay (e.g., 30ms), the participant is typically unable to identify the stimulus, but if the delay between the target and mask is longer (e.g., 200 ms), then the stimulus is readily identified.

One strategy used to explore consciousness in animals assumes that if an organism can solve complex problems behaviorally, it has complex mental capacities and therefore mental state consciousness.[32] But this approach conflates cognitive capacities with consciousness, which we've seen are not the same.[33] Animals are not, as Descartes characterized them, simple beast machines that only react reflexively to the world. They use internal (cognitive) processing of external events to help them pursue goals, make decisions, and solve problems. But because the human brain can often carry out these same tasks nonconsciously, the mere existence of such cognitive capacities in animals can't be used as evidence that consciousness was involved.

A more direct way of assessing consciousness in animals is to place them in situations that are indicative of consciousness in humans and see whether they respond in a similar way. One procedure that has been used extensively is the mirror test of self-awareness. Young children before the age of two fail to notice a change in their appearance when they view themselves in a mirror.[34] For example, if a red spot is put on a child's face, she will ignore it before roughly eighteen to twenty-four months of age, but will try to remove it at the latter age and after. This phenomenon is thought to reflect the emergence of conscious self-awareness in early life. Positive results on this test have been achieved with chimpanzees and have been used to argue that these animals have mental state consciousness. (Other animals that have passed the mirror test include monkeys, whales, dolphins, elephants, and even one species of bird.[35]) But the value of these tests has been challenged by Celia Heyes.[36] She points out that in chimps the ability to notice the spot is highest in infancy and decreases progressively with age, which is the opposite of what happens in human children, and the opposite of what would be expected if it reflects the developmental emergence of conscious awareness. Heyes's main criticism, though, is that approaches like this, which are based on a simple analogy with human behavior, don't rule out the possibility that the behavior is controlled nonconsciously (that the animal detects the spot, but the detection process does not depend on conscious awareness). Remember, to demonstrate mental state consciousness in animals, one has to not only produce evidence consistent with consciousness but also evidence that alternative, nonconscious explanations are not viable.[37] As a result, the mirror test, as typically conducted, does not provide conclusive scientific evidence for mental state consciousness in other animals.

A more stringent approach was proposed by Larry Weiskrantz.[38] He argued that, just as human studies are able to show the difference between conscious and nonconscious processing by contrasting reportable states of awareness versus nonreportable events that control behavior, animal studies also have to be able to distinguish behavioral performance from an awareness of the stimulus. His solution was the use of a commentary key. For example, a monkey is placed in a situation where button pressing will lead to a reward: If it presses button "A" when a visible light flash is present, or presses button "B" when the light flash is absent, the reward will be given. There is also a third button, "C," that produces a reward only 75 percent of the time, but independent of the light. It is assumed that if the monkey is confident about whether or not the light has flashed, it has no reason to choose "C," because it is likely to be rewarded close to 100 percent of the time for selecting "A" or "B" accurately when the light is clearly visible. But if the game rules change such that the difference between light and no light is made less clear, and is only detectable about 50 percent of the time, then pressing "C" is a better bet, because it will result in a reward 75 percent of the time, again regardless of whether the light flashed. The monkey's choice of button "C" can thus be used as an indicator of how confident it was about what was seen,[39] and this is viewed by some as reflecting conscious mental states. Although the commentary approach is, according to Weiskrantz, essential to demonstrate mental state consciousness in animals, it is not sufficient, because behavior can improve with practice and become habitual, in which case it would not necessarily depend on mental state consciousness.[40]

Studies of metacognition in animals, some of which have used commentary keys (as suggested by Weiskrantz) and tested alternative hypotheses (as suggested by Heyes), illustrate the difficulties in research in this area.[41] In simple terms, metacognition refers to cognition about cognition (thought about thought), the ability to monitor or control cognitive processes.[42] In human studies, commentary keys are unnecessary, as subjects can be verbally instructed about what to do. For example, participants can be exposed to pictures of fruits under degraded presentation conditions (using masking, for example). They are told to press a button if a fruit is shown. On each trial they are asked how confident they are—whether they are guessing or not—that a fruit picture occurred.[43] If they perform better when they say they are not guessing, this is viewed as evidence that they had some thought (a metacognition) about the accuracy of their response. A related approach, called postdecision wagering, requires the subjects to place a wager on whether their behavioral response was correct.[44] If they bet more money on trials in which they were correct, it is viewed as an indication of metacognition. Although some argue the wagering approach provides a direct, objective measure of consciousness,[45] concerns have been raised about this conclusion.[46]

In animal studies of metacognition, sophisticated training procedures involving the commentary key approach have been used to teach the subjects how to express their confidence in the accuracy of their responses[47] (Weiskrantz's monkey study described above is an example). But whether these studies actually demonstrate the existence of thoughts about thoughts in animals is debated, with some being skeptical of the claims.[48] Others, while accepting that some studies have indeed shown metacognitive ability in animals, are still hesitant about the leap from metacognition to consciousness. For example, J. David Smith, a prominent researcher in this area, is an enthusiastic supporter of the existence of metacognition in animals. He argues that research with this methodology enables comparisons of cognitive capacities between humans and other species[49] (it also provides a way to study cognitive capacities in humans who cannot communicate through language for various reasons, such as preverbal children or people with brain damage or congenital disorders leading to severe autism or mental retardation). But Smith also notes that although this kind of research can reveal significant insights into animal cognition, it does not demonstrate the existence of animal consciousness. The bar is considerably higher for the latter.

There's an additional body of work that is relevant to this topic, which involves efforts to demonstrate that animals have episodic memory, first-person conscious memories of the experiences of one's life.[50] Some claim that this kind of memory depends on consciousness and is unique to humans.[51] If animals could also be shown to possess episodic memory, it might constitute evidence that they have bona fide conscious experiences. But, as we'll see in the next chapter, the issue is more complicated than it might seem at first blush. Mammals and some birds do have some of the cognitive components of episodic memory, but whether they truly have episodic memory itself is harder to establish. For that reason most scientists who study this topic speak of episodic-like memory rather than episodic memory per se.

Although it is possible to conduct sophisticated studies of animal cognition, it is much more challenging, and maybe even impossible, to venture into another animal's mind and know what, if anything, it is experiencing and thus whether it has mental state consciousness. As Guilio Tononi and Christof Koch state, "The lessons learnt from studying the behavioural (BCC) and neuronal correlates of consciousness in people must make us cautious about inferring its presence in creatures very different from us, no matter how sophisticated their behaviour and how complicated their brain."[52] The cognitive neuroscientist and consciousness researcher Chris Frith and his colleagues stated the problem succinctly: Although they believe that monkeys do have conscious mental representations, it's very difficult to prove this assumption; even though monkeys can be trained to report their perceptions behaviorally, the inability to validate these behavioral reports with verbal ones precludes the conclusion that they consciously experienced the stimulus.[53] If we cannot conclusively establish that other primates are consciously aware of activities in their brains, it does not bode well for the possibility of proving consciousness in other, nonprimate animals.

Actions may speak louder than words in many situations. But for consciousness, even a faint whisper says more than any movement possibly can. So let's look at language and consciousness in more detail.

LANGUAGE AND CONSCIOUSNESS

As part of our daily lives we use language to label and describe our perceptions, memories, thoughts, beliefs, desires, and feelings. As we've seen, this capacity to talk about our inner states makes it relatively easy for us to study human consciousness scientifically. But the contribution of language goes far beyond simply providing a tool for assessing consciousness. Language, Daniel Dennett says, lays down tracks on which thoughts can travel.[54] Many other philosophers of mind and scientists have argued for a strong relation between language and consciousness.[55]

Language, of course, primarily involves using words to label external objects and events and inner experiences. But we also have syntax, or grammar, which structures our mental processes and guides their operation when we are thinking, planning, and deciding. As noted by the cognitive neuroscientist Edmund Rolls, syntax enables us to plan actions through many stages and evaluate their consequences without having actually to perform the actions. By contrast, non-linguistic behavior in both humans and other animals, according to Rolls, is driven by innate programs, reinforcement history, habits, and rules, but not by the capacity humans have to anticipate many steps ahead.[56]

Although there are many significant physical differences between the brains of primates and other mammals,[57] and between humans and other primates,[58] the most fundamental one is functional and involves the contributions of language to cognition. The relatively impoverished cognitive capacities of the language-deficient right hemisphere of a split-brain patient compared with the language-rich left hemisphere makes the point.[59] Some may counter that people who are born deaf and mute or who lose language abilities due to brain damage are not without higher cognition and consciousness.[60] But the issue is not about the ability to understand language and to speak, but rather how language enables the human brain to process information in ways that are not possible without it.[61]

We don't know scientifically whether other animals have mental state consciousness. But if they do have such states, and were it possible for a human to somehow experience those states in the way the animal does, the experience would likely be very different from what the human normally experiences in a comparable situation. Language is not just a system for talking and reading. Talking and reading reflect the cognitive elaborations brought into the brain by language.

The philosopher Ludwig Wittgenstein famously said that if a lion could speak, we wouldn't understand what it was saying.[62] He was probably commenting on the consequences of humans and lions having different environments and different life experiences in those environments. My take is that although the lion brain, being mammalian, is similar to the human brain in many respects, it is fundamentally different, particularly with regard to the neocortex, which, as we will see, is especially important for human consciousness. In other words, putting the ability to speak into a lion brain would still leave the lion with a lion brain. It would be a more sophisticated lion brain, but it would still be a lion brain.

WHAT IT IS TO BE: THE QUALIA PROBLEM

"Qualia" is a term about consciousness that often appears in academic discussions. In New York, it has also been a popular topic in coffeehouses, cafés, and bars in certain neighborhoods, especially around NYU. These days, New York has become a hotbed of consciousness. Some of the major philosophers of mind are based at NYU and other academic institutions in the city, but the proliferation of qualia talk in the Big Apple is also due to the New York Consciousness Collective,[63] a group of younger philosophy professors and graduate students and some neuroscientists as well that, among other activities, hosts an annual music event called the Qualia Fest, where consciousness is sung about, danced to, and altered in unmentionable ways. My band, The Amygdaloids, often plays some of our songs about mind and brain and mental disorders at the fest. By the way, The Amygdaloids' CD, also called *Anxious,* can be downloaded by scanning the barcode in the preface. The 2012 event was featured in the *New York Times,* which certainly drew broader attention to qualia around town.[64] So what are qualia?

The philosopher Tom Nagel, an NYU colleague, wrote a famous paper in 1974 titled "What Is It like to Be a Bat?"[65] His answer was that being a bat was like being a bat, and it was something a human could never truly comprehend because our experiences—our qualia—are different. Though similar to the point that Wittgenstein raised about lions, Nagel's argument was that consciousness has a subjective quality—that it is "like something" to be in a conscious state. It is now often said that when we consciously experience some inner state, what we are experiencing are its phenomenal qualities, its "qualia." We don't really know if it is actually like anything to be a bat. But we do know that if it is like something, it is nothing like what it is to be a human, in part because of language and its contributions to our brain.

In the 1990s David Chalmers, another NYU philosopher, proposed an influential distinction between the hard and easy problems of consciousness.[66] The easy problems are the ones that neuroscientists typically study—for example, how sleep and wakefulness come about, and how sensory processing, perception, motor control, learning and memory, attention, and other aspects of cognition work (aspects of creature consciousness). Explaining the phenomenal content of mental state consciousness is the hard problem.

For example, it is a fairly straightforward process to try to figure out how the brain processes the colors red, orange, and pink. We can even learn things about how cognitive processes make it possible for us to put colors and shapes together to create visual representations of the sunset. But understanding how we experience these colors in a sunset is much more difficult. As put by Ned Block, another leading consciousness philosopher at NYU, the hard problem is explaining why the physical basis of a given phenomenal or subjective experience is the basis of that experience and not of some other experience or of no experience at all.[67]

The designation "hard problem" reflects, in part, the fact that it is scientifically difficult to address. But there's more. Chalmers and Nagel believe that mind is not simply something the brain does. Although the mind does, of course, depend on the brain, its essence ultimately belongs to a nonphysical realm that is distinct from the physical world inhabited by the brain and body. In other words, Chalmers and Nagel are dualists (Chalmers calls himself a naturalistic or scientific dualist, presumably to distinguish his position from theological dualism). For them, understanding the relation of the brain to consciousness is not just a hard problem, it's an impossible one, because it's the wrong conception of the problem. Consciousness is something beyond the brain; the brain is merely a vehicle in which consciousness rides in the physical world. And because consciousness itself is not a physical event, studying the brain will not reveal the essence of phenomenal experience. Brain research will reveal neural correlates of consciousness, but not consciousness itself.[68]

Other leading philosophers argue that conscious states are brain states—that understanding conscious experience in neural terms, though difficult, is in principle possible. As a neuroscientist I pursue the mind from this physicalist perspective and assume that the brain mechanisms that contribute to consciousness, whatever they end up being, are all that is needed to explain consciousness.[69] There is nothing mental that exists independent of, and in addition to, the brain mechanisms of consciousness. When I use the term "mental," as in "mental state consciousness," I am referring to the name we give to brain states that have phenomenal properties, states that we are aware of and that we attribute to our own brain and to our mind, but where mind is a material product of the brain.

PHYSICALIST THEORIES OF CONSCIOUSNESS

Within the physicalist approach to mental state consciousness, there is much debate about how the brain might generate conscious experiences. The standard, most widely accepted view is that consciousness is a brain state in which the person in possession of that brain is aware of what is occurring within that brain and can report on the experience to others. This state is typically viewed as a product of the brain's cognitive capacities. Given that these capacities are largely dependent on the neocortex, most work on consciousness has focused on the neocortical areas. (The subcortical theories of conscious feelings by Panksepp and Damasio discussed in the previous chapters are notable exceptions.)

We will consider key theories that view consciousness in relation to cognitive processes below. These have typically addressed the issue of perceptual consciousness—how we come to consciously experience external stimuli, especially visual stimuli. Because it is not possible to review all theories of consciousness,[70] only those directly relevant to the discussion in the following chapters will be covered.

INFORMATION-PROCESSING THEORIES

Most physicalist theories of consciousness today are based on the idea that the brain is an information-processing device, and that consciousness is a result of the most advanced of its processing functions. A number of prominent psychologists and philosophers assume that a particular process called working memory plays a key role in consciousness.[71]

Working memory is a special information-processing function that consists of two main components: a temporary information storage system (the workspace) and a control system that performs executive functions (Figure 6.2). A key executive function is attention, which controls the flow of information into the workspace from sensory and long-term memory systems.

We use working memory in thought and action control. For example, suppose you are at a wine tasting. In order to compare a wine you are now sampling with others you drank a few minutes earlier, you keep a record of their individual tastes in the temporary storage component of working memory. To do this, you have to exercise control over what is and is not represented in working memory by using attention and other executive functions. But you might also be comparing the wines on their aroma and appearance, in addition to simply on the basis of their taste. Working memory thus enables not only integration over time, but also across sources, such as sensory modalities. You might also bring long-term memories from the more distant past to bear on your evaluations, experiences that enable you to know that big, bold-tasting wines have a deep, dark appearance, as compared with wines that are more subtle in taste and lighter to the eye. Executive functions are used to retrieve all these memories and maintain them in temporary storage. Executive functions also come into play if you then have to decide, on the basis of all this mental work, whether to buy one of the wines, and if so, which one.

The ability of working memory to hold disparate kinds of information in our minds while we think, decide, and plan actions is relevant to consciousness because the content of working memory is assumed to constitute the information about which we can be conscious. What was called the System 2 kind of decision-making process in a previous chapter crucially depends on working memory. But working memory depends on implicit processes that are not consciously accessible, and not all information that makes it into working memory becomes conscious content[72] (this is, in part, why there are debates about whether System 2 decisions are necessarily conscious decisions). Although working memory plays an important role, it does not, on its own, completely account for conscious experience.

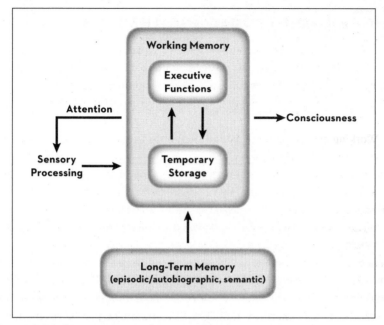

Figure 6.2: Working Memory and Consciousness.

Working memory is typically described as a mental workspace in which information can be temporarily stored while it is operated on by executive functions. A key executive function is attention, which determines which of the many sensory cues available at any one moment enter working memory. Attention also contributes to the retrieval of long-term memories that allow the stimulus to be related to similar stimuli and to experiences one has had with such stimuli. Working memory and attention are viewed by most information-processing theories as crucial (maybe necessary), though not necessarily sufficient, for the emergence of a conscious experience. Thus, while we are usually conscious of the information that is attended to and in working memory, not all information that engages attention and working memory *necessarily* makes it into conscious awareness.

HIGHER-ORDER THEORIES

Most information-processing theories implicitly assume that perceptual consciousness involves more than mere sensory processing. This idea is made explicit in higher-order theories, which assume that at least two steps are required: a first-order representation, which is not consciously experienced, and a higher-order representation, which is[73] (Figure 6.3). Without the higher-order representation, the lower-order information remains unaccessed in the cognitive unconscious and is thus not consciously experienced.

A leading proponent of higher-order theory is David Rosenthal of the City

Figure 6.3: Higher-Order Theory.

This theory proposes that in order to be conscious of some stimulus or event we must have a cognition about that stimulus or event. This allows us to be conscious of the stimulus or event, but not of the cognition about the stimulus or event. An additional higher-order event is necessary to be conscious that you are conscious of something. Working memory is necessary for these higher-order cognitions.

University of New York Graduate Center.[74] Most often the higher-order representation is viewed as a thought about a thought. Rosenthal, for example, argues that in order to be conscious, one must have the ability to reflect upon his thoughts.[75] Unless a thought is the subject of another thought, the first thought is not conscious. Importantly, even though the second thought enables information in the first-order experience to make it into consciousness, the second-order state is not itself a consciously experienced thought. For it to be consciously experienced, it has to then be the object of yet another thought (this is why metacognition, which is a second-order thought process, is not the same as consciousness). Rosenthal, in short, argues that we are not aware of the higher-order thought itself, but only of the information that it is about. Thus, nonconscious cognitive processes give rise to conscious experiences. To make this concrete, consider the steps necessary to be conscious that YOU are seeing an apple. First, the apple must be represented as a perceptual object. Second, the perceptual object has to be placed in working memory. Third, you have to have a thought about the perceptual object ("That's an apple"). But to be conscious that you are having that thought, you would need an additional thought ("I am seeing an apple"), and so on.

When I first learned of higher-order theory, I was reminded of a book on meditation that I had read some years earlier, called *Zen Training*,[76] which described the author's Buddhist ideas about thought and consciousness. He notes that "man thinks and acts unconsciously" and then proceeds to offer an explanation of how consciousness comes about. To do this, he posits three nens (thought impulses). The first nen is a primitive representation of the world. The second nen is a recognition that the first nen exists. It allows us to experience the first nen, but it knows nothing about itself. The third nen is a conscious experience of the second nen, a recognition that the experience is happening to oneself (the involvement of the self in consciousness will be discussed in the next chapter, when we examine the relation of memory to consciousness, especially the role of memories of one's personal experiences in life). The three nens are similar to the three steps posited by Rosenthal as necessary for you to know that you are conscious of something. It's interesting how very different traditions can reach similar conclusions.

Studies of metacognition essentially measure higher-order thoughts, because people are asked to think about what is in their mind.[77] Axel Cleeremans's "radical plasticity hypothesis" explicitly combines metacognition and higher-order theory.[78] He proposes that higher-order representations do not happen automatically but have to be learned: Through experience, certain unconscious states are accompanied by a learned metarepresentation that is consciously experienced. This does not mean, however, that everything measured under the umbrella of metacognition equals consciousness.

Variants on higher-order theory stress the importance of an inner narrative about experience—that consciousness is, in part, talking to ourselves. Daniel Dennett's multiple drafts theory, which draws upon Rosenthal's ideas, views conscious states as drafts of a narrative.[79] Mike Gazzaniga's interpreter theory also views conscious states as reflecting an inner narrative, one that results from the interpretation of experience.[80] Larry Weiskrantz's commentary theory emphasizes the importance of the ability to report about one's experience,[81] noting "being aware means being able to make a commentary and . . . such a capacity may be endowing and not merely enabling, which in turn bears a formal, if not literal, relation to David Rosenthal's view that consciousness entails a thought about a thought."[82] According to Rosenthal, the existence of higher-order thoughts is what allows one to give self-reports about experience.

GLOBAL WORKSPACE THEORY

Global workspace theory is another variant of the information-processing view (Figure 6.4). Originally proposed by Bernard Baars,[83] it has also been forcefully championed by Stan Dehaene, Lionel Naccache, and Jean-Pierre Changeux.[84] As in other information-processing theories, it holds that various systems compete for access to a cognitive workspace (essentially working memory), and attention selects what reaches the workspace. Information in the workspace can then be used in thinking, planning, decision making, and behavioral control. But according to global workspace theory, the executive attention and workspace of working memory alone are not sufficient for a conscious experience. The information at hand has to be broadcast widely in the brain, sent back to the workspace, and rebroadcast over and over. In global workspace theory, it is the broadcasting and rebroadcasting from the workspace that create phenomenal conscious experience.

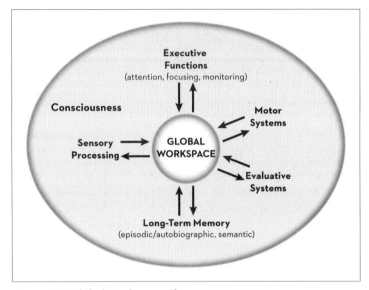

Figure 6.4: Global Workspace Theory.

In this theory, consciousness arises from the broadcasting of information out of the workspace (essentially working memory) and the return of the broadcasted information to the workspace, and rebroadcasting and return. Broadcasting and re-rebroadcasting amplify processing in such a way that produces a conscious experience. Consciousness is thus a product of the global activity in these theories.

Global workspace theory is in some respect similar to the higher-order theory of consciousness: A single level of processing is said to be insufficient for conscious experience, cognitive access is viewed as necessary for phenomenal experience, and verbal reports are taken as a mark of experience. But global workspace theory does not explicitly require that the experience be the subject of a thought or perception in order to be consciously experienced. The information simply has to be placed in the workspace and broadcast and rebroadcast.[85] One can imagine that higher-order representations are being created as part of the broadcasting and rebroadcasting, but this idea is not part of global workspace theory.

FIRST-ORDER THEORY

The simplest information-processing theory of perceptual consciousness is first-order theory[86] (Figure 6.5). It assumes that the representation of a perceptual object (e.g., a visual stimulus) is all that is needed to have a conscious experience

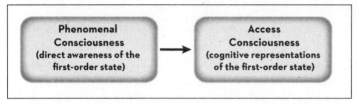

Figure 6.5: First-Order Theory.

This theory stands in contrast to the other theories described so far in that it assumes that all that is needed for the awareness of a stimulus is the processing of that stimulus itself. Awareness occurs as part of stimulus processing. Working memory, higher-order cognition, and broadcasting and rebroadcasting, according to this view, simply allow amplification of the representation and give access to the experience (access consciousness), but the experience itself (phenomenal consciousness) is independent of these operations that give rise to cognitive access.

of that stimulus. For first-order theorists, an essential part of a conscious state is its awareness of itself.[87] First-order theory thus stands in opposition to all of the theories discussed above. When I describe it as "the simplest," I refer to the fact that it requires fewer processes than the others. But it is not simple to understand.

Ned Block, a colleague at NYU, is the leading proponent of this view.[88] Block prefers the term "same-order theory" to capture the reflexive nature of consciousness—that it is a self-contained state about itself, and no other state is needed to experience it. His theory is based on a distinction between access consciousness and phenomenal consciousness.[89] Block's idea is that phenomenal experience (mental state consciousness) can exist without the cognitive access that allows you to know you are experiencing it.[90] To help understand the nature of unaccessed phenomenal consciousness, Block offers the example of what happens when you are sitting in a quiet room, reading or thinking or daydreaming, and suddenly something changes in your sensory world—for example, the motor of the refrigerator in the nearby kitchen shuts off. At that point you seem to notice that you had been conscious of the sound. This unnoticed prior experience is the essence of Block's phenomenal consciousness. It is a conscious state that you are not aware of and could not report on until you accessed it, yet, according to Block, you were phenomenally conscious of it before you accessed it. Access, Block says, is simply what allows us to become aware of the content of our subjective phenomenal experiences.[91]

In a commentary on Block's theory, Lionel Naccache and Stanislas Dehaene point out the difficulty in experimentally separating nonaccessed phenomenal consciousness from the cognitive access that is required for reporting on conscious experience. The idea of first-order phenomenal consciousness, like access consciousness, is thus intractably dependent upon the need to obtain subjective reports about the state.[92] So how do you distinguish an unaccessed state of phenomenal consciousness of which you are not aware from a nonconscious state of which you are not aware? Awareness in each case depends on access. So what is unaccessed phenomenal consciousness?

Every time I notice the motor of a refrigerator shut down, I struggle to figure out if I was aware of it or not. In that sense, Block's example is ingenious. But I think what is going on has more to do with short-term sensory memory than with phenomenal consciousness. Each sensory system can store information in a preconscious (unaccessed) state for some seconds.[93] Thus, when the motor shuts down, attention is directed to the preconscious sensory memory and pulls it into working memory. We then have a conscious experience of having been listening to the sound, but what we experience and remember is what was in our preconscious short-term memory buffer rather than some previously unaccessed phenomenal state.

Adam Zeman, a British neurologist who has written extensively about consciousness, notes that the conflict that exists between theories that use self-report to distinguish the transition from unconscious processing to conscious awareness, and theories that assume that unaccessed phenomenal states that are not reportable are states of consciousness, is a major challenge to the field.[94] How, then, are we to choose among the various theories of consciousness to try to understand fear and anxiety as consciously experienced states? Maybe the brain can help.

USING THE BRAIN TO HELP UNDERSTAND PERCEPTUAL CONSCIOUSNESS

Until recently, philosophers interested in consciousness paid little attention to the brain. Even those who ascribed to physicalist explanations of consciousness did not necessarily think that facts about brain function could contribute anything useful.[95] For example, a philosophical school called functionalism argued that trying to understand consciousness in terms of the brain is like trying to understand how a computer can play chess by analyzing its electronic components.[96] The function (playing chess) is made possible by the software that runs on the computer and is not strictly determined by the hardware itself—the same chess program can run on lots of different kinds of computer hardware. Thus, in this traditional functionalist view, consciousness is a physical event that depends on the brain, but neurons, synapses, action potentials, and neurotransmitters cannot explain how conscious experience comes about. Physicalist philosophers today, by contrast, are much more open to the idea that brain research can provide useful evidence to test philosophical theories.

Much of the discussion about how the brain makes consciousness possible, like the discussion of consciousness itself, is focused on visual consciousness.[97] This emphasis reflects the fact that research on the visual system is one of the most advanced areas of neuroscience.[98] Let's briefly review the organization of the visual system, especially its neocortical components, as most of the discussion of consciousness involves neocortically based cognitive processes.

We know that "seeing" requires the reception of electromagnetic energy (light) by the retina, whose neurons generate neural impulses that represent a stimulus in the external visual world. These impulses are transmitted to areas of the visual thalamus, a subcortical way station that processes the signals and relays the results to the visual cortex. Circuits in the earliest stage of the visual cortex (the primary visual cortex) create an initial representation of the stimulus in terms of its lines, angles, borders, degrees of luminance, and color. These representations then combine in complex ways in later stages of cortical visual processing (the secondary and tertiary areas of visual cortex) to build shape and motion information into the stimulus. We then use these representations in behavioral actions and thoughts. Identifying these stages of visual processing, and understanding their function, is part of the so-called easy problem. The harder question is how these neural representations are consciously experienced as objects and scenes in the world by the individual's brain.

One way that visual consciousness has been studied involves research on patients with brain damage. For example, it has long been known that following damage to the visual cortex in the right hemisphere, patients are blind to the left side of visual space.[99] This is due to the wiring of inputs from the eyes to the visual cortex—recall that information seen to the left of the center of visual space goes mainly to the right hemisphere (see Chapter 2). When visual stimuli are presented in this "blind" area of space, the patient has no conscious awareness of them. Building on the observation that monkeys with visual cortex lesions could respond in primitive ways to visual stimuli in the blind field,[100] Larry Weiskrantz,[101] David Milner and Mel Goodale,[102] and others have shown that patients with visual cortex damage can, if coaxed, respond to stimuli that appear in this blind area of space, in spite of denying seeing the stimuli. To assess whether they process "unseen" stimuli, they are typically given two or more choices and have to pick the one related to the stimulus. Sometimes commentary keys are used. Other studies have presented a Pavlovian threat CS (see Chapters 2 and 3) to the blind area, and use autonomic nervous system responses elicited by the CS to show that the stimulus was registered by the brain. Such patients are said to have blindsight.[103]

The reason that blindsight patients can respond this way is due to the fact that the visual cortex has two processing streams, both of which originate in the primary visual cortex.[104] The ventral stream is responsible for object recognition; the dorsal stream is responsible for processing the location of a stimulus and determining whether it is moving so that the stimulus can be acted upon behaviorally. Because the visual inputs to the ventral stream come from the primary visual cortex, damage to the primary visual area prevents the ability to identify a given stimulus. But because the dorsal stream receives visual inputs from other visual areas, especially in the visual thalamus and brain stem, it can process a visual stimulus and respond to it without being conscious of what the stimulus actually is (Figure 6.6). The question to ponder is whether this blindsight is unaccessed phenomenal consciousness (à la Block) or, instead, simply a reflection of processes that control behavior nonconsciously. Before addressing this question, let's consider another way to study visual consciousness in humans.

Other studies that have helped in understanding the brain mechanisms of consciousness are those in which visual awareness is disrupted in healthy subjects with "normal" brains using subliminal stimulus presentations, such as masking and other approaches. By using these methods in conjunction with brain imaging, it has become possible to observe what takes place in the brain when people have reportable conscious experiences as opposed to when they deny seeing stimuli (Figure 6.7). For example, studies using masking have found that when healthy subjects are able to consciously report seeing a visual stimulus, areas of the visual cortex are activated, as are areas that have been implicated in attention and working memory, such as regions in the prefrontal cortex and posterior parietal cortex.[105] With masked presentations, in which subjects deny having seen the visual stimulus, only the visual cortex is active—frontal and parietal areas are not. The same general pattern holds for other sensory systems. For example, conscious awareness of auditory stimuli requires auditory cortical processing and also requires prefrontal and parietal areas.[106] Also, imaging has been performed in blindsight patients to examine the brain activity in conditions when they say they do and do not "see." Consistent with the findings described so far, the prefrontal and parietal cortices are engaged when they report seeing the stimulus, but not when they report failing to do so.[107] The conclusion to be drawn from these various studies is that prefrontal and parietal cortices are necessary for consciousness, a conclusion that is supported by findings demonstrating that conscious awareness of stimuli is impaired when neural activity in the prefrontal or parietal cortex is disrupted.[108]

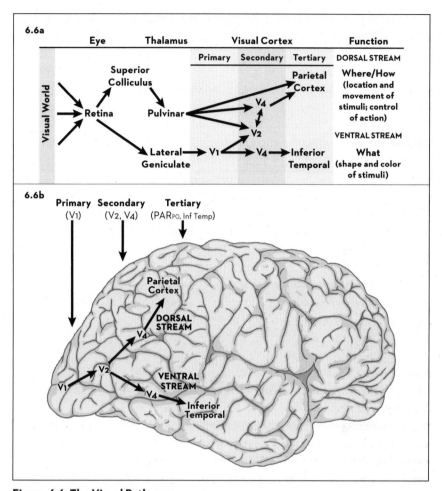

Figure 6.6: The Visual Pathways.

Much of the recent research on consciousness has involved questions about how we become aware of visual stimuli. A review of visual processing is thus in order. The visual world is projected onto the retina, which then sends visual information to the brain through several pathways. The two of most relevance here are the pathways to the lateral geniculate nucleus of the thalamus, and to the superior colliculus of the midbrain. The lateral geniculate transmits signals to the primary visual cortex (V1), which in turn connects with secondary (V2, V4) areas that then connect with tertiary areas (inferior temporal cortex). This pathway allows us to see the shape and color of objects, and is thus sometimes called the ventral or "what" processing stream. The superior colliculus connects with the pulvinar region of the thalamus, which in turn connects with V2, V4, and a visual area in parietal cortex. This pathway constructs information about the location of the stimulus and whether it is moving or stationary, and is used to guide movements in relation to the stimulus. It is known as the dorsal or "where"/"how" processing stream. In general, it is believed that we are consciously aware of processing by the "what" but not by the "where/action" stream; the latter thus controls actions without conscious awareness.

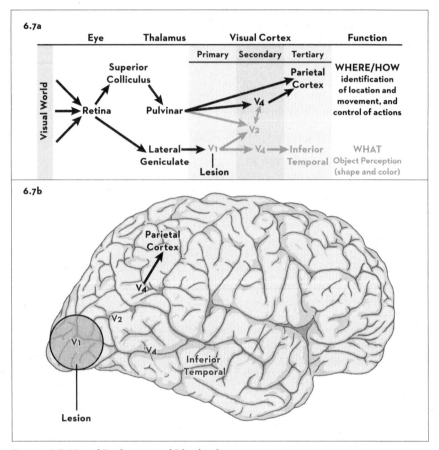

Figure 6.7: Visual Pathways and Blindsight.
Following damage to the primary visual cortex (V1) in the right hemisphere, patients do not report seeing objects in the left visual field (which connects with the right hemisphere: see Figure 2.1). Because such lesions eliminate the flow of visual information through the "what" processing stream to the inferior temporal cortex (see Figure 6.6), the "what" stream is viewed as necessary for conscious awareness of visual stimuli. These patients can, however, reach toward objects and respond in various other ways to stimuli that they claim not to "see." This capacity depends on the collicular, pulvinar, V4 pathway to the parietal cortex (see the top row of the diagram). Because these patients are consciously blind but can still respond to some extent, they are said to have blindsight.

In the remainder of this book, I will build on these observations about differences in brain activation when people can verbally report on their experiences as opposed to when they cannot. This does not mean that I believe that consciousness resides in a brain area such as the prefrontal cortex or parietal cortex. Brain

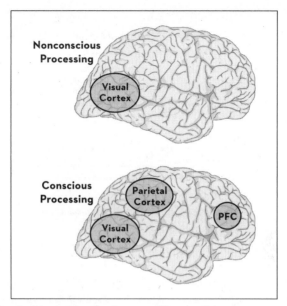

Figure 6.8: Cortical Activation for Seen and Unseen Stimuli.

Functional imaging studies show that when healthy experimental participants are exposed to a stimulus that is masked, and thus not consciously experienced and reportable (see Figure 6.1), the visual cortex is active. By contrast, when the participants are allowed to be consciously aware of stimuli and can report on the stimulus (because no mask occurred or the mask occurs at a long delay), then in addition to the visual cortex prefrontal cortex (PFC) and parietal cortex are also active. This has led to the view that the prefrontal and parietal cortex play a key role in allowing us to be aware of visual stimuli.

functions are products of circuits and systems, not brain areas. When I show illustrations in which neural activity is related to consciousness, I am emphasizing circuits within and between the depicted brain areas rather than the areas themselves.

WHAT DO THE BRAIN DATA MEAN FOR UNDERSTANDING PERCEPTUAL CONSCIOUSNESS?

The involvement of the prefrontal and posterior parietal cortices in conscious perceptual experience is an especially tantalizing discovery because these areas have been implicated in attention and other aspects of working memory,[109] processes that play a key role in theories of consciousness discussed above (other than first-order theory). These areas also have other attractive features, one of which is that they are reciprocally connected with cortical areas involved in sensory processing. The connections between these areas are called long-range connections because they transmit information between areas that are far apart in the cortex. They are also described as reentrant connections because they are reciprocal; they allow information processing in each area to recursively influence the other.[110] Another relevant feature of the prefrontal and parietal cortices is that both are convergence zones,[111] areas where diverse information about an experience can be integrated—the way something looks, smells, tastes, and feels can be blended with memory in the process of creating an experience. The work of Joaquin Fuster and Patricia Goldman-Rakic was especially notable in revealing the importance of the convergence of long-range reentrant connections between sensory processing regions and the prefrontal cortex in enabling executive functions to maintain information about immediately present, diverse types of stimuli in working memory.[112] Because more is known about how the prefrontal cortical areas contribute to consciousness, I will focus on these, and less on parietal regions.

Theories about the brain basis of consciousness often build upon ideas about long-range reentrant connections and information convergence. So let's look at how the observation that the prefrontal cortex is active when people are conscious of a sensory stimulus, but not when consciousness fails to occur, has been interpreted by different theorists (Figure 6.9).

In the 1990s neuroscientist Christof Koch and the Nobel Prize winner who helped crack the genetic code, Francis Crick, wrote several seminal articles that helped generate enthusiasm for using the visual system as a model of conscious experience.[113] They provided the fundamental logic underlying most current theories that assume that the prefrontal cortex plays a role in visual consciousness, arguing that consciousness emerges by way of loops of long-range connections from the visual cortex to the prefrontal cortex, and from the latter back to the former. (Thus, signals that reach the prefrontal cortex's working memory circuits are sent back to the visual cortex.) The visual cortex neurons that receive the prefrontal inputs are amplified by attention, creating a coalition of amplified neurons in the visual cortex. The result is that the stimulus is consciously experienced and can continue to be experienced for a time after the stimulus itself disappears. Of particular interest is the fact that only the late stages of the visual cortex (secondary and tertiary areas—see Figure 6.6) connect with the prefrontal cortex (primary visual cortex does not). This led Crick and Koch and most others to conclude that visual consciousness typically requires the late stages of the visual cortex and the prefrontal cortex.[114] Although some have argued that information in the primary visual cortex can be accessed consciously, the bulk of evidence suggests that only information that is returned from the prefrontal cortex to the sensory cortex

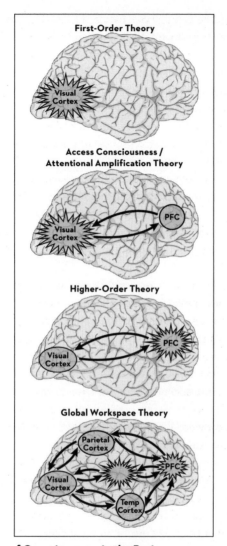

Figure 6.9. Theories of Consciousness in the Brain.

First-order theory proposes that processing in the visual cortex (especially secondary and/or tertiary areas) is all that is needed for phenomenal consciousness. Prefrontal and parietal areas simply allow cognitive access to phenomenal experience, in part by amplifying process-ing between visual and these other areas. Higher-order theory, by contrast, proposes that the prefrontal (and perhaps parietal) cortex underlies the cognitive processes that allow con-scious experience itself to occur. Global workspace theory emphasizes the amplification of processing by the broadcasting and rebroadcasting of information between areas, and assumes that consciousness emerges as a property of the global network. However, some global workspace theories give a special role to the broadcasting and rebroadcasting by the prefrontal cortex (see main text).

via reentrant connections can be consciously accessed.[115] The data described earlier showing that the prefrontal cortex is active when people can give verbal reports of visual stimuli, but not when they fail to report, are thus consistent with the Crick and Koch model. Koch is among those who take the first-person perspective of phenomenal consciousness at face value. If someone denies seeing a stimulus, Koch says, this should be accepted as the "brute fact."[116]

Like Crick and Koch, global workspace theorists accept that long-range connections from late stages of the visual cortex to the prefrontal cortex, and back to visual areas, are important, but argue that these are part of a larger network that plays a special role. That is, long-range connections are not only used to transfer information from the visual cortex into workspace areas, but also to broadcast more widely to a variety of other brain areas, each of which sends connections back to prefrontal areas, where the information from these various sources converges and can be integrated. Global (widespread) "reentrant processing"[117] is what gives rise to the amplification of broadcasting, which in turn gives rise to conscious experience in global workspace theory. Consciousness, therefore, does not reside in the visual cortex or prefrontal cortex but emerges from global reentrant broadcasting and amplification. The result of this process is a reportable conscious experience. Studies measuring neural activity in the human cortex have observed widespread patterns of activation during both consciously reportable and nonconscious stimuli.[118] However, this global activity was sustained longer (amplified) for consciously perceived stimuli. Moreover, the authors noted "special involvement" of prefrontal cortex in the conscious condition. Thus, although consciousness involves a global network sustained by reentrant processing, not all areas of the network make equal contributions to conscious awareness; the prefrontal cortex seems especially important.

Victor Lamme, a cognitive neuroscientist from the Netherlands, also argues for the importance of reentrant processing.[119] But he proposes that conscious experience does not specifically depend on the frontal cortex; instead it occurs in any cortical circuit that engages in reentrant processing. This can occur completely within the visual cortex or between the visual cortex and frontal areas. Lamme builds on Giulio Tononi's integrated information theory of consciousness, which proposes that a requirement of consciousness is integration of information by mutually interacting elements.[120]

David Rosenthal also cites the findings about prefrontal activation during conscious states. He argues that the visual cortex creates first-order visual representations, whereas prefrontal areas are needed to create higher-order representations that are accessed and phenomenally experienced.[121] Without prefrontal cortex higher-order representation, he argues, there is no phenomenal awareness.

Rosenthal challenges global workspace theory, saying that it fails to distinguish between broadcasted signals that do and do not become experienced. Consciousness, according to Rosenthal, only results from higher-order representation. In defense of global workspace theory, Stanislas Dehaene argues that preconscious processes (processes that are potentially conscious but not being accessed at the moment) can explain the middle ground between broadcasted signals that do and do not become conscious.[122]

Ned Block is a pioneer in using findings from experimental psychology and neuroscience to evaluate philosophical ideas about consciousness.[123] He argues that phenomenal visual experience occurs in the visual cortex—not the primary visual cortex, but in later stages.[124] To him, the fact that frontal and parietal areas are activated during conscious perception shows how cognitive access makes it possible to report on the phenomena occurring in the visual cortex, but lack of access does not demonstrate lack of awareness.[125] Block draws upon research involving a patient with damage to the right parietal cortex,[126] which produces a condition known as unilateral neglect.[127] Like blindsight patients, patients with unilateral neglect fail to report visual stimuli that appear to the left of center. But unlike blindsight patients, neglect patients are not blind (there is no damage to the visual cortex per se); instead, they are unable to direct attention to the left side of space due to damage to parietal attention networks. The patient in question underwent brain imaging while she was shown pictures of faces. As expected, she claimed not to have seen the faces when they were shown on the left side of space (the area of space processed and attended to by the right hemisphere). The key finding was that even though the patient failed to report seeing the face, the late visual cortex (in particular, the special region of the visual cortex involved in processing faces) was activated in her right hemisphere. Block suggests that this provides evidence that the face area of the visual cortex is the key area for phenomenal experience of faces, and that parietal cortex damage simply prevented attention to, and thus cognitive access of, the phenomenal experience. He thus maintains that although the patient could not report on the stimulus, she had phenomenal experience of it, because the face area was active.

Christof Koch notes that under some conditions it might be possible to have brief phenomenal awareness without attentional amplification from prefrontal and other areas, but he does not believe that this is what is happening when Block describes phenomenal consciousness without access. Koch's model with Crick proposes that the prefrontal cortex and visual cortex are both necessary for normal phenomenal consciousness, not just access.[128] Other critics point out that although the face cortex may be necessary for the phenomenal experience of faces, it may not be sufficient.[129] Thus, if stimulus-driven neural activity is the mark of consciousness, then consciousness could be present in any and all areas of the brain. Unaccessed states are only known from failures to access and report, rather than direct assessment of the state. If the relation between phenomenal awareness and access, which makes reporting possible, is not fundamental, how can we ever distinguish conscious from unconscious states?

Hakwan Lau and Richard Brown have challenged Block at his own game. They used findings about visual hallucinations occurring in a neurological condition to challenge first-order theory in support of higher-order theory.[130] The condition in question is a rare variant of Charles Bonnet Syndrome that can occur following damage to the visual cortex. Patients who have this condition presumably lack the ability to have a first-order phenomenal experience (due to visual cortex damage) yet can describe the experience of visual hallucinations in detail; they thus have access to phenomenal experience without the brain area that is supposed to underlie phenomenal consciousness. Lau and Brown argue that this is an example of a higher-order visual experience in the absence of a first-order experience, and for them it shows that consciousness depends on higher-order rather than simply on first-order representation.

The philosopher Martin Davies has attempted to reconcile theories of phenomenal and access consciousness.[131] He suggests that phenomenal consciousness may well be part of the causal explanation of access consciousness, noting there could be states of phenomenal consciousness without access, but not states of access without phenomenal consciousness.[132] However, the findings described by Lau and Brown seem to show just that.

First-order theorists have a difficult job[133]—probably the most difficult job in the consciousness business. They have to explain how it is possible to have a conscious experience that you do not know you are experiencing.[134] To me, it seems, the evidence suggests what makes sense intuitively: If you don't know you are experiencing something, you aren't consciously aware of it.

SUBCORTICAL THEORIES OF CONSCIOUSNESS

The views of consciousness discussed above are highly corticocentric. Some argue against this approach.[135] For example, it is known that decortication does not eliminate purposeful goal-directed behavior in animals. But one could just as easily argue that this means that consciousness is not required for goal-directed behavior. Indeed, as discussed in previous chapters, consciousness is not a requirement for using incentive stimuli to guide instrumental (goal-directed) behavior, or for behavior to be reinforced by its consequences.[136] Another point that is used to argue against cortical views of consciousness is the fact that children born without a cortex can still exhibit conscious awareness.[137] However, much evidence has demonstrated that malfunctions of brain development can be compensated for, and when this happens, all rules are off in terms of what goes where in the brain. The genetic program that builds the brain typically follows a plan that puts functional circuits in assigned places. But when that plan is disrupted, key functions are wired into alternative locations. If the visual cortex is damaged, for example, vision is handled by what is normally the auditory cortex.[138] If the left hemisphere (the language hemisphere in most people) fails to develop, the right hemisphere takes over many language functions.[139] The survival of consciousness in the absence of a normal cortex does not mean that consciousness is normally managed by subcortical areas.

In this context we should also revisit the theories of emotional consciousness by Damasio and Panksepp discussed in the previous chapter. Recall that they distinguished a primitive form of consciousness and a cognitive form. The primitive forms they postulate are in essence subcortical hypotheses of first-order phenomenal consciousness since they propose that these subcortical states do not have to be cognitively accessed in order to be consciously experienced as emotions. Then, through cognitive consciousness and its tools (such as working memory, attention, memory, and language), these primitive states can be elaborated and accessed, and thus consciously experienced, as full-blown emotions.

Panksepp and his collaborator Marie Vandekerckhove describe subcortical states of affective consciousness as "implicit procedural (perhaps truly unconscious), sensory-perceptual and affective states organized at subcortical neuronal levels."[140] But they also argue that subcortical emotional states "give us a specific feeling of personal identity and continuity without explicit reflective awareness or understanding of what is happening."[141] The states are thus implicit ("truly unconscious" and lack "reflective awareness") and, at the same time, are also consciously experienced ("give us a specific feeling"). It's hard to know what the conscious experience of a "truly unconscious" emotional state that does not enter reflective awareness might be like, but in arguing that the states are a "prereflective" form of "unknowing . . . consciousness," they are presumably referring to something like Block's unaccessed phenomenal consciousness.

Consciousness in the conventional sense (the sense in which we are aware of experiencing something) seems to depend on cortical processes. This is assumed by Block's first-order theory as well as the other information-processing theories discussed above. The processes under discussion in these theories are part of the same general cortical information-processing system. The well-established role of the visual cortex in working memory, including attention, and other cognitive functions thus provide a framework for testing where in the cortical system conscious awareness emerges from information processing. Thus, the processes are grounded in well-established circuit interactions between the visual cortex and the prefrontal and parietal cortices, and the debate is fundamentally about where in that cortical processing system consciousness emerges.

Less clear is how subcortical circuits give rise to conscious states. Why does activity in body-sensing or command system circuits give rise to conscious states, but activity in adjacent areas that control breathing, heartbeat, or reflex movements to pain or loud noise or sudden visual stimuli does not? One could make a case that the subcortical body-sensing circuits in Damasio's theory and subcortical emotion command circuits in Panksepp's theory stand in a somewhat similar relationship to cognitive consciousness as the visual cortex does. That is, the subcortical areas create first-order phenomenal experiences and then, by way of connections from the subcortical areas to cortical areas, cognitive access to the subcortical processes could be possible. But that's the easy part. The hard part in any first-order theory is explaining how the first-order state, independent of cognitive access, is consciously experienced on its own, something that has proven difficult in visual cortex and that is likely to be even harder to nail down in the brainstem.

Even if it could be shown in humans that some sort of primitive consciousness can be sustained by the brainstem, demonstrating that such states of consciousness exist in animals would face all the hurdles discussed so far. As we've seen, in animals hypothetical conscious states have to be tested by nonverbal responses, which leads to a formidable measurement problem: It is very difficult to distinguish whether nonverbal responses are based on conscious versus nonconscious processes without a verbal response as a contrast. The use of commentary keys and other clever experimental wizardry can generate evidence consistent with the idea of metacognition in animals, but even those conducting the studies acknowledge that a gap remains between establishing metacognition and proving animal consciousness.[142]

ATTENTION AND CONSCIOUSNESS

I generally side with information-processing theories in which attention controls what information is represented in working memory, and that assumes that representation in working memory is necessary for that information to become conscious content.[143] But there is a caveat: Attention may be necessary but not sufficient for mental state consciousness.

Attention does many things. Its most commonly acknowledged task is selection of the information of which we become conscious. At any given moment many stimuli are present, and we are able to attend to (be conscious of) very few of them. Prefrontal and parietal executive networks that are reciprocally connected with sensory processing areas exert top-down attentional control over the sensory cortex, selecting which inputs will be maintained in working memory. But we also know that some stimuli reach working memory by commanding attention in a bottom-up fashion.[144] Emotionally salient stimuli are particularly effective in doing so. Once a stimulus is in working memory, attention and other executive functions can then help keep it in the spotlight by suppressing competing inputs. But it would be incorrect to think of attention as only operating on information about the external environment. We also attend to signals arising from inside our body, and inside our brain (e.g., memories).

Attention is often viewed as the gateway to consciousness.[145] A group of well-regarded attention researchers put it this way: "We believe that attention is necessary for a stimulus to reach conscious awareness because attention stabilizes representations and holds them 'on line' long enough to be accessed by a variety of cortical networks and functions. Attention is the mechanism that selects certain bits of information, allowing those bits to be processed more thoroughly and reach consciousness. . . . [T]here is no evidence of stimuli reaching consciousness without some form of attentional amplification."[146] At the same time, merely because information has been attended to and has made its way into working memory does not guarantee that consciousness of the stimulus occurs.[147] In other words, although attention may be required for consciousness, it may not be sufficient.[148]

Something else may therefore be needed to turn information that is attended to and processed in working memory into conscious content. That additional something is what global workspace, higher-order, commentary, and other theories try to explain.

THE COMPLEX ROLE OF THE HUMAN NEOCORTEX IN CONSCIOUSNESS

We should not leave this discussion of the brain and consciousness with the sense that prefrontal (and/or parietal) cortex activity is a telltale sign of consciousness. You can't just look at brain imaging data and conclude that if the prefrontal cortex is active, then consciousness occurred there, because activity in the prefrontal cortex is also correlated with many unconscious processes.[149] Further, the prefrontal cortex includes many brain regions that are complexly related. And, to reiterate a point made earlier, consciousness is not something that happens in any brain region. It is, like any brain function, a product of circuits and systems.[150]

Some brain areas, such as the prefrontal and parietal areas, play a crucial role, but consciousness is not literally located in these areas.

The classic area implicated in working memory and consciousness is the lateral prefrontal cortex. Although its dorsal area (the dorsolateral prefrontal cortex, PFCDL)[151] is the region most commonly associated with working memory, its ventral part (PFCVL) has also been implicated.[152] Other areas that also contribute include the ventromedial prefrontal region (PFCVM), the anterior cingulate cortex, the orbitofrontal cortex, regions of the insular cortex, and the claustrum.[153] Damage to any one prefrontal region, or even to several, might not disrupt conscious awareness.[154] Given that the parietal cortex is also part of the consciousness picture, damage to all of the prefrontal and parietal cortices may be required to actually disrupt consciousness, but other areas (e.g., the hippocampus, basal ganglia, and cerebellum) have been implicated as well.[155] Even so, when a subset of these cortical and/or subcortical areas is damaged, others may compensate. Similarly, the fact that stimulation of the prefrontal cortex fails to trigger conscious experiences[156] does not challenge the idea that the prefrontal cortex plays a key role in consciousness, because only a limited subset of neurons in a small region is stimulated at any one time.

The general view that prefrontal and parietal networks are required for conscious perceptual experiences is supported by studies of patients as they begin to recover from coma.[157] They first transition to a vegetative state in which the brain stem and basal forebrain networks of arousal are functionally active, but the frontal and parietal networks are not. Although their eyes are open, the patients are unresponsive to sensory stimulation. When they transition to a state of minimal consciousness, where they are responsive to sensory stimuli and verbal commands, the frontal and parietal networks are also active. These findings nicely illustrate differences between the brain mechanisms underlying creature versus mental state consciousness. The same network is suppressed in states of hypnosis, in which the person is fully awake and responsive, but external awareness is altered (by way of hypnotic suggestion, attention to certain stimuli can be diminished).[158]

It is common in research on human consciousness to refer to the brain functions implicated in consciousness as the neural correlates of consciousness.[159] The subset of these correlates that I focus on in this book, arising from frontal and parietal areas, can be referred to as cortical consciousness networks (CCNs) (Figure 6.10). These circuits are key components of the global workspace, which also includes some subcortical forebrain areas connected with the cortical network, such as areas of the thalamus (especially the midline thalamus) and basal ganglia. While the CCNs are necessary for mental state consciousness, the

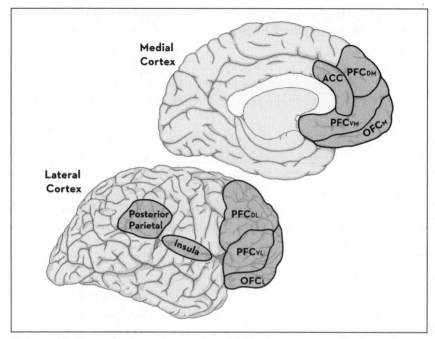

Figure 6.10: Cortical Consciousness Networks (CCNs).
Although this chapter has focused on the prefrontal, and to some extent the parietal, cortex in consciousness, a number of different regions of the prefrontal cortex have been implicated in one way or another. Abbreviations: PFC, prefrontal cortex; PFCDL, dorsal lateral PFC; PFCVL, ventral lateral PFC; PFCDM, dorsal medial PFC; PFCVM, ventral medial PFC; OFC, orbital frontal cortex; OFCL, lateral OFC; OFCM, medial OFC; ACC, anterior cingulate cortex.

subcortical areas are likely involved because of their contribution to creature consciousness and behavioral control. Specifically, the midline thalamus is a key component of the arousal/wakefulness system, and the basal ganglia are part of the system that controls instrumental behavior and reinforcement.

IS THAT ALL THERE IS?

We still don't fully understand how qualia of an experience, the "what it's like" to have that experience, comes about. But much progress has been made in understanding the brain mechanisms of mental state consciousness in the past several decades. Part of this success is due to advances in the ability to measure activity in the human brain, but equally important are conceptual advances with respect to the psychological nature of consciousness.

Although I did research on consciousness in split-brain patients several decades ago, I'm not a consciousness researcher these days. I therefore defer to the wisdom of the field to tell us how consciousness comes about. When this is figured out, we will then know how feelings like fear and anxiety come about as well, since they are states of consciousness. While emotional states of consciousness have ingredients that other states do not have, fundamentally they involve the same mechanisms as any other state in which you know that you are experiencing something.

I focused on how we become aware of a visual stimulus in this chapter. But I ignored an important part of perceptual awareness. In order to be conscious of what a stimulus is, you need conscious access to more than just its sensory properties. You also need access to memory, which gives meaning to sensory stimuli. The next chapter explores the crucial role of memory in consciousness and the contribution of different kinds of memory to different kinds of consciousness, at least one of which may be unique to humans.

CHAPTER 7

IT'S PERSONAL:
HOW MEMORY AFFECTS
CONSCIOUSNESS

"Memory is to mind as viscosity is to protoplasm, it gives a kind of
tenacity to thought."

—SAMUEL BUTLER[1]

"Every man's memory is his private literature."[2]

Your conscious experiences are personal. They are yours, and could not exist
without you. And a major fact that makes them personal is that they are
experienced and interpreted through the lens of your memories. Conscious
experiences, including experiences of fear and anxiety, are colored by memory.

CONSCIOUSNESS AND MEMORY

In Chapter 6 we discussed approaches being taken by scientists and philosophers
to grapple with how conscious perceptual experiences come about via interactions between the visual system and working memory. But when your visual system detects a round red shape in the produce section of the grocery store, and via
working memory gives you conscious access to that shape, how do you know you
are seeing an example of the fruit known as an apple?

Things like apples, chairs, spaceships, governments, and events like concerts,
weddings, and graduations are not innately wired into your brain. You have to
acquire knowledge of these through experience. So when you see that particular

shape in the store, you know it is an apple because you have learned what apples are and have stored this information in your brain in the form of a conceptual template. You can even recognize an apple in a photograph or painting, or even in a crude line drawing. This conceptual knowledge extends to all things about apples: that they are edible, and that you can eat them as is or turn them into applesauce or juice or make a pie with them; that William Tell shot an arrow through an apple resting on his son's head; that Apple is the name of a technology company and also a recording company established by the Beatles; and so on.

We have memories not only of what things are but also of experiences we've had with those things. We can revisit past experiences with a stimulus that is present, or even experiences that are indirectly related to the stimulus but that were triggered by it. For example, this past fall I was at a fruit stand looking over a pile of Macintosh apples in search of one to purchase. Suddenly I recalled dunking for apples on Halloween as a child, and this led to memories of picking apples from trees in upstate New York with my children when they were young. Recognition of a stimulus through memory can thus lead to a complex set of memories about related personal experiences.

In the 1970s the psychologist Endel Tulving was the first to establish this distinction between remembering factual information and remembering personal experiences. He called these *semantic* and *episodic memories,* respectively.[3] This remains one of the most influential distinctions in psychology today. *Semantic memories* involve knowledge you have about a thing or situation that does not personally involve you. *Episodic memories* are those in which YOU have a personal involvement. You can learn some facts about what a wedding is like by reading a guide (semantic memory), but you can only know what your own wedding was like after you experience it (episodic memory) (Figure 7.1). That "All Along the Watchtower" is a song from Bob Dylan's album *John Wesley Harding* is a semantic memory that I have stored in my brain. But that I first heard this song in the living room of Rex English's apartment on Highland Road in Baton Rouge in 1968 while I was an undergraduate at Louisiana State University is an episodic memory about a specific event in my life.

Semantic memory and episodic memory are both examples of what is called *explicit* (or *declarative*) *memory.*[4] These are memories about experiences in which YOU were consciously aware of something, and that are stored in such a way that they can be brought into your conscious awareness at later times;[5] they can be verbally reported (declared). Explicit memories contrast with implicit forms of memory, which do not require consciousness for their storage or access. We will discuss implicit memory later.

Semantic and episodic memories are closely related in that they can be

Figure 7.1: Semantic vs. Episodic Memory.
Semantic memory consists of facts, and episodic memories involve personal experiences.

consciously experienced, but they also differ in several important respects (Table 7.1). These differences can be understood by noting the distinctive features of episodic memories. First, episodic memories include information about the event that occurred (*what*), its location in space (*where*), and a time-stamp in relation to other events in your life (*when*).[6] Individually, the what, where, and when information each represents an instance of factual (semantic) knowledge. But when these are integrated into a unified representation of an event, they become the foundation for the personal memory of an experience. Episodic memory thus depends upon and builds on semantic memory. Second, episodic memories are also said to involve *mental time travel*.[7] Without episodic memory you could not recollect the past as a series of events that occurred in your life. But from an evolutionary point of view, the reason we have episodic memory is probably not simply to enable us to reminisce about the good old days. Instead, it evolved to make it possible for us to use the past to create predictions about the future so that we may benefit from those past experiences. We mentally travel to both the experienced past and the imagined future. Third, and implied by the first two characteristics, episodic memories are personal. When you mentally time travel, it is your past or present that you visit. They thus involve the self—YOU are part of the representation of the experience. As I wrote in *Synaptic Self*, there is more to the self than meets the mind's eye—much of the self is implicit or unconscious.[8] But with regard to episodic memory, "self" refers to the conscious self, those aspects of the self we consciously experience. This is key to the personal narrative created by consciousness, as proposed by a number of cognitive theories of consciousness (see previous chapter).

Though episodic and semantic memories are different kinds of explicit memory, as just noted, they do interact. In particular, episodic memory typically builds upon factual knowledge.[9] You can acquire factual knowledge about a restaurant in Istanbul by reading a guidebook before traveling (semantic memory). This semantic information can then create expectations that affect the experience you remember after you actually dine at the café (episodic memory). And remembering that you read the review of the café at home before the trip (an episodic memory) may help you remember the details of the review (a semantic memory) when you examine the menu and choose your meal. The episodic memory of the meal can also later enable you to recall the experience and anticipate returning, or help you decide never to return if you did not enjoy the experience.

Table 7.1: Comparisons Between Semantic and Episodic Memory

SEMANTIC MEMORY	EPISODIC MEMORY
Factual: "I know"	Personal: "I remember"
Consciously Accessible: you know that you know	Consciously Accessible: you know that you remember
Does not involve a unified "what," "where," and "when" representation	Involves a unified "what," "where," and "when" representation
Does not involve mental time travel	Involves mental time travel
Can be learned in a single exposure (but often benefits from repetition)	Can be fully acquired in a single exposure

Based on Table 3.1 in Gluck et al (2007).

MAKING MEMORIES CONSCIOUS

It has long been known that both semantic and episodic memory depend on the hippocampus and related areas of the medial temporal lobe (allocortical areas of the temporal lobe located on the medial side of each hemisphere).[10] In addition to the hippocampus, cortical areas surrounding the hippocampus, including the rhinal cortex (perirhinal and entorhinal areas) and parahippocampal cortex, are involved (Figure 7.2). The latter areas are way stations between neocortical

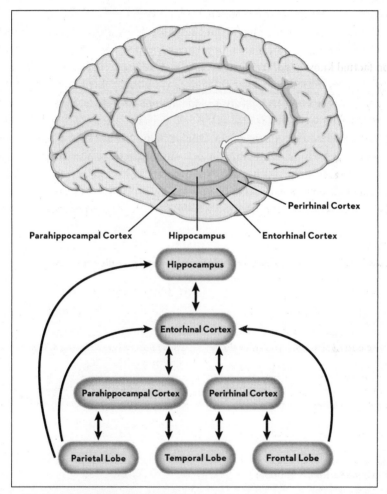

Figure 7.2: The Medial Temporal Lobe Memory System.

This system consists of the hippocampus and several surrounding cortical areas, including the entorhinal, perirhinal, and parahippocampal corticies. The perirhinal and parahippocampal areas receive information from neocortical areas and connect with the entorhinal area, which connects with the hippocampus. The connections in each case are reciprocal. In addition, the frontal cortex projects to the entorhinal cortex and the parietal lobe to the entorhinal cortex and hippocampus.

FROM NADEL AND HARDT (2011), ADAPTED WITH PERMISSION FROM MACMILLAN PUBLISHERS LTD.: *NEUROPSYCHO-PHARMACOLOGY* (VOL. 36, PP 251-273) © 2011.

systems and the hippocampus. Intricate connections between neurons in these various medial temporal lobe areas are said to constitute the *medial temporal lobe memory system*.[11] Because we can be conscious of memories stored by this system, it is common to also say that medial temporal lobe memories are conscious memories. But this is not quite accurate. Before explaining this, a brief introduction to memory terminology will be useful.

Memories are formed through a learning process called *encoding* (or *acquisition*) (Table 7.2). Memories only persist beyond a temporary state (short-term memory) if they are stored via *consolidation* processes that create a long-term memory. Once consolidated, a long-term memory has to be *retrieved* in order to be used in thought and behavior.

Table 7.2: Phases of Memory
Acquisition (learning or encoding)
Storage (consolidation of short- to long-term memory)
Use (retrieval)

Memories formed (encoded) and stored (consolidated) via the medial temporal lobe memory system in humans are potentially conscious, but are not conscious per se—they exist in a latent (preconscious) state when not being used (retrieved). Through retrieval the preconscious memory can be activated and brought into conscious awareness via placement into the temporary workspace of working memory. How many times have you tried to recall something that you know you know, but couldn't retrieve it? Then, at some point, you suddenly remembered it. That memory was encoded and stored in a way that it could be brought into consciousness, but because of a temporary inability to retrieve it, the memory was inaccessible. Once you are able to retrieve it, and not before, you become conscious of its content.[12]

The relation between long-term memory retrieval and working memory is supported by several lines of research. For example, both episodic and semantic memory retrieval are impaired by damage to areas of the prefrontal cortex that are involved in working memory, and prefrontal neural activity has been correlated with both episodic and semantic memory retrieval.[13] It should be pointed out that episodic memory also engages the parietal cortex,[14] which, as we've seen, is involved in attention and working memory. Further, retrieval is impaired by increased cognitive load on working memory (i.e., thinking about something else while trying to remember a particular detail)[15] and improved when cognitive load is reduced.[16]

Retrieval of a long-term memory from its storage sites and transfer of the stored information into working memory seem to be necessary steps for the conversion of a preconscious memory into a memory that is consciously experienced (Figure 7.3). We can thus distinguish between two states in which a potentially conscious memory can exist: preconscious (*inactive* and nonconscious at present, but potentially conscious) and conscious (*active* and presently conscious).[17]

The brain basis of conscious memory experiences can be analogized with conscious visual experience. Just as connections between the visual cortex and CCNs (e.g., prefrontal cortex circuits) enable conscious access to preconscious visual processing, connections between temporal lobe regions and areas of CCNs[18] enable inactive preconscious memories to be accessed and become active conscious memories.

More specifically, the reciprocal connections between the cortical memory storage sites and CCNs enable information about semantic facts and episodic

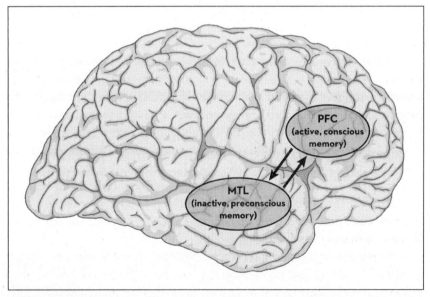

Figure 7.3: Contributions of the Prefrontal Cortex to the Conversion of Inactive (Preconscious) to Active (Conscious) Explicit Memory.

Memories stored via the medial temporal lobe (MTL) memory system are in an inactive, preconscious state until they are retrieved into active, working memory circuits involving areas of the prefrontal cortex (PFC; see Figure 6.10).

experiences to be retrieved into working memory. From there a conscious experience of the memory can occur via higher-order representation, interpretation, commentary, broadcasting, and/or some other means yet to be uncovered. As in perception, attention plays multiple roles in memory: top-down selection of which semantic/episodic memories move into working memory, bottom-up reception of memories that force themselves into working memory, and sustaining the memory, once it is in working memory.

Given that semantic and episodic memory are different kinds of explicit memory, the cortical mechanisms engaged should be different. Indeed, this is the case. In the temporal lobe, episodic memory depends on the hippocampus, whereas semantic memory involves cortical areas that are intermediate between the sensory cortex and hippocampus, especially areas of the perirhinal, entorhinal, and parahippocampal cortex,[19] but may also involve the hippocampus.[20] Further, different prefrontal circuits are involved in episodic and semantic memory retrieval.[21]

NONCONSCIOUS MEMORY

Not everything is remembered in a way that allows conscious access to it. So-called *implicit memory* does not require consciousness for either storage or retrieval (Table 7.3). Such memory is typically expressed as behavior rather than by being accessed as conscious content.

Most of our memories are of the implicit type. For example, damage to the hippocampus has no effect on the ability to undergo Pavlovian conditioning and express conditioned responses but does disrupt the ability to consciously remember being conditioned.[22] Performing a learned skill also depends on implicit memory: riding a bike, playing a musical instrument. You cannot teach these skills to another person in words. The other person's brain has to learn to do them. You can describe how you learned to ride a bike (which is an explicit memory about the experience), but that memory is not the implicit memory that

Table 7.3: Explicit vs. Implicit Memory		
PROPERTY	EXPLICIT	IMPLICIT
Consciously Accessible	yes	no
Can Be Expressed Flexibly	yes	no
Depends on Medial Temporal Lobe Memory System	yes	no

enables you to perform the skill. Although explicit memories can be expressed outwardly in a variety of ways (through speech or any of several nonverbal channels, such as drawings, sounds, whole body movements), implicit memories are typically expressed through the output modality of the system that acquired the memory. Riding a bike, for example, involves the systems that learned to maintain balance while rhythmically moving the legs in sync to propel the bike forward and depends very little on conscious semantic or episodic memories (an amnesiac patient lacking conscious memory will still remember how to ride a bike, but have no knowledge of having done so afterward).

Yet another form of implicit memory is *priming*[23] (Figure 7.4), which occurs when information stored in semantic memory that one is not consciously aware of facilitates behavioral performance. Consider the word fragment "_urse." I can bias you to complete this fragment as "nurse" or "purse" by priming you with a story about either doctors or wallets prior to showing you the fragment. You don't have to be thinking about the prime for this to occur, and you don't even have to be aware of the prime, as it can be delivered subliminally, thus bypassing

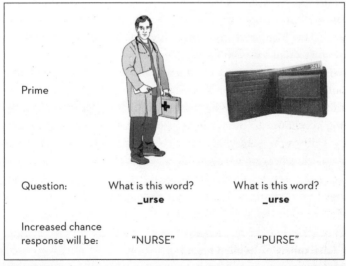

Prime

Question: What is this word?   What is this word?
                 _urse   **_urse**

Increased chance
response will be: "NURSE"   "PURSE"

Figure 7.4: Priming.
Priming is an implicit form of information processing in which some prior stimulus influences performance. In the example, seeing a picture of a doctor biases one to say "nurse" while seeing a picture of a wallet biases the saying of "purse" as a way of completing the word fragment "_urse." That priming is an implicit or nonconscious form of processing is indicated by the fact that the prime can be presented using a masking procedure that prevents awareness of the prime, and by the fact that patients with hippocampal damage can be primed even though they do not remember seeing the prime.

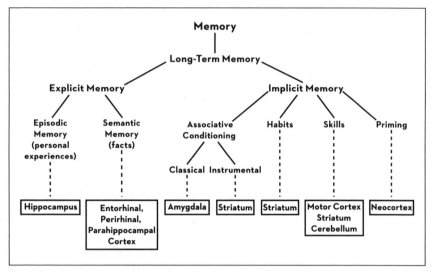

Figure 7.5: Summary of Explicit and Implicit Memory Circuits.

(MODIFIED FROM SQUIRE [1987].)

consciousness.[24] Even people who cannot consciously remember the prime because of damage to the medial temporal lobe memory system still benefit from prior exposure and exhibit the priming effect.[25]

Priming nicely illustrates a point made above—that memories are not conscious unless retrieved into working memory. For example, semantic memories, which are often described as conscious memories, can be activated subliminally and affect the speed or accuracy with which one responds nonconsciously.[26] Only when a semantic or episodic memory is consciously retrieved does it become an explicit, consciously experienced memory.

Conclusions about differences in the neural basis of explicit and implicit memory emerged from studies of the famous patient HM, who had undergone surgical removal of the hippocampus and could no longer form new, consciously retrievable memories; he could, however, learn new skills, and could be conditioned and primed.[27] He couldn't remember having done these things afterward, but he could do them. It then emerged that each type of implicit memory (conditioning, skill learning, priming, etc.) depends on its own distinct brain circuit.[28]

Just as episodic memory depends on semantic memory, semantic and episodic memory both depend on implicit memory.[29] For example, each time we consciously recognize a stimulus, we are drawing upon implicit processes operating in the medial temporal lobe memory system. Sensory cues that activate elements of a memory are stored via the medial temporal lobe system; then, through a process known as pattern completion,[30] a memory is assembled in a way that can be retrieved into working memory, where it can be consciously experienced. Although the result is a conscious memory, the processes that package it in such a way to allow it to become conscious content are not consciously accessible. Recall Karl Lashley's prescient statement that we are never aware of the processes that give rise to conscious content (see Chapter 2).

DEFINING CONSCIOUS STATES IN RELATION TO MEMORY

Endel Tulving, the psychologist who introduced the notion of semantic and episodic memory, has argued that these two types of memory depend on different forms of consciousness (Table 7.4). According to Tulving, semantic memory is an example of *noetic consciousness,* whereas episodic memory is an example of *autonoetic* consciousness. "Noetic" is derived from the Greek noun *nous* (which means mind or consciousness) and its corresponding verb *noein* (which means to notice).[31] *Noetic consciousness* thus allows conscious awareness of factual knowledge (semantic memory), whereas *autonoetic consciousness* allows knowledge about personal experiences (episodic memory).

Noetic consciousness enables you to know you are seeing an apple when you see one, whereas autonoetic consciousness enables you to attend to or become aware of knowledge based on personal experience, memory related to your self. For example, knowing that I was looking at an apple (semantic memory expressed in noetic consciousness) helped me remember the day when I went apple picking with my kids (an episodic memory expressed in autonoetic consciousness).[32] Both forms of consciousness are introspectively accessible and verbally reportable.

The difference between semantic memory and noetic consciousness, and between episodic memory and autonoetic consciousness, is that semantic and episodic memory content, until retrieved into working memory, is in a preconscious

Table 7.4: Memory in Relation to Conscious States	
KIND OF MEMORY	RELATION TO CONSCIOUSNESS
Episodic	Autonoetic Consciousness
Semantic	Noetic Consciousness
Implicit	A-Noetic State (nonconscious)

state. To be conscious of a memory, you not only need the underlying stored information, you also need to retrieve the memory into a cognitive workspace (working memory, global workspace) that allows the memory to be consciously experienced.

We are not just conscious of stimuli in our environment. We are also conscious of ourselves, which is the core of autonoetic consciousness. An autonoetic state is a kind of metacognitive consciousness, a version of higher-order consciousness—specifically, a thought about one's self. And to be conscious of who we are requires episodic memory. Any significant experience in our life is experienced personally and thus concerns the self. When we are conscious of the self, we are experiencing our self in light of a remembered self-concept. The remembered self allows us to draw upon past memories that are relevant to our self, and also to project our self into the future. It is an autonoetic experience.

The episodic and related supporting semantic memories of a given experience are what allow that experience to be remembered as part of one's personal narrative, or autobiography.[33] Although autobiographical states of consciousness involve facts or semantic memories about your self, semantic memories alone do not constitute a full-blown autobiographical memory of who you *were,* who you *are,* and who you might come to *be;* autonoetic awareness of episodic memory is also required.

Autonoetic self-consciousness is thus much more complex than noetic or factual consciousness. Chris and Uta Frith, cognitive neuroscientists at University College London, propose that our self is a social construct, and that deliberate conscious social interactions require reflective awareness of one's self.[34] The developmental psychologist Michael Lewis, like Tulving, described self-consciousness in terms of the ability to think about who are we now in terms of the past and future.[35] Gazzaniga's view of consciousness as interpreter[36] similarly treats consciousness as a means by which we make sense of our lives at any given moment by using memory, information about what we are presently doing, and information about our physical and social environment as raw material for constructing our narrative.

A-NOESIS: THE OPERATING MODE OF THE UNCONSCIOUS BRAIN

Noetic and autonoetic states both involve consciously accessible knowledge. Tulving proposed a third kind of state, also based on the Greek root *noein:* So-called *a-noetic* states occur without an individual's knowledge of their existence— without cognitive access to the state; without noticing or attending to the state;

without a phenomenal experience of the state.[37] (I will use the unconventional spelling "a-noetic" to help distinguish it from "autonoetic.")

Just as noetic and autonoetic states are associated with explicit or conscious memory, a-noetic states occur in conjunction with implicit memories, which do not require conscious access for their formation, storage, or retrieval. According to Tulving, a-noetic states are triggered automatically (involuntarily), and they "remain concealed from consciousness."[38]

Tulving used an unfortunate term, referring to a-noetic states as instances of *a-noetic consciousness*." What he had in mind was something akin to creature consciousness, rather than mental state consciousness.[39] But others have treated a-noetic states as instances of a primitive form of mental state consciousness.[40] A-noetic states will be treated here as instances of creature rather than mental state consciousness. In mental state terms, they are nonconscious events (Table 7.4). They are not directly accessible (retrievable into working memory) and do not result in conscious experiences that can be verbally reported.

A-noetic is a useful description of the vast expanse of the nonconscious brain, including mental states that are part of the cognitive unconscious. Such states arise from circuits that are not wired to generate conscious access and thus do not readily give rise to direct conscious experiences within the brain itself. At the same time, as I will explain in the next chapter, a-noetic states do have various consequences in the brain and body that can be attended to or noticed consciously, thereby indirectly contributing to conscious experiences—such as the feeling of being afraid or anxious.

What also bears reemphasizing here is that semantic knowledge can exist both noetically and a-noetically. In the priming example above, for example, stored semantic facts can prime one to respond with either "purse" or "nurse" when presented with the fragment "_urse." That the effect is nonconscious can be shown by degrading the stimulus in ways that prevent awareness,[41] or by testing patients with damage to the hippocampus who do not remember the prime but are still primed by it.[42] Although conscious semantic memory requires the medial temporal lobe memory system and working memory, and thus reflects noetic consciousness, semantic priming operates nonconsciously and as such is a form of a-noetic memory that does not require working memory.

Given that semantic knowledge can be used to guide behavior in the absence of conscious awareness of the knowledge, when an animal or person uses semantic information in carrying out some task, one cannot conclude that the stimulus involved was consciously experienced. Conscious awareness of the stimulus requires that semantic memory of what the stimulus is be made conscious by its contents being retrieved into working memory. Only then can it become a state of noetic knowledge, and only if the organism also has the additional cognitive capacities that turn the contents of working memory into consciously experienced content.

For example, a threatening stimulus—say, a snake at your feet on a path in the woods—will automatically elicit defensive responses that occur as a result of activation of a defensive survival circuit. This is an a-noetic state that does not have any necessary connection to conscious knowing or the self. However, the same stimulus that triggered the a-noetic state can, and likely will, also result in the retrieval of conscious noetic knowledge (semantic memory) about the threat (some snakes are venomous) and also result in an autonoetic state of fear and anxiety (this snake might bite me; if it's venomous, I might not be able to make it out of the woods to a hospital, and even if I do they may not have the antidote, or they might have it, but it may be too late). Further, as mentioned above, the a-noetic state has observable bodily consequences (rapid heartbeat, freezing behavior). The semantic fact that freezing and a quickened heartbeat are associated with feelings of fear and anxiety, and the episodic fact that these symptoms are happening to YOU, can interact with other semantic and episodic information and contribute to the feelings of fear and anxiety that are evolving in working memory at the moment.

ANIMAL CONSCIOUSNESS: THE EPISODIC MEMORY DEBATE

Animals obviously have creature consciousness—they are alive and respond to their environments. And some animals, especially mammals, but also birds and other vertebrates[43] and even some insects,[44] are said to use complex cognitive (mental) operations to guide their behavior. The question is not so much about whether they have mental states, but instead whether they have mental state consciousness. Hard data are sparse when it comes to this topic, because we can never truly know whether an animal consciously experiences what is going on inside its brain: We can measure information processing but not conscious content. Although much of the debate has been based on passion and speculation, in recent years some have tried to close this gap with empirical evidence. I mentioned this work in Chapter 6 but held back one important aspect of the research until after explaining what episodic memory is and how it relates to consciousness. So let's look at the very interesting studies that have asked whether evidence for episodic memory can be demonstrated in animals.

The current debate was triggered, in part, by Tulving's claim that episodic memory is a unique human adaptation,[45] which led some researchers to design experiments to test whether episodic memory is present, or at least possible, in other animals. If animals have episodic memory capacities similar to human episodic memory, they may also have autonoetic consciousness, in which case they can comprehend that what is happening now is happening to them, and can relate this to their past, and extrapolate the implications for personal well-being into the future.

In evaluating the possibility of episodic memory in animals, researchers have generally focused on one of its key features: event representations that include what, where, and when information (a representation of the event itself and its place and time of occurrence). In a seminal study conducted in the 1990s, Nicky Clayton and Anthony Dickinson discovered that birds could form memories that included what, where, and when information.[46] They studied shrub jays, a species that stores food for future use. The researchers gave the birds either worms (which decay rapidly) or peanuts (which don't) and allowed them to bury the food. A key factor was that the researchers manipulated the freshness of the worms. After a delay during which the birds were deprived of food, they were allowed to retrieve what they had stored. The studies showed that the birds clearly have the building blocks of episodic memory—they could represent *what* (worms versus nuts), *when* (older worms were retrieved before fresh worms or nuts), and *where* (the location where each kind of item was buried). Being careful not to overinterpret the data, however, Clayton and Dickinson called this *episodic-like* memory.

What, where, and when memories have since been reported in a variety of species, including nonhuman primates,[47] rodents,[48] birds,[49] and even bees.[50] But concerns have arisen about the interpretation of the data as representing genuine episodic memories.[51] One issue is whether behavioral performance in these studies depends on a unified representation of the experience that includes what, where, and when information, or whether performance might depend on separate storage of these forms of information. In other words, could these be separate semantic memories about what, where, and when? We know that what and where information is separately encoded in the medial temporal lobe (what memories involve the perirhinal cortex; where memories, the parahippocampal cortex)[52] (see Figure 7.2). Research by Howard Eichenbaum's laboratory has shown that only when information from the perirhinal and parahippocampal areas is integrated in the entorhinal cortex (the gateway into the hippocampus) do they begin to coexist as unified representations.[53] But does coexistence of representation of what, where, and when necessarily mean that an integrated episodic memory exists?

Even if animals do have unified what, where, and when representations, are they merely a complex form of semantic memory? And, if so, is this complex semantic memory consciously experienced, or does it instead control behavior nonconsciously? (Recall that semantic memory can control behavior nonconsciously or consciously.) Ultimately, the key question is whether animals consciously experience an integrated what, where, and when memory in relation to a personal sense of self with a past and future. This is far more complicated to answer, as it depends on the memory's being experienced as a state of autonoetic consciousness—a state of consciousness that involves the self.

Proponents of episodic memory in nonhuman animals bolster their case by arguing that reptiles, birds, and nonhuman mammals have brain areas similar to those that create human episodic memory—a hippocampus to make episodic memories and a prefrontal cortex to access them.[54] Specifically, areas that are precursors to the hippocampus and prefrontal cortex are present in reptiles and birds, and all mammals have bona fide versions of some structures that have been implicated in long-term memory[55] and in attention and working memory.[56]

While all this is true, there are problems with this line of reasoning. Arguing that humans and animals should have conscious memory simply because both have a hippocampus and prefrontal cortex gets two things wrong. First, it misrepresents the relation between explicit memory and consciousness in humans—as I have noted, explicit memory stored in the hippocampus is not a conscious memory until it is retrieved into working memory, where it can then be made consciously accessible. Thus, the demonstration that a memory depends on the hippocampus is not sufficient to conclude that the memory is consciously experienced, much less experienced as a state of autonoetic consciousness. Second, it gets wrong the relation between brain areas and brain functions. Merely because a rat or bird has some version of a hippocampus and prefrontal cortex does not mean that it has all the functions of the human hippocampus and prefrontal cortex.

We have to be particularly cautious about generalizing from structure to function when cortical areas are under consideration, especially when the prefrontal cortex is involved. Although all mammals have a prefrontal cortex, there are regions of it that are unique in primates.[57] Clearly, the prefrontal cortex of a mouse or rat is an impoverished version of that in monkeys or chimps. And the version of attention and working memory that a mouse or rat has pales in complexity relative to that of primates. Although nonhuman primates have brains and psychological capacities that more closely resemble ours, even these animals fail to exhibit all the characteristics of human cognition. Further, certain structural features of the human prefrontal cortex distinguish the brains of humans from the brains of even our closest primate relatives, the great apes.[58] Just because rats, or even monkeys, have a prefrontal cortex does not mean that their prefrontal cortex enables the capacities that allow autonoetic consciousness.

The key point to be made here is that having some form of attention and working memory may well be necessary for cognitive access, and thus phenomenal experience, but these do not guarantee that phenomenal experience occurs. This is why higher-order thought and global workspace theories are not simply working memory or attention theories. They attempt to specify what sort of capacities other than attention and working memory are necessary for human conscious awareness.

To form a true episodic memory requires not just what, where, and when memory, but also a self-concept: knowledge that an event being stored is something happening to YOU. Only then does an episodic memory become a state of autonoetic consciousness. One solution to the problem of establishing the existence of episodic memory in animals might be to relax the "self" requirement from the definition. This would make it easier to demonstrate episodic memory in animals but would eliminate a key aspect of episodic memory that makes it so interesting—its autonoetic quality.

True episodic memory based on self-awareness is clearly present in humans. It has also been claimed for some hominids and some marine mammals and elephants, and even birds.[59] However, conclusions about nonverbal organisms are, for measurement reasons, less certain. As discussed in Chapter 6, such studies are often based simply on analogy with humans and often begin by assuming consciousness and then looking for supporting evidence rather than testing alternative hypotheses that do not require consciousness to explain a particular behavioral response. There is no firm evidence that any animal, other than humans, has the ability to consciously experience the self as an entity with a past, present, and future. Studies based on commentary keys to indicate confidence are a significant methodological advance, but lacking language, other animals have no way to tell us what they might experience (see Chapter 6).

The gold standard of consciousness remains verbal self-report.[60] But even if particular animals do have some form of consciousness, even some form of self-awareness, the presence of language in the human brain changes the information-processing mode, and consciousness potential, of the brain. For example, the human representation of "what," via our vast semantic capability, far surpasses capacities in other animals to learn items and concepts and to group (or "chunk") information for use in thought and decision making. And one of the most sophisticated concepts that our semantic capacity makes possible is the concept of "me," or "I." But whatever nonverbal semantic skills nonhuman organisms might have, or can be trained to have, no nonhuman has anything that rivals the sophisticated linguistic computational mechanism of syntax, which enables us to relate what and where information to absolute and relative time via past, present, and future tense. For example, through syntax the concept of "me" in the present can be projected into the past and future. With a syntax-based language system in the brain, consciousness becomes self-referential and timeless. One could say that the time travel component of autonoetic consciousness is conferred, or at least greatly facilitated by, the past- and future-tense features of language.

Through our capacity for semantics, we can code experiences with labels that discriminate between present experience and others that have taken place, and relate present experience to broad categories of experience. We may know that John is a middle-aged white male with a temper and has an especially short fuse when he has been drinking. With syntax, we can make predictions about semantically labeled items—Because John is drinking, I had better steer clear so he doesn't have the opportunity to hurt me tonight. Animals can learn predictions through experience. But humans are especially adept at constructing predictions on the fly; our ability to envision possible futures sets us apart cognitively. As we will discuss in later chapters, this comes with a price—anxiety.

The question of whether animals can engage in mental time travel and incorporate self-awareness into the memory is unresolved and not easily resolvable. For now, such memories remain *episodic-like* at best. But even this description, to me, implies that these memories are closer to episodic memory than they actually are. It is usually best to adopt the simplest explanation of a phenomenon, because more complex ones tend to add features that come to be assumed to be real rather than just hypothesized characteristics. (The simplest is not always the ultimate answer but should clearly be considered and ruled out before adopting more complex explanations that are harder to evaluate.) What are called episodic-like memories might be more accurately viewed as nonverbal, semantic memories of what, where, and when. Because most animals seem to lack the hardware (brain organization) and software (cognitive processes) that underlie human autonoetic consciousness, I think episodic memory and autonoetic consciousness are the wrong targets for those who want to build a case for animal consciousness. Semantic memory and noetic consciousness might be more tractable.

Mammals, or at least some, have some of the hardware (bona fide hippocampus and prefrontal cortex) and software (attention and working memory) that could make semantic-like memory (not semantic in the linguistic sense, but semantic in the factual sense) and noetic consciousness possible. But whether these, like instances of priming described above, reflect nonconscious semantic content (a-noetic content) as opposed to semantic content that has made it into conscious awareness (noetic content) is difficult to determine.

The problem is that there may not be any way to really prove animal consciousness with data. Clever experiments can show that animals perform behaviorally in ways that people behave when they are in a particular state of phenomenal consciousness. But we can create robots that behave the way humans behave when we are having a phenomenal experience. Consciousness is, and probably always will be, an inner experience that is unobservable to anyone other than the experiencing organism. And in the absence of verbal report, there is little to measure.

If I were forced to take a guess, I would say that some other animals probably have at least noetic conscious states—momentary conscious states about facts. But this obviously has to involve facts that can be represented without language, such as the knowledge that a certain food item was stored at a particular place more recently than other items at other places, or that predators are more likely to be near a certain watering hole at the end of the day than earlier in the day.

Findings by Nikos Logothetis and colleagues are consistent with semantic consciousness in nonhuman primates. These researchers have pursued the neural correlates of consciousness with a variety of state-of-the-art imaging and neural recoding techniques.[61] They exposed monkeys to competing images rapidly presented separately to the two eyes, creating a situation of perceptual conflict that the brain has to resolve. What the animal "sees" in this test thus varies from moment to moment. These impressive studies show that neural activity in the prefrontal, but not visual, cortex resolves the difference and very well may account for which of the conflicting views dominates perception at any one moment. From this work one can conclude neural responses in areas of monkey cortex that have been implicated in human conscious perception resolve perceptual conflict in behaviorally meaningful ways and could constitute neural correlates of noetic (semantic) consciousness. But there are caveats. Just because the prefrontal cortex is active does not mean that conscious awareness is occurring (much of the activity that occurs in the prefrontal cortex is not consciously experienced), and just because a stimulus has been attended to and has made it into working memory does not mean that it is conscious. These may be necessary but not sufficient for consciousness (see Chapter 6). Further, as noted, research using verbal reporting in humans shows that cognitive processing of a stimulus is not the same as conscious awareness of the stimulus. Whether the animals are aware, in the sense that you and I are aware of what we see when we look at a visual stimulus, is thus not really known.

SO HOW DO WE KNOW HUMANS ARE CONSCIOUS?

At this point, you may well be thinking that the same argument applies to human consciousness: How do we know other people are conscious? We can, after all, know only our own states of consciousness. It was this realization that motivated

Descartes' ideas on consciousness, ideas that have shaped the philosophical debate about consciousness in the Western world ever since. But we have two advantages when it comes to humans that are lacking in studies of other species.

First of all, modern neuroscience has provided compelling evidence that psychological functions are products of brain systems. All members of a species are genetically endowed with brains that have the same general capacities, so it is safe to assume that if one person has the capacity for consciousness, other humans are very likely to as well. And because the brain circuits that play a key role in human consciousness (especially the prefrontal cortex) are different (at least to some degree) even in nonhuman primates,[62] we should tread carefully when attributing consciousness to other species.

But more important, through language we can share our inner experiences with one another. If you and I are sitting on the beach in California and watching the sun disappear over the western horizon, we can compare our qualia through language. We don't know whether we are having the exact same experience inside our heads (what I call pink, you might think of as orange), but we are having the same *type* of experience. We can thus study consciousness with some degree of confidence and with minimal assumptions in humans. The degree of confidence decreases and the assumptions increase when we consider other animals.

Human research, at least for the present, is limited in what it can reveal about detailed brain mechanisms. But it remains the best and maybe the only way to truly study consciousness itself (as opposed to information processing that may or may actually be part of conscious experience). Studies of perception and memory in humans have made significant advances in clarifying the nature of consciousness, its underlying components, and its basis in the brain. These achievements will be leveraged when we turn to the question of how the brain makes conscious feelings, in other words, emotions, and especially feelings of fear and anxiety, in the next chapter.

CONNECTING THE DOTS

In my daily life as a layperson I treat my cat, Petey, as if he has self-awareness and feelings. When he is in the kitchen meowing and rubbing my leg, and his food bowl is empty, I assume he is telling me he is hungry. If he knocks something off a cabinet, I ask him, in an annoyed tone of voice, why he did that, as if an intentional conscious motivation was responsible for his action. When I scratch his tummy and he purrs, I think he is happy. But merely because it seems that he must be experiencing these states does not mean that he is actually experiencing them. The world seems flat, but we know it is not, because scientific evidence has shown that to be the case. In the absence of some way of answering the question of whether animals have conscious experiences that bear some relation to the conscious experiences humans have, we should not simply assume they do.

While it's very difficult to do so, I think scientists should guard against bringing anthropomorphic assumptions into the laboratory. As we have seen, Lloyd Morgan urged scientists as early as the late nineteenth century to resist the temptation to view animal behavior in terms of the human mind; otherwise the study of animal behavior would lose its anchoring in empirical facts.[63] Animal psychology has struggled with this challenge in one way or another ever since.[64]

I'm surprised when scientists today simply assume that animals have conscious experiences and build this assumption into their research as if it were a hard fact. Hunger, pleasure, fear, and the like are freely called upon to explain animal behavior. The same scientist can be very rigorous in designing experiments and performing statistical tests of behavioral data and in using sophisticated neurobiological techniques to study the brain, but then become quite free with interpretations of the emotional life of animals in relation to the statistically significant neurobiologically manipulated behavioral responses. Although interpretations are by their nature speculative, problems arise when the speculations come to be treated as unquestioned facts. Part of the fun of being a scientist is speculating, imagining answers to things we do not, and may never, know, but scientists also have an obligation to keep the difference between these hypotheses and scientifically based observations separated. Otherwise, the speculations come to be thought of as the facts, and these "facts" then become part of the assumed reality that the scientists work with.

Ultimately, I think the key issue in regard to animal consciousness is what we should consider as the default condition of the nervous system. Should we assume that all animals are conscious until it is proved that they are not, or should we assume that the primordial condition of the nervous system is a-noetic (nonconscious)? This question is often derailed by failing to distinguish creature and mental state consciousness. But it can also be derailed by the assumption that mental state consciousness simply must exist in other animals. To me, the nonconscious assumption clearly seems to be preferable over the consciousness assumption. It allows us to study the a-noetic aspects of brain function in humans and other animals alike without having to make untestable assumptions about consciousness, and makes it less likely that presumptions will be treated as facts. Ultimately, the difference is whether we approach the question of animal consciousness by showing that nonconscious explanations cannot explain the data, or whether we assume that observations that seem like they might indicate consciousness constitute hard evidence of consciousness. I prefer the strategy where we climb up to consciousness rather than slide down from it.

One problem in making cross-species comparisons is that the criteria used are often a moving target. No one believes that rats or octopi have the kinds of experiences we do, so when researchers talk about consciousness in other animals, they are not really talking about what humans call consciousness. But then the evolutionary card is played—human consciousness must have evolved from similar processes in animals; therefore, behavioral evidence that might be consistent with consciousness in animals tells us how human consciousness evolved. But this only holds true if direct evidence for mental state consciousness in other animals can be obtained. Otherwise, we are talking about nonconscious brain processes that may have been precursors to human cognition, which would be relevant to consciousness but not because the precursor state was itself a state of consciousness.

Scientists should always be very careful about what constitutes data and how the data are interpreted. But the stakes are especially high when the implications go beyond the quest for knowledge about how the world works and attempt to address the core issues that affect the lives and well-being of people. When we treat the circuits that a-noetically control defense responses as if they give rise to autonoetic conscious feelings of fear in animals, we misrepresent what we are studying and mislead others who are applying our research in an effort to aid people who suffer from overwhelming feelings of fear or anxiety. Animal research can help tremendously but is most effective if we interpret the results in the most accurate way possible.

CHAPTER 8

FEELING IT: EMOTIONAL CONSCIOUSNESS

"You're not me. You can't feel like I feel."

—JOHN FOWLES[1]

A man standing next to you is aiming a gun at a bull's-eye target some distance away. Without warning he turns and points the barrel at your head. Although the stimulus elements of the physical situation remain nearly the same, the meaning of the situation changes drastically for you when the gun is turned your way. The experience is transformed into a decidedly emotional experience, most likely an experience in which your conscious mind is consumed with feelings of fear and anxiety—fear that the person might pull the trigger, and anxiety about what that might mean for your survival and well-being if he does.

The qualia of an emotional experience, the "what it's like" to be in that experience, are different from those of a nonemotional experience. This chapter explores the nature of the differences, and how they come about in the brain. Specifically, it delves into what happens in the brain and body when people encounter emotionally arousing, and especially threatening, stimuli as opposed to emotionally neutral sensory stimuli of the type we discussed in Chapters 6 and 7.

A defining feature of the feeling of fear is that you are afraid of something. Thus, a key part of the experience of being afraid of a snake or mugger, or a gun pointed at your head, is the awareness that the snake, mugger, or gun is present. I therefore start this discussion with a consideration of how the conscious awareness of a threat stimulus comes about and also consider what's different about the brain processes that are engaged by threatening versus neutral stimuli. Ultimately, these differences account for why you consciously experience the feeling of fear when you encounter threats, and anxiety when you worry about threats that are not present.

CONSCIOUS VERSUS NONCONSCIOUS PROCESSING OF THREATS

In studies of threat processing in humans, researchers typically present participants with stimuli that are either inherently threatening or that have acquired threat significance through Pavlovian conditioning (Figure 8.1). Each species is innately prepared to treat certain stimuli as threatening.[2] Innate/prepared threats include pictures of people with facial expressions of anger or fear, or pictures of venomous animals like snakes and spiders. Conditioned threats are created by pairing a benign stimulus with a mild electric shock. In some studies, the two approaches are combined (pairing an innately threatening stimulus, such as an angry face, with shock) to enhance the effects of conditioning.[3]

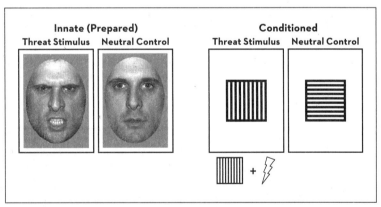

Figure 8.1: Innate (Prepared) and Conditioned Threat Stimuli in Human Studies.

(*Left*) Some stimuli function as threats for humans without any obvious prior learning. However, because not all people respond to the same degree, and it is hard to completely rule out prior learning, these stimuli are sometimes designated as prepared rather than innate stimuli. (Modified from Ewbank et al [2010].) (*Right*) Biologically neutral stimuli, when paired with an aversive unconditioned stimulus (US), such as electric shock, become conditioned threat stimuli (CSs). It is also possible to pair prepared stimuli with aversive stimulation to boost the effects of conditioning, since in human studies the intensity of the US is typically fairly weak.

The effects of conscious versus nonconscious threats are investigated much as they are in the case of neutral stimuli—by using masking or other experimental tricks to prevent awareness of threats or by studying threat processing in patients with brain damage, such as blindsight patients (see Chapter 6). Numerous studies have shown that the human brain can process threat significance without awareness of the trigger stimulus itself.[4]

As I discussed in earlier chapters, one reason we can study consciousness in humans, and have trouble doing so in animals, is that we can create situations in which people are able to verbally report whether they are consciously aware of a stimulus or not. Nonverbal responses reveal whether the brain registered the stimulus in a meaningful way but do not, on their own, reveal differences between conscious and nonconscious processing.

As we saw in Chapter 6, studies of nonconscious processing of neutral stimuli (e.g., through masking or in blindsight patients) often require that the participants be coaxed to respond in some fashion. Subjects are usually asked to choose between two or more items, even if they feel they are guessing. But threatening stimuli offer an experimental advantage over neutral stimuli, as threats automatically and involuntarily elicit autonomic nervous system responses (such as changes in blood pressure, heart rate, respiration, and perspiration) and potentiate body reflexes, like startle.[5] The participants do not have to be coaxed or made to respond to stimuli that they claim not to have experienced. Threat-elicited body responses thus provide an objectively measureable, nonverbal indication that unseen threats are meaningfully processed by the brain when verbal report fails. These responses are usually taken as evidence that threat processing does not require consciousness. It is fundamentally an a-noetic (implicit or nonconscious) form of processing.

Under normal conditions, however, we obviously can, and usually are, consciously aware of threats. Consciousness adds dimensions to threat processing that are not possible when threats are processed only a-noetically. Although much of human decision making is based on nonconscious processes that assess the value of stimuli and responses (see Chapters 3 and 4), we can also bring consciousness to bear in our choices.[6] Past learning may suggest that climbing up a tree is the best option if a pit bull is growling at you and seems poised to attack. But if you notice that the tree has no low branches, or that its low branches won't support you, then you might have to try an alternative method of escape, and you can use conscious memories and imagination (mental time travel to the future) to test various strategies and select one that seems best. In addition, though, threats, once in consciousness, can cause us to ruminate and worry—will you be severely injured if the pit bull catches up with you? Let's look at what happens in the human brain during conscious processing of threats.

BRAIN SYSTEMS ACTIVATED BY CONSCIOUSLY PROCESSED THREATS

Recall from Chapter 6 that when neutral visual stimuli are prevented from entering conscious awareness by masking or other procedures, areas of the visual cortex are activated. In the absence of such experimental trickery, when participants are able to report seeing visual stimuli, in addition to the visual cortex, frontal and parietal areas of the CCNs are activated as well.

Not surprisingly, the same basic pattern of brain activation occurs for threatening visual stimuli: Nonconsciously processed threat stimuli activate the visual cortex but not the frontal and parietal areas, whereas consciously seen visual threats activate the visual, frontal, and parietal areas[7] (Figure 8.2). We thus become consciously aware of visual threat stimuli in the same way that we become aware of any other kind of visual stimulus: by interactions between the visual cortex and cortical networks of consciousness that control attention and other executive functions and enable representation of the stimulus in the

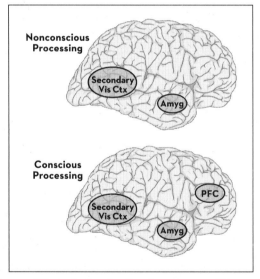

Figure 8.2: Patterns of Brain Activation for Conscious and Nonconscious Threats.
As shown in Figure 6.8, freely seen emotionally neutral stimuli activate the visual cortex as well as the prefrontal and parietal areas, but masked stimuli that are not reportable only activate the visual cortex. The same pattern of activation holds for threats. However, in addition, both seen and masked threats result in amygdala activation. The fact that the amygdala can be activated by threats that are not reportable suggests that amygdala activation occurs independent of conscious awareness of the stimulus.

cortical workspace of working memory. As noted before, this does not mean that consciousness resides in these areas, but instead that cells, molecules, synapses, and circuits in these areas help make consciousness possible.

Although the same cortical areas are activated in both consciously seen emotionally neutral stimuli and consciously seen threats, the degree of cortical activation is greater for the latter.[8] The net result is that a threat stands out in the conscious mind relative to neutral stimuli that are also vying for attention. We will consider how this boosting of cortical processing occurs. First, though, a word about memory in relation to conscious threat processing.

In Chapter 7 we discussed the role of memory in consciousness. Surprisingly, the contribution of medial temporal lobe memory systems to conscious processing of threats has not been studied much. However, given what we know about the role of memory in consciousness, it seems likely that the medial temporal lobe memory systems are called upon when we consciously process threats. For example, in order to be conscious that you are being threatened, you have to know what a threat is (have the concept of a threat stored in your brain), knowledge that requires semantic memory. You also have to know that the particular stimulus present is an example of a threat (which also requires semantic memory). In addition, past personal experiences you've had with threats in general or with this particular threat are likely to be retrieved (which requires episodic memory). If these various medial temporal lobe representations make it into working memory, the net result is that in the presence of a threat, you begin to assemble a state of consciousness that is both noetic and autonoetic in nature. And memory processing, like sensory processing, is boosted for threats relative to neutral stimuli.[9]

WHAT'S DIFFERENT ABOUT THE WAY THE BRAIN PROCESSES THREATS VERSUS EMOTIONALLY NEUTRAL STIMULI? ENTER DEFENSIVE SURVIVAL CIRCUITS

The reason that threats produce autonomic nervous system responses, but neutral stimuli do not, is that threats activate specific circuits that control these responses. The circuits involved belong to the category of defensive survival circuits, a prime example of which is the amygdala-based defensive circuit described in Chapters 2 and 4. These circuits process threats without the need for conscious processing of the stimulus.

That defensive circuits, such as those involving the amygdala, are nonconscious threat processors has been shown in several ways. For example, as discussed earlier, amygdala activation occurs in normal, healthy subjects regardless of whether the threat stimulus is freely seen or masked.[10] In addition, threats activate the amygdala in blindsight patients who deny seeing threat stimuli presented in their area of blindness.[11] Such findings are compatible with data showing that damage to the human amygdala prevents the implicit, nonconscious threat conditioning but does not affect the ability to consciously remember being conditioned.[12] In contrast, hippocampal damage prevents the ability to consciously remember being conditioned, but not the ability to be conditioned or to respond to the CS afterward.[13]

Another line of research that is relevant here involves contingency awareness during threat conditioning. The question under consideration is whether conscious awareness of the relation (contingency) between the CS and US is necessary for threat conditioning to occur. Although studies in the past suggested that conscious awareness of this relation is required for conditioning,[14] recent work shows that conditioning can indeed occur when awareness of the relation is prevented by making the conditioned stimulus information difficult to detect.[15] Moreover, amygdala activation occurs both when subjects are aware and not aware of the contingency, but hippocampal activity occurs only when the subjects are aware of the contingency.[16] Thus, we see here implicit and explicit forms of memory in action: Implicit memory underlies conditioning itself, but explicit memory (which involves the medial temporal lobe memory system and, presumably, the prefrontal/parietal areas) is required for conscious knowledge of the contingency between the CS and US (semantic memory)[17] and for conscious awareness of having been conditioned (episodic memory).[18]

Beyond interfering with the expression of implicit (nonconscious) responses to threats, damage to the amygdala has another important effect. Recall that threats boost sensory processing in the visual cortex of humans. Amygdala damage eliminates this effect, with the result being that threats and neutral stimuli produce similar degrees of cortical activation.[19]

One could criticize this emphasis on the amygdala as being too narrow, as it is not the only region of the brain that contributes to threat processing (see Chapter 4).[20] But because its role is fairly well understood, it is an excellent focal point for examining how threats affect cortical processing and thus influence how threats are consciously experienced. At the same time, an emphasis on the role of the amygdala in threat processing should not detract from the many other functions to which the amygdala contributes.[21]

The main difference, then, between what happens when the brain processes threats versus neutral stimuli is that a defensive survival circuit involving the amydgala is activated. This then has consquences for how the cortical areas process the threat. Table 8.1 summarizes the differences in brain activation for consciously seen versus masked visual threats.

Table 8.1: Brain Activation by Seen and Masked Emotional and Neutral Stimuli				
	NEUTRAL STIMULUS		THREATENING STIMULUS	
	SEEN	MASKED	SEEN	MASKED
Visual Cortex	active	active	active	active
Frontal/Parietal Cortex	active	not active	active	not active
Amygdala	not active	not active	active	active

HOW DO THREAT STIMULI REACH THE AMYGDALA?

A key part of my argument here is that threat stimuli activate amygdala-based defensive survival circuits, which then initiate a number of responses in the brain and body that change the way the threat is further processed by the brain. These circuits ultimately make significant, albeit indirect, contributions to the conscious experience of fear. As a starting point for understanding how threat processing is affected by amygdala activation, we will first consider the routes by which sensory information reaches the amygdala. The focus will be on auditory and visual threats, because most of the research has involved these sensory modalities.

It was long thought that the main way that the amygdala is activated by sensory stimuli was via pathways originating in the late stages of cortical sensory processing.[22] But research I conducted in the mid-1980s showed that sensory stimuli do not have to engage cortical processing areas in order to activate the amygdala and thereby elicit innate defensive responses (freezing) and autonomic nervous system responses in rats.[23] Specifically, these studies revealed that the amygdala receives sensory inputs not only from late stages of cortical processing but also from subcortical sensory processing areas in the thalamus. Thus, thalamic areas that provide the primary sensory cortex with its sensory inputs also originate a shortcut to the amygdala that bypasses the cortex. The thalamic and cortical sensory inputs to the amygdala came to be known as the *low road* and the *high road*, respectively[24] (Figure 8.3). Although the two roads originate in the same general areas of the thalamus, they involve different populations of neurons, with different capacities, within those areas.[25]

The thalamic cells that project to the primary visual cortex are high-fidelity processors that make possible precise representations of features of external stimuli in the primary sensory cortex. The primary sensory cortex then connects with late (secondary and tertiary) visual areas that integrate across visual features (shape, color, motion) to construct perceptual representations of objects and events. By way of connections with frontal and parietal working memory/attention networks, and with medial temporal lobe memory networks, representations created in late areas of the visual cortex can be used in cognitive processing and in creating conscious awareness of the stimulus. The late processing areas

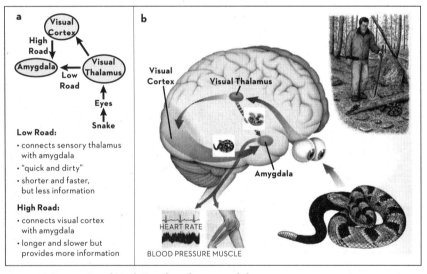

Figure 8.3: Low Road High Road to the Amygdala.

a. Sensory stimuli reach the amygdala by way of two routes. Information transmitted to the sensory thalamus is then sent to the sensory cortex, but in addition, is transmitted to the amygdala. The sensory thalamic neurons that connect with the amygdala are not part of the main system that connects the thalamus to primary sensory cortex. In the visual system, for example, the amygdala receives inputs from the collicular-pulvinar pathway rather than the geniculo-cortical pathway (see Figure 6.6). **b.** Illustration of the low road in action. A hiker is walking along and is about to step on a rattlesnake (from LeDoux [1994]). The low road can then trigger freezing before the person consciously knows, through the cortical processing of the visual stimulus and interactions between the visual cortex and the frontal and parietal areas that, together with the visual cortex, contribute to conscious visual experience (see Figure 8.2).

of the visual cortex are also the origin of the high-road connections to the amygdala.

The thalamic cells that project directly to the amygdala give rise to the low road and provide the amygdala with simple, primitive features of stimuli, such as the relative intensity, size, and speed of approach of a visual stimulus, rather than exact information about the object or event that is occurring. In the visual system, for example, the amygdala receives inputs from the collicular-pulvinar pathway, which is the origin of the "where/action" pathway that functions nonconsciously and underlies blindsight (see Chapter 6), and accounts for why blindsight patients show amygdala activity and express autonomic response threats (see above). Though the low road provides less information content, it requires fewer processing steps than the high road and is thus a faster route to the amygdala.

Compared with the high road, the low road is a "quick and dirty" route; it allows you to respond with speed rather than accuracy in a situation of danger. If you find yourself freezing to a curved shape on the ground (via thalamic inputs to the amygdala) that you then realize (via cortical processing) is a stick, the cost of this erroneous preemptive defense response is small compared with the potential cost of actually stepping on a snake. I have often written and spoken about the snake mistake, and the image in Figure 8.4 was sent to me by someone who found himself frozen by just such a curved "stick in the grass."

Figure 8.4: A Stick in the Grass.
Snake or stick? Within milliseconds, your amygdala can begin to trigger defensive responses to such a stimulus. The circuits underlying this response are shown in Figure 8.3.

IS THE AMYGDALA REALLY A NONCONSCIOUS PROCESSOR?

The low-road/high-road model is supported by studies of nonconscious processes in healthy brains,[26] blindsight patients,[27] and in patients with electrodes implanted in the amygdala as part of the treatment of epilepsy.[28] But it also became the springboard for much (perhaps too much) enthusiasm about the contribution of thalamic inputs to the amygdala in nonconscious processing—the low road came to be equated with nonconscious processing and the high road with conscious processing. It is now clear that both roads should be viewed as nonconscious inputs to the amygdala.[29] The labels "low road" and "high road" are best thought of as shorthand descriptions of sensory input routes to the amygdala from the thalamus and cortex, rather than ways of accounting for differences between nonconscious versus conscious processing of threats by the brain.

The excessive attribution of nonconscious processing to the low road to the exclusion of the high road eventually led to a backlash against the notion that nonconscious thalamic inputs might be significant in the human brain.[30] That is, because thalamic inputs had come to be viewed as nonconscious inputs, and cortical inputs as conscious inputs, the attacks on the low road became attacks against the very idea that the amygdala is a nonconscious processor.[31]

Two key pieces of evidence were used to challenge the nonconscious nature of amygdala processing.[32] First, under some conditions, the degree of amygdala activation (as measured by functional magnetic resonance imaging, or fMRI) is greater when subjects are aware of a threat. Second, amygdala activation is reduced if subjects are engaged in an attention-demanding task while processing a threat. These findings have been interpreted to mean that amygdala activity is modulated by attention and, thus, that the amygdala participates in conscious rather than nonconscious processing.[33] But there are other factors to consider.

The conclusions about attentional modulation of the amygdala are based on fMRI results, which are very crude. The fMRI technique, which estimates neural activity from oxygen use in the brain, can only measure relevant changes over the course of several seconds. This is a serious limitation, because we know from animal studies that amygdala cells respond to stimuli within milliseconds.[34] More recent studies in humans have used techniques that overcome the poor temporal resolution of fMRI and have shown that early responses elicited by threats are unaffected by attention, but later responses are.[35] The fast responses, moreover, occur in the lateral nucleus of the amygdala (LA), where the sensory inputs arrive from both the low and high roads.[36] Thus, top-down attention does not affect the quick responses in the LA. But for reasons described below, neither the early nor the late responses should be discussed in terms of top-down attentional modulation.

Why might amygdala activity under some conditions be altered during tasks that involve attention? Consider an example. When participants in experiments are asked to focus on a difficult visual discrimination task that requires considerable attention (e.g., judging whether two lines have the same angle of tilt), the amygdala is less activated by masked ("unseen") visual stimuli with emotional significance (e.g., faces showing expressions of fear or anger). Although amygdala activity is affected in such situations, this may not be for the reason that has been proposed (i.e., attention modulates the amygdala). If the water supply of New York City is shut down by a mechanical failure, the water supply to my apartment building in Brooklyn will be affected, but not because my building was specifically targeted. In other words, the amygdala may respond less to an angry face when attention is focused on something else because the normal amplification of visual processing that would occur when attention is focused on the face is lacking.[37] The absence of attentional amplification reduces activity in the visual cortex, and this weakens the signals that this cortex sends to the amygdala. The amygdala responses may therefore be affected by attention but not because attention controls the amygdala but instead because it affects activity in cortical areas that connect with the amygdala.[38]

It should also be noted that increased attentional load does not eliminate amygdala activity but only reduces it.[39] The remaining threat-triggered amygdala activity may well be driven, at least in part, by thalamic inputs.

Consciousness is an intrinsic feature of a neural network with unique information representation capacities made possible by unique patterns of connectivity. Specifically, conscious awareness of visual stimuli occurs by way of reciprocal connections between visual and prefrontal/parietal areas that enable the information to be represented in working memory, amplified by attention, broadcast, and/or entered into a higher-order representation. Consciousness is thus not passed onto the amygdala simply because it is connected, via the high road, with the late stages of the visual cortex. The high road, like the low road, is a nonconscious processing channel;[40] the amygdala is a nonconscious processor of information from both roads.

The elephant in the room of human laboratory studies of threat processing is that the stimuli used—static pictures of people looking afraid or angry, or neutral pictures paired with very weak electric shocks—are hardly threatening by real-world standards, where one's well-being or even life may be on the line, or even by the standards of animal research, where the stimuli used predict predators or electric shocks that are considerably more aversive than those used in human studies. The shocks we use in our rat research are brief and in many of our studies are given only once or a small number of times, but they are set to a level that is physically uncomfortable (as determined by behavioral and physiological responses) and uncontrollable. In human studies the participants adjust the shock themselves to a level that they find tolerable; granting subjects some degree of control over the situation further reduces the threatening nature of the stimulus so that no one in these studies feels that he is in actual danger. Fortunately, in spite of the mild nature of the threats used, human studies have nonetheless been able to confirm the basic brain circuits discovered in the animal work. Still, one has to be careful not to overstate the implications of what the use of weak aversive stimuli can tell us. Attentional load, in other words, may be less effective in reducing human amygdala activity in a real-life situation of danger. In fact, as we will see below, even when attention is focused on some task, threats have the ability to interrupt the focus of attention and redirect it to the threat. Otherwise, we could be harmed every time a threat suddenly appears.

One additional point should be made about the interpretation of the results based on methods that prevent (or reduce) conscious awareness of stimuli. Subliminal and masked stimuli are not as effective as freely seen stimuli in supporting cognitive processing (e.g., semantic processing)[41] and in driving brain activity.[42] Such results are sometimes used to argue that nonconscious processing is limited. However, these findings reveal more about the limits placed on the brain when the stimulus input is degraded by brief exposures to stimuli than they do about the limits of nonconscious processing per se. Some studies use more sophisticated procedures to prevent awareness that do not require brief inputs but still show some processing limitations; still, these, like other human studies, are constrained by the use of static CSs and weak USs.[43]

In daily life naturally occurring cues that are fully visible or audible can bias our behavior in complex ways that we have no knowledge of or conscious control over.[44] This is what the psychologist John Bargh calls "the automaticity of everyday life."[45] The biases can be as benign as effects on what you choose to eat or as insidious as the subtle way you respond to members of other races.[46] The nonconscious prowess of the brain is likely much more powerful than artificial laboratory studies using degraded stimuli can reveal.

CHANGING AMYGDALA ACTIVITY BY REAPPRAISING THREATS

Research on emotion regulation is also relevant to the question of whether top-down cognition can influence the amygdala. We know from personal experience that it is hard to intentionally control our emotions; they seem to be better at controlling us. In my earlier books I suggested that the reason for this is related to the paucity of connections from the lateral prefrontal cortex working memory circuits to the amygdala.[47] However, seminal studies from the laboratories of James Gross and Kevin Ochsner show that teaching people to reappraise emotional stimuli can reduce the subjective reports of stimulus salience, and also reduce amygdala activity.[48] For example, participants who are instructed to think about something pleasant when a negatively charged stimulus is presented rate the stimulus as less arousing. The key finding by these researchers is that areas of the lateral prefrontal cortex that have been implicated in working memory and executive control functions are involved in this form of cognitive regulation of the amygdala (Figure 8.5).

Does this mean that attention and working memory functions of the lateral prefrontal cortex directly control the amygdala? Not necessarily. The extent to which attention is a major factor in this kind of cognitive reappraisal is debated.[49] Also, the amygdala does not seem to be directly affected by the lateral prefrontal cortex, which is not surprising because there are no known connections from the former to the latter.[50] Instead, the effects observed seem to occur indirectly by way of connections from the lateral prefrontal regions to other areas. One candidate is the medial prefrontal cortex; however, Ochsner and colleagues have obtained evidence suggesting that lateral prefrontal connections to posterior regions involved in semantic processing of visual stimuli, which then connect to the amygdala, are key.[51] When the semantic meaning of a stimulus is reinterpreted from a threat to

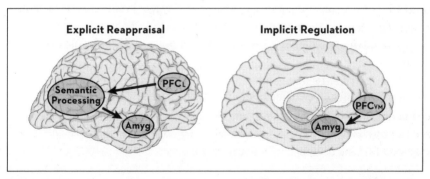

Figure 8.5: Two Forms of Cognitive Reappraisal Affect the Amygdala Differently.
(*Left*) Explicit reappraisal (using reappraisal to change self-reported emotional experience). Primarily involved are interactions between the lateral prefrontal cortex (PFCL) and the amygdala via mediation by semantic processing in the posterior neocortex. (*Right*) Implicit reappraisal (using reappraisal to change autonomic responses controlled by the amygdala). Primarily involved are interactions between the medial prefrontal cortex (PFCvm) and amygdala.

a nonthreat, the cortical signal that reaches the amygdala is weaker, and less amygdala activity occurs—as above, not because lateral prefrontal cortex and top-down attention regulate the amygdala but, instead, because the amygdala receives a weaker input from other cortical areas that are directly influenced by the lateral prefrontal cortex. Thus, executive functions don't directly change the amygdala in this form of reappraisal for the same reasons discussed above about attention and consciousness in relation to the high road.

Studies by Mauricio Delgado, Liz Phelps, me, and other colleagues took a different approach to cognitive regulation of the amygdala.[52] We were specifically concerned with whether the ability of conditioned threat stimuli to activate the amygdala and elicit autonomic nervous system responses (as opposed to self-reported emotion) could be regulated by reappraisal. In our task participants were instructed that they would sometimes see a visual stimulus and a shock would follow. They then underwent threat conditioning. In addition, they were also trained to imagine a pleasant scene from nature when the visual CS appeared. Once they were well trained in the use of the regulation strategy, they were placed in the brain scanner, reminded of the regulation strategy, and exposed to the conditioned threat stimuli. The results indicated that the ventromedial prefrontal cortex was involved in reducing amygdala activity, which in turn resulted in a decrease in the autonomic nervous system responses elicited by the CS. The same ventromedial prefrontal cortex region regulates the amygdala in another form of emotion regulation, extinction.

The cortical circuits of working memory and its executive control functions not only involve lateral prefrontal areas, which do not connect with the amygdala, but also several prefrontal cortical areas (e.g., ventromedial, anterior cingulate cortex, orbital) that do connect with the amygdala (see Chapter 6). So it is possible that via connections from medial cortical areas, top-down regulation of the amygdala might be achieved.

But let's consider the Delgado study in more detail. Emotion regulation in this experiment started out in the form of instructed, explicit cognition and involved the lateral prefrontal cortex. Once the training was established, though, the reappraisal process was automatically performed to control the amygdala-dependent ANS response. Ultimate control by the medial prefrontal cortex in this regulation strategy may therefore, like extinction, be a form of new implicit learning that enables the medial prefrontal cortex to exert control over the expression of the CS-US association stored in the amygdala and weaken the expression of autonomic responses.

In sum, the reappraisal approach used in the Gross and Ochsner studies changed self-reported conscious experience, but in the Delgado study, reappraisal was used to change autonomic responses. In both cases the reappraisal process involves explicit cognition, but the changes involved explicit control in one case and implicit control in the other.

THREATS NONCONSCIOUSLY PROCESSED BY THE AMYGDALA DIRECTLY INFLUENCE CORTICAL PROCESSING AND GRAB ATTENTION

So far we've looked at how attention affects, or does not affect, threat processing. Here we turn to the flip side of this issue—how threat processing grabs attention. You have probably noticed that if something significant happens when you are focused on some task, your attention can shift off the task and to the new event. This alone suggests that attention isn't necessary for threat processing; otherwise, threat processing would always require attention first. Threats instead capture attention and direct it to the threat. This realization led Herbert Simon, a pioneer in cognitive science and a Nobel Prize winner in economics, to propose in the 1960s that an efficient cognitive system must not only be able to focus attention on specific tasks, but must also have an interrupt mechanism, a means of redirecting attention when a high-priority event arises unexpectedly.[53]

I became aware of Simon's notion in the late 1990s, when I was working with computational modelers, scientists who tested psychological and brain theories with computer simulations.[54] One of the ideas that came out of this collaboration was that the rapid amygdala activation that takes place outside of awareness could be a means by which attention is shifted to dangerous events that suddenly occur.[55] Such a process may be at work in Michael Eysenck's *attention control theory of anxiety*, which distinguishes two attention systems for processing threats: one that is goal directed, and one that is stimulus driven.[56] When the goal-directed system is busy working on some task, the stimulus-driven system can still detect threats and redirect attention in a nonconscious, bottom-up way.

Numerous laboratory studies confirm the intuition that threats or other "emotional" stimuli can seize attention.[57] This body of work paved the way for studies of the brain mechanisms involved. Not surprisingly, the amygdala has been implicated in the allocation of attention, including the involuntary redirection of attention.[58] For example, if two stimuli are presented in rapid succession, the first one is noticed but the second tends not to be,[59] a phenomenon known as *attentional blink*. It's as if the first stimulus leads to a mental blink during which the second stimulus is not registered. But if the second stimulus is a threat, it has a good chance of overriding the blink effect and being seen.[60] In patients with amygdala damage, this override does not take place, suggesting that threat-driven amygdala activity is normally responsible for the ability of threats to persist and invade consciousness.[61]

How exactly does nonconscious threat processing by the amygdala exert bottom-up influences on cortical processing? When the lateral amygdala (LA) detects that a threat (innate or conditioned) is present, it sends signals to other areas of the amygdala to control behavioral and physiological adjustments in the brain and body (see Chapter 4). One of the targets of the LA is the basal amygdala (BA), which is particularly well connected with cortical areas, including areas of frontal and parietal attention networks.[62] Thus, via outputs of the BA to prefrontal and parietal areas, circuits that exert top-down attentional control over sensory processing can be influenced (Figure 8.6). In particular,

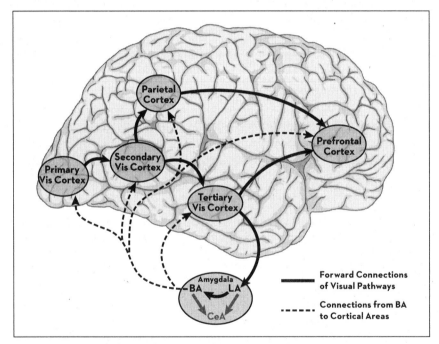

Figure 8.6: Direct Effects of the Amygdala on Cortical Processing.

Using visual processing as an example, the figure illustrates how the amygdala primarily receives visual inputs from the tertiary stage of processing but sends connections back to the earliest stages of cortical visual processing. Within the amygdala, the lateral nucleus (LA) is the main recipient of visual inputs. It then connects with the basal amygdala (BA), which is the origin of most of the connections to sensory and other cortical areas. While only the visual cortex is shown, auditory and somtosensory cortical systems have a similar arrangement (taste and olfaction are different). The main connections from the amygdala to the nonsensory areas shown include the prefrontal cortex and parietal cortex because of their prominent role in the ideas developed here. The prefrontal cortex in this case refers to both lateral and medial areas.

bottom-up amygdala influences on cortical attention networks can disrupt top-down attention focused on nonthreatening targets, allowing redirection of attention toward threats.

The BA also connects directly with sensory areas of the cortex and can directly affect processing there as well. Although only the late stages of visual processing send inputs to the LA (this is the high road), the BA connects back to all areas of the visual cortex, including the earliest stages of processing (the primary visual cortex).[63] This anatomical fact is consistent with the observation that threat stimuli activate both primary and secondary areas more strongly than do neutral stimuli and also with studies showing that the processing of very primitive visual features (such as differences in color or brightness) that depend on the primary visual cortex is enhanced for threat stimulus relative to neutral stimuli.[64] Thus, once the LA is activated by a threat, the BA can begin to influence all aspects of cortical visual processing.

Another way that threats may alter attention is by engaging areas of the medial prefrontal cortex involved in amygdala regulation.[65] Earlier I discussed how activity in these areas might dampen amygdala activity to reduce unwanted responses to threats. But in addition, the medial prefrontal cortex may be able to elevate amygdala output activity and thereby boost cortical processing of threats.[66] This would facilitate reentrant processing between CCNs involving prefrontal/parietal areas and sensory processing and enhance conscious awareness of the threat.

NONCONSCIOUS PROCESSING BY THE AMYGDALA ALSO INFLUENCES ATTENTION AND SENSORY PROCESSING INDIRECTLY BY CHANGING AROUSAL LEVELS

Another very significant consequence of threat detection by the amygdala is an increase in overall brain arousal. Creature consciousness depends in part on arousal, as do alertness, attention, and vigilance (sustained attention).

Threats are very effective at raising arousal levels globally in the brain.[67] This effect, sometimes called *generalized arousal*,[68] is achieved by populations of neurons that make and release the neuromodulators (e.g., norepinephrine, dopamine, serotonin, acetylcholine, orexins, and other chemicals).[69] Though the cell bodies of these neurons are restricted to specific brain areas, their axons are widely distributed, enabling them to exert broad influences on brain activity (see Chapter 4, and especially Figure 4.4).

Although the term "generalized arousal" describes the global nature of the

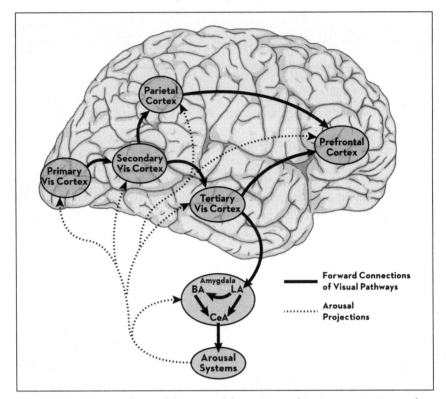

Figure 8.7: Indirect Effects of the Amygdala on Cortical Processing via Arousal.
Sensory information transmitted to the lateral amygdala (LA) reaches the central amygdala (CeA) directly and via intra-amygdala connections in the basal amygdala (BA) and other areas not shown (such as the intercalated nuclei). We have discussed the role of the CeA in controlling defensive behavior and supporting physiological responses of the body (not shown in this diagram). Another important set of outputs of the CeA is to arousal systems. These contain neurons that manufacture various neuromodulators (such as norepinephrine, serotonin, dopamine, acetylcholine, and others) and send axons widely throughout the brain (only connections to key areas being discussed are shown). When activated by the CeA, the arousal systems release their chemicals from their axons and change information processing in those areas.

effects of neuromodulators on the brain, the behavioral and cognitive consequences ultimately reflect the local consequences of modulator chemicals on information processing in specific circuits when cells with receptors for a specific modulator bind the chemical. For example, the binding of neuromodulators to their specific receptors facilitates processing between the sensory thalamus and sensory cortex, within the sensory cortex, in the prefrontal and parietal cortices, in areas of the amygdala, in the hippocampus, and in many other brain regions.[70]

The local specificity of modulatory action is illustrated by results showing that whereas the modulator acetylcholine affects sensory processing and attention in cortical areas, arousal-induced changes in sensory processing areas of the cortex can be achieved without changing processing in executive control areas responsible for attention.[71] Thus, in addition to influencing attention and thus conscious sensory processing, neuromodulators can also separately affect non-conscious sensory processing, which can influence other nonconscious processes that occur outside the influence of attention and consciousness.

Neuromodulators are most effective in modulating the activity of neurons that are already active,[72] which accounts for how their nonspecific release can selectively affect particular neurons. Thus, neurons that are actively processing a visual stimulus or that are controlling attention to visual stimuli will be affected, but neurons that are not engaged will be less affected, or even unaffected.

A major way that threats change arousal is via outputs of the CeA to neuro-modulatory systems (Figure 8.7).[73] (By the way, the amygdala also processes appetitive stimuli and the CeA also activates neuromodulatory systems in their presence[74]). The consequence of CeA activation of neuromodulatory systems is an increase in attention and vigilance, which may be achieved by lowering the threshold to detect sensory stimuli.

The opportunities for arousal to influence information processing in the brain are manifold. The fact that the amygdala itself is a recipient of neuromodu-latory inputs means that its processing is also boosted during arousal. As the amygdala drives arousal and arousal in turn drives the amygdala, a self-sustaining reentrant loop is engaged that helps keep the brain and body revved up as long as the threat remains.[75]

The amygdala thus facilitates sensory processing in two complementary ways. Sensory and working memory cortical networks, including attentional networks, are directly influenced by BA outputs and are also affected by neuromodulators released by way of outputs of the CeA. The net effect is that connections between sensory and working memory/attention networks are doubly facilitated, which sharpens and sustains focus on the threat, causing the threat to stand out in conscious awareness relative to neutral stimuli that are also present. Thus, cortical reentrant processing occurs above and beyond that achieved in perceptual states of awareness that lack emotional significance (Figure 8.8).

This helps account for why threat detection results in risk assessment and heightened sensitivity to the environment. If your brain has detected a potential source of harm, and arousal has been triggered, through focused attention you begin to monitor the environment in search of other possible harmful things. The mechanisms involved are similar to those mentioned above, in which the amygdala

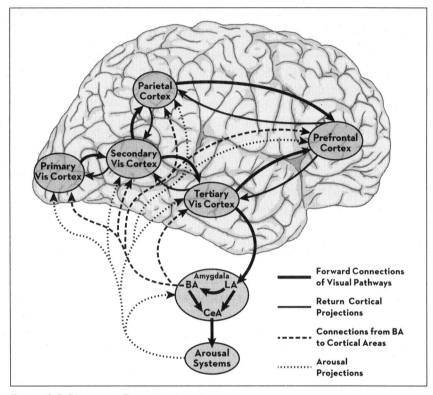

Figure 8.8: Recurrent Perpetuation.

Reciprocal connections between brain areas (or more precisely, between neurons in those areas) create processing loops. Recurrent activity in these loops is believed to give rise to the amplification of processing and the emergence of conscious experience. While several levels of processing loops are illustrated here, note two points. First, once the amygdala is activated by a sensory stimulus, it can, via basal amygdala (BA) outputs, not only influence ongoing sensory processing but also influence cortical areas involved in attention and working memory that also are reciprocally connected with sensory processing areas. Second, via the central nucleus (CeA), in addition to initiating defensive responses and supporting physiological changes in the body (not shown here), arousal systems within the brain are activated and release their chemicals and modulate processing in all of the areas mentioned above, but also release in the amygdala and modulate its processing. Thus, so long as the threat persists, a massive system of feed-back and feed-forward amplification of processing occurs via multiple layers or recurrent connectivity, keeping the organism engaged and energized to cope with the threat.

exerts bottom-up influences on sensory processing and attention; once attention is captured, top-down executive attention biases sensory processing. People with anxiety disorders have this in the extreme.[76] Through arousal and reentrant processing, their brains are on high alert. They are hyperaroused, thus hyperattentive to threats, and hypervigilant even when threats are not present. What might be a safety signal for most people can be perceived as a warning of danger for them.

PROCESSING OF UNCERTAIN THREATS INVOLVES THE BED NUCLEUS

By virtue of extensive connectivity with sensory processing areas, the amygdala-based defensive circuits described above are especially well suited to control behavioral and physiological responses to immediately present stimuli. These circuits are thus particularly relevant to understanding nonconscious processes that contribute to feelings of fear. But what about feelings of anxiety, which occur when we are worried about threats that are not present and may never occur? As we saw in earlier chapters, the bed nucleus of the stria terminalis (BNST) comple-ments the amygdala in the control of behavior in situations in which threats are possible but uncertain. The BNST, unlike the amygdala, is not well connected to sensory systems but instead has strong inputs from the prefrontal cortex and hippocampus, as well as from the amygdala (see Chapter 4). In short, the BNST can be triggered by the cognitive representation of events, including the ability to predict and worry about events that your future self might experience.

Many situations of danger may involve both actual threats and uncertainty. When this occurs, the amygdala and BNST may well both be engaged; the amyg-dala to respond to the certain threat and the BNST to initiate risk assessment of the less certain elements. When novelty or uncertainty is primarily involved, then the BNST does most of the work.

NOTICING THE UNNOTICEABLE

Defensive motivational states are nonconscious, a-noetic states. We cannot sim-ply turn our attention to defensive survival circuits and come to know exactly what they are doing. But we can come to know about defensive survival circuits and their accompanying motivational states indirectly by monitoring their observable consequences.

When split-brain patients fabricate verbal (left hemisphere based) explana-tions for behaviors that were produced by the right hemisphere, the left hemi-sphere is generating explanations of behaviors produced by nonconscious systems and does so in the maintenance of a sense of self. That is, our behavior is an important way we come to know who we are. This is the essence of Gazzaniga's interpreter theory of consciousness (see Chapter 6).

The most obvious way that defensive motivational states make themselves known to us is, in fact, through our own behavior. The ability to observe one's behavior and thus create representations of behavior in working memory is called *monitoring*.[77] By directing our attention to our behavioral output, we can acquire information about what we are doing and intentionally adjust our behavior in light of thoughts, memories, and feelings. As an executive function of working memory, monitoring, not surprisingly, involves circuits in the prefrontal cortex.[78] We use observations of our own behavior to regulate how we act in social situations.[79] If you become aware that your behavior is negatively affecting others, you can make adjustments as a social situation evolves. Or if you notice you are acting in a biased way toward some group, you can make corrections. In addition, through monitoring one can observe undesirable habits and seek to change these through therapy or other means. Not everyone is equally adept at using monitoring to improve self-awareness. The field of emotional intelligence is all about how people differ in such abilities and how one can be trained to do better.[80]

We can also monitor signals coming from within our body. This is, again, the foundation of Damasio's somatic marker hypothesis (see Chapter 5). The most specific of these are signals from the somatosensory system, which transmits information about touch, temperature, irritation, and pain from skin and muscles to cortical processing areas, much like the visual or auditory systems do. When you have a headache, backache, sore muscles, an itch, feel warmth or cool air on your skin, or are feverish, you become aware of somatosensory information being processed in cortical areas.

We can also notice some signals from our inner organs, many of which have nerves that transmit signals to the brain for processing. For example, you notice when your bladder is full, your stomach is empty or burning from acid indigestion, or your heart is racing. Most other signals that come from the body, however, especially those from the inner organs, are more amorphous, more difficult to pinpoint. We know little about the state of our gallbladder, appendix, pancreas, liver, kidneys, and most other organs unless they malfunction and result in pain or other unexpected consequences. However, because sensory nerves in these organs do send messages to the brain, they could have some indirect and implicit influences on perception, attention, memory, and emotion. Also, some hormones released by various organs in the body can affect consciousness indirectly by binding to receptors in many areas of the brain, including the amygdala and all other regions that control defensive motivational states, as well as cognitive processing areas of the cortex that contribute to sensory processing, attention, long-term memory (semantic and episodic), and working memory.

Brain arousal is also noticeable, at least to some degree, if only indirectly. When arousal is high, we feel alert, energetic, and vigilant. When arousal is low, we are sluggish and inattentive. Drugs like amphetamines raise brain arousal levels and artificially increase alertness and the ability to concentrate by mimicking the arousing effects of neuromodulators.[81] Although sudden changes in arousal can be detected, the informational content associated with these is weak: Arousal can tell us something important has happened but not what it is or what it means. In the absence of other disambiguating information, we turn to our outer senses to monitor our behavior and our surroundings.[82] Arousal also plays an important role in most contemporary theories of how we experience emotions.[83]

Thus, through monitoring, the various components of a defensive motivational state can, via their consequences, affect conscious experience. Monitoring is a terrific work-around of our inability to be able to directly access the nonconscious brain, but it is not risk free. The interpretation of the nonconscious sources and motivations underlying body responses might or might not be accurate.[84] When we are monitoring vague, imprecise physiological signals in the brain or body that are not expressed in behavior, the opportunity for misattribution of the motivational significance of the signals is high. When people make decisions based on "gut feelings," they are using nonconscious processing, but the fact that we do make many decisions this way does not mean we should intentionally avoid conscious decision making. Gut feelings can sometimes be useful but can also cause problems (see Chapter 3) and should not, out of mental laziness, be relied on as a routine mode of decision making.

REMEMBERING WHAT IT IS LIKE TO FEEL FEAR OR ANXIETY

In a previous chapter I argued that we are not innately endowed with feelings. A similar point was made in a very interesting article titled "The Language of Feeling and the Feeling of Anxiety."[85] In it the authors note: "It is unlikely that we are born knowing what our feelings are any more than we are born knowing the world and reality. Instead, we learn what our feelings are and this learning has a social-verbal experiential basis." In other words, as you go through life you learn what words like "fear" and "anxiety" mean, and you attach these words to experiences that you monitor in your brain and body. When a child has been in a situation of danger, a parent may tell her, "You must have been really afraid." Or if the child seems nervous about being in a school play, comfort from parents may come in the form of, "Don't be worried, it's natural to feel a little anxious when people are watching you." The child also hears others talking about fear or anxiety and sees examples of these states played out on TV and in the movies. Children's movies these days often involve quests in which characters, often animals, have to face many fear-arousing challenges and are anxious for prolonged periods until they finally reach their des-tination, and their fear and anxiety are replaced with jubilation. As the psychologist Michael Lewis points out, children act afraid and anxious long before they can feel these emotions.[86] Words like "fear" and "anxiety" have established relations with propositions such as "I am afraid of X" or "I am anxious about X." Once one cogni-tively learns these, it is easy to know *what it means* to be afraid of or anxious about X without necessarily fully understanding *what it is like* to feel fear or anxiety.[87]

Children thus build up a catalog of what canonical examples of different emo-tions look like in others and feel like in themselves. The Swiss developmental psy-chologist Jean Piaget used the term "schema" to describe an organized collection of information about a topic or situation that children acquire and then use in thought and action.[88] Emotional schemas are stored in semantic and episodic memory as emotional concepts.[89] These schemas are used to categorize situations as threaten-ing or safe. In threatening situations, signals in brain and body are monitored, as is one's behavior. If a match occurs (through a process called *pattern recognition*) between a stored schema and the present situation and/or state, the present condi-tion is cognitively conceptualized (interpreted) as the schematized emotion and labeled with the emotion word assigned to that state. Sometimes only some of the components of the emotional state will be present, in which case *pattern comple-tion* is called upon. As one becomes more emotionally experienced, the states become more differentiated. Fright comes to be distinguished from startle, panic, and terror, and dread distinguished from concern, wariness, and edginess. Because the labeling process is imprecise and depends on individual learning and interpre-tation, each person may use the terms a little differently.

Psychologists Lisa Barrett and James Russell refer to these schematic and interpretive processes underlying emotional experiences as *conceptual acts* that contribute to the psychological construction of emotions.[90] As Assaf Kron and colleagues have noted, "Feelings don't come easy."[91] They don't simply happen; they require considerable mental work.

Although the memory of monitoring helps you label experiences, the words we use as labels are not the states. They are the effort of consciousness to classify and make sense of what is being experienced and enable one to report on the experience if questioned about it. So how, then, does the experience of emotion itself come about?

FEELING IT

It's useful to think about how emotional feelings emerge in consciousness by way of analogy with the way the flavor of a soup is the product of its ingredients.[92] For example, salt, pepper, garlic, and water are common ingredients that go into a chicken soup. The amount of salt and pepper added can intensify the taste of the soup without radically changing its nature. You can add other ingredients, like celery, green peppers, and parsley, and have a variant of a chicken soup. Add roux and it becomes gumbo, whereas curry paste pushes it in a different direction. Substitute shrimp for chicken, and the character again changes. None of these individual items are soup ingredients per se: They are things that exist independent of soup and that would exist if a soup had never been made.

The idea that emotions are psychologically constructed states is related to Claude Levi-Strauss's notion of "bricolage."[93] This is the French word referring to something put together (constructed) from items that happen to be available. Levi-Strauss emphasized the importance of the individual, the "bricoleur," and his social context, in the construction process. Building on this idea, Shirley Prendergast and Simon Forrest note that "maybe persons, objects, contexts, the sequence and fabric of everyday life are the medium through which emotions come into being, day to day, a kind of emotional bricolage."[94] In the brain, working memory can be thought of as the "bricoleur," and the content of emotional consciousness resulting from the construction process as the bricolage.

Similarly, fear, anxiety, and other emotions arise from intrinsically nonemotional ingredients, things that exist in the brain for other reasons but that create feelings when they coalesce in consciousness. The pot in which the ingredients of conscious feelings are cooked is working memory (Figure 8.9). Different ingredients, or varying amounts of the same ingredients, account for differences between fear and anxiety, and for variations within each category. Although my soup analogy is new, I've been promoting the basic idea that conscious feelings are assembled from nonemotional ingredients for quite some time.[95]

FEELING FEARFUL

Typically, you are afraid of something that is present. It is the awareness of this thing that ultimately, with the aid of other ingredients, gives rise to the feeling of fear. So the first ingredient of a fearful experience is a representation of a particular sensory object or event in the brain.

The second ingredient is defensive survival circuit activation, by way of thalamic and cortical inputs. This initiates the expression of defense responses and supporting physiological changes.

The third ingredient is attention/working memory. In order to consciously know that a stimulus is present, you have to attend to it. Attention delivers the stimulus to working memory. This involves interactions between sensory areas of the cortex and prefrontal and parietal circuits.

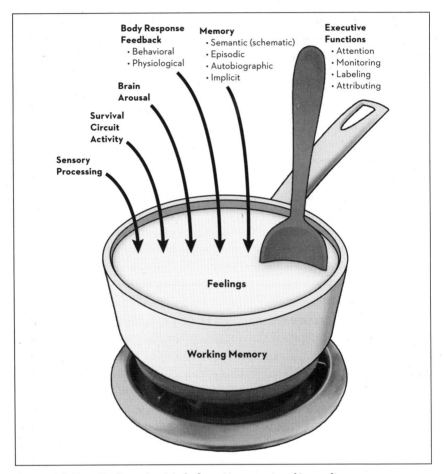

Figure 8.9: How Feelings Are Made from Nonemotional Ingredients.

The way conscious emotional feelings emerge from the brain can be thought of by way of analogy with the way the flavor of a soup emerges from nonsoup ingredients. In the case of chicken soup, water, onions, garlic, carrots, chicken, salt, pepper, bay leaves, parsley, and other ingredients are not soup ingredients per se. They are just things in nature that can be used to make soup. Change the ingredients and the flavor of the soup changes, either subtly (adding salt and pepper can intensify the flavor) or qualitatively (substituting fish for chicken or adding roux or curry paste). Emotions are like this in that their characteristic quality (the way they feel) is determined by the combination of nonemotional ingredients. Some of the nonemotional ingredients that combine to make emotions are sensory processing, survival circuit activity, brain arousal, body feedback, and memory. Working memory is the pot in which emotions are cooked. While the character of the emotion can be greatly influenced in a bottom-up fashion (i.e., by the survival circuit that is active and the degree of arousal that occurs), executive functions can contribute to the particular mixture of ingredients that enter working memory from the sensory systems and from long-term memory, and to the interpretation of the experience and the labeling of its quality (worry, concern, dread, anxiety, fear, panic, terror, horror).

The fourth ingredient is semantic memory, which enables you to apprehend what the particular thing is (object recognition) and to distinguish it from other objects (discrimination). This kind of information makes it possible for you to consciously know that a stimulus of the type present now is potentially harmful (this is a snake, some snakes are venomous, venomous snakes can kill you). The integration of semantic memory with stimulus information involves interactions between cortical sensory processing circuits, prefrontal and parietal attention/working memory circuits, and medial temporal lobe semantic memory circuits. The merging, by way of executive control functions such as attention, of sensory information with semantic knowledge results in a kind of factual consciousness about the stimulus—what Tulving calls noetic consciousness.

The next ingredient is episodic memory. Episodic memories are about YOU; they involve your past but are also the foundation for predictions about your future self. Your conscious experience of an episodic memory is, in Tulving's terms, an autonoetic state of consciousness. This involves interactions between sensory cortex circuits, medial temporal lobe episodic memory circuits, and prefrontal and parietal circuits that create a contemporaneous representation in working memory involving the stimulus (what), its location and yours (where), and its time of occurrence (when). In addition, working memory must represent the fact that all of this is happening to YOU—the self has to be involved. Although these are the basic ingredients of an autonoetic experience, they are not sufficient to constitute an autonoetic state in which you feel fear or anxiety.

The activation of defense circuits not only triggers body responses but also changes the way cortical circuits process information. Amygdala outputs trigger an increase in brain arousal, potentiating processing by active neurons throughout the brain. Amygdala outputs to cortical areas grab attention and boost sensory processing of the threat via reentrant processing loops. Connections from the amygdala to long-term memory and working memory and attention circuits are also reentrantly engaged, facilitating retrieval of semantic and episodic memories of the threat. The fact that the amygdala itself is a target of modulators means that it drives all of these processes more strongly. Although the defensive motivational state that results is a nonconscious (a-noetic) state, it, or components of it, can come to be part of the conscious experience of fear when you monitor, through executive capacities, noticeable consequences of the state.

You are now in a state that is very close to being a conscious state of fear. It might even be a primitive, poorly differentiated one. But one more step is needed. By way of past personal experiences with danger, you have learned what ingredients contribute to fear. As monitorable aspects of the state build up in consciousness, you begin to recognize, via memory and pattern completion, that ingredients that are present are the indicators of fear, and you label and categorize the state. The cognitive templates, schemas, and conceptual acts of constructivist theories proposed by Barrett, Russell, Clore, Ortony, and others do the work.[96] Whether you feel concern, alarm, fright, panic, or terror depends on the particular blend of ingredients and the way they are cognitively interpreted via stored schemas. Fear and other emotions are based on assumptions, presuppositions, and expectations; they are constructed in the brain from nonemotional ingredients.[97]

What distinguishes the various kinds of fears is the combination and amount of the raw materials involved. *What ties together all instances of fear* is the awareness that a threat to well-being is present or is soon very likely to occur.

In short, in order to be felt as fear, components of a nonconscious defensive motivational state have to invade and become a presence[98] in conscious awareness. This can only happen in organisms that have the capacity to both be aware of brain representations of internal and external events and to know in a personal, autobiographical sense that the event is happening to them—someone has to be home in the brain in order to feel fear when the defensive state knocks on the door.[99]

Defensive circuits organized in subcortical brain areas mature earlier than cortical circuits involved in cognition and language. This is why, as I noted above, infants can react "emotionally" long before they can actually feel emotion,[100] and why adult humans and nonhuman animals can react "emotionally" without necessarily feeling emotional. Unless the brain has the wherewithal to be conscious of its own activities, noetic states of fear cannot exist. In the absence of the ability of the brain to apprehend that the event is happening to itself, there can be no autonoetic experiences of fear. Autonoetic fear is likely very limited in the animal kingdom, and may, as Tulving argued, be exclusive to humans. Although noetic fear could, in principle, be experienced by other animals, measurement problems have so far precluded a clear demonstration of noetic consciousness, even in nonhuman primates. The challenges will be even greater in other species with brains in which the cerebral cortex differs even more from ours.

The human brain's compulsion to organize experiences into categories is unique and depends, in part, on natural language as well as other cognitive capacities. As pointed out earlier, we have English words to distinguish more than three dozen variants of fear-related kinds of experience.[101] The idea that language and culture shape experience,[102] including emotional experience, is currently thriving in psychology.[103] Culture and experience, for example, contribute to how the amygdala responds to threats.[104] Whatever limited self-awareness is possible without language, it is clear that its presence changes the self-and-consciousness game in the brain (see the discussion of animal consciousness in Chapters 6 and 7). Among the things that language does for the human brain is enable the symbolic representation of experiences of fear or anxiety without the actual exposure to the stimuli that normally elicit these emotions.[105] In certain circumstances, this helps keep us safe but can also become a vehicle for excessive rumination and can have unwanted consequences in the form of debilitating anxiety.

I have emphasized here how fearful feelings related to defensive circuits arise, which makes it easy to slip back into the idea that the defensive survival circuit is a fear circuit. But the fact is, we can feel fear in response to activity in other survival circuits (fear of starving, dehydrating, or freezing to death). I call these "fears" because they are triggered by specific stimuli and interpreted in terms of schemas related to danger and harm to well-being. But as soon as you interpret these signals as sources of harm, you begin to contemplate, via autonoetic consciousness, what the consequences might be, and at that point anxiety schemas are activated, and worry takes over.

FEELING ANXIOUS

Like fear, anxiety can be initiated from the outside, as when stimuli reliably predict that harm will occur but do not predict exactly when it will take place. Anxiety can also be triggered by stimuli that are only weakly associated with danger, so harm may or may not actually follow. Novel situations in which one does not know exactly what to expect are also triggers of anxiety. In addition, anxiety can develop when particular memories or thoughts, independent of external events, lead to worry. In each of these cases, once the stimulus, situation, thought, or memory is attended to and enters working memory, it is interpreted in light of cognitive templates and schemas based on semantic and episodic memory and gives rise to noetic and autonoetic states of consciousness that include implications for the self. During this process the representations of stimuli and situations and of thoughts and memories, by way of cortical connections to the amygdala and/or bed nucleus of the stria terminalis (see Chapter 4), activate defensive survival circuits, leading to arousal and other physiological consequences in the brain and body and other aspects of defensive motivational states that sustain attention to the anxiety-provoking stimulus, thought, or memory.

Like fear, anxiety often involves defensive circuit activation. But as we've just seen, anxiety can arise from the consequences of activity in other survival circuits. Anxiety, again like fear, is not directly the result of the activation of a survival circuit. It is a cognitive interpretation that sometimes, but not always, depends on survival circuit activity in generating autonoetic conscious feelings. Existential anxieties, such as worries about leading a more meaningful life or of the eventuality of death, are not dependent on survival circuits but are lofty concerns that live in autonoetic consciousness. They may have an indirect impact on survival circuit activity but primarily involve abstract concepts related to choices and potential consequences in anticipated future situations, all centered on the conscious self.

WRAPPING UP

The psychotherapist Mark Epstein says that trauma is an indivisible part of human life.[106] It lacks logic, but it connects the person to the world at a fundamental level. Looking at trauma through the lens of fear and anxiety, its lack of logic can be seen as a consequence of the storage of memory of the event in part through implicit systems that cannot be accessed by consciousness and its linguistic analytic tools, and thus cannot be directly monitored. And the fundamental connection to life Epstein refers to can be seen as arising from the storage of these implicit memories in universal survival circuits that exist to keep the organism alive and well, and that contribute, albeit indirectly, to the assemblage of feelings in each human brain. This nonconscious connection to life is, at a primitive level, a major factor that connects us to all members of our species, and to other species. It makes it possible to relate to people and animals in ways that defy words and logic. We are empathic and anthropomorphic, not necessarily because we share feelings with others, but because nonconscious interactions between our a-noetic brains. Under such conditions, feelings are fostered in us that we naturally presume exists in others, including other species. As I noted before, just because something is natural (genetically encoded) does not mean it is scientifically correct.

Feelings like fear and anxiety are icing on the evolutionary cake, a layer added long after the cake was baked.[107] Once you have a mental schema of a certain kind of cake—say, devil's food cake with vanilla icing—it's difficult to imagine an example of that kind of cake without also thinking about the icing. Similarly, once you feel fear when responding to danger, it's hard to imagine that other responses that occur at the same time are not causally tied to the feeling and also hard to imagine that other animals might respond to danger without this same kind of feeling.

Fear and anxiety are not biologically wired. They do not erupt from a brain circuit in a prepackaged way as a fully formed conscious experience. They are a consequence of the cognitive processing of nonemotional ingredients. They come about in the brain the same way any other conscious experience comes about but have ingredients that nonemotional experiences lack.[108]

CHAPTER 9

FORTY MILLION
ANXIOUS BRAINS

"We live only a few conscious decades, and we fret ourselves enough
for several lifetimes."

—CHRISTOPHER HITCHENS[1]

The biological building blocks we share with bacteria, plants, sponges, worms, bees, fish, frogs, dinosaurs, rats, mice, cats, monkeys, and chimps make it possible for us to survive, but we are ultimately more than mere survival machines. Although we live *in* the present, we humans live *for* the future, a mental quality that is rare, and maybe unique, in the animal kingdom. This takes a special kind of brain, one that can be self-aware and aware of the relation of self to time. Autonoetic consciousness is our blessing and curse. It enables us to strive to achieve, but also to worry that we will fail.

Restating Kierkegaard, the cultural historian Louis Menand noted, "Anxiety is the price tag on human freedom."[2] Kierkegaard believed that we are free to choose our future actions, and this defines who we are.[3] But modern science has concluded that our freedom is often more illusory than we think.[4] Regardless of how free our will really is, though, the fact that we believe it is free makes us anxious when we perceive that we do not actually have control, when we face risky options in situations of uncertainty, or when we ruminate over how the present and future might be different if we had acted differently in the past.

Though anxiety is fundamental to human nature, for some people it has debilitating consequences. According to the National Institute of Mental Health, forty million people in the United States alone have some form of an anxiety disorder.

One could quibble over whether every person with a diagnosis of anxiety qualifies as having a disorder as opposed to just a greater amount of anxiety than he or she cares to have.[5] But just as there are surely people who may not warrant the "disorder" label, there are likely others who are effectively incapacitated but not officially diagnosed. Also, as noted in Chapter 1, because generalized anxiety is considered an impairing cocondition in most other psychiatric disorders, including depression, schizophrenia, autism, and others, and occurs in people with serious, and sometimes not so serious, medical problems, the reach of anxiety extends far beyond the 40 million people with a specifically diagnosed anxiety disorder.

DEFINING ANXIETY SCIENTIFICALLY

The traditional view of anxiety handed down from Freud and Kierkegaard is that it is an unpleasant conscious experience. Indeed, we each recognize anxiety by the way it makes us feel personally, and people typically seek help for anxiety because they want to feel better. The psychologist Richard McNally, a leading anxiety researcher, highlights the importance of conscious experience in anxiety, stating that the notion of "unconscious anxiety" is oxymoronic.[6] At the same time, because feelings are private and not easily measured in people, and even more difficult to study rigorously in animals, scientists have sought ways to redefine fear and anxiety in nonconscious (nonfeeling) terms to make the condition more tractable as a research topic. The central motive state was a solution for dealing with animals. Another approach became popular in human research.

In the 1960s anxiety was still typically thought of in Freudian terms by clinicians—as a conscious feeling with hidden causes in the unconscious. But the behaviorist movement had suggested that fear and anxiety problems could be treated using ideas derived from learning theory, such as principles underlying extinction (see Chapter 10). Peter Lang, a young researcher interested in this approach, recognized the need to have objective measures to validate whether such treatments were effective. He argued that emotions such as fear and anxiety could be assessed using three response domains: (1) *language behavior* (what people say about their situation); (2) *behavioral acts* (such as escape and avoidance); and (3) *physiological reactions* (including changes in blood pressure, heart rate, sweating, and muscle tension; startle responses; later he added physiological changes in the brain, such as increased arousal). Successful treatment, he proposed, would result in significant and persistent changes in all three response systems.[7] Lang thus sought a way to retreat from an exclusive focus on the "hidden phenomenology" of fear and anxiety that had been the main concern of psychoanalysis and instead focus on objectively measureable responses.[8]

A word about language behavior as one of Lang's three response systems. Behaviorists considered thought a covert form of behavior and viewed speech (verbal behavior) as a way to measure this covert behavior. Language behavior, according to B. F. Skinner, is subject to the same rules of reinforcement as any other behavior.[9] Behaviorists have even said that the most powerful way to change behavior is through language.[10] By including language behavior as a response system in his fear and anxiety model, Lang was not using language as a means of obtaining a verbal report of "hidden phenomenology" but instead as objective behavior.

Lang's three-response model not only redefined anxiety in objective terms for research purposes but also influenced therapy, suggesting that the response systems themselves be targeted rather than the ineffable phenomenology inside a patient's head. But in solving one problem, this approach created new ones.[11] In particular, it continued the behaviorist trend of marginalizing what most people conceived of as the essence of anxiety—namely, its subjective (hidden) phenomenology—the way it feels.

Interestingly, studies by Lang found that behavioral and physiological measures of fear and anxiety were often discordant with language behavior.[12] In therapeutic settings, people might improve behaviorally (a claustrophobic patient might be able to take the subway) or physiologically (a spider phobic may show less arousal when exposed to spider pictures) but may state that they still feel anxious and worry about their condition.[13] In fact, laboratory studies in which multiple measures of each response system are taken have found discordance among different behavioral responses or different physiological measures, whereas the verbal measures are more consistent.[14] On the basis of this observation, Stanley Rachman, another influential figure in the field of human anxiety, concluded that "verbal report is definitional and essential."[15]

When Lang first offered his proposal, consciousness still had a bad scientific reputation due to the lingering effects of behaviorism—most cognitive scientists were more interested in how information processing works than in how conscious experiences come about. Verbal behavior was, as noted, viewed in behaviorist terms. But, as discussed in previous chapters, the field is now more comfortable with the idea that verbal self-reports are windows into conscious experience. Self-report is often used in research on consciousness in general,[16] as well as in studies of fear and anxiety in particular,[17] including fear and anxiety disorders.[18] Again, as Rachman noted, self-report is "definitional and essential." Thus, feelings, being conscious experiences, can be studied experimentally using information-processing approaches such as those that explore the role of attention and working memory in consciousness (Lang, in fact, turned to an information-processing perspective in his later work[19]).

We also know that the discordance between what one says about one's feelings and how one's body is reacting in the face of a threat is a natural consequence of brain organization. Bodily responses are products of survival circuits that operate nonconsciously, and working memory, which is crucial to self-reports about consciousness, does not have direct, inside-the-brain access to the implicit systems that control these responses.[20] Working memory acquires information about these states indirectly, by monitoring their noticeable consequences.

A verbal report about the way one feels is thus not just another measure. The essence of an emotion like anxiety is, as Freud said, "that we should feel it" (see Chapter 1) and conscious states are best assessed via verbal reports. As I've argued throughout this book, nonreportable nonconscious factors that *contribute to* anxiety should *not* be *equated with* the conscious experience of anxiety. They are part of the brain's means of coping with challenges and opportunities, a capacity that, as I've noted, has ancient biological roots and is not in the brain to make conscious fear or anxiety.

Efforts to redefine fear and anxiety in nonsubjective terms (e.g., as behavioral or physiological responses or nonconscious central states) have complicated and, in fact, impeded the goal of understanding what fear and anxiety really are.[21] The default meanings of mental-state terms like "fear" and "anxiety," which are borrowed from common usage, will always be the mental state (i.e., phenomenal, subjective, conscious) meaning—the actual feeling of fear or anxiety. When such terms are used as labels for nonconscious states or responses for scientific purposes, it is all too easy to unwittingly slip from talking about fear and anxiety in nonconscious terms to talking about conscious feelings. This leads to confusion about what is meant when fear and anxiety are discussed among scientists and also when scientists communicate with nonscientists; scientists have to guard against these conceptual slippages, which are so natural that they often go unnoticed.[22] More important, they can also lead to research that purports to be about anxiety but turns out to be about something controlled separately in the brain (e.g., defense responses). Scientists have an obligation to communicate their work in a clear and accurate way, even if it is less sexy. We aren't selling a product; we are trying to understand and explain how things work.

THE LOGIC OF DRUG DISCOVERY RESEARCH

To illustrate some of the problems that arise from viewing anxiety as something other than what it is for scientific purposes, we will examine how researchers seeking biologically based treatments through animal research conceptualized anxiety. As we will see, the success of such approaches has been limited. I argue that a big part of the problem is the way fear and anxiety have been conceived.

Research attempting to identify new drugs to treat problems involving fear and/or anxiety has traditionally measured behaviors (including innate responses, such as freezing or flight, and learned responses, such as escape and avoidance) and/or physiological changes in the body (including autonomic nervous system responses and endocrine responses) and brain (brain arousal or more specific brain activity). This approach fit well with Lang's response-based concept of anxiety because it focused on objectively accessible behavioral and physiological responses rather than thorny questions about feelings. But in contrast to Lang's emphasis on anxiety as a collation of response measures rather than an actual entity in the brain (Lang noted that anxiety does not reflect some single entity in the brain that can be manipulated), drug discovery work has often treated anxiety as a singular central state that could be pharmaceutically controlled and assumed that the effects of the treatment could be measured by behavioral and physiological responses. And because the central state was, for many, a physiological one, rather than a conscious feeling, efforts to find drugs that would alter the central state could be pursued through animal studies without having to wrestle with the consciousness problem. Recall Jeffrey Gray's notion of anxiety as a central state of behavioral inhibition, which was very influential in drug discovery research.

But here's the problem: The central state itself is never actually measured in such research—it is simply assumed to exist. Then a further assumption, even more troubling, is made: The central physiological state of anxiety, though viewed as a way of studying anxiety without having to solve the problem of consciousness in animals, is assumed to be one and the same as the conscious feeling of anxiety. Therefore, drugs that reduce behavioral or physiological responses thought to be indicative of the central state should make rats or mice—and, by implication, people—feel less anxious. It is in fact common for researchers to describe animals as being made less anxious by such drugs. To make the data of a drug study in animals relevant to human feelings, assumptions about what the data mean for animals are thus piled up, one on top of the other, and then further assumptions about humans are piled on top of these. This conflation of implicit processes that nonconsciously detect threats and control responses in the brain and body with processes that give rise to conscious feelings is bound to lead to disappointing results, which, in spite of a tremendous amount of money, effort, and time, is exactly what happened in drug discovery studies.[23]

Anxiolytics, drugs that help reduce anxiety, *have* been discovered, and help some people feel less anxious, at least to some extent. But the fact remains that current medications are not really viewed as ideal by either the treatment community or those taking the drugs. I maintain that part of the problem is the conceptual foundation of the research and its influence on the way drugs have been used in treatment. Before turning to these issues, though, let's look at anxiolytic drug discovery research in more depth.

SEARCHING FOR ANXIOLYTICS

Psychoactive drugs are believed to change behavior, physiology, thoughts, and feelings by altering the neurochemistry of the brain. This approach to treatment took off in the 1950s with the discovery of medications that help people with schizophrenia and depression.[24] These drugs were found mainly to affect monoamine neurotransmitters (dopamine in schizophrenia; norepinephrine in depression) and led to the chemical imbalance hypothesis that still dominates today.[25] In this view, mental illness is due to a disruption in the balance of neurotransmitters in the brain, so restoring this balance should restore mental health.

I summarized the history of antianxiety treatments in *Synaptic Self*. Scott Stossel's *My Age of Anxiety*[26] and Elliot Valenstein's *Blaming the Brain*[27] also have insightful comments on this history. I will highlight only a few key points most relevant here.

Until the development of pharmaceuticals, alcohol was the most commonly used anxiolytic. In the middle of the twentieth century, barbiturates and mepobromate became the first prescription drugs to be used to treat anxiety but were found to be highly sedating and addictive. In the 1960s they were replaced by a new drug class, benzodiazepines, which include Valium, Librium, Klonopin, and Xanax. In contrast to medications for schizophrenia and depression, benzodiazepines do not involve monoamine transmission but, instead, the inhibitory transmitter GABA (gamma-aminobutyric acid). Specifically, benzodiazepines have a special binding site on the GABA receptor, and when they occupy that site, GABA inhibition increases, reducing the ability of the affected circuits to process information.[28]

For decades benzodiazepines have been the most widely prescribed medications in the United States, and the number of prescriptions continues to rise.[29] Unlike many other psychiatric drugs, benzodiazepines act quickly after a single dose, which makes it possible to get a bit of relief on an as-needed basis. These drugs came to be used not only by people with a psychiatric diagnosis of anxiety, but by those who just wanted to feel a little less anxious. But benzodiazepines also have unwanted side effects, including sedation and muscle relaxation, memory impairment, and the potential for addiction, and withdrawal symptoms when discontinued.

———

The early success in the pharmaceutical treatment of schizophrenia, depression, and anxiety led to great enthusiasm for studies of the chemistry of the brain in relation to mental illness. The treatments available were somewhat helpful but not ideal. Could a concerted effort in drug discovery find the magic bullet for these conditions? This idea caught on. In the 1960s, federal funding became available for studies of the neurotransmitter basis of mental illness, and major pharmaceutical companies created CNS (central nervous system) divisions to develop new drugs for psychiatric treatment. Animal models began to be used to screen drugs for antianxiety properties in the hope of finding drugs with more clinical efficacy and fewer side effects.[30]

In a typical study, rats or mice are put through one or more behavioral tests in which they face some sort of challenging, often threatening, situation (some examples are shown in Figure 9.1). These involve electric shock; being in an open, unprotected area; exposure to cues related to a predator; encountering a situation of motivational conflict; or being placed alone with an unknown fellow species member. Drugs that enhance the animal's ability to deal with the situation are viewed as having anxiolytic properties.

In the literature today, there are more than one hundred behavioral tests used to model human anxiety in animals, and most of these focus on generalized anxiety.[31] But new treatments have, for the most part, become available as the result of coincidental observations, often about the effects of other drugs in people rather than from specific research testing hypotheses about anxiolytics in animals.[32]

For example, because certain antidepressant drugs, such as the tricyclic antidepressants (e.g., Tofranil) and monoamine oxidase inhibitors (e.g., Nardil), were also found to have anxiolytic effects in people, anxiolytic properties of the newer and better-tolerated antidepressants called selective serotonin reuptake inhibitors (SSRIs) were tested as well. SSRIs, which include drugs like Prozac and Zoloft, did turn out to be useful for some anxiety disorders but had their own problems as a treatment solution (slow onset, gastrointestinal and other physical symptoms, and also tolerance and withdrawal). Other antidepressant medications that affect serotonin receptors (specifically, the 5HT1A receptor), or that alter norepinephrine levels (e.g., selective norepinephrine reuptake inhibitors, or SNRIs) are also somewhat useful for some people with anxiety. Again, these were not new drugs discovered through research but rather a new use for existing drugs.

Today anxiety is most often treated first with psychotherapy, and if drugs are used the most common choices are still benzodiazepines or SSRIs, or other drugs that target monoamine (serotonin or norepinephrine) systems. Things have not

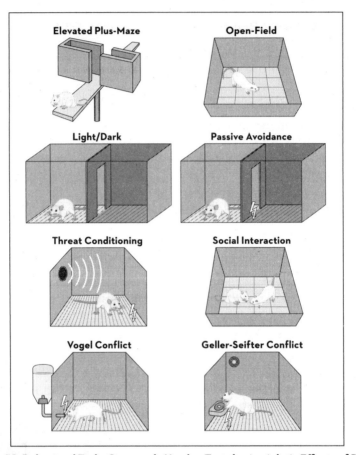

Figure 9.1: Behavioral Tasks Commonly Used to Test the Anxiolytic Effects of Drugs.

Anxiolytic drugs are screened by their ability to make animals more inclined to spend time in brightly lit, open, unprotected areas (elevated plus maze, open-field, light/dark, passive avoidance), freeze less to stimuli that predict shock (threat conditioning), interact more with other species members (social interaction test), and be more willing to undergo distress (shock) to receive a reward (Vogel and Geller-Seifter conflict tests).

FROM GRIEBEL AND HOLMES (2013), ADAPTED WITH PERMISSION FROM MACMILLAN PUBLISHERS LTD: *NATURE REVIEWS DRUG DISCOVERY* (VOL. 12, PP. 667–87), © 2013.

changed much in spite of years of research. Table 9.1 lists drugs typically recommended for different conditions.

Drug discovery research continues. Because benzodiazepines act via GABA receptors, much attention has focused on other ways to alter GABA transmission and perhaps make the drugs more effective and tolerable. Because SSRIs act via serotonin, efforts continue to find new drugs that alter this system. Norepinephrine

Table 9.1: Drug Classes Recommended for Treatment of DSM-IV Anxiety Disorders*

DRUG CLASS	DSM-IV ANXIETY DISORDER				
	GAD	PANIC	SAD	PTSD	OCD
Selective Serotonin Reuptake Inhibitors (SSRIs)	√	√	√	√	√
Serotonin Norepinephrine Reuptake Inhibitors (SNRIs)	√	√	√	√	
Tricyclic Antidepressants		√		√	√
Calcium Channel Modulators	√		√		
MAO Inhibitors		√	√	√	√
Reversible MAO Inhibitors			√		
Benzodiazepines	√	√	√		
Atypical Antipsychotics	√			√	
Tricyclic Anxiolytics	√	√			
Noradrenergic and Specific Serotonergic Inhibitors	√			√	√

*For each category of drug, different drugs may have different applications. For example, different SSRIs have different effects on different disorders; the recommended SSRI depends on the disorder.

Based on Table III in Bandelow et al (2012)

continues to be a popular target for efficacy improvement as well. Calcium channel modulators are used for some conditions, and research on these is ongoing. The excitatory transmitter glutamate has been studied, as have drugs that change hormone or peptide levels in the brain. For example, one promising target is the peptide oxytocin, which has been reported to reduce anxiety, promote affiliation, attachment, and affection,[33] and facilitate extinction of threat conditioning.[34] Although the jury is still out,[35] a recent study shows that treatment of GAD patients with oxytocin resulted in lower levels of anxiety and stronger connectivity between the medial prefrontal cortex and amygdala, suggesting enhanced cortical control of amygdala activity.[36] Another target is the endocannabinoid system, one of the newer neurotransmitter systems to be discovered. It involves a lipid molecule, called anandamide, that bind to specific receptors and has been implicated in a variety of behaviors in animals, most notably extinction of threat responses.[37] Cannabinoid receptors also bind marijuana, *Cannabis*, and contribute to its psychoactive properties. As with some other drugs with anxiolytic properties, cannabinoids interact with GABA to achieve their effects. The endocannabinoid system has also been implicated in anxiety disorders.[38]

WHAT'S THE PROBLEM WITH ANTIANXIETY DRUG RESEARCH?

Much has been written about the failures of anxiolytic drug discovery. I will summarize some of the main criticisms here.[39]

One major problem is that the behavioral approach used in antianxiety research typically only assesses a temporary condition elicited by a threat or other challenge. Such passing episodes connected to a specific stimulus are called examples of *state anxiety*. Anxious people typically have what is called *trait anxiety*,[40] which is a chronic condition they want to be rid of. Putting a random group of rats through a test in which a momentary defensive response is elicited is somewhat useful but has limits. For some tests, the same animal can show strong responses one day and weak responses the next. Also, the strength of the response can vary across multiple tests. Compounding the problem is the fact that in most studies the animals are not selected because they exhibit especially high levels of anxiety-like behaviors but are randomly chosen from the existing supply in the colony. One consequence of trait anxiety in people is that trigger stimuli are more effective at inducing strong state anxiety.[41] Unless the rats also have a chronic condition, tests will not be especially informative about pathological trait anxiety.

To surmount the problem of testing temporary states in relatively normal animals, researchers have used several strategies. One is to expose the animals to stressors for some time prior to testing in an effort to get them to develop a more chronic condition, and maybe even a pathological one. Another alternative is to use animals with genetically modified brains that might capture some features of pathological anxiety.[42] Animals have also been selectively bred to perform strongly or weakly on behavioral tests viewed as anxiety assays. One can also simply take advantage of individual differences between animals. For example, in our studies of Pavlovian threat conditioning, we have found that randomly selected rats show a wide range of conditioned freezing responses.[43] Studying those rats that exhibit exaggerated defense responses in such tests might be a productive approach.

Another problem is that often the behavioral tests that are used are selected because past studies have shown that drugs known to be anxiolytic in humans (most often benzodiazepines) weaken the behavioral response in the tests. The behavioral tasks are then used to test other drugs for antianxiety properties. But this turns out to be useful mainly in finding more benzodiazepine-like drugs, because tasks sensitive to benzodiazepines are not necessarily sensitive to other known anxiolytics.[44] Novel anxiolytics that work in a different and perhaps more effective way are not likely to be identified with this kind of strategy.

Yet another problem is that drugs are usually administered to animals once for a given test. This is useful when studying benzodiazepines, which, within an hour of the first dose, can have positive effects in people. But many psychiatric drugs take weeks to achieve a therapeutic effect. This may explain why, in the majority of animal studies that use a single SSRI treatment, the drug has no effect or may even increase defensive and risk assessment behaviors. For example, in studies of the effects of SSRIs on threat conditioning, we in fact found that a single treatment increased freezing to a tone paired with shock—an effect that would typically be called anxiogenic (anxiety producing) in the literature. But when the rats were given twenty-one days of treatment, they froze much less.[45] This is consistent with the fact that in people there is often an initial period of anxiety/agitation/depression that precedes any therapeutic effect, with the latter requiring two to three weeks of treatment. Testing drugs after a single dose makes the drug screening process easier but ultimately compromises what can be learned.

Gender is another significant factor: Women are far more likely than men to develop anxiety disorders,[46] but animal models focus mainly on males, a situation that is true of animal research in general.[47] (Ignoring gender has always been recognized as problematic, but accounting for it increases the number of animals needed for each study and adds complications to the design; although this needs to change, the funding has to be available to accommodate the change.)

Although the results of all this work have not lived up to the hopes, this by no means should be taken to mean that it was useless. Research is a process of learning from mistakes and false starts. Some of the problems mentioned would not have been identified had the work not been done. Clearly, some simple fixes can help make future studies more useful. For example, efforts could focus more on the effects of chronic rather than acute pharmaceutical treatments in both male and female subjects that exhibit exaggerated behavioral responses on the tasks used. Measures could be repeated on multiple days to assess reliability. Multiple tasks could be used that are known to give reliable responses and that are thought to be valid measures of the behavior in question. Finally, one might also require an additional standard. In the 1980s, Jeffrey Gray proposed that studies of anxiety in animals should use behavioral tests that are sensitive to multiple classes of drugs known to be anxiolytic in humans.[48] The assumption was that the brain system underlying the common effect of multiple antianxiety drugs on behavior would be the anxiety system. At the time, the drug options were different from those today, as we currently have more options. It could be of interest to revisit Gray's suggestion with contemporary drugs and determine whether they overlap in terms of anatomical, cellular, molecular, and/or genetic factors. If so, this might be a way to search for new pharmaceutical options for treating behavioral aspects of anxiety.

Even if all the corrections described above were made, though, there would still be a problem. As described later in this chapter, it's a conceptual problem.

LOOKING FOR ANXIOUS GENES

Alongside drug research, effort has been directed to finding a genetic basis for psychiatric disorders. If faulty genes could be identified, drugs that compensate for the malfunctioning genes might be useful in treatment.

The search for anxious genes proceeds on two fronts. I mentioned earlier efforts using selective breeding or gene targeting to produce animals that show anxiety-like behaviors. The other approach is to search for genes that correlate with anxiety symptoms in people with these disorders. If anxious genes can be identified in people, they can in turn be targeted in animals, which will, in principle, make it possible to conduct mechanistic studies of the malfunctioning of those genes in the genesis of pathological anxiety.

We don't need scientific evidence to tell us that some people are more anxious than others, and anecdotal evidence also suggests that nervousness runs in families. The latter suggests that individual differences in anxiety may have a genetic component, which is supported by studies that have shown that anxious tendencies in early life tend to be carried into adulthood, as though anxiety were a stable (and therefore perhaps genetically inherited) characteristic of the individual.[49]

The traditional approach to relating genes to psychiatric disorders starts with comparing the trait in question in people with similar and different genetic backgrounds. Such studies are most powerful when comparisons are made between identical and fraternal twins raised together, and between identical twins raised apart. Because identical twins have identical genes but fraternal twins do not, such studies allow the estimation of the influence of genetic versus nongenetic (especially environmental) factors on a given trait. For example, twin studies of anxiety have revealed that genetic factors account for roughly 30 percent to 50 percent of an individual's tendency to be generally anxious or to have a specific anxiety disorder.[50] Once a genetic component has been established, the search for the genes involved can begin. This is a time-consuming and complex process that has recently been greatly facilitated by the information obtained by the Human Genome Project.[51]

The success of genetic studies in neurological diseases, such as Huntington's disease, the familial form of Parkinson's disease, and a few others, led to hopes that similar advances could be made for psychiatric conditions. But unlike these neurological diseases, psychiatric disorders are not inherited following the simple laws of Mendelian genetics, in which traits are controlled by a single gene and result in a few standard inheritance patterns of being dominant or recessive.[52] Heritability in psychiatric disorders typically involves complex inheritance patterns controlled by multiple genes that interact with environmental factors to produce their results.

With the rise of molecular genetics, it has become possible to search for possible changes (mutations, polymorphisms) in target genes. Much effort has gone into investigating variations in genes that contribute to serotonin transmission, because serotonin-related drugs have antidepressant and anxiolytic properties. This assumes, however, that the treatment mechanism is the same mechanism that gives rise to the disorder.[53] Although this is consistent with the old chemical imbalance hypothesis, it is not a conclusion that should simply be accepted without careful assessment. Nevertheless, studies of the genetic control of serotonin have found interesting results. For example, people with a certain variant (polymorphism) of a gene controlling a protein involved in serotonin transmission are more reactive to threatening stimuli, and this hyperreactivity is associated with increased amygdala activity during the threat.[54] Further, it has been reported that this variant of the gene can account for 7 percent to 9 percent of the inheritance of anxiety.[55]

Recent studies grabbing lots of attention involve a polymorphism of the gene for an enzyme that breaks down anandamide. As we've seen, it is a naturally occurring substance in the brain that binds to endocannabinoid receptors, and mice lacking the receptor fail to extinguish. Further, animals and people with a gene variant that results in less of the enzyme, and thus more anandamide, exhibit less "anxiety-like" behavior and increased functional connectivity between the prefrontal cortex and the amygdala.[56] Though the authors describe the results in terms of "anxiety-like" behavior, writing in the *New York Times Sunday Review* psychiatrist Richard Friedman summarizes the work in an article titled "The Feel Good Gene."[57] This is yet another example of how research on behavior is freely generalized to conscious feelings, which I believe is inappropriate for several reasons. First, the key data in the study are behavioral; although the study reports changes in self-report of anxiety in a figure, these differences are very small and not mentioned in the text of the paper. Second, even if the gene variant biases people to feel less anxious, that is not the same as feeling good. Third, correlation is not causation, and there is no evidence that the gene is the cause of the slight reduction in anxiety. But, dopamine fans, don't fret. The article goes on to also describe dopamine as giving rise to the "sense of pleasure."

Recently, much enthusiasm has been generated by the discovery of epigenetic mechanisms,[58] which refers to the fact that the function of genes can be regulated by environmental influences. This doesn't mean that the environment is mutating our DNA; rather, what is affected is how genes carry out their function—making proteins. Epigenetics is the brave new world of biology and has already provided new insights into the biology of processes that are important in anxiety, such as threat processing, risk taking, and stress, and other conditions, such as addiction and eating disorders, as well.[59]

TAKING STOCK OF DRUG AND GENE STUDIES

Elucidation of the role of drugs and genes in anxiety or other psychiatric disorders would constitute a major development. The way these drugs and genes act on the brain could then be pursued in detail in animal models in an effort to develop better treatments for humans. But this effort would only be as good as the conception of anxiety that underlies it.

For example, if some set of genes that is correlated with uncontrolled feelings of anxiety is found in humans and then also discovered in rats and shown to affect their behavior in anxiety tests, that would surely make headlines, much like the story about anxious crayfish did (see Chapter 2). It would also likely lead to studies trying to find drugs that could normalize the behavioral performance, and then to studies that attempt to figure out where in the brain the circuits are that are affected by the maladaptive genes, and how these circuits contribute to the maladaptive behavior. But what would we have if we found this?

We would certainly know something about the role of the genes in the function of the circuit and of the circuit in the behavior. But we would not necessarily have found the key to how anxious feelings arise. Three other considerations are important. First, the gene would have to be shown to have a causal role in anxious feeling in the human brain and not simply be correlated with anxiety. Second, the causal effect would have to be shown to be direct and not due instead to an effect such as revving up survival circuit activity and/or inducing a hyperactive defensive motivational state, both of which could indirectly contribute to anxious feelings. Third, and perhaps most important, unless the rat brain can be shown to have the capacity for autonoetic consciousness, the studies would not tell us how the genes contribute to the feeling of uncontrollable dread that consumes the pathologically anxious human mind.

DIAGNOSTIC CATEGORIES

When researchers attempt to understand the pharmacological, neural circuit, cellular, molecular, and/or genetic basis of a DSM-defined psychiatric disorder, they accept the idea that the disorder, as so defined, is a biological entity that can be related to a specific mechanism and treated by altering the function of the disordered mechanism. The researchers thus assemble a sample of people with a given diagnosis—say, PTSD or panic disorder—and then attempt to relate the severity of their symptoms to brain function or to genes, and to assess the effects of drugs in improving the symptoms. Because the diagnosis is assumed to be biologically meaningful, one simply has to pinpoint the dysfunctional mechanism in order to figure out how to treat it.

This point of view has come under attack from social scientists Allan Horwitz and Jerome Wakefield. In their 2012 book *All We Have to Fear: Psychiatry's Transformation of Natural Anxieties into Mental Disorders*,[60] they argue that anxiety is often a normal response of the brain to the typical challenges that life presents rather than a pathological state. Thus, fear of heights and snakes, fear of being judged by others, and fear of being reminded of past trauma reflect the operation of a fear system in the brain that is functioning as it was genetically designed to and do not represent disorders that need medical drug treatment. In a review of the book, Ken Kendler, a leading geneticist who studies anxiety, praises its efforts to redefine the boundaries of disorders but thinks its authors fail to appreciate the significance of the fact that sometimes our genetic endowment is out of sync with the present environment.[61] He notes, for example, that in the "age of McDonald's" our fat storage system is responding to the environment in its genetically programmed way, but the result is an epidemic of type 2 diabetes. Do we, then, really want to argue that "individuals with type 2 diabetes are not disordered and hence should not be eligible for insurance coverage because their metabolism is doing what it was evolved to do"?

But the DSM approach has not just been the target of criticism by social scientists. It has also come under attack from biologically oriented psychiatrists. Tom Insel, director of the National Institute of Mental Health (NIMH), the Vatican of biological psychiatry, has been a particularly vocal critic. He points out that in other fields of medicine, early descriptive diagnoses that are not based on an understanding of the pathological biology have often run into trouble.[62] Disorders may seem unitary, but as more is learned about biology they turn out to be heterogeneous. This is the case, he argues, with mental and behavioral problems:

> Diagnostic categories based on clinical consensus fail to align with findings emerging from clinical neuroscience and genetics. The boundaries of these categories have not been predictive of treatment response. And, perhaps most important, these categories, based upon presenting signs and symptoms, may not capture the underlying mechanisms of dysfunction.

Let's dig deeper into the problem by considering how a patient typically receives a diagnosis. On the basis of the person's response to a series of questions (verbal self-report), a determination is made as to how many symptoms of a certain type he has. For example, in order to qualify as having PTSD under DSM-IV, a person must have one of five reexperiencing (recurrent memory) symptoms, three of seven avoidance/numbing symptoms, and two of five hyperarousal symptoms. My colleague Isaac Galatzer-Levy calculated that this means that there are more than 70,000 permutations under which one can be diagnosed to have PTSD under DSM-IV (and over 600,000 under DSM-5 because it lists even more potential symptoms that can be combined).[63] Further, because there are a large number of symptoms of PTSD, but also rigid rules about specific symptom requirements in multiple domains, one individual with the right combination of six symptoms may be diagnosed with the disorder, whereas someone who has eighteen symptoms of PTSD that do not meet the specific rules of symptom combination for diagnosis would be considered healthy.[64]

The DSM system does not identify a single disorder but a range of factors that may well depend on different brain systems (threat processing, attention, memory, arousal, avoidance, etc.). This does not mean that there are no consistent biological dysfunctions in mental and behavioral problems, but rather that the dysfunctions may not be categorized in a biologically meaningful way by the DSM approach.

DSM categories clearly have proved useful to some extent.[65] They provide a common language for clinicians and researchers that can be used to assess symptoms across cultures. The DSM system also offers broad guidelines for treatment that are somewhat effective. At the same time, patients don't always fall neatly into one of the DSM categories, and a large proportion end up with multiple diagnoses. (Most people with major depression, for example, also have generalized anxiety.) The latter could indicate that one condition is a risk factor for the other, that the diagnosis makes a distinction between conditions that are effectively the same, or that the diagnostic categories are just wrong.

Clinicians acknowledge that the DSM system is not perfect. A number of therapists with whom I've spoken agree that the categories should be considered rough guides that seldom truly capture a given individual's problems. Further, they say, using labels runs the risk of causing the therapist, family members, or social relations to make unwarranted suppositions about the person's condition and/or cause the person to start acting or feeling a certain way once he or she is so labeled. Nevertheless, in order to be reimbursed by insurers, therapists are required to use the labels provided by the DSM system; benefits to veterans are likewise often dependent on these diagnostic labels.

It should be noted that it was not the intent of the people who devised the DSM categories to provide a roadmap for brain researchers.[66] It's hard enough to understand the brain when we make a concerted effort to do so; we shouldn't expect that the DSM creators, seeking ways to organize diagnoses for the purpose of treating problems but not being especially focused on or knowledgeable about the brain, would have somehow come up with categories that accurately reflect the fundamental biological organization of the brain.

RESEARCH DOMAIN CRITERIA

In 2007 Steve Hyman, former director of the National Institute of Mental Health, noted:

> Although the central role of the brain in [mental] disorders is no longer in doubt, the identification of the precise neural abnormalities that underlie the different mental disorders has stubbornly defied investigative efforts.[67]

Hyman goes on to argue that part of the problem is that early on in the DSM history, a fairly arbitrary decision was made to split symptoms to establish many categories of disorders rather than grouping them into a smaller number of conditions. He also questioned the DSM view of disorders as qualitatively different from a state of well-being and instead suggested that mental disorders might be better viewed as involving traits that are on a continuum with the "normal" state. Thus, changes in the way the function of one or more neural circuits is altered in an individual might lead to deviations from normal.

The NIMH, under Insel, followed Hyman's lead. In 2010 the agency outlined a new approach to research on psychiatric disorders. Its Research Domain Criteria (RDoC) project rests on three concepts:[68]

1. Mental and behavioral problems are brain problems.
2. The tools of neuroscience can identify the brain dysfunctions underlying behavioral and mental problems.
3. Biological markers of the brain dysfunctions can be discovered and be used to guide diagnosis and treatment of mental and behavioral problems.

The basic concept underlying these criteria is that a problem like anxiety or depression does not emerge from a core depression or anxiety system in the brain. Rather, mental and behavioral problems reflect changes in specific brain mechanisms that operate at different levels and that perform basic psychological and behavioral functions. Blair Simpson, a leading anxiety researcher, summarized the RDoC approach[69] as a framework that specifies psychological constructs within key neural domains and delineates different units of analysis that pertain to these constructs, independent of traditional diagnostic categories. As shown in Table 9.2, the psychological constructs fall under five broad functional domains

Table 9.2: Research Domain Criteria: The RDoC Matrix

| | UNITS OF ANALYSIS | | | | | | | |
FUNCTIONAL DOMAINS	GENES	MOLECULES	CELLS	CIRCUITS	PHYSIOLOGY	BEHAVIOR	SELF-REPORTS	PARADIGMS
Negative Valence Systems (fear, anxiety, loss)								
Positive Valence Systems (reward, learning, habit)								
Cognitive Systems (attention, perception, memory, working memory)								
Arousal and Regulatory Systems (arousal, circadian rhythm, motivation)								
Social Process Systems (attachment, communication, perception of self and others)								

Based on http://www.nimh.nih.gov/research-priorities/rdoc/research-domain-criteria-matrix.shtml

or systems: negative valence systems (e.g., threat processing), positive valence systems (reward processing), cognitive systems (e.g., attention, perception, memory, working memory, executive function), arousal and regulatory systems (e.g., brain arousal, circadian rhythm, motivation), and social processing systems (e.g., attachment, separation). For each domain, data from a set of objective measurements at a different level of analysis (genetic, molecular, cellular, physiological, systems, behavioral, introspective self-report, etc.) are to be supplied on the basis of either existing or future research. Each of the five domains includes several lower-order factors for which the measurements would also be obtained. For example, under negative valence are circuits involved in acute threat, future threat, and sustained threat, among others.

The RDoC approach promoted by the NIMH is well suited to guiding research on basic mechanisms, many of which can be studied similarly in animals and humans.[70] Because particular types of symptoms and signs (self-reported feelings of fear and anxiety, hyperarousal, enhanced attention to threats, diminished safety detection, excessive avoidance and risk assessment, etc.) depend on particular circuits, they may well be vulnerable to specific predisposing factors. Different symptoms may also be treatable with different approaches that target the underlying circuits. Thus, identifying circuits underlying specific cognitive and behavioral processes that malfunction in relation to various symptoms offers a novel approach to understanding and treating anxiety and other mental and behavioral problems.

The RDoC model can't immediately replace the DSM system. It will take time to compile enough information to offer new ways to categorize mental and behavioral problems. However, existing data show how this might be effected. For example, symptom *grouping* seems required because amygdala-based threat-processing circuitry has been implicated in most if not all anxiety disorders[71] as well as in schizophrenia, depression, borderline personality disorder, autism spectrum disorders, and other conditions.[72] But given that the categories are not likely to go away anytime soon, we can provisionally improve things by *splitting* current categories on the basis of research data: For example, different processes (RDoC dimensions) are altered in PTSD caused by single versus multiple trauma episodes.[73]

Recall from Chapter 5 that six processes characterize people who suffer from anxiety, regardless of their particular disorder:[74] (1) increased attention to threats; (2) failure to discriminate threat and safety; (3) increased avoidance; (4) heightened reactivity to unpredictable threats; (5) overestimation of threat significance and likelihood; and (6) maladaptive behavioral and cognitive control. Circuits involving the amygdala, nucleus accumbens, bed nucleus of the stria terminalis, lateral prefrontal cortex, ventromedial prefrontal cortex, orbitofrontal cortex, anterior cingulate cortex, hippocampus, insula cortex, and arousal systems contribute to these processes. Of great interest would be comparisons of the involvement of the processes and the specific circuits and molecular mechanisms underlying each of them across anxiety disorders, and also between anxiety disorders and other mental disorders in which increased anxiety is a factor (e.g., depression, schizophrenia, autism).

Unlike DSM-guided drug discovery, which has often succeeded through accident, the RDoC approach allows facts about the brain to suggest how a given problem should be understood, researched, and treated. This leads to a far more complex view of anxiety and challenges the simple idea that there might be a magic pharmaceutical bullet that can solve the problem. Although the RDoC approach adds complexity to the therapeutic challenge, when considered in light of the views I have developed in this book it also helps explain why current treatments are not more successful.

Drugs that are developed by testing their effects on defensive reactions (freezing and accompanying physiological responses) and actions (avoidance) are essentially targeting defensive survival circuits and defensive motivational states and thus only indirectly change anxious feelings. This is likely why, after treatment, people may be less physiologically aroused by threats and less prone to avoid stressful situations but still feel anxious. It is no small achievement to be able to change the way survival circuits operate and contribute to behavioral reactions, physiological responses, and actions. But in order to end up with drugs that specifically make people feel less anxious by directly changing feeling, brain systems that make conscious feelings would have to be targeted.

Benzodiazepines are an interesting case. They do make people feel subjectively less anxious, and also affect animal behavior in certain "anxiety" tests. But before concluding that these drugs make rats feel less anxious, and therefore that drugs that affect behavior in anxiety tests can be expected to make people feel less anxious, we need to consider the benzodiazepines further. This class of drugs did not result from drug discovery research in animals but from findings in humans. Their receptors are components of GABA receptors, and activation of these receptors increases inhibition. This has profound effects on neural activity in the brain, producing widespread and often nonspecific effects like sedation. But let's ignore the general effects of increased inhibitory drive and just focus on the role of inhibition in functions relevant to anxiety. In the bed nucleus and hippocampus, increased inhibition reduces risk assessment behavior in situations of uncertainty; in the hippocampus, they also impair memory, which may reduce the impact of past dangerous situations and further reduce risk perception.

Benzodiazepine receptors are also present in the prefrontal cortex, and in the human brain changes in benzodiazepine receptor function occur in areas involved in working memory, attention, and consciousness in people with anxiety disorders.[75] Thus, effects of benzoidazepines on defensive behavior and subjective feelings of anxiety may well depend on the fact that the receptors are present in both working memory circuits and threat processing/risk assessment circuits. Consequently, we can't simply assume that because a drug has an effect on "anxiety-like" behavior in animals that it will also affect feelings of anxiety in people unless, like benzodiazepines, the drug affects circuits underlying both kinds of processes. Animal behavior tests are excellent tools for exploring effect on risk assessment, but human studies are required to determine effects on subjective well-being. As we've seen repeatedly, these two effects do not necessarily go hand in hand.

Evaluation of treatment effects thus needs to be based on realistic expectations about what treatments do, and this depends on an understanding of the brain mechanisms that underlie the tasks that are used to make the assessment. Anxiolytic drugs that are judged to be only somewhat successful, or even unsuccessful, because they don't do a great job of making people feel less anxious may actually be doing as much as they can reasonably be expected to do given that they are based on studies that measure survival circuit activity in animals rather than feelings in people. I am not saying that people don't sometimes feel better with drug treatment. The issue instead is whether better outcomes might be achieved if we recognized the difference between treatments that affect implicit and explicit processes in the brain.

PUTTING CONSCIOUS EXPERIENCE FRONT AND CENTER IN THE SCIENCE OF ANXIETY

As I have argued, the essence of anxiety is the unpleasant feeling—the apprehension, dread, angst, and worry—that one experiences when one perceives a lack of control in situations of uncertainty and risk. It is a by-product of our unique ability to envision our future self and especially to anticipate unpleasant, or even catastrophic, scenarios regardless of their likelihood.[76] Earlier in the chapter, I quoted Menand's restatement of Kierkegaard: Anxiety is the price humans pay for freedom. My restatement of Kierkegaard and Menand is that anxiety is the price humans pay for autonoetic consciousness.

The conscious experience of anxiety is the anxious individual's point of contact with the altered brain functions that are causing problems. It is also what often leads him to seek help and is what he reports on when being diagnosed or treated. Clinicians spend their time interacting with the conscious minds, including the conscious feelings, of their clients and thus know that consciousness is critical. A person's troubles may stem from deep down, but consciousness, via verbal self-report, is the main vehicle by which to assess what the person is thinking and feeling.

Some scientists have avoided consciousness in conceptualizing fear and anxiety. This puts a gap between scientific research and the actual problem of understanding feelings of fear and anxiety. Others go too far in the other direction, assuming that behavioral tests in animals directly reveal mechanisms of conscious anxiety. This connects the wrong brain mechanisms to fear and anxiety and causes interpretative problems in understanding the implications of the research. Still others, while avoiding consciousness in conceptualizing their research, have called upon conscious feelings when interpreting their behavioral data from animals. This, too, causes confusion when the implications of the work are considered.

As discussed in earlier chapters, much progress has been made in the science of human consciousness in recent years through studies of humans, in which it is possible to clearly separate conscious and nonconscious processes and understand their distinct contributions to mental life. To be clear, conscious experience is still private. In spite of some claims, scientists haven't figured out how to read the content of people's minds with imaging machines. What has changed is that some of the component processes that contribute to conscious experiences have been identified—such as working memory, attention, monitoring, other executive functions, and long-term semantic (noetic) and episodic (autonoetic) memory.

The RDoC approach provides a framework that could give the phenomenal experiences a more prominent role in the scientific picture of anxiety. It includes all the functions necessary to characterize global defensive motivational states (arousal, threat processing, risk assessment, avoidance, etc.), as well as functions that cognitively process signals related to these activities (sensory processing, long-term explicit memory, attention, working memory, monitoring, self-evaluation, verbal self-report, etc.). Unfortunately, it stops short of emphasizing the centrality of the phenomenal experience itself and instead focuses on the role of cognitive processes in information processing—leaving out the role of these processes in generating the content that constitutes the conscious experience of anxiety. It includes verbal self-report, but as just another level of analysis (refer to Table 9.2). All the other RDoC measures provide information about how various nonconscious ingredients contribute to feelings of anxiety but are not themselves measures of anxiety. The conscious experience of anxiety, the way it feels, is *not* just another level of analysis. It is what anxiety is.

But not all conscious states are the same. There are many kinds of cognitive processes and representations that can give rise to conscious states. For example, anxious people worry about a potential negative event even if they also know that the event has a very low likelihood of occurring.[77] Estimates of the likelihood of an event and worry about the event are different cognitions, and both involve conscious evaluations of the future;[78] however, the facts about the likelihood of the event are not sufficient to short-circuit the worry. Also, you may know you are anxious but not know why. It is thus important to distinguish cognitions that contribute to the interpretation and naming of states happening to one's self (I'm feeling anxious or afraid) from cognitions that attribute causes to such experienced states (my feeling is due to the situation I am in or a result of what happened to me in the past). Both are autonoetic states, but they are different. One is a felt experience, and the other is a speculation about the nature and origin of the felt experience. Such speculations can then feed anxious feelings.

Often one does not know why one is anxious, and this uncertainty increases anxiety. Misattribution of the motivational cause of feelings or actions is a source of further anxiety, as it fails to adequately explain the state and results in cognitive dissonance.[79] People are compelled to reduce such states by further attributions,[80] which creates the opportunity for additional misattribution and anxiety. Attribution (interpretation) of experience is a key factor in creating the mental continuity of the conscious self,[81] but it can also enable anxiety.

FOUR WAYS TO BE ANXIOUS

The ideas discussed in this and previous chapters attempt to bring phenomenal experience into a brain-based understanding of fear and anxiety. The key concepts are summarized below by considering four scenarios (Table 9.3).

Table 9.3: Four Ways to Be Anxious
1. In the presence of an existing or imminent external threat, you worry about the event and its implications for your physical and/or psychological well-being
2. When you notice body sensations, you worry about what they might mean for your physical and/or psychological well-being
3. Thoughts and memories may lead to you to worry about your physical and/or psychological well-being
4. Thoughts and memories may result in existential dread, such as worry about leading a meaningful life or the eventuality of death

Scenario 1. When a threat signal occurs, it signifies either that danger is present or near in space and time or that it might be coming in the future. Nonconscious threat processing by the brain activates defensive survival circuits, resulting in changes in information processing in the brain, controlled in part by increases in arousal and behavioral and physiological responses in the body that then produce signals that feed back to the brain and complement the physiological changes there, intensifying them and extending their duration. Collectively, these give rise to a defensive motivational state. When the state itself, or components of it, grab attention and enter working memory, a representation of the experience is created. The representation includes not just the defensive motivational state information (including noticeable responses like rapid heartbeat and behavioral avoidance) but also information about external stimuli (the threat and other stimuli that are present) and memories about the semantic meaning of the stimuli and past episodic experiences with such stimuli. The result is a variant of the conscious feeling of fear or anxiety, depending on whether the initial threat signal is itself a clear and present danger or a warning about a potential future danger. But even if the threat is present, the feeling of fear quickly gives way to anxiety. These conscious feelings do not simply present themselves but have to be assembled via interpretation. Indeed, a leading contemporary theory assumes that conscious emotional feelings are psychological constructions in which schemas stored in memory are matched with present cues (brain arousal, body feedback, memories, etc.) in working memory to give rise to the experience.[82]

Scenario 2. The trigger stimulus does not have to be an external stimulus but can be an internal one, as some people are particularly sensitive to body signals. The slightest twinge in the gut, or muscle spasm, is sufficient to trigger a health concern in those prone to hypochondria. People who suffer from panic attacks are especially attuned to body sensations. These sensations become conditioned triggers that activate (much like an external stimulus) the defensive circuit and have many of the same consequences. The cognitive biases of the person, based on past experiences stored as episodic and semantic memories in the form of schemas, are what cause him or her to worry about illness or the imminence of a panic attack when such symptoms occur and match stored schemas. Note that I am not proposing that conditioned sensations are causes of panic attacks but, instead, that these sensations initiate processes that lead to anxiety, dread, and worry that a panic attack might be coming and may therefore indirectly prime the brain in such a way as to lower the threshold for an attack. (For an excellent overview of a modern learning theory approach to panic disorders, see the paper by Mark Bouton, Susan Mineka, and David Barlow, who each come from a different background in psychology.[83])

Scenario 3. Anxiety can be triggered by thoughts and memories as well. We do not need to be in the presence of an external or internal stimulus to be anxious. An episodic memory of a past trauma or of a panic attack in the past is sufficient to activate the defense circuit and produce all of the typical consequences, which then match with stored schemas to give rise to the feeling.

Scenario 4. A thought or memory may also produce a different kind of anxiety, what is often called existential dread. Examples are the contemplation of whether one's life has been meaningful, the inevitability of death, or the difficulty of making decisions that have moral value. These do not necessarily activate defensive systems; they are more or less pure forms of cognitive anxiety. If such contemplations become threatening, though, they can activate defensive circuits and give rise to the more typical form of anxiety associated with body tension and physiological arousal.

Anxiety, in short, is a conscious feeling. It can arise in a bottom-up way, driven by activity in defensive circuits or from higher processes that conceptualize worry, either about an uncertain future or about existence itself. In each case, anxiety is, like fear, dependent on cortical processes that enable sensory information and memories, along with consequences of survival circuits activity, if present, to be represented in working memory and available for conscious thought.

Anxiety (worry, dread, apprehension, trepidation, angst, and worry) involves a particular kind of conscious thought. It is all about the self. Yes, we worry about our loved ones, but it is because they are part of us. I'm not speaking here of the "blood is thicker than water," "selfish gene," or "maternal instinct" varieties of biological explanation. What I'm referring to is the type of bond that requires an episodic, autonoetic self, a self that can be projected into the future, a contemplation of what that future self will be like if bad things happen—not just to it but also to those the self cares about, whether they are biologically related or not, whether they are a person or a pet, whether they are known personally or only as an idol or hero, as these are all psychologically part of our extended self. In the words of William James, "a man's Self is the sum total of all that he *can* call his, not only his body and his psychic powers, but his clothes and his house, his wife and children, his ancestors and friends, his reputation and works, his lands and horses, and yacht and bank-account. All these things give him the same emotions. If they wax and prosper, he feels triumphant; if they dwindle and die away, he feels cast down—not necessarily in the same degree for each thing, but in much the same way for all."[84]

CHAPTER 10

CHANGING THE ANXIOUS BRAIN

"Anxiety's like a rocking chair. It gives you something to do, but it doesn't get you very far."

—JODI PICOULT[1]

How can one feel less afraid and worried? How can anxieties be made to disappear or at least come to be controlled so they are less debilitating?

Mental and behavioral problems are most often treated with either psychotherapy or medication, and sometimes with a combination of the two. In the previous chapter I discussed the use of drugs to relieve fear and anxiety, and the difficulties involved in finding new and better pharmaceutical treatments. But the fact is that for many problems related to anxiety and fear, psychotherapy is a viable—and in fact the best—option. In this and the following chapter we will consider psychotherapy as a way to change the anxious brain.[2] Although I am neither a therapist nor a physician and have had little experience inside a therapist's office, I have learned some things about what goes on in the brain when organisms are threatened and will discuss the question of therapy from this perspective.[3]

PSYCHOTHERAPEUTIC APPROACHES

The American Psychological Association lists several categories of psychotherapy, including psychoanalytic and psychodynamic, humanistic, behavioral, and cognitive, as well as integrative and eclectic therapies, which are blends of two or more of the other approaches[4] (Table 10.1). Classic psychodynamic treatments based on Freud's psychoanalytic method use verbal free association and introspection to seek the root cause of mental and behavioral problems buried in repressed (unconscious) memories, especially of early trauma or socially unacceptable desires.[5] Newer psychodynamic approaches tend to place more emphasis on insights into present interpersonal conflicts.[6] Humanistic therapies (existential, Gestalt, and client-centered) help people make rational choices and realize their potential in life while showing care and concern for others.[7] Behavior therapy assumes that many problems are due to learning and uses principles of Pavlovian and instrumental conditioning to change maladaptive behaviors.[8] Particularly important in the behavioral treatment of fear and anxiety is the method of exposure, which was inspired by the principle of extinction and involves repeated encounters with objects or situations that make one anxious or afraid. Cognitive therapy is based on the assumption that dysfunctional cognitions (beliefs) underlie pathological emotional states (such as anxiety) and behaviors (such as avoidance).[9] By changing these beliefs, the fears, anxieties, and behaviors associated with them can be changed as well. Cognitive-behavioral therapy combines cognitive interventions with methods that seek to reduce fear and anxiety through exposure to the threats. Acceptance and commitment therapy, a variant on cognitive therapy, attempts to teach people to accept rather than change their emotions and make decisions within the context of what they value, as opposed to letting negative feelings control their behavior.[10] The various cognitive therapies are by far the most common psychotherapeutic approaches in use today.

Table 10.1: Some Common Types of Psychotherapy

Psychoanalysis and Psychodynamic Therapies
Humanistic Therapy
Behavior Therapy
Cognitive Therapy
Integrative/Eclectic Therapies
Alternative/Complementary Therapies

Based on http://www.apa.org/topics/therapy/psychotherapy-approaches.aspx

There is growing interest in alternative approaches to the treatment of anxiety, though their effectiveness has not been evaluated in all cases. Mindfulness-based methods use relaxation, breathing exercises, meditation, yoga, and other techniques to focus on the present and reduce tension and worry.[11] Although each of these can be used on its own to reduce stress and anxiety, they can also be incorporated into other psychotherapeutic approaches. For example, behavioral and cognitive therapies often include relaxation training, whereas acceptance and commitment cognitive therapy employ mindfulness and meditation.[12] Hypnosis, one of the first methods used by Freud but later dismissed by him, is gaining popularity.[13] Another approach, called eye movement desynchronization reprocessing (EMDR), uses visual stimuli to induce patterns of eye movements as part of a treatment that helps the client reprocess troubling events and acquire new coping skills.[14]

PSYCHOTHERAPY AND THE BRAIN, CIRCA 2002

In *Synaptic Self* I drew a broad distinction between therapies based on talk and those based on exposure therapy (Figure 10.1). I went on to argue that these approaches are fundamentally different because they depend on different brain circuits. Talk therapy requires conscious retrieval of memories and thinking about their origins and/or implications and thus depends on working memory circuits of the lateral prefrontal cortex. By contrast, therapies involving exposure depend on medial prefrontal areas that contribute to extinction, the process on

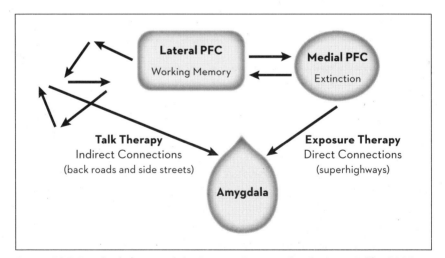

Figure 10.1: Psychotherapy and the Brain as Portrayed in *Synaptic Self* in 2002.

which exposure is modeled. I suggested that the fact that the medial frontal areas connect with the amygdala, whereas the lateral areas do not, might account for why it is easier and faster to treat fears, phobias, and anxiety with exposure-based approaches (behavioral or cognitive-behavioral therapy) than with psychoanalytic or humanist approaches based on talk.

In retrospect, this neural hypothesis got some things right but was overly simplistic in other respects. For example, working memory and its executive functions, such as attention and other cognitive control functions, involve both lateral and medial areas of the prefrontal cortex. And while the medial (but not lateral) prefrontal cortex is strongly connected with the amygdala, medial and lateral prefrontal areas are also interconnected.[15] Research on emotion regulation and reappraisal discussed in Chapter 8 illustrates the complex ways that these areas can affect amygdala activity. But more important, my idea that talk therapy depends on cognition and consciousness, but exposure therapy does not, was naïve. All forms of psychotherapy depend on verbal exchanges between the client and therapist and engage cognitive processes,[16] including those that contribute to consciousness.

For example, exposure therapy, whether conducted by pure behavioral therapists or cognitive behavior therapists, requires language for clients/patients to discuss their problems and concerns, to comprehend a treatment plan involving exposure, to process verbal instructions about how to deal with stress during exposure, and to acquire new coping skills and implement them during the session and also when feeling anxious or afraid later. Some behaviorists even argue that language is the most powerful way to change fearful and anxious behavior.[17] All of these activities also involve working memory. Thus, like traditional psychotherapy, exposure therapy depends on talk and engages working memory circuits,[18] including both lateral and medial prefrontal circuits and probably parietal ones as well.

In the material that follows, I will explore exposure therapy in some detail to present a more accurate view of how it works. Although it is obviously not the only psychotherapeutic approach available to treat problems with fear and anxiety, it is the most effective and widely used treatment today.[19] Because extinction is a key factor that enables exposure to help anxious people,[20] and much progress has been made in understanding its neural basis, insights into how extinction works in the brain can be leveraged to understand why and how exposure helps people.[21] But as we will see below, much more goes into exposure therapy than the stimulus repetition processes that induce extinction. And by separating the role of extinction from other processes that contribute to exposure, we can develop a more nuanced approach to what takes place in the brain when people are treated by encounters with the things about which they are fearful or anxious.

Exposure is thus an excellent starting point if one wants to understand how psychotherapy works in the brain. This does not mean that I think therapy *has* to be understood in terms of neural processes in the brain, nor does it mean that I think that other therapeutic approaches lack value. I am simply focusing on exposure because of its connection to extinction, which we understand neuroscientifically.

IN THE BEGINNING

As the old saw goes, if you get thrown off a horse, the best way to get over your fear is to hop back in the saddle. The German poet Goethe, who suffered from fear of heights, figured this out.[22] He would force himself to slowly ascend to the top of his local cathedral and then stand on a small platform overlooking the city with nothing to hold on to until his fear dissipated. He repeated this exercise often and was eventually able to go on enjoyable field trips to the mountains.

It is generally accepted, at least by behavioral and cognitive therapists, that exposure is fairly effective in reducing fear and anxiety across a variety of disorders.[23] Although exposure is most often thought of as a treatment for fears or anxieties about specific stimuli or situations (animals, heights, germs, tests, public speaking, social encounters, past traumas), it is, as I will explain later, also an important method for treating the excessive worry that typifies generalized anxiety.[24]

Freud considered having his patients confront the objects or places they feared[25] (Figure 10.2), but the use of the exposure approach did not become formalized as a treatment for fear and anxiety until much later. In the mid-twentieth century, behaviorist principles based on Pavlovian and instrumental conditioning began to be used to reconceptualize the dominant Freudian view of anxiety and its treatment.[26] In contrast to dynamic therapies, which sought to reveal the root cause of adjustment problems, behavior therapies ignored the cause and attacked the symptoms.[27]

Exposure therapy flowed naturally from the interpretation of anxiety championed in the mid-twentieth century by O. Hobart Mowrer and Neal Miller in their famous two-factor theory of avoidance behavior.[28] As discussed in Chapter 3, Mowrer and Miller viewed avoidance learning as a combination of Pavlovian and instrumental conditioning. First, through Pavlovian conditioning, neutral stimuli acquire the ability to elicit fear, and then, through instrumental conditioning, responses are learned that enable escape and then avoidance of the fear-evoking situation. But if in the future the stimulus loses its predictive relation to harm, a person never has the opportunity to extinguish his fears, because

Figure 10.2: If Freud Did Exposure.

successful avoidance prevents the harm from being experienced. To be rid of fear, the theory goes, one has to overcome habitual avoidance and be reexposed to the fear-arousing stimulus, experience the fear, and then learn, via extinction, that the stimulus is not really the portent of a harmful outcome. The logic of the Mowrer-Miller theory still contributes to the rationale for the use of exposure as a treatment for problems of fear and anxiety.[29]

Thus, the basic idea underlying exposure therapy is that facing your fears will condition you, via extinction, to be less responsive to the trigger stimuli. If you are afraid of elevators, for example, a therapist can show you pictures of elevators and weaken your responses that way. Or the therapist might ask you to imagine being in an elevator and encourage you to not lose your concentration on that thought, because thinking about something else will allow you to mentally escape from your fears and worries and reduce the effects of the mental exposure. To add a more realistic element, the therapist might take you on elevator rides; by forcing you to stay on the elevator, avoidance is prevented and extinction can do its work. Once some success is achieved through initial exposures, the client is then given instructions on how to conduct the exposure alone, especially in daily life situations, in order to strengthen and maintain the beneficial effects of the exposure.

The first form of psychotherapy explicitly based on exposure was *systematic desensitization*, introduced by Joseph Wolpe in the late 1950s.[30] This approach used repeated gradual exposures to imagined threats accompanied by the use of relaxation exercises. A number of variants of exposure therapy emerged in the years that followed.[31] *Graded retraining* also involves gradual exposure, but to real-life, anxiety-provoking situations rather than imagined stimuli.[32] In contrast to gradual exposure approaches, *flooding*[33] (also called *implosive therapy*) induces and maintains a high level of fear during imagined exposures, preventing escape and avoidance until fear levels dissipate. In some forms of implosive therapy, the therapist guides the imagined exposures in such a way as to ensure that a high level of fear is maintained. *Prolonged exposure therapy*, a variant of flooding, attempts to maintain a high level of fear arousal, but its key premise is that all aspects of fear, as defined by Lang's three response systems (behavioral avoidance, physiological responses, and verbal behavior), have to be reduced in order for exposure to be effective.[34] A different approach focuses on low levels of exposure to real-life threats in a highly structured situation and uses verbal reinforcement to motivate the individual to stay with the procedure even if it becomes somewhat stressful.[35] Still other approaches use *vicarious exposure* through observation of social situations to reduce fear and anxiety.[36] Virtual reality technologies are also used as a vehicle for exposure therapy in a semirealistic context.[37]

On the whole, exposure approaches work fairly well, helping roughly 70 percent of those who are treated.[38] Clearly, though, there is room for improvement, as discussed in the following chapter.

THE COGNITIVE MAKEOVER OF EXPOSURE

Exposure thus arose in behavior therapy as extinction plus muscle relaxation training, breathing exercises, and some cognitive support (such as instruction, verbal reinforcement, and/or social modeling). But with the rise of the cognitive movement in psychology, cognitive principles began to infiltrate theories of conditioning and extinction[39] and also made their way into behavior therapies influenced by such theories. The result was that standard therapeutic procedures began to become more and more cognitively weighted.[40] Cognitive therapy was the result.[41]

Cognitive therapy essentially consumed behavior therapy from within, originally calling itself cognitive-behavior therapy.[42] (The terms "cognitive-behavioral therapy" and "cognitive therapy" are essentially synonymous today.) Aaron Beck, the founder of cognitive-behavior therapy, argued that cognitive change is a pre-requisite for sustainable emotional and behavioral change.[43] Exposure was viewed as an adjunct to procedures that sought to alter the way people think about their problems rather than as a vehicle of change on its own. A cognitive therapist may choose to use approaches that include exposure as part of a cognitive change strat-egy or approaches weighted toward changing cognition without using exposure.

When done today by cognitive therapists, exposure is somewhat different than that performed by early behaviorists. Although verbal exchanges and instruction are involved in both approaches, a change in cognitive (especially maladaptive) beliefs is crucial to the cognitive version. Thus, to the cognitive therapist, explicit cognition, working memory, and executive control processes are as important, if not more so, than extinction processes engaged by exposure.

Donald Levis, a behavioral psychologist and anxiety theoretician, described the fundamental difference between cognitive and behavioral therapy in terms of scientific philosophy: Behavior therapy is committed to a focus on observable fac-tors and avoids making reference to inner thoughts and feelings.[44] Indeed, accord-ing to Beck, the key difference is that the cognitive therapist attempts to modify the mental content (maladaptive thoughts or beliefs) associated with emotional distress (feelings) or behavioral problems, whereas the behavior therapist tries to change the overt behaviors (such as avoidance responses) themselves.[45]

For Beck, the goal of cognitive therapy is to change streams of negative thoughts based on *core beliefs* (or *schemas*) that give rise to automatic (noncon-scious) maladaptive cognitive appraisals of situations and lead to feelings of anx-iety and cognitive and behavioral avoidance. In order to eliminate negative emotions and avoidance and secure enduring therapeutic changes, Beck argues that it is necessary to identify and evaluate the maladaptive beliefs (some of which are unconscious) and replace them with more realistic thought patterns, which will result in healthier thinking, behaviors, and feelings.[46]

In parallel with Beck, Albert Ellis developed a cognitive approach called *rational emotive therapy*.[47] In Ellis's ABC model, A is an antecedent stimulus (a noise), B is a belief (that the noise signifies danger), and C is the consequence (fearful feelings and avoidance responses). Because anxious people are predis-posed to have beliefs that lead them to construe harmless things as dangerous, the therapist's job is to help identify the beliefs that connect antecedent events to their consequences so that beliefs can be changed.

David Clark, a leading cognitive therapist, calls maladaptive beliefs "cata-strophic misinterpretations."[48] Writing with Beck, Clark notes that exposure therapy is useful as part of the therapeutic process because it enables a deeper activation of the threat schema and provides opportunities to disconfirm the catastrophic misinterpretations sustained by avoidance.[49] Again, in this cognitive model, exposure is not the main vehicle of treatment but a method to get at beliefs that can then be disconfirmed.

The work of Anke Ehlers and Clark can be used to illustrate how cognitive therapists view PTSD and its treatment.[50] Ehlers and Clark assume that PTSD develops after a trauma if a person ends up appraising stimuli associated with the trauma as threatening in their present or future circumstances. If so, the stimuli will serve as triggers that end up producing memory distortions about threats and their consequences. This, in turn, leads to hyperarousal, intrusions of anxious thoughts into consciousness that can result in the reexperiencing of mental or physical symptoms that occurred in connection with the past trauma. The appraised threat also motivates cognitive and behavioral avoidance, which maintains the problems rather than reducing them (due to the prevention of exposure and extinction). The therapy thus starts by identifying negative appraisals, memories, trigger stimuli, and cognitive and behavioral factors that preserve symptoms. Then the therapist helps the client modify the excessive negative appraisals, elaborate the memories, discriminate triggers that lead to reexperiencing them, and eliminate cognitive and behavioral avoidance.

The version of cognitive therapy called acceptance and commitment therapy takes a slightly different approach, focusing on thought acceptance, via mindfulness training, rather than thought or belief change, to counteract cognitive avoidance strategies.[51] Although some view this as a new wave of cognitive method therapy, others regard it as simply an additional tool in the cognitive therapy arsenal.[52]

Although maladaptive beliefs may become habitual and carried out automatically by nonconscious processes, the process of belief change engaged in cognitive therapy often involves explicit cognition and working memory. Several top-down cognitive tools are used.[53] The therapist helps the person introspectively recognize the automatic thoughts and uncover the beliefs they reflect. Through a process of reevaluation, or reappraisal, the beliefs come to be seen from a different perspective. When exposure is involved, the client is strongly encouraged to test whether the pathological thoughts and behaviors being avoided are actually harmful. Relaxation and other stress-reduction techniques, as well as mindfulness training and thought acceptance, depend on also top-down processes because they are initiated by an intentional effort to control body physiology (relaxation) and mental states (mindfulness).

Beck, as noted, asserted that cognitive change is a prerequisite to enduring behavioral change.[54] But although cognitive change may be directed by explicit cognition—that is, by working memory and its executive functions—and may be part of conscious experience, as Beck also notes, nonconscious cognitions (automatic thoughts arising from nonconscious beliefs, or schemas) also need to be changed. In the field of social psychology, compelling evidence has demonstrated that nonconscious beliefs, known as biases, can have profound effects on thought and behavior.[55] As discussed in the previous chapter, anxious people have biases for detecting and responding to threats and overvalue the significance of potential threats. These kinds of biases and valuations, even when processed unconsciously, can control behavior and influence conscious thought. Successful therapy thus involves changes in explicit (conscious) and implicit (nonconscious) cognition.

DID COGNITIVE EXPOSURE THERAPY IMPROVE BEHAVIORAL EXPOSURE THERAPY?

The infusion of cognitive ideas into behavior therapy reframed the purpose of exposure. It became less directed toward the extinction of conditioned responses by overcoming avoidance and more focused on cognitive change. Did this achieve anything?

In 1987 the British fear expert Isaac Marks reviewed the literature and concluded that the various supplements that were added to simple exposure (including relaxation, breathing, changing of false beliefs about threats, etc.) were redundant—that exposure alone was sufficient.[56] More recent findings confirmed this result.[57] One might, therefore, be tempted to conclude that contrary to the principles of cognitive therapy, cognitive processes play no role in exposure therapy. But this would be the wrong inference.

The cognitive-behavioral therapist Stefan Hofmann proposed that the addition of cognitive treatment to exposure does not add to the therapeutic effect because of cognitive overlap across the procedures.[58] In other words, cognition is not just the basis of cognitive therapy but also contributes to exposure therapy and even to extinction. In particular, he argues that changes in cognitive expectations about the source of harm constitute a common currency that underlies extinction, exposure therapy, and cognitive therapy.

To evaluate Hofmann's hypothesis, we need to examine the role of cognition in extinction and exposure therapy more deeply. Specifically, we have to consider four questions. First, what is the nature of the cognitive functions that contribute to extinction? Second, to what degree do these cognitive functions overlap with the cognitive functions that contribute to exposure therapy? Third, to what degree do the therapeutic effects of exposure therapy depend on extinction (stimulus repetition) as opposed to the other therapeutic procedures that are also used (belief change, etc.)? And fourth, to what degree does the cognitive engagement in extinction and exposure therapy overlap with the cognitive processes that contribute to cognitive therapy that lacks an exposure component? Further, each of these questions has to be considered in light of the distinction between implicit cognition, which operates nonconsciously, and explicit cognition, which involves working memory and its executive functions and enables conscious awareness. Resolution of these issues will form the basis for a more sophisticated view of the neural underpinnings of psychotherapy than I achieved in *Synaptic Self.*

COGNITION IN EXTINCTION

Extinction in a laboratory setting is sometimes called *experimental extinction* and involves relatively pure stimulus repetitions for the purpose of scientific research as opposed to treating people with anxiety. When humans are the subjects of extinction studies, there is, of course, some instruction involved, but the procedure itself is mainly based on stimulus repetition; in studies of animals, it's stimulus repetition all the way.

When rats or people are being given stimulus repetitions in an extinction procedure, their brains are learning. Learning, including extinction learning, is a cognitive process, as it involves information processing to create internal representation of events.[59] The difference between exposure and extinction is not, therefore, simply that one involves cognition and one does not. As Hofmann suggested, they both do. At the same time, although some of the cognitive processes may overlap between extinction and exposure therapy,[60] others clearly differ, because there is more explicit cognition involved in the interaction between the therapist and client in exposure therapy than is typically the case when experimental participants simply undergo extinction procedures in the laboratory.

Let's revert to the language of conditioning to discuss the contribution of cognition to extinction. The simplest view of extinction is that the CS-US association acquired during Pavlovian threat conditioning is weakened when the CS no longer predicts the US. From the purely behaviorist point of view, all that is necessary for acquisition of threat conditioning is the co-occurrence of the CS with the US, and all that is required for extinction is the repeated presentation of the CS without the US.

During the initial conditioning (CS-US pairing) the organism learns that the CS predicts the US; during extinction, it learns instead that the CS predicts the absence of the US (a "CS–no US" association is involved). In effect, as a result of extinction training, the CS comes to predict safety. For example, if you are shocked when you turn on a lamp because of a wiring defect in the switch, the lamp-shock association (a CS-US association) will cause you to avoid touching the lamp. Then, if after getting the lamp repaired you cautiously turn it on and find you are not shocked, you can proceed with abandon in using it. You've formed a new association—a "lamp–no shock" association (CS–no US association)—that overrides or suppresses the original association.[61]

With the infusion of cognitive ideas into the study of learning and memory, predictions in terms of stimulus and responses came to be supplemented with cognitive mediators.[62] Specifically, the occurrence of the CS came to be thought of as triggering a "representation" of the CS-US association, such that the occurrence of the CS led to an "expectation" of the US, and it was this expectation that caused the response. Then, during extinction, the expectation established during conditioning is replaced by a new one that indicates that the CS is now safe.

For example, an influential psychological theory of conditioning by Robert Rescorla and Allan Wagner proposed that during conditioning the "surprising" (unexpected) outcome of a shock after a tone leads the brain to learn—that learning, in effect, occurs when new information is present.[63] Because a seemingly meaningless event like a tone is not expected to have any bad event follow it, the occurrence of the shock violates this prediction and leads to learning of the tone-shock association. Then, in extinction, the absence of shock conflicts with the learned expectation, and this prediction error triggers new learning. Prediction errors have been shown to be a significant factor in a variety of forms of learning, including Pavlovian threat conditioning and extinction as well as in the reinforcement of new instrumental responses in both animals and humans.[64]

In humans, when expectancies control behavior, it is common to regard predictions, beliefs, and decisions as explicit forms of conscious cognition. But as we've seen in previous chapters, research in animals and humans shows that behaviors that appear to be based on conscious beliefs and decisions can often be accounted for in terms of nonconscious information processing. Mark Bouton, arguably the leading expert in the study of extinction in animals, and a strong proponent of a cognitive view of extinction, views consciousness as irrelevant to extinction in animals.[65] If he is right, and I think he is, why should we suppose that it is necessary to account for extinction in people? We can, of course, be conscious of the absence of the US when the CS alone occurs in the course of extinction. Although conscious awareness of the absence of the US when the CS occurs alone may well underlie changes in your conscious memory of the CS-US relation, it is unlikely to account for the creation of the implicit memory that suppresses defense responses in extinction. Explicit and implicit memories of the same situation are formed and stored separately. As noted in Chapter 8, conscious awareness of the CS-US relation is not necessary for conditioning and presumably is not needed for extinction, either.

There is compelling evidence that interactions between the medial prefrontal cortex and the amygdala underlie threat extinction in animals[66] and humans.[67] The rat undergoing extinction freezes less to the CS because interactions between the medial prefrontal cortex and the amygdala change the ability of the CS to drive the lateral-to-central amygdala circuit and its outputs to PAG (the conditioning circuitry will be elaborated on in the next chapter). The rat is not freezing less because it is consciously thinking, *Oh, the tone no longer predicts the shock, so I don't need to freeze.* Stimuli are evaluated for their threat significance by retrieving memories (expectations based on past learning), but these are implicit memories stored as associations (CS-US, CS–no US) within the amygdala and do not require working memory and executive control functions or the conscious content that these enable. Similarly, the reason that a person who has successfully undergone exposure therapy for spider phobia can now look at spiders in magazines is that, as a result of extinction, the sight of the spider has become less capable of flowing through amygdala circuits and activating defense responses. This is probably not all that is needed to treat the phobia (the person probably has beliefs about spiders that need to be dealt with so that they do not cognitively reinstate the extinguished responses). But extinction of the implicit memory of the threat as an activator of defense circuits is what underlies the extinction of behavioral responses to phobic stimuli.

The fact that conscious decisions and the beliefs and values on which such decisions depend do not fully account for our reactions, actions, and habits does not mean that explicit, conscious thought plays no role in behavior, nor does it mean that changing explicit conscious thoughts is useless as an approach to therapy. It just means that implicit processes also play an important role, and consciousness should not be assumed to account for the observed effects until implicit cognition has been ruled out and explicit cognition has been directly implicated.

COGNITION IN EXPOSURE THERAPY

An important difference between exposure and extinction that needs special highlighting is that in exposure therapy the measure most used to gauge treatment effects is the client's self-report of a reduction in feelings of fear (or anxiety) in the presence of the threat stimulus or situation. By contrast, in a typical study of threat extinction, the effects of stimulus repetition are measured as a reduction in behavioral or physiological responses. Although it is common for researchers studying either humans or animals to state that extinction reduced "fear," the fact is that the goal of the study is usually to determine whether behavioral or physiological responses have been affected. A decrease in freezing in a rat or a change in some physiological response, such as skin conductance (a measure of perspiration) in humans, does not indicate that the conscious feeling of fear has been reduced—as I've pointed out, studies in humans have shown that the degree of fear measured behaviorally or physiologically often does not coincide with self-reports of subjective feelings.[68]

Because extinction alters the propensity of the threat CS to activate defensive circuits, it is tilted toward changing implicit processes. Exposure therapy adds layers of top-down cognition, such as reappraisal, to this process and assesses progress based on self-report. Experiments that have examined the neural basis of emotion regulation in healthy human participants are revealing. For example, as noted in the last chapter, studies by Liz Phelps and colleagues using extinction or other emotion regulation training techniques to implicitly change physiological responses implicated the medial prefrontal cortex,[69] whereas studies by James Gross, Kevin Ochsner, and their colleagues using top-down reappraisal strategies to change self-reported emotion found that the lateral prefrontal cortex played a more significant role.[70]

I will explain why these and other differences between extinction and exposure are significant by discussing a popular form of exposure therapy called *prolonged exposure*.[71] This method is based on *emotional processing theory*, as proposed by Edna Foa and Michael Kozak.[72] The basic concept of prolonged exposure therapy is that fearful feelings must be elicited and sustained during repeated exposure until fear reduction occurs, allowing disconfirmation of false belief about the irrationally feared object or situation's ability to cause harm. If fear is not fully activated, it will not fully extinguish, and problems will continue.

Foa and Kozak built on Peter Lang's idea that fear is represented in the brain in the form of *fear structures*, or schemas,[73] similar to Aaron Beck's notion of automatic thoughts and beliefs being represented in schemas. (Recall that schemas are also part of the psychological construction theory of emotion, and my theory described in Chapter 8.) This was an elaboration of Lang's earlier idea that fear and anxiety could be accounted for in terms of response systems. The fear structure is viewed as a program (in the sense of a computer program) for escaping or avoiding danger and includes several kinds of stored propositions: It includes propositions about threats—when the threat signal (CS) occurs, a bad thing (US) follows; propositions about physiological changes—when the CS occurs, I sweat and my heart beats faster; propositions about actions—if I perform a certain response when the CS occurs, the US will be avoided; propositions about the meaning of stimuli and responses—the CS makes me feel afraid, and avoiding the

CS prevents fear.[74] Input stimuli that match the stimulus information stored in the structure activate the fear program and produce behavioral, physiological, and verbal responses that were the centerpiece of Lang's original three-response system theory. Differences in the way fear structures represent threats distinguish healthy from pathological fear structures, or fear structures in people with different forms of pathological fear and anxiety.

According to emotional processing theory, the two key factors required for a therapeutic reduction in fear and anxiety are complete activation of the fear structure by fear-arousing stimuli and the insertion of new information into the fear structure that is incompatible with the pathological information. Both of these are thought to be accomplished by prolonged exposure. By forcing the person to approach distressing but safe objects or situations and eliciting fearful feelings, and finding out that harmful consequences do not occur, corrective information is added to the fear structure, and fear and avoidance are reduced.

In line with contemporary learning theory, emotional processing theory holds that new information does not replace old information in the fear structure but instead creates a competing memory that suppresses the old memory.[75] The new memory (new fear structure) lacks the pathological associations that trigger fear and that support and sustain avoidance and prevent extinction; therefore, negative emotions (fear and anxiety) and avoidance are reduced. Also in line with learning theory, the emotional processing approach assumes that learning results when there is a discrepancy between what is expected and what actually happens, leading to a modification of future expectations.[76] But unlike pure extinction, in which implicit expectancies are changed directly by the stimulus repetition procedure, in implosive therapy and other therapies, expectancy change is as much, if not more, a top-down process.

As I have noted many times now, learning, whether it involves Pavlovian or instrumental conditioning or higher cognitive forms of learning, is a process by which synapses are changed. Synaptic change is a physiological process sustained by molecular events. We are not consciously privy to these activities. In the case of explicit memory, we can be conscious of the content that is stored but not of the processes that enabled the storage. But in implicit systems, the changes are nonconscious all the way. We can, in other words, explain a change in defensive behavior and physiological responses without assuming that a change in a subjective feeling of fear mediates the behavioral change. Indeed, as noted, subjective feelings of fear do not always correlate well with physiological and behavioral responses elicited by threats.[77] Emotional processing theory would explain this as the result of an incomplete activation of the fear structure, and so the exposure only partly extinguishes the fear structure's control over overall fear response, which includes behavioral, physiological, and cognitive (as assessed verbal responses) components.

But this assumes that behavioral, physiological, and verbal responses are all products of a single system in the brain (the fear structure or program). I have argued against such notions of a unitary fear system throughout this book. Although emotional processing theory has been very influential, it has also been challenged on certain points.[78]

Although Lang stated that fear is not a lump in the brain that can be palpated the concept of a fear structure that has to be changed in its entirety seems to lend credence to the idea of a system or module in the brain dedicated to making all things related to fear. I support the idea of a defensive system that detects and responds to threats, but I don't believe that fear is a direct product of this system. The defense system works implicitly, whereas fear is constructed as conscious feeling via cognitive systems responsible for any kind of conscious awareness. For this reason, I believe that the implicit and explicit processes have to be targeted separately with different therapeutic strategies. A similar idea has been proposed by Chris Brewin and Tim Dalgleish in their multiple representation theory, in which they suggest that both verbally accessible and automatic implicit processes underlie problems with fear and anxiety and should be treated separately.[79]

Therapeutic procedures that target systems that work implicitly are best at changing implicit memories, whereas procedures that engage explicit processes and working memory are best at changing explicit processes. In that light, the specific aspects of exposure therapy that depend on extinction or other emotion regulation functions contributed to by medial prefrontal–amygdala connections and related components of defensive control circuits are best for changing how a stimulus activates defensive circuits and controls defensive behaviors, physiological responses, and avoidant behaviors. However, the aspects of exposure therapy that change maladaptive beliefs and other cognitions that lead to cognitive avoidance, and that aim to store new explicit memories that compete with irrational, pathological memories brought into consciousness through therapy, are best approached through talk, instruction, reappraisal, verbal reinforcement, and so on. It would be of interest to be able to identify the therapeutic effects of a pure extinction treatment (minimal verbal exchanges or instruction, with stimulus repetition emphasized) versus a more traditional exposure treatment, as this makes it possible to more effectively distinguish the effects due to nonconscious extinction from those due to explicit cognitive change. Also of great interest would be studies that use nonconscious (masked) stimulus presentations as a way of conducting exposure therapy.

As far as I know, an experimental comparison of the benefits of exposure therapy (as usually practiced) and pure stimulus repetition (as typically carried out in research on extinction) has not been made. Nor has nonconscious extinction (presentation of the to-be-extinguished stimulus with masking or other techniques that bypass consciousness) been attempted in an experimental or a clinical setting. However, very interesting results have been reported in studies that have examined the effects of a single masked exposure versus a prolonged, freely seen exposure to a phobic stimulus in phobic patients.[80] This is not extinction per se, or is a very limited version of it, because only a single trial was used, but the effects were dramatic. In one study the single masked (nonconscious) stimulus reduced avoidance behavior, whereas the freely seen stimulus did not. In a second study the effects on both avoidance and subjective distress were compared for masked and freely seen stimuli. The masked stimulus reduced avoidance but did not affect subjective distress. The freely seen stimulus had the opposite effect—it did not reduce avoidance but did reduce distress. Such results illustrate the importance of recognizing the different contributions of implicit and explicit systems in controlling behavioral responses and subjective feelings. I will return to these studies in the next chapter, where I discuss evidence that a single exposure to a stimulus can induce persistent changes via processes related to memory reconsolidation, also to be discussed in the next chapter.

Under normal viewing conditions and repeated exposures, changes in both defensive behavior and conscious feelings in people may result from, but for different reasons. The effect on the defensive behavior is a direct result of stimulus repetition on implicit circuits. The effects on feelings can come about in two ways.

First, feelings may change secondary to the control of the defensive circuits. If the CS-US association in the amygdala is suppressed by extinction, threat-elicited responses in the brain and body will be reduced. To the extent that they contribute to nonconscious defensive motivational states, components of which feed the construction of a feeling of fear via cortical areas, extinction can reduce feelings of fear or anxiety.

Second, stimulus repetition without consequence can also change the cognitive representation of the CS-US association in explicit memory. One of the advantages of human cognition is that it can make decisions on the fly rather than having to rely on new learning. When explicit cognition is in control, if you observe that something that was previously a danger is no longer so, you can quickly reappraise it and start to think about it differently and act differently toward it. This is why overcoming avoidance helps change conscious beliefs. With conscious expectations changed, previously threatening stimuli can be viewed in a new light: They no longer trigger top-down activation of fear schemas that normally lead to the conclusion, "I am in danger and feel afraid," and also normally activate defense circuits and their physiological consequences that then support the cognitive construction of the feeling. The greatest benefit is likely to be obtained if both the defensive circuit and the explicit representations are changed. Although my conclusion is similar to that of prolonged exposure proponents, I reach it from a different perspective.

In sum, if only the explicit or only the implicit system is treated, the untreated system can reinvigorate the fear. The implicit system can seize attention, and attention can retrieve past memories about the danger of the stimulus, triggering new feelings of fear and reestablishing the belief that the stimulus is dangerous. For its part, the explicit system can lead to worry and avoidance and create feelings of fear in an abstract cognitive sense that can then release stress hormones that revitalize the CS-US association, reestablishing threat sensitivity, hyperarousal, behavioral avoidance, and other consequences of amygdala-based threat conditioning. Indeed, stress has long been recognized to be a potent trigger that reestablishes extinguished defensive reactions in animals and phobic fears in people.[81]

It is well known to anxiety researchers that people's fear and anxieties are often irrational and not particularly modifiable by relying upon logical reasoning alone.[82] People afraid of riding in elevators or flying in planes know rationally that the likelihood of being harmed in them is very low, yet this conscious knowledge does not override the implicit control of behavior, and the feeling of fear and anxiety that results.

Foa, Kozak, Lang, and others who argue that incomplete extinction is why therapy fails may be right. But not because there is a unitary fear structure that has to be completely activated. It is, I think, because exposure procedures, as usually carried out, do not attempt to treat implicit and explicit processes separately, and this leads to incomplete extinction and/or incomplete thought change and competition for cognitive resources, as described in a later section. First, let's talk about worry.

TREATING WORRY WITH EXPOSURE THERAPY

The chief feature of pathological anxiety is chronic worry.[83] People with GAD, for example, are not specifically worried about spiders, elevators, or social situations—they just worry. When a person is worried, working memory is occupied and decreases one's ability to perform effectively and efficiently.[84] Given that there is no specific object or situation to extinguish, you may wonder how exposure therapy can be used to treat generalized anxiety. This will be clearer after we consider the nature of worry in more detail.

Thomas Borkovec, who specializes in the science of worry,[85] notes that it usually appears in the form of internal verbalization—thoughts in the form of words. Inner discussion with one's self makes threat more abstract, and allows disengagement. This helps avoid deeper, more concrete processing that more effectively triggers emotional arousal and prevents disconfirmation about the actual degree of danger posed. Worry is the cognitive equivalent of behavioral avoidance. As noted by Borkovec, "Despite creating conditions that restrict the individual's life and/or generate other kinds of disturbance, s/he can reduce some distressing experiences by avoiding their source."[86] Worriers, Borkovec explains, escape from fearful images by thinking about the future in abstract verbal terms (if the event worried about comes to pass, awful things might happen).

We know that anxious people are highly sensitive to threats, so if they perceive one, whether real or imagined, semantic and episodic threat-related memories will be retrieved in the form of thoughts and images. This leads to catastrophic thinking about negative scenarios and gives rise to cognitive coping strategies that anticipate negative outcomes and attempt to prevent them from occurring. Because worriers worry about so many things, the worst-case scenario seldom happens, and through negative reinforcement, their worrying is perpetuated by its perceived ability to avoid the worst-case outcome. Worry occurs to a significant degree in all anxiety disorders (Table 10.2).

Michael Eysenck's *attention control theory of anxiety* conceptualizes the cognitive basis of worry.[87] He distinguishes two attention systems: one that is goal directed and one that is stimulus driven. His theory proposes that anxiety and worry disrupt the goal-directed system, allowing the stimulus-driven system to dominate. Worry produces thoughts that use the executive attention resources of working memory, making it less available for dealing with one's professional, personal, and social obligations. Executive functions are also engaged by the consequent efforts to avoid thinking about threats and their consequences. One becomes more distractible because attention is being competed for by worry, and it is harder to stay on task. With working memory focused on threats, the stimulus-driven form of attention allows threatening stimuli to more easily seize attention. In addition, as noted earlier, anxious people treat nonthreatening stimuli as dangerous because their ability to discriminate between threat and safety is weakened, and they overvalue the danger of weak threats. Threats come to play a larger role in their lives than would otherwise be the case.

With this background, let's consider how exposure therapy might be useful in treating worry in GAD.[88] The idea is to use anxious thoughts as instances of exposure and then to apply treatment strategies to them. In initial sessions,

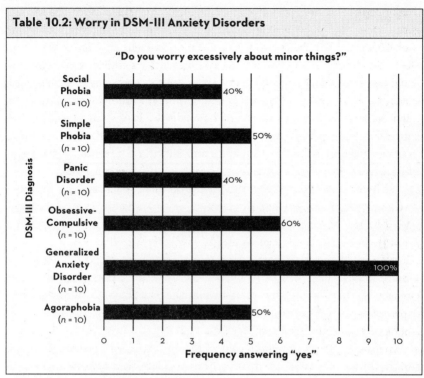

Table 10.2: Worry in DSM-III Anxiety Disorders

"Do you worry excessively about minor things?"

Based on Figure 4.3 in Sanderson and Barlow (1990)

clients discuss their concerns and are taught coping strategies to deal with their anxieties when they arise. Such strategies include relaxation training (breathing, muscle relaxation, meditation), self-control desensitization (frequent rehearsal of relaxation when worrisome thinking or imagery-induced anxiety occurs), and cognitive restructuring (identification of frequent automatic thoughts and beliefs, development of multiple alternative perspectives, behavioral testing of predictions, reappraisal, and decatastrophizing). They are then encouraged to attend to any shifts in their level of anxious feelings and to note whether they occur in conjunction with worrisome thinking, imagined threats or their future danger-ous outcomes, physiological responses, behavioral avoidance, and any external cues that might be associated with these symptoms. By becoming aware of the anxiety cues, the patient can use coping strategies early and prevent the escala-tion of anxiety.

FROM THE ANIMAL LABORATORY TO THE THERAPIST'S OFFICE: TARGETING IMPLICIT PROCESSES

It's common these days for medical practitioners to speak about the need to treat the whole person, whether her problem involves heart disease, cancer, anxiety, or depression. Obviously, it's important for a therapist to understand a patient's particular issues and their relation to her overall life. In that respect, verbal exchanges based on explicit cognition are essential. But while the therapist should take the whole person into account, the best way to do so might *not* be to treat the whole person all at once. Specifically, I propose that behavioral and physiological responses that are controlled by implicit processes should be, to the extent possible, treated separately from thoughts and behaviors that are controlled by explicit processes and top-down cognition.

The idea of targeting specific processes is not a novel idea. Lang's three-response approach suggested that different components of the fear structure all needed to be treated. Cognitive therapists likewise use a family of approaches to target specific symptoms associated with particular disorders.[89] Barlow's *unified cognitive therapy* uses antecedent reappraisals, avoidance prevention strategies, and modification of emotionally driven behaviors to target different dysfunctional emotion regulation processes.[90] The multiple representation theory of Brewin and Dalgleish, mentioned earlier, also proposes this approach. What I have in mind is not a replacement for these ideas but instead a complement to them.

A large part of the reason that animal research is so successful in relating behavioral and cognitive functions to brain mechanisms is that specific processes can be targeted in relative isolation. You don't have to ask rats to think about their pasts or their concerns for the future; you simply present them with tones and shocks together and then just the tone without the shock. The rat may be "thinking" in a rat kind of way about what is happening, but that is not what changes the brain.

In therapy, by contrast, thinking is usually assumed to be part of the process. But what I am suggesting is that the therapeutic goals of exposure therapy might be more effectively achieved if the procedure were conducted more like laboratory extinction, by stripping away some of the cognitive processes that are typically involved. The studies mentioned above showing a dissociation between masked (nonconscious) and freely seen (conscious) exposure of phobic stimuli suggest that this might work.

Now you may be asking yourself, why would this need to be done if implicit and explicit processes are independent? Shouldn't extinction just proceed automatically even if the conscious mind is otherwise engaged? The fact is, although we can isolate implicit processes in the laboratory, especially in animal studies, in real life, as we have seen, the systems interact. Implicit sensory processing contributes to explicit perception, implicit memory processing contributes to explicit memory, and implicit cognitive processing contributes to working memory and consciousness. But it also works the other way around: Explicit processes can initiate implicit processes that carry out a task. For example, when people are told to reappraise some stimulus, the task is initiated with verbal instruction and voluntary control, but these processes then set up other processes that proceed implicitly and interact with the amygdala to change its activity.

What this means is that although certain brain chores may ultimately depend on explicit or implicit systems, in real time in real life these systems work together, and may well utilize common resources. So if you are attempting to change beliefs in the same session in which you are extinguishing, you are asking the brain to learn and store memories in a way that might not be ideal. The psychologist Michelle Craske has similarly argued that extinction and cognitive interventions should be separated.[91]

As an example, let's review reappraisal. Although belief change through reappraisal is initiated by top-down cognition and thus involves the lateral prefrontal cortex, it indirectly affects the amygdala. One way it does so is by way of interactions between the medial prefrontal cortex and the amygdala,[92] which are the same areas that are involved in extinction.[93] However, we don't know if exactly the same circuits, cells, and synapses in the prefrontal cortex and amygdala are involved in reappraisal and extinction. But even if they only partially overlap, instructed cognitive reappraisal during extinction might compete with the processes that are attempting to change behavioral and physiological responses to the threats via extinction, and vice versa.

Further, as the Eysenck model suggests, in an anxious individual, working memory can be distracted by task-irrelevant threat information. So if the person is talking and processing instructions while being exposed to threats, and his brain has to undergo extinction learning and change beliefs simultaneously, the cognitive systems required for belief change may be distracted. In other words, the fact that exposure therapy, as is typically practiced, almost always involves top-down cognition that engages the lateral and medial prefrontal cortex may mean that cognitive change and the implicit behavioral change are both less effective.

Nonconscious exposure therapy may be a useful option, either as a supplement to or a replacement for traditional exposure. Whether nonconscious extinction is a feasible method or not has not been explored, but Liz Phelps and I are currently testing it. The data may not come fast enough to appear in this book, but consult my laboratory's website, which includes my frequently updated publication list (www.cns.nyu.edu/ledoux).

Recognizing the distinction between explicit and implicit processes helps understand why research in animals can provide useful insight that can potentially improve people's lives. Explicit processes are important, but animal research, such as the kind of work I do, is especially useful in helping us understand the implicit processes. In the last chapter, I will review some findings from animal research that have either contributed to new treatment strategies or that may be useful to consider in the development of new approaches to psychotherapy.

CHAPTER 11

THERAPY: LESSONS FROM THE LABORATORY

"How happy is the blameless vestal's lot!
The world forgetting, by the world forgot.
Eternal sunshine of the spotless mind!
Each pray'r accepted, and each wish resign'd."

—ALEXANDER POPE[1]

"Changing the content of our memories or altering their emotional tonalities, however desirable to alleviate guilty or painful consciousness, could subtly reshape who we are, at least to ourselves. With altered memories we might feel better about ourselves, but it is not clear that the better-feeling 'we' remain the same as before."

—PRESIDENT'S COUNCIL ON BIOETHICS[2]

In the fall of 2000 I started getting calls and emails from people asking me to erase their memories. Karim Nader, Glenn Schafe, and I had recently published a paper in the journal *Nature* with a rather technical title, "Fear Memories Require Protein Synthesis in the Lateral Amygdala for Reconsolidation after Retrieval."[3] In this study we conditioned rats with a tone and shock and then later presented them with the tone alone after a drug that blocks protein synthesis had been infused in the lateral amygdala (LA), a key area of the amygdala where the tone-shock association is stored. When tested the following day, or at any time afterward, the rats behaved as though they had never been conditioned. The procedure, in other words, seemed to erase the memory that the tone was a signal of danger. Toward the end of the short piece, we proposed that it might be possible to use a technique like this (but without having to inject a drug directly into the amygdala) to dampen traumatic memory in people with PTSD.

The *New York Times* published a summary of the piece,[4] and letters to the editor ensued; some of the correspondents were fascinated, but others appalled, thinking that it might be possible to delete memories. A trauma therapist, for example, implied we were playing with fire, as the experience of trauma becomes part of one's self and needs to be remembered: "The idea may be appealing to a culture that uses avoidance and denial to deal with painful events. Yet despite the pain associated with traumatic events, they represent our personal, social and political realities. . . . Would we really be better off, for example, if Holocaust survivors forgot what happened? As a culture, would we be doomed to repeat our less desirable behaviors? What an endless nightmare that would be."[5] President George W. Bush's science advisory panel had a similar outlook, essentially declaring that memory is sacrosanct and should not be played with by scientists, even if the goal is to make people feel better. (See the epigraph at the head of the chapter.)

People who were actually suffering with troubling memories of horrible experiences, however, desperately wanted the treatment, no matter how unproven in humans, even if it did mean that they would lose part of their self (which is, in fact, precisely what they desired). These individuals wanted to have their conscious memories erased, though what we actually did in the study was dampen implicit memories controlled by defense circuits.

We continued to publish in this area, and each new paper led to more press and more phone calls and emails. I'll return to the details of this research area, and its implications, later. For now, I've introduced the subject briefly as a way to consider the question of whether memories should be off limits as a therapeutic target.

The fact is, every exchange between two people involves a process of memory retrieval and memory storage. Without memory, social relations would not be possible—for example, when you interact socially you draw upon remembered details about the other person, shared interests and conflicts, and the facts and/or experiences you want to convey. But more to the point for the purposes of this discussion is that many—and maybe every—form of psychotherapy engages and changes memory in some way. Psychoanalysis was based on extracting repressed memories and bringing them into consciousness, where they could be reinterpreted. Cognitive therapy is a process of changing beliefs, which are memories, and learning and remembering new coping skills. In exposure therapy, extinction creates new implicit memories that compete with existing ones to prevent the latter from causing trouble. Fundamentally, our study simply offered another way to achieve what therapy often tries to accomplish—to prevent troubling memories from causing trouble.

There are a number of new approaches that have been developed to treat fears and anxieties, many by changing memories. These mostly involve methods that change the brain behaviorally through learning rather than by way of pharmacological treatments. In some cases drugs are used acutely to boost the learning effect, rather than as a prolonged treatment, and once the change is established, drugs are no longer needed (it's better thought of as temporary use of pharmaceutical agents to facilitate the learning of coping skills rather than as drug therapy).

The procedures I am going to describe in the following sections are targeted primarily at implicit memory processes. Many are focused on ways to improve extinction and thus improve exposure therapy. But I will also describe alternatives to extinction, especially procedures that might literally erase threat memory, rather than simply inhibit it the way extinction does. None of the treatments, however, is a panacea that can, on its own, cure problems with fear and anxiety. For therapy to have maximally beneficial and persistent effects, it may well be necessary to change both conscious and nonconscious memories that contribute to distress.[6] Therefore, the emphasis of this chapter on the neuroscience of implicit learning does not mean that I believe there is no role for explicit processes such as talk, insight, compassion, and person-to-person interactions. As I argued in the last chapter, both conscious and unconscious processes have to change, and this might be best achieved by targeting explicit conscious memories and implicit memories separately.

HOW EXTINCTION WORKS IN THE BRAIN

I've already spent a good deal of time on the nature of extinction and its role in exposure therapy. But because it has limitations as a therapeutic tool, much effort has gone into trying to make extinction, and exposure, work more effectively. Before discussing its drawbacks, I want to expand on the underlying circuitry of extinction in more detail than I did in Chapter 4 because this information will help in understanding how its limits are manifested in the brain and how extinction might be modified to overcome them.

In order for extinction to reduce the ability of a Pavlovian CS to elicit defense responses, the amygdala's control over these responses has to be changed. Key to this process is the ventromedial prefrontal cortex (PFCvM) and its ability to regulate the amygdala-housed circuits that store the CS-US memory, which when activated results in the expression of defense responses. The idea that this is how extinction works emerged from studies done in my laboratory by Maria Morgan in the early 1990s.[7] We had previously found that animals with lesions of the visual cortex seemed to be unable to stop freezing to a light paired with shock, even after extensive extinction training.[8] We didn't think that the visual cortex itself was responsible for extinction but thought instead that damage to the visual cortex might have prevented information about the visual CS from reaching extinction circuits, rendering the CS-US memory indelible (resistant to extinction).

In trying to understand how these irreversible, or what we called indelible, amygdala memories might be sustained, we considered the fact that once animals and people with prefrontal cortex damage have acquired some cognitive or behavioral response, they tend to repeat this behavior (i.e., perseverate), even when the response is no longer useful.[9] Perhaps lesions of the prefrontal cortex in rats had given rise to an "emotional perseveration" of CS-elicited freezing. Morgan then examined the effects of PFCvm lesions in rats and found that, again, the animals could not stop freezing in response to the CS. It was as if removal of PFCvm influences resulted in an out-of-control amygdala, one that responded to stimuli that were, objectively speaking, no longer threatening. This immediately suggested that the type of unregulated fear and anxiety that occurs in people with anxiety disorders might involve some dysregulation of prefrontal-amygdala circuits. It was as if the amygdala was the accelerator of defensive reactions and the prefrontal cortex the brake upon them (Figure 11.1). Malfunction of the brake makes the expression of the reactions hard to control. This idea has since been supported by research in animals and humans and is now commonly accepted.[10]

At the time Morgan was doing these studies, Greg Quirk, another member of the laboratory, became fascinated with extinction and soon went on to become a leading investigator of the prefrontal cortex control of extinction.[11] His work pushed the exploration of how the PFCvm regulates the amygdala to a new level and inspired many others to enter the field.

A simple diagram of the extinction circuitry is shown in Figure 11.2. The three major players are the amygdala (which stores the CS-US association), the PFCvm (which regulates the amygdala), and the hippocampus (which encodes the context in which the CS-US association is learned initially and extinguished).

Anything that changes the ability of the CS to activate the CS-US association in the lateral amygdala (LA) or the ability of the CS-US association to control CeA outputs will affect the expression of defensive behaviors and physiological responses in the presence of a threat CS. Morgan's studies in rats showed that two regions of the PFCvm make different contributions to amygdala regulation.[12] The prelimbic region modulates the ability of an individual CS presentation to control

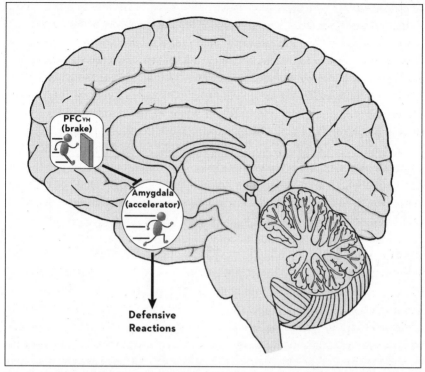

Figure 11.1: The Accelerator and Brake on Defensive Reactions in the Brain.
The amygdala is the driving force (accelerator) of defensive responses and supporting physiological responses in the brain and body. The ventromedial prefrontal cortex (PFCvm) regulates (applies the brake to) the amygdala, thus adjusting the occurrence and intensity of defensive
reactions as situations change. This mechanism is often impaired in people who suffer from
problems with fear and anxiety (not because the amygdala is the source of feelings of fear or
anxiety but because amygdala-dependent brain and body responses contribute ingredients
that are assembled into a feeling of fear or anxiety).

amygdala outputs for that CS event. By contrast, the infralimbic region of the
PFCvm, via extinction training involving multiple CS presentations, shapes a
new association (a CS–no US association) that overrides the CS-US association
stored in the LA, engendering a more lasting change that applies to later occurrences of the CS. In other words, the prelimbic region regulates the expression of
CS-elicited defense responses controlled by the CeA on an exposure-by-exposure
(in the laboratory, trial-by-trial) basis but does not produce persistent changes,
whereas the infralimbic cortex engenders changes during extinction learning
that persist over exposures and into the future. The hippocampus contributes by
encoding the context in which both initial learning[13] and extinction[14] occur. As

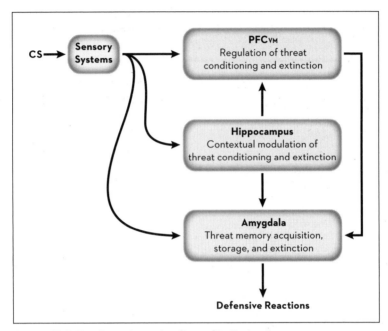

Figure 11.2. Key Brain Areas Implicated in Extinction.

The amygdala plays an important role in the acquisition, storage, expression, and extinction of threat memories. The ventromedial prefrontal cortex (PFCvm) regulates the acquisition, storage, expression, and extinction of threat memories by the amygdala. The hippocampus learns about the context of acquisition and modulates the expression and extinction of threat memories in relation to context.

we will see, extinction is highly dependent on context, which limits the degree to which the beneficial effects of extinction generalize beyond the therapeutic situation.

So how does the new learning that occurs in extinction (learning of the CS–no US association) prevent the expression of the original memory (CS-US association) and the defense responses it controls? To understand this we have to consider the cellular and molecular mechanisms in the amygdala threat-learning[15] and extinction[16] circuitry in a bit more detail than I have provided so far, mainly highlighting key points rather than the details of their workings. To do this I will focus on the contribution of the amygdala and PFCvm to initial learning and to subsequent extinction learning (Figure 11.3).

Meaningless stimuli are prevented from activating LA cells and triggering the defense circuitry by means of a strong network of GABA inhibitory cells.[17] During threat learning the CS and US converge in the LA, creating the CS-US

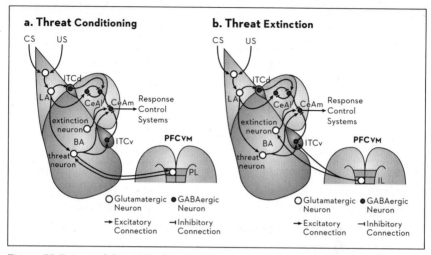

Figure 11.3: Amygdala and Prefrontal Circuits Underlying the Conditioning and Extinction of Defense Responses.

Threat conditioning involves convergence of the conditioned stimulus (CS) and the unconditioned stimulus (US) in the lateral amygdala (LA). LA sends connections to the intercalated cells (ITC), lateral part of the central nucleus (CeAL), and threat neurons in the basal amygdala (BA). BA threat neurons connect with ITC and the medial part of the central nucleus (CeAm). Within CeAl different cells receive inputs from LA and ITC, but the two cell populations are interconnected. The prelimbic region of the prefrontal cortex (PFCPL) connects and regulates threat neurons in BA. Outputs of CeAl connect to CeAM, which connects with response control areas. Similar circuits are involved in extinction except that the PFCIL connects with extinction rather than threat neurons in BA. Extinction neurons in BA then regulate CeAM.

BASED ON LEE ET AL (2013).

association and establishing the ability of the CS, when it later occurs, to shut down GABA inhibition so that the CS can activate the LA, and, in turn, lead to the expression of defensive behaviors and physiological responses by way of CeA outputs.[18] As shown in Figure 11.3, the LA connects with the CeA in several ways: (1) by direct connections from the LA to the CeA; (2) by connections from the LA to the basal amygdala (BA), and from there to the CeA; and (3) by connection from the LA and BA to a group of GABA inhibitory cells called the intercalated masses, which then connect to the CeA.[19] Information flow through these various connections is regulated by complex interactions between excitatory and inhibitory cells.[20] The result is that the pattern of excitation and inhibition within the amygdala enables the CS to drive CeA outputs. But there is more. Within the BA, evidence suggests that there are "threat neurons" that represent the CS-US association and that link the LA to CeA (either directly or by way of the intercalated cells)[21] and "extinction neurons" that prevent the threat neurons from activating the

CeA.[22] Within the CeA, interactions between two subareas, the lateral and medial CeA, are involved.[23] The lateral CeA receives the inputs from the other amygdala areas (LA, BA, and intercalated cells), and the medial CeA is the source of outputs to the hypothalamus and brain stem that control the various consequences of defense circuit activation (see Chapter 4).

The expression of CeA-controlled defense responses, once learned, is regulated by connections from the prelimbic area of the medial prefrontal cortex to the BA, which then connects with the CeA directly and by way of the intercalated cells.[24] The extinction circuitry is similar but with one important difference. Connections from the infralimbic region, rather than the prelimbic region, of the medial prefrontal cortex to the basal and intercalated cells of the amygdala are involved.[25] Furthermore, connections from the hippocampus to the BA (not shown in Figure 11.3) are important for the discrimination of different contexts and their relation to the danger signaled by the CS.[26] All of these neural interactions underlying the expression and extinction of defense responses are further regulated by neuromodulators (such as norepinephrine and dopamine) and peptides (such as endocannabinoids, enkephalin, substance P, oxytocin, and brain derived nerve growth factor, among others).[27]

There are at least three components to the process that prevent the original CS-US association from being expressed after extinction.[28] One is that the CS–no US association exerts inhibitory control over LA processing (thus weakening the ability of the CS to activate the CS-US association). Another is that connections from the infralimbic cortex to inhibitory cells in the basal nucleus and the intercalated area further suppress the ability of the CS to flow from the LA to the CeA. Finally, extinction interferes with the balance between the lateral and medial areas of the CeA and prevents the expression of outputs of the medial CeA to response control circuits.

Like all forms of learning,[29] extinction requires the synthesis of proteins in neurons that are learning and storing the new information. In this case, protein synthesis is required in both the infralimbic cortex[30] and the amygdala[31] for the effects of extinction to persist as a long-term memory.

For most if not all forms of memory, in most if not all organisms, the protein synthesis process underlying memory storage is triggered by the activation of certain genes within the neurons forming memory. A key activator is the gene transcription factor, cyclic AMP response binding element protein (CREB).[32] Extinction learning is no exception because it, too, involves CREB-dependent protein synthesis.[33] CREB activity is regulated by neuromodulators, such as norepinephrine and dopamine, that are released by way of CeA outputs. Many other molecules and steps come into play as well.[34] The result is that during extinction learning, recently active synapses on those neurons that undergo protein synthesis are strengthened, and new patterns of synaptic connections across the various neurons in the network constitute the memory. Reactivation of those synapses results in retrieval of the extinction memory, which suppresses the original CS-US association.

The detail in which these circuits have come to be understood provides a strong foundation upon which to understand extinction and ways that it might be improved to enhance exposure therapy. First, though, let's consider some of the limitations of extinction that need to be overcome.

LIMITATIONS OF EXPOSURE THERAPY ARE REVEALED BY LABORATORY STUDIES OF EXTINCTION

In spite of the wide use of exposure-based techniques, exposure is limited in what it can achieve. Many of the limits are limitations of extinction itself. Table 11.1 lists several, which are also graphically depicted in Figure 11.4.

Extinction is strongly dependent on the context in which it occurs.[35] If a rat is conditioned with a tone and shock in one chamber, and then the effects of the tone are extinguished in a novel chamber, the rat will cease freezing to the tone in the novel extinction chamber but will still freeze to the tone in the original conditioning chamber or in another chamber in which extinction has not occurred.[36] The contextual control of extinction depends on the hippocampus in

---

Table 11.1: Limits of Extinction

1. Context Dependence: extinction in one context does not generalize well to other contexts

2. Spontaneous Recovery: the effects of extinction often dissipate over time such that the conditioned stimulus becomes threatening again

3. Renewal: exposure to the original conditioning context can also reverse the effects of extinction and revive the threat-inducing potential of the conditioned stimulus

4. Reinstatement: re-exposure to the unconditioned stimulus can also reverse the effects of extinction

5. Stress-Induced Reversal: stressful experiences completely unrelated to the original threat learning can also undo the effects of extinction

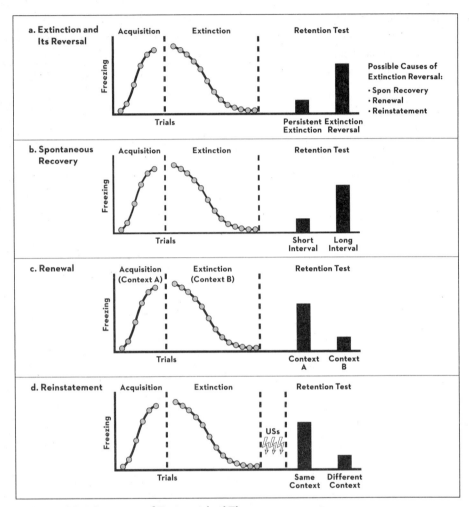

a. Extinction and Its Reversal

Acquisition   Extinction   Retention Test

Freezing

Trials

Persistent Extinction
Extinction Reversal

Possible Causes of
Extinction Reversal:
• Spon Recovery
• Renewal
• Reinstatement

b. Spontaneous Recovery

Acquisition   Extinction   Retention Test

Freezing

Trials

Short   Long
Interval Interval

c. Renewal

Acquisition   Extinction   Retention Test
(Context A)   (Context B)

Freezing

Trials

Context Context
A   B

d. Reinstatement

Acquisition   Extinction   Retention Test

Freezing

USs

Trials

Same   Different
Context Context

Figure 11.4: Recovery of Extinguished Threats.

a. Through extinction, the ability of learned threats to control defense responses is weakened. Successful extinction results in low levels of defensive responses like freezing. However, the original threat memory can be revived, suggesting that extinction is a form of inhibitory learning that suppresses the original threat memory. Recovery can occur due to the mere passage of time (spontaneous recovery), exposure to the context in which the original memory was formed (renewal), or by exposure to the unconditioned stimulus (reinstatement). **b. Spontaneous Recovery** occurs when there is a long delay between conditioning and extinction; it is less likely to occur with a short delay. **c. Renewal** occurs in the training context but not in novel context. **d. Reinstatement** involves the delivery of the unconditioned stimulus (US) after extinction. If the subsequent test occurs in the same context, then freezing is reinstated.

FROM QUIRK AND MUELLER (2008), ADAPTED WITH PERMISSION FROM MACMILLAN PUBLISHERS LTD.: *NEUROPSYCHOPHAR-MACOLOGY* (VOL. 33, PP. 56–72, © 2008, AND MYERS AND DAVIS (2007), ADAPTED WITH PERMISSION FROM MACMILLAN PUBLISHERS LTD.: *MOLECULAR PSYCHIATRY* (VOL. 12, PP. 120–50)), © 2007.

animals and humans.[37] This context specificity restricts the beneficial effects of exposure therapy done solely in the therapist's office and has led to the idea that exposure should be done in as many different situations as possible,[38] especially in the real world, where genuine threats occur.[39] Accordingly, as part of cognitive behavior therapy, people are sometimes taught to follow exposure protocols to use when they encounter threats in daily life.[40]

Another limit is that the effects of extinction can be reversed, restoring the original CS-US association and, in effect, overriding the CS–no US association established during extinction learning. Pavlov discovered one version of this in his pioneering studies of conditioned salivation in dogs.[41] Once the dogs learned to salivate to the sound of a bell, the response could be extinguished by ringing the bell repeatedly without presenting food. But when the bell was rung again after a few days with no exposure to it, the dogs salivated as they had before. Pavlov called this phenomenon *spontaneous recovery*—the reversal of extinction by the passage of time. Later work showed that spontaneous reappearance is a general feature of extinction, as it occurs with threat[42] as well as appetitive conditioning,[43] and affects animals and people alike.[44] Unfortunately, the therapeutic effects of exposure therapy, because they depend on extinction, are also subject to being compromised by spontaneous recovery. (Note that in this situation recovery does not imply healing, but instead more suffering).[45]

Extinction can also be reversed by processes known as *reinstatement* and *renewal*.[46] In renewal, exposure to the context in which conditioning occurred is sufficient to revive the CS-US association.[47] Renewal depends on contextual processing by the hippocampus in rats[48] and humans.[49] Reinstatement, by contrast, occurs when one is exposed to the aversive US by itself, which, like the context or the CS itself, can activate the original CS-US association and revive the conditioned responses.[50] If a rat that has been completely extinguished after threat conditioning is exposed to a single presentation of the US, it will again freeze to the CS even though the CS and US are never re-paired. The fact that after extinction the effects of the original conditioning experience can spontaneously recover, or be reinstated or renewed, demonstrates that extinction is not memory erasure.[51]

Stressful or painful events unrelated to the original learning can also reverse the effects of extinction and reinstate the original conditioned responses. Phobias successfully treated with exposure therapy, for example, can reappear after stressful events that have nothing to do with the phobia[52] (a fear of heights may return after the death of a family member or after an automobile accident). Stress can also prevent extinction from taking place.[53] This is thought to be due to the fact that stressful events release the hormone cortisol via the pituitary-adrenal system (see Chapter 3); this hormone has impairing effects on PFCvm function.[54] Thus, the very factor needed to induce extinction—exposure to stressful threats—can prevent extinction. This is an argument against the use of flooding and related exposure procedures that elicit high levels of "fear." However, the hormones released during stress that impair extinction have complex and sometimes opposing effects on different phases of learning (acquisition, memory consolidation, memory retrieval, extinction, and memory reconsolidation).[55] Relatively high stress may either help or impair, depending on which phase of learning is in play.

Some argue that a further limit of extinction, and thus of exposure therapy, is that its value is restricted to situations in which the responses one is trying to treat were established through learning; you can't extinguish an association that you have not learned. The key observation used to support this point of view is that people cannot always point to some harmful experience as the source of their problems.[56] However, the fact that one doesn't remember having been conditioned does not mean that the conditioning did not take place.[57] In fact, the more stressful a situation, the more likely the individual involved will fail to remember it because cortisol released during stress can attack the hippocampus and produce amnesia with respect to the event, leaving no conscious memory, or an impaired one, of it.[58] The same hormones actually potentiate the nonconscious threat-conditioning process in the amygdala.[59] The anxiety expert David Barlow notes that all that is needed for the conditioning of anxiety responses to external stimuli or to inner thoughts or sensations from the body is activation of the fight-flight system (that is, defense circuit activation)—external stimuli can become associated with the state rather than to an external US.[60] Again, this conditioning process need not reach conscious awareness. Evidence that nonconscious conditioning effects can be weaker than conscious conditioning effects[61] does not discredit the idea that nonconscious conditioning may be important; masking and other methods used to induce lack of awareness severely limit the depth of information processing (this is a technical limit of the presentation method, as discussed in previous chapters). In real life, many events, though freely visible, do not enter conscious awareness yet still influence cognition and behavior.[62]

Still, it is not necessarily the case that all threats faced by humans are due to specific learning experiences. Some threats may well have an innate component. For example, in Chapter 3 we discussed evolutionarily prepared threats. Not everyone has a pathological fear of snakes, spiders, heights, or social situations. But these are conditions that we are inherently sensitive to as a species. And although some people are more sensitive than others, we may all, under certain conditions, have some susceptibility to them.

Technically, it is correct that extinction can't be used to help people cope with problems that are not based on learning because what it does is induce new learning that competes with the original learning. But the stimulus repetition procedure of extinction can still be used. In this case, the stimulus exposure procedure is called *habituation*.[63] Habituation is a form of *nonassociative learning* because it involves a single stimulus that has an innate or otherwise preexisting capacity to affect behavior. For example, a loud noise elicits a startle reflex the first time it occurs, but this ability weakens with repetition. Extinction, by contrast, is associative because it is based on a stimulus that acquired its potency through learning.[64] Like extinction, habituation can be undone by stressful stimuli (electric shocks, for example, lead to the dishabituation—reversal of habituation).[65] Habituation is a key idea in prolonged exposure therapy, in which threatening stimuli are presented over and over in a session until the person's reactions subside.[66] During exposure therapy, habituation and extinction may both contribute of the effects of stimulus repetition.[67]

The reversibility of extinction and habituation, though troubling in a therapeutic situation, is a natural, useful feature of the brain. When things change, the brain has to adapt. Spontaneous recovery allows a testing of the waters after the passage of time. Renewal is a way to be put on alert when one returns to the specific situation in which danger previously occurred. Reinstatement induces a full-throttle defensive mode and brings back learning from the past if a stressful, painful, or otherwise harmful event actually occurs. The limitations of extinction are, in some sense, inseparable from the advantages that it naturally endows.

CAN THE EXTINCTION PROCEDURE BE REFINED TO MAKE EXPOSURE THERAPY MORE EFFECTIVE?

Although exposure is very useful, the fact that it is so easily reversed has prompted the search for alternatives, and ways to make it more effective. This section will consider behavioral refinements, and the next will turn to neurobiological proposals.

The psychologist Michelle Craske and her colleagues have been leaders in the effort to improve exposure therapy by enhancing extinction.[68] They have made several specific suggestions, based mostly on extinction research in animals, aimed at making exposure stronger and more resistant to being reversed.

1. Increase expectancy violation during extinction.

As noted in Chapter 10, extinction learning depends on a prediction error, a mismatch between expectations and experiences.[69] Creating exposure situations that have greater mismatch should therefore improve the therapeutic outcome. Craske and her colleagues argue that strategies that attempt to change beliefs about threats, thereby reducing the tendency to overestimate the potential harm that a stimulus will cause, will decrease the mismatch and thereby weaken extinction learning. Cognitive interventions, by this logic, should be used only after exposure; otherwise, they might weaken rather than improve exposure therapy effects. This is certainly true with regard to aspects of exposure that involve explicit cognition. But it may also apply, at least to some extent, to implicit learning in extinction—recall my point in the last chapter about separating implicit and explicit aspects of therapy as much as possible to allow each to be maximally effective.

2. Use only occasional reinforcement during extinction.

The CS is usually presented alone in extinction. But in the "occasional reinforcement" procedure, it is sometimes followed by the US.[70] This presumably increases the expectancy violation, improves the effects of extinction, and also increases the salience of the CS, which stimulates new learning. Additionally, such reinforcement would give a person repeated experience with fear reduction and fear revival, so that the return of fearful feelings in real-life situations would be less troubling and discouraging.

3. Deepen the extinction process.

In the "deepened extinction" procedure, multiple-threat CSs are separately extinguished and then combined in a composite extinction session.[71] This approach reduces spontaneous recovery in both animals[72] and humans.[73]

4. Remove safety signals and safety behaviors.

Safety signals and behaviors (see Chapter 3) are crutches that give short-term relief but interfere with inhibitory learning and ultimately are unproductive. Elimination of these may lead to better outcomes.

5. Use multiple exposure contexts.

Because extinction is context dependent, conducting exposure in different contexts is a way to reduce renewal. The exposures should thus be done in different situations, at different times of the day, in the presence and in the absence of the therapist, using imagination and stimuli, and in safe situations (home, therapist's office) as well as in the real world.

6. In Addition . . .

These and other suggestions made by Craske and colleagues seem well founded and have the advantage of adhering closely to the approaches used in laboratory studies of extinction and also of keeping cognitive and behavioral manipulations separate. I will make some additional suggestions, although implementing some of them will, from the point of view of the status quo of psychotherapy, seem impractical. But simply considering why they might, in principle, improve exposure therapy is useful.

First, because explicit and implicit learning can interfere with each other, a single type of learning experience should be used in a given session (see previous chapter). If it is an implicit procedure, like extinction, then talk, instruction, and other cognitive engagements that may interfere and compete for resources should be minimized (also emphasized by Craske in the discussion above). This limits what can be accomplished in a single session, but it may make for a better result.

Second, don't try to extinguish an entire complex event. The scene of an accident consists of many cues, only some of which are hot points for traumatic memory retrieval. Break down the event into specific triggers and work on these separately.

Third, consider giving exposures under conditions that bypass consciousness by using masking or other approaches. This may be the best method to address implicit processes. Though not easy to achieve, it may be particularly helpful.

Fourth, within the session, the learning should be spaced out. It is well known that massed training is much less effective than spaced training in creating enduring memories,[74] including implicit memories of extinction.[75] The explanation at the molecular level involves CREB, the transcription factor that initiates gene expression and protein synthesis in the conversion of short-term to long-term memory.[76] Massed training depletes CREB, and once used up about sixty minutes of recovery is needed to replenish the supply, so additional training within that period only interferes with the resupply process.[77] It has been shown that CREB-dependent protein synthesis in the PFCvm [78] and amygdala[79] is required for the long-term retention of extinction. So if one is going to do twenty-five exposures, they should be done in blocks of five, with breaks between, rather than all twenty-five at once. Temporal spacing, in short, could make the effects of extinction and exposure more persistent.

Fifth, after any learning experience, effort should be made to minimize activities that might disrupt memory consolidation. For both explicit and implicit memory, consolidation depends on gene expression and protein synthesis.[80] This process takes a minimum of four to six hours; events that happen molecularly or behaviorally during this time can interfere with the consolidation process and make the memory weaker.[81] Isolating patients after exposure might be impractical, as they usually need to get back to life's activities; but it may be needed for optimal benefit. One could envision overnight clinics where such a procedure could be done.[82] An overnight sequester would also take advantage of the fact that important aspects of memory consolidation occur during sleep.[83] But perhaps a few hours would do. Recent work shows that a "nap" after a therapy session enhances the therapeutic benefit.[84] But it is also possible that a nap right after therapy helps eliminate rumination and exposure to trigger cues, and thus prevents memory strengthening through reconsolidation.[85] Regardless of the reason why the nap works, it appears to be of practical value. Obviously, people in controlled environments can still engage in thought that could interfere with consolidation, so establishment of structured activities that support rather than interfere with therapy consolidation during these extended sessions would have to be carefully considered.

Putting these suggestions into practice would not be easy. For one thing, several of my suggestions cannot be implemented within the confines of the accepted therapeutic format of the fifty-minute session. But if ease of doing things was the guiding principle of medical treatment, people with heart disease could not benefit from pacemakers or people with Parkinson's disease from deep-brain stimulation or gene therapy. I'm not suggesting that psychotherapy should follow a medical model. I'm saying that treating people most effectively may require a change in the way psychotherapy sessions are done.

I have one more important suggestion that, unlike some of those listed above, is very simple to implement. It has to do with the relation of the first trial of extinction to successive ones. We'll consider this proposal after we examine the topic of memory reconsolidation.

USING NEUROBIOLOGICAL ENHANCEMENT OF EXTINCTION TO IMPROVE EXPOSURE THERAPY

Until now I've emphasized methods of enhancing extinction via procedural refinements. Another approach combines standard extinction procedures with brain manipulations, such as drugs. It is important to emphasize, though, that these are not drug therapies: the patient is not "on the drug" for an extended period of time. The drug or biological manipulation is only used acutely to enhance the effects of extinction learning.

Drug Enhancement of Extinction

A promising way to improve extinction emerged from studies of the neural basis of threat learning in rats. In the 1990s a number of studies by my laboratory and others demonstrated that the synaptic plasticity underlying acquisition of the CS-US association by the amygdala depends on a subcategory of glutamate receptors called NMDA (*N*-methyl-D-aspartate) receptors.[86] The mechanism of these receptors was described in *Synaptic Self,* but the precise details are not important for us here. The relevant point for our purposes is that when NMDA receptors are blocked in the LA, threat conditioning is disrupted. This basic finding led Michael Davis to assume that facilitation of NMDA receptor function would enhance learning, and indeed this proved to be the case. When rats were given the NMDA facilitating drug D-cycloserine (DCS), the memory created by conditioning was stronger.

Recognizing that extinction itself is a form of learning, Davis hypothesized that facilitation of NMDA receptor function would likewise enhance the effects of extinction. Davis was then inspired to collaborate with psychiatric colleagues, most notably Barbara Rothbaum and Kerry Ressler, who tested whether DCS would improve exposure therapy.[87] Initial results supported Davis's hypothesis, although several follow-up studies had mixed results.[88] Overall, though, the results in animal and human studies suggest that DCS does strengthen the effects of extinction and exposure, at least in some conditions.[89]

Inspired by the success of the DCS work, researchers have begun to explore other means of chemically enhancing exposure. After it was demonstrated in rodents that the adrenal cortex hormone cortisol (or synthetic variants of it) facilitated extinction,[90] the effects of administering cortisol prior to exposure therapy was tested in phobic patients.[91] The studies showed that cortisol treatment reduced self-reports of anxiety and physiological responses during exposure to the phobic stimuli and enhanced the persistence of these effects. Cortisol thus seems to affect both explicit and implicit processing systems, consistent with the widespread nature of its receptors in the brain in neocortex and in subcortical survival circuits.[92]

Rob Sears in my laboratory found that blocking the neuromodulator orexin disrupted threat conditioning.[93] This effect was shown to involve interactions between norepinephrine neurons in the locus coeruleus and the hypothalamic orexin system. Hypothalamic neurons release orexin in locus coeruleus and the latter releases norepinephrine in the amygdala. Blocking orexin receptors in the locus coeruleus thus reduces norepinephrine in the LA and thereby interferes with conditioning. Other recent studies have demonstrated that blockade of a particular receptor of this modulator facilitates extinction when infused directly into the amygdala but not into the PFCVM or hippocampus.[94] Orexins have also been implicated in human anxiety, especially panic disorder.[95] These studies make orexins a potential target for enhancing exposure therapy.

An exciting new body of research on acid-sensing receptors[96] in the brain holds great promise, though I don't think they have yet been studied in relation to extinction. These receptors detect pH levels in the cerebrospinal fluid (CSF), which is the substance that surrounds neurons in the brain and spinal cord. A low pH reading is an indicator that acid levels are too high; this is a consequence of the diffusion of carbon dioxide (CO_2) from blood to the CSF. In the brain, CO_2 is broken down, and the result is an increase in acidity. Levels of both CO_2 and acidity are detected by special sensors on the respiratory neurons in the brain stem, which send signals to the diaphragm muscles to increase the respiration rate to bring in more oxygen to balance elevated CO_2. Recent studies of rodents have also found acid sensors located on neurons in the amygdala (LA and BA) and in the BNST.[97] A rise in acidity increases the excitability of neurons in these areas, making them more responsive to threats.[98] Hypersensitivity to acid in the brain has been proposed as a genetic predisposition to panic disorder,[99] consistent with Klein's suffocation alarm theory of panic (see Chapter 3).[100] Given that the rodent research suggests that changing acid levels in the amygdala and BNST alters the response to external as well as internal stimuli, research on acid-sensing receptors might be pertinent to a broader range of conditions involving fear and anxiety. New pharmacological tools are becoming available for altering acid levels, and these may offer yet another approach to treating problems with fear and anxiety in people;[101] studies of extinction in animals would be an ideal place to explore this possibility.

Promising results have also been found using drugs that augment the function of the endocannabinoid system of the brain to enhance extinction in animals and exposure therapy in humans.[102] Other drugs that have been tested for this purpose target GABA, serotonin, dopamine, acetylcholine, and other transmitters, as well as hormones such as oxytocin.[103] Recently, hallucinogens have also been reported to have a beneficial effect on end of life anxiety,[104] and research on how these chemicals interact with survival circuits and neocortical circuits implicated in working memory, attention, and other cognitive functions that contribute to consciousness will be interesting to pursue.

Using Brain Stimulation to Enhance Extinction

A number of procedures are available for stimulating the brain or peripheral nervous system for therapeutic purposes. These include deep brain stimulation, transcranial magnetic stimulation, and vagus nerve stimulation.[105]

Deep brain stimulation (DBS) is quite invasive and involves the delivery of electric current through electrodes implanted in the brain. This technique has been used with some success in Parkinson's disease, Tourette syndrome, depression, anorexia, and anxiety disorders.[106] Studies in rats have shown that DBS can enhance extinction.[107] Although relatively few studies have examined whether DBS can enhance exposure therapy in humans, the evidence that does exist is supportive.[108] The mechanisms underlying the therapeutic effect, however, are poorly understood.

A less invasive procedure is *transcranial direct current stimulation,* a technique in which low levels of electric current are delivered to the surface of the skull. The current flows through the skull and into the superficial parts of the underlying cortex and changes the excitability of neurons in the affected area. This procedure has been used to modify information processing in a variety of laboratory cognitive tasks and has been helpful in treating depression but does not appear to have been applied to anxiety disorders.[109] A related procedure uses *transcranial magnetic stimulation* to manipulate brain activity. A pilot study in people with PTSD found that it led to enhancement of exposure therapy.[110]

Another stimulation approach targets the vagus nerve. The descending vagus nerve is the main pathway by which the brain controls the parasympathetic nervous system and thus counters the sympathetic nervous system (fight-flight system), whereas the ascending vagus nerve carries signals about body states to the brain and is responsible for regulating arousal systems in the brain stem. That vagal stimulation might be useful in treating anxiety and depression was first suggested by studies in which epileptic patients receiving such stimulations displayed improved mood.[111] People with anxiety disorders were then tested and were indeed helped by this treatment.[112] In rats, vagal stimulation also enhanced extinction.[113] Stephen Porges has proposed that the descending vagus has two distinction components: an evolutionarily older one that elicits immobilization and feigns death, and a newer one that promotes a sense of calm and facilitates social interaction.[114] Targeting the vagal system may be useful for enhancing exposure therapy.

Stimulation techniques carry various degrees of risk. All but the transcranial method are invasive. The application of some is relatively expensive, and effectiveness is as yet not well established for all procedures. There are no well-accepted criteria for determining which patients are most appropriate candidates for such invasive treatments for anxiety. Layers of ethical questions are raised by these and other emerging neuroscientific technologies.[115] The next approach, gene therapy, is no exception.

Extinction and Gene Therapy

Perhaps the most radical approach under consideration is gene therapy. This has been used with some success in the treatment of Parkinson's disease, which results from the loss of neurons that make dopamine.[116] Areas responsible for motor control depend on dopamine, and its absence leads to dysregulation of the associated circuits and results in tremors. In gene therapy for Parkinson's, genes are attached to viruses that are injected into the motor control regions of the basal ganglia and carry the new genes into neurons. The genes then reprogram nondopamine neurons so that they are able to produce dopamine. Though the procedure has been applied only in a small sample, the results were viewed as successful and offer hope that this method might be able to be used more widely.

Application of gene therapy to conditions affecting the brain is predicated on knowledge of the pathological factors and the brain circuits that underlie them. We have learned a great deal about the circuits that contribute to defensive survival circuits, so we have some sense of what locations in the brain might be targeted to address different symptoms arising from malfunction in them. For example, Robert Sapolsky's laboratory started with the observation that the receptors that bind cortisol are highly concentrated in the LA and BA.[117] These receptors contribute to the acquisition and expression of defense responses by increasing the excitability of neurons. Noting that estrogen receptors inhibit activity in these same neurons, the researchers created a chimeric gene that coded aspects of both cortisol and estrogen. When infused into a rat's amygdala before conditioning with a tone and shock, the consolidation of the threat memory to the tone was impaired such that the animal expressed a reduced degree of freezing to the tone.

Although such findings are promising, don't expect genes to be injected into the brain anytime soon for the purpose of treating people with anxiety disorders. Even if we did have the knowledge to pinpoint the circuit that is giving rise to exact symptoms in a given individual, this would involve an indirect approach to changing the way the fearful or anxious person feels. Further, questions would arise as to whether it is justifiable to use such an invasive approach, which would be extremely expensive and carries the potential for infections and other side effects. There is also the very concrete risk of altering not only the target function but other functions in which the amygdala participates, such as appetitive motivation. We know a good deal about both appetitive and aversive circuits, but they seem to overlap to a great extent.[118] Reducing anxiety at the expense of losing the ability to benefit from positively reinforcing experiences would hardly be considered a desirable outcome.

The rise of nanobots may someday make gene therapy and drug delivery in general easier.[119] These tiny (nanometer-scale) molecular robots could potentially deliver drugs to designated target areas and even to specific kinds of neurons in those areas. Researchers are currently attempting to use this approach with cancer drugs. Such technology would, like other cutting-edge advances, raise questions regarding safety,[120] access, and affordability. And as is often the case with brain drugs, the line between therapeutic and recreational uses would be blurry— someone would surely figure out how to get the best high by delivering street drugs to just the right sweet spot in our synaptic sea.

CAN MEMORIES BE ERASED INSTEAD OF JUST INHIBITED?

Extinction creates new memories that override or inhibit threat memories.[121] As we've seen, this method is effective but not perfect, because the original memory can reappear. The paper by Karim Nader discussed at the opening of this chapter suggested the possibility of a different approach.[122] Could, perhaps, the original memory be more effectively controlled and, in fact, even erased by manipulating it after it was retrieved? A wave of research on memory erasure resulted.

Reconsolidation Blockade

In Michel Gondry's film *Eternal Sunshine of the Spotless Mind*, Clementine leaves Joel. Overcome with sorrow and loneliness and recurrent thoughts of Clementine, Joel seeks the help of a memory erasure company that claims to be able to wipe all traces of her from his brain. To do this, they zap Joel's brain when memories of Clementine are active. While this sounds like a scenario from science fiction, and in some respects it was (the machine in the film could identify, monitor, and pinpoint specific memories and destroy them, which is not possible at present), one part of it was not so far-fetched. The film came out four years after Nader's publication showed how infusing a drug into the amygdala before a CS-US memory was retrieved eliminated the ability of the CS to later activate the amygdala-based memory and elicit freezing in rats.

Why did Nader undertake that study? Beginning in the 1960s, a long line of research had shown that certain drugs, especially protein synthesis inhibitors, administered immediately after learning disrupted consolidation (conversion of temporary short-term memory into persistent long-term memory).[123] The basic idea behind these studies was that memory is in a fragile or labile state and subject to disruption until it is stabilized by protein synthesis. The window during which a memory could be disrupted was four to six hours from the time it was acquired; after that, it became stable and persistent. This led to the standard view that a memory is stored once; then each time some stimulus appears that is relevant to it, the original memory is activated and expressed.

But there had also been studies that found that memory became labile and disruptable after it was retrieved.[124] It was as if retrieval reopened the consolidation process, and in order for the memory to persist after retrieval it had to be restored or reconsolidated. Conceptual differences between the consolidation and reconsolidation views of memory, and procedural differences in the way studies are conducted, are depicted in Figure 11.5. Because this idea did not fit with the standard and very widely accepted consolidation theory, it was rejected by leading

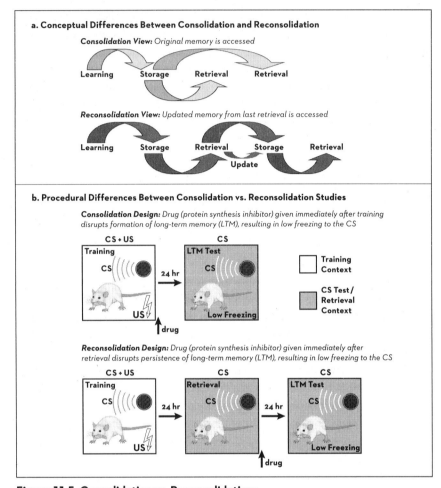

Figure 11.5: Consolidation vs. Reconsolidation:

a. Conceptual Differences Between Consolidation and Reconsolidation. According to consolidation theory each time we retrieve a memory we retrieve the original memory. Reconsolidation theory, on the other hand, suggests that each time we retrieve a memory, the memory is potentially changed (updated); thus, you retrieve the memory you stored after the last retrieval rather than the original memory. **b. Procedural Differences Between Consolidation vs. Reconsolidation Studies.** In studies of consolidation, typically a protein synthesis inhibitor (a drug known to block consolidation) is given immediately after training. Then short-term memory (STM) is tested. The next day, long-term memory is tested. The typical finding is that STM is intact (showing that the memory was formed) but LTM is impaired (indicating that STM was not consolidated into a persistent LTM). In reconsolidation studies, the drug is given after retrieval of a fully consolidated memory. STM and LTM are then tested. The typical finding is that post-retrieval STM is intact (showing that the memory was present during retrieval) but LTM is impaired. The conclusion is that during retrieval, memory is destabilized and has to be reconsolidated via protein synthesis to persist as LTM.

researchers[125] and fell by the wayside. The so-called reconsolidation hypothesis enjoyed a bit of a revival in the 1990s, through the work of Susan Sara,[126] but did not gain traction.

If you take a drug orally, by injection, or by intravenous infusion, it reaches the brain via the bloodstream. This means, though, that the entire body and the entire brain are potentially affected by the drug. Most of the early work on consolidation and reconsolidation was carried out this way. My laboratory's contribution to consolidation and reconsolidation was to target drug infusions to the specific brain region that we had implicated in threat memory. Glenn Schafe showed that blockade of protein synthesis in the LA immediately after conditioning did not disrupt the conditioned freezing response elicited by the CS but that the same rats failed to freeze to the CS when tested the next day. The drug, in other words, had no effect on short-term memory but prevented the conversion (consolidation) of a long-term memory that persisted.[127] Nader then built on this. He demonstrated that blockade of protein synthesis in the LA during memory retrieval had no effect on responses elicited by the tone shortly after conditioning but disrupted the memory when tested the next day.[128]

Because we had targeted a specific brain region with a known role in memory, our study attracted the attention of other researchers in a way that the earlier studies using systemic injections did not. Hundreds of related studies followed.[129] Our basic finding was demonstrated for memories that depend on the amygdala, hippocampus, neocortex, basal ganglia, and other regions of the brain; memories based on threat conditioning, or appetitive reinforcement with food or addictive drugs, were shown to be susceptible; and animals as diverse as worms, bees, snails, and a variety of mammals also showed the effects. Reconsolidation has also been demonstrated in humans.[130] A comprehensive volume covering the topic of reconsolidation, edited by my NYU colleague Cristina Alberini, who has done much important work in this area,[131] was published in 2013.[132]

Why would the brain have such a strange mechanism that allows memory to be disrupted when it is retrieved? Actually, it's not strange at all. The purpose of reconsolidation is not to make memory subject to disruption but to enable it to be updated.[133] An example of updating comes from a study that Lorenzo Diaz-Mataix and Valérie Doyère did in my laboratory.[134] To understand this study, though, we need to consider something else first.

Initially, we thought that all memories might be susceptible to disruption by reconsolidation blockage. But Nader, who is now a professor at McGill University, found that strongly conditioned memories (those conditioned with an especially intense US) were protected from reconsolidation blockade.[135] This was bad news for those who had hoped that reconsolidation blockade could be used as a treatment for PTSD, because this disorder usually results from very strong memories having been created in response to horrific situations. But Diaz-Mataix and Doyère found that strong memories can indeed undergo reconsolidation if new information is incorporated into that memory—in other words, if the memory is updated.

What they did was condition rats with a CS paired with a strong US. Then they later reactivated the memory using a trial in which both the CS and US were presented. But they used two different scenarios. For one group, the US occurred at the same time in relation to the CS (at the end of the CS) during both conditioning and during retrieval. Thus, there was no new information about the relation of the CS to the US. For the other group, the CS occurred at a different time during retrieval than during conditioning, which means something different occurred. All animals were then tested again the next day. For animals in which the retrieval trial introduced new information, treatment with the reconsolidation blocking drug during retrieval disrupted memory in the later test (freezing to the tone was reduced); for the animals in which the retrieval trial did not introduce new information, reconsolidation blockade had no effect, just as Nader had found. This is another example of how expectation violations (prediction errors) trigger new learning.

In summary, memory disruption is a useful trick for scientists, but not the function that nature provided; the purpose of reconsolidation is memory updating.[136] Although memory updating often works to our advantage, because it allows memory to be dynamically changed to accommodate new information that comes along, it can also give us trouble. Consider the case of a person who witnesses a sensational crime and gives a report to the police on the spot. Then, at the trial, he gives a different account of what he had seen. Between those two events the witness had read about the crime in the newspaper and had also become aware of new details and rumors concerning it while surfing the Internet. These informational encounters retrieved the original memory and updated it. At the trial the testimony morphed into a composite memory that included some of the new information that had not actually been experienced.

It is hard for one to realize when memories are accurate and when they are not. For example, Liz Phelps, Bill Hirst, and others conducted studies of memories of the World Trade Center attacks.[137] They found that their subjects had very vivid, unshakeable memories of that day. But because the facts could be checked, the investigators were able to show that some of the details that people were very confident about were actually inaccurate.

As Elizabeth Loftus[138] and Daniel Schacter[139] have noted, memory can be unreliable for a variety of reasons. It usually works fairly well, all things considered, but is clearly not a carbon copy of experience. Eyewitness testimony is often all the evidence that juries have on which to base decisions, but, as in the example above, testimony can be inaccurate. I think that unless there is corroborating evidence, a single eyewitness's testimony should never be treated as irrefutable evidence, no matter how certain the witness feels about what he witnessed.

In the late 1990s a major controversy erupted when, through therapy, some people claimed to have recovered lost memories of sexual abuse.[140] Some vividly recalled sexual assault by family members, others being imprisoned and abused in demonic rituals. While not taking a side in any particular case, in light of reconsolidation research we can see how it is possible for potentially false memories to arise in such accounts. For example, rogue revelations of this type could result from the retrieval of memories of benign childhood experiences while a person is being asked to think about the possibility of having been sexually abused. If the discussion also involves family members, or demonic rituals, these could be incorporated into the reconsolidated memory. Sadly, many real cases of rape and childhood sexual abuse do exist, but it must be acknowledged that not everything people remember actually happened to them.

As I noted at the beginning of this chapter, some objected to the idea that it might be possible to help people with PTSD by blocking the reconsolidation of trauma-related memories, arguing that it is important for trauma to be remembered. We took these criticisms to heart and tried to test whether a memory of a complex experience (as opposed to a single stimulus, like a conditioned tone) could be wiped out by reconsolidation blockade. Rather than just administering a single tone and shock, we used various combinations of multiple CSs and USs to create relations between multiple elements of an experience. Then we activated just one part of the complex memory by using a single CS or US. These studies, done by Jacek Dębiec, Lorenzo Diax-Mataix, Valerie Doyère, Karim Nader, and others, showed that only the specific component that was reactivated was affected;[141] other parts of the memory remained intact. These results suggest that individual triggers might be worked on by the patient and therapist, and when the memories are addressed to the extent at which they are both comfortable, the process can stop. Of course, this would be a judgment based on explicit cognition and might lead to incomplete extinction of the implicit memory.

Reconsolidation disruption has also generated excitement as a treatment for other disorders. For example, research done by Barry Everitt and Trevor Robbins, and by Jane Taylor, has shown that reconsolidation blockade can prevent drug relapse in addicted rats.[142]

Promising results have been obtained in some clinical studies of anxiety disorders,[143] but as of now the findings have not been as dramatic as in the associated animal work, and the verdict is still out on how effective reconsolidation will be in clinical settings.[144] One problem is that many of the drugs that are most effective in blocking reconsolidation in animals are not safe or approved for human use. Time will tell whether suitable drugs can be found to make reconsolidation a valuable therapeutic tool in humans.

Although most of the clinical interest has focused on reconsolidation blockade, Jacek Dębiec found that he could not only disrupt memories but could also make them stronger. To do this, he used drugs that facilitated rather than interfered with the protein synthesis process.[145] The fact that memory can be strengthened with this procedure suggests that triggers of anxiety or depression could be transformed in such a way as to represent a pleasant or exhilarating event. This method could function as a biological tool to induce reappraisal, perspective change, cognitive restructuring, and the like.

The prolonged effects of a single-stimulus exposure in reconsolidation studies are reminiscent of the interesting findings discussed above involving the ability of a single stimulus to alter avoidance or subjective feelings. Those procedures may have, in fact, somehow tapped into the mechanisms of reconsolidation. Also, the studies described earlier showing that cortisol strengthens the effects of exposure therapy could well be a reconsolidation, rather than a consolidation, effect.[146]

Given the power of the reconsolidation approach in experimental studies in animals, it seems likely that it will work in clinical settings under the right conditions. Regardless, though, as a normal process that occurs in the brain, reconsolidation probably contributes to all therapeutic situations in which memories are retrieved and changed. In other words, because retrieval makes memory labile and subject to change, reconsolidation is taking place constantly—in fact, potentially every time we remember something.

Disentangling Reconsolidation and Extinction

Extinction and reconsolidation have a complex relationship. The first trial of extinction is, in effect, a reconsolidation trial, because the CS results in the retrieval of the CS-US association. Furthermore, many of the same molecular players involved in extinction are also involved in reconsolidation (protein synthesis, CREB, glutamate receptors, various kinases).[147] So how are these to be distinguished?[148]

Recall that both extinction and reconsolidation depend on protein synthesis for long-term retention of a new memory. Building on this, Yadin Dudai and colleagues conditioned groups of animals in one of two ways on day one, and then presented the CS the next day to retrieve the memory on day two.[149] The memory was then tested on day three. For one of the groups, the conditioning procedure was such that the CS presentation during the retrieval trial on day two produced extinction learning, as judged by the absence of the conditioned response the third day. For the other, CS presentation during the retrieval trial on day two did not produce extinction, because the conditioned responses were strongly expressed the next day. The key issue is what happened to the conditioned responses in these two groups on day three when protein synthesis was blocked on day two. For the group that normally extinguished on day two, strong conditioned responses were expressed on day three (protein synthesis inhibition on day two thus blocked the consolidation of extinction memory, and the conditioned response acquired on day one persisted on day three). For the other group, however, conditioned responses were not expressed on day three (protein synthesis inhibition on day two thus blocked reconsolidation and thus prevented the expression of conditioned responses on day three). Thus, whether protein synthesis blockade on a given exposure to a CS disrupts extinction consolidation or instead disrupts reconsolidation of the original threat memory depends on the dominance of extinction versus reconsolidation processes during retrieval; these compete to determine how a memory is expressed.[150] This is Dudai's theory of the dominant trace.[151]

From a therapeutic point of view, this interaction between extinction and reconsolidation could complicate potential treatments, especially because the same drugs affect both extinction and reconsolidation.[152] It might be necessary, though tricky, to coordinate drug delivery in relation to the timing of exposure in order to target the exact memory process and achieve the intended therapeutic effect rather than an unintended consequence.

Reconsolidation Without Drugs

Many important findings are discovered accidentally, as was the case with one of Marie Monfils's studies in my laboratory[153] (Figure 11.6). For reasons unrelated to the conditions of the actual experiment, she inserted a short break between the first and second trial of an extinction procedure. When she then tested for spontaneous recovery and renewal, she found that they did not occur. This led to much discussion in the laboratory, and the idea arose that inserting a gap between the first and second trials may have led the brain to effectively treat the first trial as a reconsolidation trial.[154] In other words, it made the threat memory vulnerable to change for the next four to six hours, so that doing extinction during that time period changed the stimulus from a predictor of danger to a predictor of safety. Indeed, if the space between the first trial and second trial was between ten minutes and about four hours, the threat memory never returned. But if the window

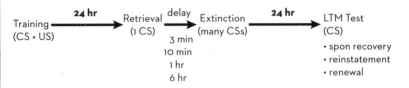

a. Reconsolidation/Extinction Design

Training (CS + US) → **24 hr** → Retrieval (1 CS) → delay / 3 min / 10 min / 1 hr / 6 hr → Extinction (many CSs) → **24 hr** → LTM Test (CS)
• spon recovery
• reinstatement
• renewal

b. Effects on Extinction of Different Delays Between the Retrieval Trial and Extinction in Both Rats and Humans

Delay	Effects on Recovery of Initial Threat Learning
3 min	extinction subject to spontaneous recovery, renewal, and reinstatement
10 min	extinction protected from spontaneous recovery, renewal, and reinstatement
1 hr	extinction protected from spontaneous recovery, renewal, and reinstatement
6 hr	extinction subject to spontaneous recovery, renewal, and reinstatement

c. Conclusions: A single retrieval of a threat memory destabilizes the memory by inducing reconsolidation. If extinction is then carried out within the reconsolidation window (10 min–4 hrs) the CS becomes a safety signal rather than a threat and extinction is protected from spontaneous recovery, renewal, and reinstatement.

Figure 11.6: Improving Extinction by Combining Extinction and Reconsolidation in Rats and Humans.

(BASED ON FINDINGS REPORTED BY MONFILS ET AL [2009] AND SCHILLER ET AL [2010].)

was less than ten minutes or longer than six hours, it did return. It thus seems that some rapid molecular mechanism is engaged and opens the reconsolidation window, and then lasts several hours. Monfils's initial study implicated a subgroup of glutamate receptors; later, in Richard Huganir's laboratory at Johns Hopkins, Roger Clem clarified in detail how glutamate receptors participate in this process of triggering memory lability and stabilizing a new (updated) version of the memory by using state-of-the-art molecular genetic techniques.[155]

We then set up a collaboration with Liz Phelps's laboratory, with Daniela Schiller taking the lead. The studies confirmed the rat findings in college students who underwent conditioning and then received one retrieval trial followed by an extinction procedure ten minutes, one hour, or six or more hours later. The ten-minute and one-hour time periods prevented recovery even when tested a year later, but the later time period had no effect.[156] Studies by Schiller and others in the Phelps laboratory extended the psychological findings but also implicated the PFCvm—in fact, the same region involved in extinction and implicit reappraisal—in this use of extinction to update the memory of the threat stimulus as safe rather than dangerous.[157]

Based on the studies by Monfils and Schiller, extinction during the reconsolidation window has been proposed as a mechanism of action underlying a novel approach for treating combat PTSD.[158] Addiction researchers also picked up on Monfils and Schiller's work, and tested the effects on drug cue–induced relapse in addicted rats and people.[159] In both rats and humans, the procedure produced a persistent prevention of relapse. This is an impressive application of a simple but powerful procedural change.

It should be stressed that all that was done in these studies was to change the amount of time between the first and second trial of extinction. No drug manipulation was involved, but only a procedural change that inadvertently took advantage of the reconsolidation window opened by a single retrieval trial. It's very exciting that a very simple adjustment in the timing of stimulus exposures has the potential to greatly improve the effectiveness of exposure therapy. Although not all studies have found the effect as described,[160] there have been a number of successful replications in different species and different kinds of tests. Additional research will have to clarify the conditions under which the effect can be expected to occur.

Any future clinical applications of these methods should follow the laboratory procedures as closely as possible, including the minimization of explicit cognition. This will make it easier to replicate the animal findings in humans and may also be significant for preventing interference when multiple processes compete for brain resources.

Zipping Memory

Another approach to memory erasure has come from work on an enzyme called PKMzeta. Todd Sacktor at SUNY Downstate, Brooklyn, discovered that he could enhance synaptic plasticity in the hippocampus by using this enzyme, and could disrupt plasticity with a chemical called ZIP (zeta inhibitory peptide).[161] This led him to explore, with Andre Fenton (now a colleague at NYU), the role of PKMzeta in hippocampal-based memory.[162] They found that ZIP, given long after learning, eliminated the conditioned memory. Many studies followed in the amygdala, neocortex, and other areas, confirming the effect.[163] Although these findings suggest that ZIP may be a potentially powerful tool for changing memory, there is a catch. Procedures based on memory reactivation and reconsolidation can target specific memories through their individual reactivation—injection of a protein synthesis inhibitor into the LA affects only the reactivated memory, leaving other memories stored in the LA intact. But ZIP interferes with *all* memories that are stored in the region where the drug is infused. In order for PKMzeta manipulations to be useful therapeutically, some means of selectively affecting specific memories would be needed.

TRUMPING EXTINCTION WITH PROACTIVE AVOIDANCE

After 9/11 many people in New York and elsewhere seemed to cope with the consequences of the terror attacks by staying close to home and avoiding the routines of daily life, such as work, school, and social interactions.[164] Indeed, they seemed pathologically glued to their televisions, watching the planes fly into the towers over and over again as one news program transitioned to the next.

Avoidance, as we've seen, is generally viewed in a negative light by the mental health community. But I teamed up with a psychiatrist, Jack Gorman, to write an editorial in the *American Journal of Psychiatry* suggesting that some forms of avoidance can be thought of as an adaptive and useful strategy, as a means of actively coping with and gaining mastery over anxiety and its triggers.[165] The editorial was based on studies that we had done in the laboratory using the escape-from-threat procedure (see Chapters 3 and 4) in rats.[166] In brief, we conditioned the rats in the usual way to a tone and shock and then moved them into a new situation in which the tone was again presented. If they made any movement at all, the tone was terminated. These movements were thus reinforced by the fact that they allowed escape from the threatening tone. Over the course of a few trials, the rats learned that if they scampered to the opposite side of the chamber as soon as they were placed in it, the tone never appeared. The rats, in short, learned to control their environment and its threat triggers by taking action.

A key part of the study was that it also involved a separate group of rats, called yoked controls.[167] Each went through the same procedure as the experimental animals, but in a separate chamber. They received the same stimulus presentations as the experimental rats, but only the experimental rats' behavior was allowed to control the CS. By the end of training neither group exhibited freezing to the CS. Freezing in the experimental group was eliminated by giving them active control over the CS occurrence, but in the yoked rats freezing was eliminated passively by extinction since the CS was not followed by the US (remember, the yoked animals received the same stimuli as the other group, the behavior of which prevented the US from occurring). When we tested spontaneous recovery and reinstatement in both sets of rats, the group that stopped freezing because of extinction again froze (recovered the CS-US association), but the group that stopped freezing because they learned to control the CS did not. Active behavioral engagement and control seem to be more effective than extinction in preventing the ability of the threat to start triggering defensive reactions again.

As described in Chapter 4, the way that active control seems to work is by preventing the CS from driving the LA to the CeA pathway, and instead directing the LA outputs to the BA, and the BA outputs to the nucleus accumbens. Through these connections, the CS functions as a negative reinforcer, a stimulus that strengthens the behavior that eliminates the aversive stimulus (see Chapters 3 and 4).

In our editorial Gorman and I argued that in the wake of 9/11 or other traumatic situations, each time people went to work or met with friends, they moved away from being frozen in place and passively avoiding life and took a step toward active coping. Our ideas on active coping resonated with the work of the trauma therapist Bessel van der Kolk, who found that in traumatized individuals training in active coping could help overcome the dominant tendency for exaggerated freeze-flight-fight reactions.[168]

The use of such strategies may help explain why resilient individuals adapt quickly following a trauma. Work by George Bonanno has demonstrated that resilient individuals tend to have a large repertoire of active coping options, are adept at choosing ones that are appropriate to a particular context, and are good at using feedback from their environment to adapt their strategies as needed.[169] Training with active coping strategies could help traumatized individuals learn behaviors that resilient individuals use naturally.

Our rats were avoiding the tone and the adverse consequences it warned about. When avoidance involves behaviors and thoughts that directly engage with stress-related cues and events in order to change their impact and enable the organism to exert control over them, it is a useful form of avoidance, a form of *active coping.*[170] I wrote a series of pieces for the *New York Times* Opinionator series on anxiety, the last of which was about active coping.[171] In this article, I used the term *"proactive avoidance"* to describe behaviors and thoughts that directly engage with anxiety triggers in order to change, through learning, their impact and thereby help the organism exert control. (The term *"agency,"* as used in the therapeutic community, seems to have a similar meaning.) This kind of strategy combines self-exposure to anxiety-provoking situations with strategies for gaining control over the trigger cues. Michael Rogan,[172] a former researcher in my laboratory and currently a therapist who specializes in social anxiety, suggests that rather than forcing oneself to ride out anxiety at a party (flooding), it's more effective to use anxiety control strategies, such as relaxation and active coping (e.g., trips to the bathroom or stepping out to make a call), that enable regrouping before reexposure. Exposure can therefore be achieved in a way that allows instrumental learning to be reinforced by the successful regulation of defensive responses (Figure 11.7). This does not contradict my earlier warning about not attempting to accomplish too much at once, because the reduction of defense responses triggered by the threatening social stimulation is necessary to reinforce

Figure 11.7: Active Coping.

(BASED ON IDEAS DEVELOPED BY LEDOUX AND GORMAN [2001] AND LEDOUX [2013].

instrumental learning. This strategy, it should be noted, circumvents the question of whether the triggers have been learned or not (see above), because the goal is to reinforce active control behaviors that reduce the responses elicited by the cues, regardless of their origin.

For the most part, animals have to learn avoidance behaviors via trial and error. We humans likewise learn through instrumental reinforcement processes (implicit learning), but we're also able to use observation and instruction to explicitly learn to avoid.[173] Through these approaches, or by sheer imagination, we create avoidance concepts or schemas, and when in danger we draw upon these stored action plans. Then, when we encounter threats, they can trigger the avoidance schema and motivate performance of the response. Given the hypersensitivity of anxious people to threats, learned or schematized avoidance can be easily triggered and drive behavior in pathological ways, but proactive avoidance schemas can function as part of one's repertoire of coping skills. Shifting the balance from pathological to adaptive (proactive) avoidance is the key, and though it's not easy to accomplish, understanding the difference between the two is a valuable first step.

BREATHING AWAY ANXIETY

Your brain usually takes care of breathing for you,[174] so that you don't normally have to think about it at all. For example, if you exert physical effort—say, while jogging—your respiratory rate speeds up to take in more air, allowing more oxygen to be extracted and brought into the bloodstream to help with the metabolic processes that generate energy by using oxygen to break down glucose.

Breathing on autopilot is controlled by respiratory circuits in the medulla and pons of the hindbrain that connect to the muscles of the lungs.[175] Neurons in these are sensitive to CO_2 and acidity (see earlier discussion), and this plays a key role in their control over the contraction of the diaphragm muscles, which in turn control the amount of air we take in to balance CO_2 and oxygen in our body. In addition to this automatic breathing control, we also have voluntary control of air intake and how fast or slow we breathe. Singing involves voluntary breath control, as does playing a flute, saxophone, or harmonica. Conscious control of this process is achieved by interactions between the executive control functions of the neocortex and the neurons in the medulla–spinal cord that control breathing.[176]

When someone is stressed, it is common to advise, "Just take a deep breath." This folk wisdom has a grain of truth to it. During stress the sympathetic nervous system dominates, overshadowing the parasympathetic system. The result is fast heart rate, but low heart rate variability, and shallow breathing.[177] When one breathes in the slow, measured way that is commonly taught in meditation, yoga, and relaxation training, the vagus nerve, which controls the parasympathetic nervous system, becomes more active, and the balance between the sympathetic and parasympathetic system improves. As a result, heart rate variability increases, and the times when it is somewhat slower provide windows of opportunity exists for automatic processes to drive heart rate down and thus reduce elevated blood pressure and other sympathetic responses.[178]

Because it is an easy and free way to have some power over anxiety, everyone should learn to do this. In fact, I believe that instruction in the use of controlled breathing should be a major part of early education, something that children can be trained to do to the point that it becomes a habit that is simply expressed when any sign of tension arises. This simple trick, if built into one's life early, before major problems arise, could greatly reduce the adverse consequences of uncontrolled stress in childhood.[179]

MAKING WORKING MEMORY "SELF-LESS"

In the 1960s meditation was often thought of as a hippie fad, just another example of the fascination with all things Eastern and/or mystical. But eventually meditation became a mainstream phenomenon. Along with relaxation, reappraisal, exposure, and coping strategies, some cognitive therapists today use meditation, or what is called mindfulness, as part of the therapeutic program. In the cognitive therapy called acceptance and commitment therapy,[180] the patient is encouraged to be mindful (present in immediate experience) and to accept rather than react to, judge, and change thoughts and experiences.

So what goes on in the brain in meditation? James Austin, author of *Zen and the Brain,* describes meditation as "a relaxed attentive state" that "helps . . . relieve us from self-inflicted trains of thought."[181] Discussions of meditation often mention the removal of the "the self," as exemplified in expressions such as "no mind" or "no self."[182] But this does not mean a blank mind so much as a mind that is free from what Austin calls "thought pollution."[183] When the "incessant chatter" drops out, what remains is "the present moment."[184]

Although Austin's ideas are speculative, in recent years meditation has become the subject of a great deal of research that is grounded in basic principles of modern neuroscience and cognitive psychology. According to Richard Davidson and Antoine Lutz, leaders in this field, meditation is a "family of complex emotional and attentional regulatory strategies for . . . the cultivation of well-being and emotional balance."[185] One style of meditation called *focused attention* involves sustained concentration on some object or thought, whereas another, *open monitoring,* involves observing experiences moment to moment and nonre-actively. Often both styles are used in an individual's practice.

It is significant that meditation training often starts with training in breath-ing. Control of breathing, as we have just seen, can have a palliative effect on anxiety,[186] which helps prepare the mind for the more intensive activity of trying to be "in the present." In the book *Zen Training,* which I have mentioned previ-ously, Katsuki Sekida points out that it is easy to concentrate while holding one's breath, because tension in the respiratory muscles sustains attention. Although we can hold our breath for only a short time, by learning to breathe using the methods employed by Zen masters, Sekida explains, breathing can be controlled in succes-sive repeated cycles that make sustained attention possible. This, he proposes, occurs by way of the influence of the act of breathing on the reticular formation (what we now call arousal systems). As noted in other chapters, by releasing neu-romodulators, arousal systems regulate attention and vigilance functions con-trolled by cortical areas. Interestingly, controlled breathing also sends messages to arousal systems by way of the ascending vagus nerve, doubling up on the ability of breathing to affect arousal.

In light of the fact that attention has appeared repeatedly in this discussion of meditation, it is interesting to consider findings from studies that have mea-sured brain activity using fMRI while people meditate. This is challenging research to carry out, because being in an fMRI machine is about as far removed as possible from the kind of serene, quiet environment in which meditation is often practiced. Yet a number of studies have been conducted in people with various degrees of training in meditation, from skilled monks to novices. Brain areas of the CCNs involved in attention and working memory, including various areas of the frontal (lateral, medial, orbital, cingulate, insular) and parietal corti-ces, are often implicated.[187] In addition, the brain's so-called "default network,"[188] which is engaged when the brain is doing nothing in particular (such as day-dreaming), is sometimes also involved. A leading researcher in this area, Peter Malinowski, has developed a model of the brain and meditation in which five cognitive processes, each related to a different brain circuit, are involved: orient-ing, alerting, salience, executive, and default mode.[189] This seems to be a useful scheme to guide future research.

Now for an exercise in pure speculation: Through controlled breathing, arousal systems are engaged in such a way as to enable sustained attention via working memory networks. Because the breathing process can be trained to the point that it becomes habitual and does not require executive control to be carried out, executive control can then be used solely for attentional control of the con-tents of working memory. Although we have discussed attention in the context of selecting what enters working memory, selection, by definition, also includes exclusion. Executive functions might therefore be capable of preventing information from entering working memory. Studies have indeed shown that people can be trained to ignore particular stimuli or memories.[190] By isolating working memory from external stimuli and memories about the self (episodic memory and

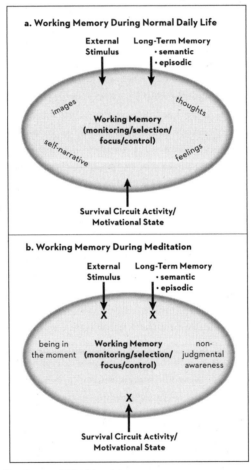

Figure 11.8: Meditation and Working Memory.

The hypothesis depicted here builds on the established role of working memory in cognition (monitoring, selecting, focusing, controlling). These processes determine what occupies working memory and constitutes the content of immediate conscious experience (images, thoughts, feelings) and drives one's self-narrative. During meditation, working memory uses these same cognitive processes to prevent inputs from entering working memory, allowing the mind to be present in the moment and nonjudgmental, and free of an ongoing self-narrative.

autonoetic consciousness), a sustained focus on the free flow of unselected thought, supported by breathing-induced control of the arousal system, might be maintained. This would be a kind of pure state of "self-less" working memory (Figure 11.8).

How might this scenario of self-less working memory help relieve fear and anxiety? Feelings of fear and anxiety, I've argued, are states of autonoetic consciousness, and are thus about the self. If indeed the neural circuits that provide working memory with the ingredients required to experience these feelings have effectively been controlled via meditation, this "self-less" mind *cannot* feel fear or worry in the sense of a personal experience. With extensive training, one might learn to call upon this self-less mind state when the possibility of threat or worry arises, and thus short-circuit the cognitive construction of feelings of fear or anxiety. By using this kind of mental posture, we are better able to think and act in ways that are more "aware," "nonjudgmental," and "in the present," and to achieve the beneficial effects of meditation on physical and mental well-being.[191]

Buddhist teachings have long emphasized the importance of being less "self-involved." When we are fearful and anxious, we are concerned about our self and its well-being. These are autonoetic thoughts about one's health, family, friends, wealth, life, death, and so on. Our conscious self, according to Mark Epstein, who is a Buddhist psychoanalyst, will do almost anything to maintain the independence, power, control, or success that it has achieved, even if to do so other people, other cultures, or the world has to suffer.[192] A healthier approach, Epstein says, is to let go of the "absolute self" that we construct and recognize our broader role in life.

Learning to meditate can be a challenging task, but it is obviously a skill that is within the capacity of our species. Individuals who are inherently relaxed and low in anxiety may have a particular penchant for preventing the outer world or inner worries from dominating their thoughts. Perhaps those who first discovered this practice and its benefits were naturally endowed with the propensity and learned to pass it on to others through training. Not everyone has the time and motivation to spend years attaining the highest levels of training possible. However, learning simple relaxation and breathing strategies that are a key part of mindfulness and meditation practice is not especially difficult or time consuming and likely can benefit most if not everyone who puts in a minimal amount of effort.

THE ANXIETY OF EVERYDAY LIFE

Autonoetic consciousness is our best friend and worst enemy. It enables us to write and revise our narrative, our self-story, as we live each moment of each day.

It also enables us to fill in the blanks of our future self. The way we fill in those blanks is an important element in our overall outlook on life. Fearful, anxious people see troubles ahead and lead their lives dwelling on worst-case scenarios that often do not come to pass. They believe that worrying gave them the power to execute plans that prevented the bad things from happening in the past. But just as the brain can learn to be anxious, it can also learn to not be that way. Although some people are by their natures more anxious than others, ever-increasing anxiety doesn't have to be their destiny. Change is difficult, and for a variety of reasons it's more difficult for some people to change than others. But the brain is adaptable. It's just a matter of being able to make those changes happen. That's where the science of fear and anxiety can hopefully help. Although we've come a long way in this effort, we still have a ways to go. But with clear, empirically grounded concepts, new scientific tools, good ideas, and future generations may be less inclined to think of theirs as the age of anxiety.

NOTES

CHAPTER 1: THE TANGLED WEB OF ANXIETY AND FEAR

1. Montaigne (1993).
2. Dickinson (1993).
3. Kagan (1994); Eysenck (1995).
4. Kagan (1994).
5. LeDoux (2002).
6. In this section I was greatly aided in summarizing the history of anxiety by several authors. Particularly important were books by Zeidner and Matthews (2011) and Freeman and Freeman (2012). These authors consulted with me in writing their books, and I, in turn, benefited by reading their texts. Also very useful were articles and books by Menand (2014), Smith (2012), and Stossel (2013), http://blogs.hbr.org/2014/01/the-relationship -between-anxiety-and-performance/ (retrieved Nov. 20, 2014). I came across Stossel's excellent book *My Age of Anxiety* and Menand's succinct and informative review of it in *The New Yorker* relatively late in the writing of this book. But these, nevertheless, became useful resources for some information not covered in other sources I was using.
7. This etymological history is based on: Lewis (1970); Rachman (1998); Zeidner and Matthews (2011); Freeman and Freeman (2012), and *The Online Etymology Dictionary* (http://www.etymonline.com). Stossel (2013) covers additional aspects of the etymology.
8. Peter Meineck, a classicist at NYU, pointed out to me that a transliteration of *angh* is *ankhô*, which refers to choking. My son, Milo LeDoux, who studied classics at the University of Oxford, was also helpful.
9. Freeman and Freeman (2012).
10. Laocoön, a priest of Apollo in Troy, was supposedly being punished by the Greek gods Athena and Poseidon for having warned the Trojans that the equestrian Greek gift was a plot. Boardman (1993); Laocoön, cat. 1059, Pio Clementino Museum, Octagonal Court. Retrieved Sept. 21, 2014, from mv.vatican.va.
11. Retrieved Sept. 19, 2014, from http://www.theoi.com/Daimon/Deimos.html.
12. St. Thomas Aquinas, *The Summa Theologica*.
13. Makari (2012).
14. Kierkegaard (1980).

15. Freud was translated into English by a number of authors, but the definitive translation, called *The Complete Psychological Works of Sigmund Freud (the Standard Edition)*, is by James Strachey.
16. Klein (2002).
17. Zeidner and Matthews (2011); Freeman and Freeman (2012).
18. Breuer and Freud (1893–1895).
19. Freud (1917), p. 393.
20. Spielberger (1966), Chapter 1, p. 9.
21. Freud (1917). Quoted by Zeidner and Matthews (2011).
22. Freud (1959).
23. Heidegger (1927).
24. Sartre (1943).
25. Freeman and Freeman (2012).
26. Kierkegaard (1980).
27. "Existentialism," *Stanford Encyclopedia of Anxiety*, http://plato.stanford.edu/entries/existentialism/#AnxNotAbs.
28. Tauber (2010).
29. Kierkegaard (1980), p. 156.
30. Epstein (1972), p. 313.
31. Yerkes and Dodson (1908); McGaugh (2003).
32. David Barlow, quoted by Scott Stossel. Retrieved Nov. 20, 2014, from http://blogs.hbr.org/2014/01/the-relationship-between-anxiety-and-performance/.
33. Kandel (1999).
34. However, within the analytic community, some have attempted to forge a connection between neuroscience and psychoanalysis (http://neuropsa.org.uk/). Mark Solms (an analyst) and Jaak Panksepp (a neuroscientist) have been particularly active proponents of this. See Solms (2014) and Panksepp and Solms (2012).
35. Freeman and Freeman (2012); Menand (2014); Stossel (2014).
36. Auden (1947).
37. From the Introduction in Auden (2011).
38. http://www.laphil.com/philpedia/music/symphony-no-2-age-of-anxiety-leonard-bernstein.
39. Smith (2012).
40. I remembered this as a scene from Paul Mazursky's film *An Unmarried Woman* (1977), but thanks to Robin Marantz Henig's *New York Times* "Opinion" piece, I have correctly attributed it to Pakula's *Starting Over*. Both starred Jill Clayburgh, which may have contributed to my confusion. http://www.nytimes.com/2012/09/30/sunday-review/valium-and-the-new-normal.html.
41. May (1950); Menand (2014).
42. Quoted in Smith (2012).
43. While fear and anxiety are thus distinguishable states, the words "fear" and "anxiety" are sometimes used interchangeably and sometimes inconsistently. For example, while Freud's translator, Strachey, interpreted *Angst* as anxiety, in German *Angst* can refer to a state with a particular object (fear) as well as to a more general state of worry and dread (anxiety). Strachey was well aware of this, but he felt anxiety was the state Freud generally had in mind when he used *Angst* (Freeman and Freeman, 2012). Strachey admitted the possible confusion between *Angst* and *Furcht* in some of Freud's writings. In English, we readily use fear to refer to situations involving anxiety or worry (e.g., "I fear I will let you down" or "I'm afraid to tell him the truth"). The fungibility of the terms is also shown by

the fact that Freud and Kierkegaard both viewed anxiety as a kind of fear (free-floating fear for Freud; fear of nothingness for Kierkegaard). But Freud also said fear was a kind of anxiety (primary anxiety). Further conflation of fear and anxiety can be found in other expressions used by Freud. He talked about *expectant fear* and *anxious expectation,* which are seemingly identical in emphasizing "expectation" when referring to worry, dread, and apprehension about unpredictable future events. Freud's expression *free-floating fear* is most often called *free-floating anxiety* today.

44. Marks (1987).
45. See Smith (2012).
46. Wenger et al (1956) proposed such an approach. The states listed in the text are my interpretation of what would be low, medium, and high variants on this scale.
47. Hofmann et al (2012); Barlow (2002).
48. Barlow (2002); Rachman (1998, 2004); Zeidner and Matthews (2011); Stein et al (2009); Beck and Clark (1997); Anxiety Disorders Association of America (ADAA): http://www.adaa.org/understanding-anxiety; National Institute of Mental Health: http://www.nimh.nih.gov/health/topics/anxiety-disorders/index.shtml; http://www.psychiatry.org/dsm5.
49. http://en.wikipedia.org/wiki/Diagnostic_and_Statistical_Manual_of_Mental_Disorders.
50. http://apps.who.int/classifications/icd10/browse/2010/en#/V.
51. The cognitive-behavior therapist and researcher Stefan Hofmann of Boston University was very generous in helping me understand the history of anxiety disorders. If there are mistakes, they are due to my interpretation of his comments.
52. This history was summarized to me by Stefan Hofmann, who referred me to Richard McNally's excellent summary of the history of panic attacks (McNally, 1994). See also Klein (1964, 1981, 1993, 2002); Klein and Fink (1962); Barlow (1988); Marks (1987).
53. Meuret and Hofmann (2005).
54. The relation to respiratory distress (shortness of breath) suggested to some that hyperventilation might be the root of panic, depriving the person of oxygen; however, Donald Klein argued instead that an increase in carbon dioxide in the blood sets off an alarm system in the brain that falsely leads the person to believe that he or she is about to suffocate (Klein, 1993; also see Roth, 2005). The cause of panic remains unresolved (Ley, 1994; Stein, 2008).
55. The term "nostalgia" was originally used to refer to debilitating homesickness in soliders, and considered a disturbance rather than a quaint longing for the good old days.
56. A case could be made for including OCD as well, but I have chosen not to go in that direction.
57. Anxiety Disorders Association of America (ADAA): http://www.adaa.org/understanding-anxiety; National Institute of Mental Health: http://www.nimh.nih.gov/health/topics/anxiety-disorders/index.shtml.
58. Anxiety Disorders Association of America (ADAA): http://www.adaa.org/understanding-anxiety; National Institute of Mental Health: http://www.nimh.nih.gov/health/topics/anxiety-disorders/index.shtml.
59. Lim et al (2000).
60. Galea et al (2005); Kessler et al (1995).
61. Barlow (2002). This summary from Barlow is based on Meuret and Hofmann (2005).
62. Hettema et al (2001a, 2001b, 2008); Kendler (1996); Kendler et al (2008, 2011).
63. Horwitz and Wakefield (2012).
64. Wakefield (1998).
65. Epstein (1972), p. 313.
66. Grupe and Nitschke (2013); Meuret and Hofmann (2005); Hofmann (2011); Dillon et al (2014); Bar-Hamin et al (2007). The model proposed by Grupe and Nitschke will be discussed in detail in Chapter 4.

67. In Chapter 9 I describe how the National Institute of Mental Health has moved in this direction, deemphasizing DSM categories as guides to brain research on the causes and treatment of mental and behavioral problems.

68. LeDoux (1984, 1987, 1996, 2002, 2008, 2012, 2014, 2015).

69. LeDoux (2012, 2014, 2015).

70. This summary is based on Winkielman et al (2005).

71. James (1884, 1890).

72. Freud (1915), p. 109.

73. Barrett (2006a, 2006b, 2009); Barrett and Russell (2015); Russell (2003); Russell and Barrett (1999); Lindquist et al (2006); Barrett et al (2007); Lindquist and Barrett (2008); Clore and Ortony (2013).

74. Clore (1994).

75. Watson (1913, 1919, 1925, 1938); Skinner (1938, 1950, 1953, 1974).

76. Tolman (1932, 1935); Hull (1943, 1952).

77. Morgan (1943); Hebb (1955); Stellar (1954); Bindra (1969, 1974); Rescorla and Solomon (1967); Bolles and Fanselow (1980); McAllister and McAllister (1971); Masterson and Crawford (1982); Gray (1982, 1987); Gray and McNaughton (2000); Bouton (2005).

78. Scherer (1984, 2000, 2012).

79. Tomkins (1962); Ekman (1972, 1977, 1984, 1992a, 1992b, 1993, 1999); Izard (1971, 1992, 2007); Panksepp (1982, 1998, 2000, 2005); Panksepp et al (1991); Vandekerckhove and Panksepp (2009, 2011); Damasio (1994, 1996, 1999, 2010); Damasio and Carvalho (2013); Damasio et al (2000); Prinz (2004); Scarantino (2009).

80. LeDoux (1984, 1987, 1996, 2002, 2008, 2012, 2014, 2015).

81. Schachter and Singer (1962); Arnold (1960); Smith and Ellsworth (1985); Scherer (1984, 2000, 2012); Lazarus (1991a, 1991b); Ortony and Clore (1989); Ortony et al (1988); Clore (1994); Clore and Ketalaar (1997); Clore and Ortony (2013); Johnson-Laird (1988); Johnson-Laird and Oatley (1989, 1992); Levenson, Soto, and Pole (2007).

82. Barrett (2006a, 2006b, 2009); Barrett and Russell (2015); Russell (2003); Russell and Barrett (1999); Lindquist et al (2006); Barrett et al (2007); Lindquist and Barrett (2008); Clore and Ortony (2013).

CHAPTER 2: RETHINKING THE EMOTIONAL BRAIN

1. Kagan (2003).

2. LeDoux (2012, 2014).

3. MacLean (1949, 1952, 1970).

4. The limbic system theory was based on the concept of brain evolution proposed in the early twentieth century by Ludwig Edinger (Edinger, 1908) and his followers (Arien Kappers et al, 1936; Herrick, 1933, 1948; Papez, 1929). This evolutionary theory has been criticized by numerous scholars (Nauta and Karten, 1970; Butler and Hodos, 2005; Northcutt, 2001; Reiner, 1990; Jarvis et al, 2005; Striedter, 2005). The limbic system theory itself has also been highly criticized (Brodal, 1982; Swanson, 1983; Reiner, 1990; Kotter and Meyer, 1992; LeDoux, 1991, 1996, 2012b).

5. Gazzaniga and LeDoux (1978).

6. Gazzaniga (1970).

7. Watson (1925); Skinner (1938).

8. Neisser (1967); Gardner (1987).

9. Hirst et al (1984); LeDoux et al (1983); Volpe et al (1979).

10. Gazzaniga and LeDoux (1978).

11. The Society for Neuroscience was founded in 1969, and the first meeting of this group was held in Washington, D.C., in 1971.

12. Kandel and Spencer (1968); Kandel (1976); Kandel and Schwartz (1982); Hawkins et al (2006); Kandel (2001, 2006).

13. Pavlov (1927).

14. Thorndike (1913).

15. Skinner (1938).

16. Skinner (1953).

17. Carew et al (1972, 1981); Pinsker et al (1973); Walters et al (1979); Kandel et al (1983); Hawkins et al (1983).

18. Cohen (1975, 1984); Schneiderman et al (1974); Berger et al (1976); Thompson et al (1983); Woody (1982); Ryugo and Weinberger (1978); Berthier and Moore (1980).

19. Blanchard and Blanchard (1969); Bolles and Fanselow (1980); Bouton and Bolles (1980); Brown and Farber (1951); McAllister and McAllister (1971); Brady and Hunt (1955).

20. Blanchard and Blanchard (1969); Bolles and Fanselow (1980); Bouton and Bolles (1980).

21. Blanchard and Blanchard (1969); Bolles and Fanselow (1980); Bouton and Bolles (1979); Gray (1987); Edmunds (1974); Brain et al (1990).

22. Schneiderman et al (1974); Kapp et al (1979); Smith et al (1980); Cohen (1984); Gray et al (1989); LeDoux et al (1982); Sakaguchi et al (1983).

23. For discussion of some of the issues, see Lorenz (1950); Tinbergen (1951); Beach (1955); Lehrman (1961); Elman et al (1997); Blumberg (2013).

24. The Blanchards had done a study of amygdala lesions on conditioned fear in the early 1970s (Blanchard and Blanchard, 1972). Although Bruce Kapp was about to publish a paper on the role of the central nucleus of the amygdala in fear conditioning when I was just starting my research (Kapp et al, 1979), I did not learn about Kapp's work until after my work was well under way.

25. Weiskrantz (1956); Goddard (1964); Sarter and Markowitsch (1985).

26. For a summary of my work on fear conditioning, see LeDoux (1987, 1992, 1996, 2000, 2002, 2007, 2008, 2012a, 2014); Quirk et al (1996); LeDoux and Phelps (2008); Johansen et al (2011); Rodrigues et al (2004).

27. Kandel (1997; 2012); Byrne et al (1991); Glanzman (2010).

28. See list in preface.

29. Kapp et al (1984, 1992); Davis (1992).

30. Fanselow and Lester (1988).

31. Kim et al (1993); Maren and Fanselow (1996).

32. Sample trainees from different laboratories. *Kapp laboratory:* Paul Whalen, Michaela Gallagher; *Davis laboratory:* David Walker, Jeff Rosen, Serge Campeau, Katherine Myers, Shenna Josslyn; *Fanselow laboratory:* Jeansok Kim, Fred Helmstetter, Steve Maren.

33. Other researchers who have made significant contributions include Denis Paré, Andreas Luthi, Chris Pape, Pankaj Sah, and Vadim Bolshakov. Many others have entered the field in the last several years. While they are too numerous to mention, a number of them are cited in various places in the book.

34. LeDoux (1987, 1992, 1996, 2002, 2007); Rodrigues et al (2004); Johansen et al (2011); Fanselow and Poulos (2005); Davis (1992); Paré et al (2004); Pape and Paré (2010); Sah et al (2008).

35. LeDoux (1996), p. 128.

36. James (1884, 1890).

37. Darwin (1872).

38. Panksepp (1998).

39. Mowrer (1939, 1940, 1947); Mowrer and Lamoreaux (1946).

40. Miller (1941, 1948, 1951); Brady and Hunt (1955); Rescorla and Solomon (1967); McAllister and McAllister (1971); Masterson and Crawford (1982); Bolles and Fanselow (1980).

41. These researchers typically claimed that the state of fear was not meant to refer to subjective feelings; yet they often wrote in a way that failed to make this case, often talking about rats "freezing in fear," for example. However, the intellectual leader of this field, O. Herbert Mowrer (1960), explicitly claimed that conscious feelings of fear are what motivate rats to avoid shock. The standard position of most theorists, though, was that fear was a nonsubjective motivational state.

42. McAllister and McAllister (1971); Masterson and Crawford (1982); Bolles and Fanselow (1980).

43. Central motivational states like this were proposed to exist starting in the 1940s (Beach, 1942; Hull, 1943; Mowrer and Lamoreaux, 1946; Morgan, 1943, 1957; Stellar, 1954; Hebb, 1955; Bindra, 1969, 1974). At the time little was known about brain function, and it was proposed that these states be talked about as components of a conceptual nervous system rather than the actual central nervous system (Hebb, 1955).

44. Tolman (1932); Hull (1943); MacCorquodale and Meehl (1948); Marx (1951).

45. Early proponents of the fear circuit as mediating a central state of fear were Michael Davis, Peter Lang, and Michael Fanselow (Davis, 1992; Lang, 1995; Fanselow, 1989; Fanselow and Lester, 1988). Other researchers also adopted this view as well (e.g., Rosen and Schulkin, 1998; Adolphs, 2013).

46. Personal communication via email.

47. Bolles (1967).

48. Personal communication via email.

49. Summarized in Gazzaniga and LeDoux (1978).

50. Gazzaniga and LeDoux (1978).

51. Gazzaniga (1998).

52. LeDoux (1984).

53. LeDoux (1996), p. 267.

54. Olsson and Phelps (2004); Bornemann et al (2012); Mineka and Ohman (2002); Vuilleumier et al (2002); Knight et al (2005); Whalen et al (1998); Liddell et al (2005); Luo et al (2010); Morris et al (1998); Pourtois et al (2013).

55. Bornemann et al (2012).

56. Kahneman (2011).

57. Fletcher (1995); Churchland (1988).

58. Bacon (1620), p. 68; Arturo Rosenblueth and Norbert Wiener, quoted in Lewontin (2001), p. 1264.

59. Panksepp (1998, 2000); Ekman (1992a, 1992b, 1999); Tomkins (1962); Izard (1992, 2007). For a critique of the natural kind view of emotion, see Barrett (2006a, 2006b, 2013); Barrett et al (2007); LeDoux (2012).

60. Panksepp (1998, 2000, 2005, 2011); Adolphs (2013); Anderson and Adolphs (2014).

61. Ekman (1992a, 1992b, 1999); Tomkins (1962); Izard (1992, 2007); Scarantino (2009); Prinz (2004); Panksepp (1998); Damasio (1994).

62. Feinstein et al (2013).

63. Gray and McNaughton (2000).

64. Fossat et al (2014).

65. Headline from the website PsychCentral, http://psychcentral.com/news/2014/06/17/fear-center-in-brain-larger-among-anxious-kids/71325.html. Retrieved Jul. 20, 2014.

66. Qin et al (2014).

67. Ekman (1992a, 1992b, 1999); Tomkins (1962); Izard (1992, 2007); Panksepp (1998); Damasio (1994).

68. Kelley (1992); Fletcher (1995); Mandler and Kessen (1964).

69. Fletcher (1995).
70. Mandler and Kessen (1964).
71. Kelley (1992).
72. Mandler and Kessen (1964).
73. Churchland, P.M. (1984, 1988); Churchland, P.S. (1986, 1988); Graziano (2013); Graziano (2014).
74. Fletcher (1995).
75. The idea of a central defense system was originally an offshoot of central state theories of Morgan, Konorski, Hebb, and Bindra (Morgan, 1943; Bindra 1969; Hebb, 1955; Konorski, 1967). And while the central defense system was initially thought of in terms of defensive motivation, the terms "defense system" and "fear system" often came to be used interchangeably. Some of the ideas about defensive motivational systems are illustrated in these various publications: Konorski (1967); Masterson and Crawford (1982); Bolles and Fanselow (1980); McAllister and McAllister (1971); Fanselow and Lester (1988); Cardinal et al (2002); Blanchard and Blanchard (1988); Davis (1992); Rosen and Schulkin (1998); Adolphs (2013); Bouton (2007); Lang et al (1998); Mineka (1979).
76. Gazzaniga and LeDoux (1978); Gazzaniga (1998, 2008, 2012).
77. Ekman (1992a, 1992b, 1999); Tomkins (1962); Izard (1992, 2007); Scarantino (2009); Prinz (2004); Panksepp (1998); Damasio (1994).
78. LeDoux (2012, 2014).
79. LeDoux (2012, 2014).
80. LeDoux (2012); Sternson (2013); Giske et al (2013).
81. Wang et al (2011); Lebetsky et al (2009); Dickson (2008); McGrath et al (2009); Pirri and Alkema (2012); Garrity et al (2010); Bendesky et al (2011); Kupfermann (1974, 1994); Kupfermann et al (1992).
82. Macnab and Koshland (1972); Hennessey et al (1979); Fernando et al (2009); Berg (1975, 2000); Harshey (1994); Eriksson et al (2002); Helmstetter et al (1968); Rothfield et al (1999).
83. Emes and Grant (2012).
84. LeDoux (2012).
85. LeDoux (2012, 2014).
86. LeDoux (2012); Giske et al (2013).
87. Beach (1942); Morgan (1943, 1957); Stellar (1954); Hebb (1955); Bindra (1969, 1974).
88. Bargmann (2006, 2012); Galliot (2012); Lebetsky et al (2009); Bendesky et al (2011); Dickson (2008); Pirri and Alkema (2012); Garrity et al (2010); Kupfermann (1974, 1994); Kupfermann et al (1992).
89. Sara and Bouret (2012); Bouret and Sara (2005); Foote et al (1983); Aston-Jones and Cohen (2005); Saper et al (2005); Nadim and Bucher (2014); Luchicchi et al (2014).
90. Konorski (1967); Masterson and Crawford (1982); Bolles and Fanselow (1980); McAllister and McAllister (1971); Fanselow and Lester (1988); Cardinal et al (2002); Blanchard and Blanchard (1988); Davis (1992); Rosen and Schulkin (1998); Adolphs (2013); Bouton (2007); Lang et al (1998); Mineka (1979).
91. Barrett (2006, 2009, 2012); Barrett et al (2007); Lindquist and Barrett (2008); Wilson-Mendenhall et al (2011); Russell (2003, 2009); Russell and Barrett (1999); Wilson-Mendenhall et al (2013).
92. Russell (1991, 1994, 2003, 2009; 2012, 2014); Russell and Barrett (1999); Barrett (2006a, 2006b); Barrett and Russell (2014); Lindquist and Barrett, L.F. (2008); Clore and Ortony (2013); Levenson, Soto, and Pole (2007).
93. Lashley (1950).
94. Kihlstrom (1987).

95. Dickinson (2008); LeDoux (2008, 2012a); Winkielman and Berridge (2004).

96. Balleine and Dickinson (1998); Dickinson (2008); Heyes (2008).

97. Chamberlain (1890).

98. Heyes (2008); Rosenthal (1990).

99. Hatkoff (2009).

100. Goodall's Introduction in Hatkoff (2009).

101. Goodall, quoted in "Should Apes Have Legal Rights?" *The Week*, August 3, 2013. http://theweek.com/article/index/247763/should-apes-have-legal-rights. Retrieved Nov. 5, 2014.

102. Arguments for legal rights in animals are often based more on moral than scientific grounds. See "Should Apes Have Legal Rights?" *The Week*, August 3, 2013. http://theweek.com/article/index/247763/should-apes-have-legal-rights. Retrieved Nov. 5, 2014.

103. Caporael and Heyes (1997).

104. Frans de Waal, interviewed by Edwin Rutsch at the Center for Building a Culture of Empathy. http://cultureofempathy.com/references/Experts/Frans-de-Waal.htm. Retrieved Nov. 6, 2014.

105. Frans de Waal, interview for Wonderlance.com. http://www.wonderlance.com/february2011_scientech_fransdewaal.html. Retrieved Nov. 6, 2014.

106. Barbey et al (2012).

107. Semendeferi et al (2011).

108. Preuss (1995, 2001); Wise (2008).

109. Dennett (1991); Jackendoff (2007); Weiskrantz (1997); Frith et al (1999); Naccache and Dehaene (2007); Dehaene et al (2003); Dehaene and Changeux (2004); Koch and Tsuchiya (2007); Sergent and Rees (2007); Alanen (2003).

110. Weiskrantz (1997); Heyes (2008).

111. Mitchell et al (1996); Kennedy (1992).

112. Decety (2002).

113. Fletcher (1995); Churchland (1988).

114. Heider and Simmel (1944); Heberlein and Adolphs (2004); Greene and Cohen (2004).

115. Greene and Cohen (2004).

CHAPTER 3: LIFE IS DANGEROUS

1. Emerson (1870).

2. Moyer (1976).

3. LeDoux (2012).

4. Gallistel (1980); Godsil and Fansleow (2013); LeDoux (2012).

5. Cannon (1929).

6. Darwin (1872).

7. Miller (1948); Hunt and Brady (1951); Blanchard and Blanchard (1969); Bouton and Bolles (1980); Bolles and Fanselow (1980).

8. Suarez and Gallup (1981).

9. Edmunds (1974); Blanchard and Blanchard (1969); Bracha et al (2004); Ratner (1967, 1975).

10. This paragraph is based on Edmunds (1974); Ratner (1967, 1975); Langerhans (2007); Pinel and Treit (1978).

11. Edmunds (1974).

12. Rosen (2004); Takahashi et al (2005); Gross and Canteras (2012); Dielenberg et al (2001); Hubbard et al (2004).

13. Breviglieri et al (2013); Zanette et al (2011).

14. Litvin et al (2007).

15. Vermeij (1987); Dawkins and Krebs (1979); Mougi (2010); Edmunds (1974).

16. Edmund (1974); Dawkins and Krebs (1979).

17. Langerhans (2007).
18. This paragraph is based on: Benison and Barger (1978); Fleming (1973); Brown and Fee (2002).
19. Bernard (1865/1957); Langley (1903); Cannon (1929).
20. Blessing (1997); Porges (2001).
21. The value of the term "innate" was discussed in Chapter 2.
22. Lang (1968, 1978, 1979).
23. Selye (1956).
24. Rodrigues et al (2009).
25. McEwen and Lasley (2002); Sapolsky (1998); McGaugh (2000); de Quervain et al (2009).
26. Klein (1993); Preter and Klein (2008); Roth (2005).
27. Freire et al (2010); Johnson et al (2014); Wemmie (2011).
28. Ley (1994); Vickers and McNally (2005).
29. See Blanchard and Blanchard (1988); Gray (1982); Bolles and Fanselow (1980); Fanselow and Lester (1988); Fanselow (1989).
30. Edmunds (1974); Ratner (1967, 1975).
31. Tolman (1932); Blanchard et al (1976); Blanchard and Blanchard (1988); Bolles and Fanselow (1980); Adams (1979).
32. Bolles and Collier (1976); Bolles and Fanselow (1980); Blanchard et al (1976); Blanchard and Blanchard (1988).
33. Fanselow and Lester (1988).
34. Fanselow calls this the postencounter stage, but the simpler term "encounter stage" is more straightforward.
35. Bolles (1970); Bolles and Fanselow (1980); Fanselow (1989); Fanselow (1986); Fanselow and Lester (1988).
36. See quotes from these authors in Chapter 1.
37. Brain et al (1990), p. 420.
38. Rosen (2004); Takahashi et al (2005).
39. Rosen (2004).
40. Hebb (1949); Magee and Johnston (1997); Bliss and Collingridge (1993); Martin et al (2000); Johansen et al (2010); Kelso et al (1986).
41. Pavlov (1927); Myers and Davis (2002); Milad and Quirk (2012); Bouton (2002); Sotres-Bayon et al (2004, 2006).
42. Jacobs and Nadel (1985); Bouton (1993, 2002, 2004); Bouton et al (2006).
43. Wolpe (1969); Rachman (1967); Eysenck (1987); Kazdin and Wilson (1978); Hofmann et al (2013); Beck (1991); Foa (2011); Marks and Tobena (1990); Barlow (1990); Barlow (2002).
44. Williams (2001); Beck et al (2011); Genud-Gabai et al (2013).
45. See Grupe and Nitschke (2013).
46. Rogan et al (1997); Rogan et al (2005); Etkin et al (2004); Walasek et al (1995).
47. Demertzis and Kraske (2005).
48. Ohman (1988, 2002, 2005, 2007, 2009); Phelps (2006); Phelps and LeDoux (2005); Dolan and Vuilleumier (2003); Buchel and Dolan (2000); Armony and Dolan (2002); Dunsmoor et al (2014); Schiller et al (2008); Pine et al (2001); Olsson et al (2007); Delgado et al (2008); Lau et al (2011); Grillon (2008).
49. Bandura (1977); Rachman (1990).
50. Mineka and Cook (1993); Berger (1962); Hygge and Öhman (1978); Olsson and Phelps (2004); Olsson et al (2007); Olsson and Phelps (2007).
51. Litvin et al (2007); Jones et al (2014); Masuda et al (2013); Kim et al (2010); Chivers et al (1996); Gibson and Pickett (1983); Flower et al (2014).
52. Olsson and Phelps (2004); Raes et al (2014); Dymond et al (2012).

53. Mineka and Cook (1993); Berger (1962); Hygge and Öhman (1978); Olsson and Phelps (2004); Olsson et al (2007); Olsson and Phelps (2007).

54. Miller (1948); McAllister and McAllister (1971); Mineka (1979); Moscarello and LeDoux (2013); Choi et al (2010); Cain and LeDoux (2007); LeDoux et al (2009); Cain et al (2010); Cain and LeDoux (2008).

55. Balleine and Dickinson (1998); Cardinal et al (2002).

56. Miller (1948); Choi et al (2010).

57. Robert Bolles, a powerful figure in this field in the 1960s and 1970s, was a strong critic of the instrumental nature of avoidance (Bolles, 1970, 1972; Bolles and Fanselow, 1980). He argued that avoidance responses simply reflect species-specific responses and are not learned themselves. His vocal negative attitude inhibited research on this topic.

58. LeDoux (2014).

59. McAllister and McAllister (1971); Miller (1941, 1948, 1951); Mowrer and Lamoreaux (1946); Miller (1948); Amorapanth et al (2000); Coover et al (1978); Daly (1968); Dinsmoor (1962); Esmoris-Arranz et al (2003); Goldstein (1960); Kalish (1954); McAllister and McAllister (1991); Desiderato (1964); Kent et al (1960); McAllister et al (1972, 1980).

60. McAllister and McAllister (1971); Mineka (1979); Levis (1989); Cain and LeDoux (2007); LeDoux et al (2009); Cain et al (2010); Cain and LeDoux (2008).

61. Cain and LeDoux (2007).

62. Skinner (1938, 1950, 1953); Kanazawa, S. (2010). Common Misconceptions About Science VI: "Negative Reinforcement." *Psychology Today*. Retrieved Oct. 29, 2014, from http://www.psychologytoday.com/blog/the-scientific-fundamentalist/201001/common-misconceptions-about-science-vi-negative-reinforcem.

63. Mowrer and Lamoreaux (1946); Miller (1948); McAllister and McAllister (1971); Masterson and Crawford (1982); Levis (1989); Gray (1987).

64. Mowrer and Lamoreaux (1946); Miller (1941, 1948, 1951); Miller (1948); McAllister and McAllister (1971); Levis (1989); Masterson and Crawford (1982).

65. Thorndike (1898, 1913); Olds (1956, 1958, 1977); Olds and Milner (1954); Panksepp (1998).

66. Rescorla and Solomon (1967); Bolles (1975); Bolles and Fanselow (1980); Masterson and Crawford (1982).

67. Ricard and Lauterbach (2007); Hofmann (2008); Dymond and Roche (2009).

68. Schultz (2013); Tully and Bolshakov (2010).

69. Grupe and Nitschke (2013).

70. Borkovec et al (1999).

71. See Thorndike (1913); Cardinal et al (2002); Balleine and Dickinson (1998); Balleine and O'Doherty (2010).

72. Church et al (1966); Solomon (1980).

73. LeDoux (2013).

74. Barlow (2002).

75. Dymond and Roche (2009).

76. For summaries of research on decision making, see Glimcher (2003); Bechara et al (1997); Levy and Glimcher (2012); Sugrue et al (2005); Rorie and Newsome (2005); Shadlen and Kiani (2013); Rangel et al (2008); Dolan and Dayan (2013); Balleine and Dickinson (1998); Cardinal et al (2002); Balleine (2011); Delgado et al (2008); Delgado and Dickerson (2012); Hartley and Phelps (2012); Dayan and Daw (2008); Rolls (2014).

77. Corbit and Balleine (2005); Holmes et al (2010); Holland (2004); Rescorla (1994). Two motivational processes underlie these value transfer effects. The CS triggers a general motivational process that energizes behavior in a nonspecific way. The CS can also trigger a motivational process that is specific to the value of the Pavlovian US. This US-specific form of motivation underlies the greater facilitation by a CS that has a motivational match with

the outcome of the instrumental response (food in both cases or footshock in both cases). What remains after the specific effect is accounted for is the general motivational effect.

78. Campese et al (2013).
79. Holmes et al (2010); Volkow et al (2008); Robinson and Berridge (2008).
80. Grupe and Nitschke (2013); Beck and Emery (1985); Barlow (2002).
81. Bindra (1968); Cofer (1972).
82. Gray and McNaughton (2000); Grupe and Nitschke (2013).
83. Kendler et al (2003); Bell (2009); Bevilacqua and Goldman (2013); Gorwood et al (2012); Pavlov et al (2012); Nemoda et al (2011); Mitchell (2011); Congdon and Canli (2008); Casey et al (2011).
84. Blanchard and Blanchard (1988).
85. Gray and McNaughton (2000); Grupe and Nitschke (2013).
86. File et al (2004); Campos et al (2013); Sudakov et al (2013); Davis et al (1997); Davis et al (2010); Belzung and Griebel (2001); Clément et al (2002); Crawley and Paylor (1997); Griebel and Holmes (2013); Kumar et al (2013); Millan (2003); File (1993, 1995, 2001); File and Seth (2003).
87. Erlich et al (2012).
88. Waddell et al (2006); Walker and Davis (1997, 2002, 2008).
89. Millan and Brocco (2003).
90. Gray (1982, 1987); Gray and McNaughton (2000); McNaughton and Corr (2004); McNaughton (1989).
91. Blanchard and Blanchard (1988); Gray and McNaughton (2000).
92. Loewenstein et al (2001). Drawing from a broad range of psychological and behavioral economics research, the authors propose the risk-as-feelings hypothesis that emphasizes the impact of emotions on apparently suboptimal decision making. I disagree with how they use the words "emotion" and "cognition." In contrast to them, I would say that what they call emotional systems are survival circuits and what they call cognition subsumes both cognitive and emotional (feeling) processes. Nevertheless, I agree with their point that different brain systems deal with risk differently.
93. This expression is borrowed from Michael Gazzaniga's book of the same name.
94. Evans (2008); Kahneman (2011); Newell and Shanks (2014).
95. Kahneman (2011); Tversky and Kahneman (1974); Kahneman et al (1982).
96. Tversky and Kahneman (1974); Kahneman et al (1982); Kahneman (2011).
97. Park et al (2014); Redelmeier (2005); Minué et al (2014); but see Marewski and Gigerenzer (2012).
98. Evans (2010); Evans (2014).
99. Nisbett and Wilson (1977); Wilson (2002); Wilson et al (1993); Bargh (1997); Kihlstrom (1987).
100. Gazzaniga and LeDoux (1978); Gazzaniga (2012); Wilson (2002); Evans (2014).
101. Wegner (2002); Velmans (2000); Bargh and Ferguson (2000); Evans (2010); Greene and Cohen (2004); Gazzaniga (2012).
102. This issue was discussed extensively in Newell and Shanks (2014). Although they minimized the importance of unconscious factors, others provided strong critiques of their view and compelling support for the role of unconscious factors in decision making. See commentary at the end of the article by Evans et al, Coppin et al, Ingram and Prochownik, Ogilvie and Carruthers, and Finkbeiner and Coltheart.
103. Gazzaniga (2012); Jones (2004); Zeki and Goodenough (2004).

CHAPTER 4: THE DEFENSIVE BRAIN

1. Tolkien (1955).
2. Barres (2008)
3. Behrmann and Plaut (2013).
4. The *neocortex* was so named because it was thought to be a new addition with the evolution of mammals (Edinger, 1908; Ariëns Kappers et al, 1936), a view that has since been challenged (Nauta and Karten, 1970; Butler and Hodos, 2005; Northcutt, 2001; Reiner, 1990; Jarvis et al, 2005; Striedter, 2005). Some now prefer the more neutral term *"isocortex"* to avoid this evolutionary implication. But because the term "neocortex" is in wide use, I adopt it here.
5. Some medial cortical areas with five layers are considered transitional between the neocortex and allocortex (Mesulam and Mufson, 1982; Allman et al, 2001), but for simplicity I will refer to both allocortex and transitional areas as the medial cortex.
6. A portion of the neocortex curls around to the medial side, but most neocortical tissue is located laterally.
7. I have long been a critic of the limbic system theory of emotion. It is based on Edinger's theory of brain evolution (Edinger, 1908; Ariëns Kappers et al, 1936), which has since been discredited (see Nauta and Karten, 1970; Butler and Hodos, 2005; Northcutt, 2001; Reiner, 1990; Jarvis et al, 2005; Striedter, 2005). For other critiques of the limbic system theory, see Brodal (1982); Swanson (1983); Kotter and Meyer (1992); Reiner (1990).
8. Goltz (1892).
9. Cannon (1929); Cannon and Britton (1925); Bard (1928).
10. Karplus and Kreidl (1909).
11. Cannon (1929, 1936).
12. Bard (1928); Bard and Rioch (1937).
13. Ranson and Magoun (1939); Eliasson et al (1951); Uvnas (1960); Eliasson et al (1951); Grant et al (1958).
14. Hess and Brugger (1943); Hess (1949); Hunsperger (1956); Fernandez de Molina and Hunsperger (1959); Hoebel (1979); Vaughan and Fisher (1962).
15. Abrahams et al (1960).
16. Hilton and Zbrozyna (1963); Hilton (1979); Fernandez de Molina and Hunsperger (1962).
17. Kluver and Bucy (1937); Weiskrantz (1956); MacLean (1949, 1952).
18. See the discussion of innateness in Chapter 2.
19. Hilton (1982).
20. Lindsley (1951).
21. Moruzzi and Magoun (1949).
22. Lindsley (1951).
23. Saper (1987).
24. Flynn (1967); Siegel and Edinger (1981); Panksepp (1971); Zanchetti et al (1972).
25. Sternson (2013); Wise (1969); Valenstein (1970).
26. Bandler and Carrive (1988).
27. Deisseroth (2012); Boyden et al (2005); Sternson (2013); Lin et al (2011).
28. Lin et al (2011).
29. Sternson (2013).
30. Although a good deal has been learned about how innate odors related to predators elicit defensive behaviors in rodents (Gross and Canteras, 2012; Rosen, 2004; Blanchard et al, 1989), I will not cover this work in any detail.
31. Hebb (1949).
32. Johansen et al (2011); Maren (2005).

33. Hebb (1949); Brown et al (1990); Magee and Johnston (1997); Bliss and Collingridge (1993); Martin et al (2000); Johansen et al (2010).

34. For review see Quirk et al (1996); LeDoux (2002); Maren (2005); Johansen et al (2011); Rogan et al (2001); Paré and Collins (2000); Paré et al (2004).

35. Rodrigues et al (2004); Johansen et al (2011); Maren (2005); Tully and Bolshakov (2010); Sah et al (2008); Rogan et al (2001); Nguyen (2001); Josselyn (2010); Fanselow and Poulos (2005); Schafe and LeDoux (2008).

36. Pitkanen et al (1997); Pitkanen (2000); Amaral et al (1992).

37. Quirk et al (1995, 1997); Repa et al (2001).

38. Paré and Smith (1993, 1994); Royer and Paré (2002).

39. Haubensak et al (2010); Ciocchi et al (2010).

40. Price and Amaral (1981); Hopkins and Holstege (1978); da Costa Gomez and Behbehani (1995).

41. LeDoux et al (1988); Amorapanth et al (1999); Fanselow et al (1995); Kim et al (1993); De Oca et al (1998).

42. Gross and Canteras (2012).

43. LeDoux (1992, 1996); Davis (1992).

44. LeDoux et al (1988); Amorapanth et al (1999).

45. LeDoux et al (1988).

46. Morrison and Reis (1991); Cravo et al (1991); Saha (2005); Macefield et al (2013); Reis and LeDoux (1987).

47. This contrasts with results described above showing that electrical stimulation of the PAG elicits autonomic responses. However, electrical stimulation is a crude method that is prone to giving false positive results: Just because a response can be elicited by artificial stimulation of the brain does not mean that under more natural conditions the brain functions this way.

48. Kapp et al (1979, 1984); Schwaber et al (1982); Danielsen et al (1989); Pitkanen et al (1997); Pitkanen (2000); Liubashina et al (2002); Veening et al (1984); van der Kooy et al (1984); Takeuchi et al (1983); Higgins and Schwaber (1983).

49. Gray and Bingaman (1996); Gray et al (1989, 1993); Rodrigues et al (2009); Sullivan et al (2004).

50. Sara and Bouret (2012); Bouret and Sara (2005); Foote et al (1983); Aston-Jones and Cohen (2005); Saper et al (2005); Nadim and Bucher (2014); Luchicchi et al (2014); Holland and Gallagher (1999); Whalen (1998); Weinberger (1982, 1995); Lindsley (1951); Aston-Jones et al (1991); Sears et al (2013); Davis and Whalen (2001).

51. Kapp et al (1992); Weinberger (1995); Sears et al (2013); Davis and Whalen (2001); Holland and Gallagher (1999); Gallagher and Holland (1994); Lee et al (2010); Wallace et al (1989); Van Bockstaele et al (1996); Luppi et al (1995); Bouret et al (2003); Spannuth et al (2011); Samuels and Szabadi (2008).

52. Holland and Gallagher (1999); Whalen (1998); Weinberger (1982, 1995); Lindsley (1951); Aston-Jones et al (1991); Sears et al (2013); Davis and Whalen (2001).

53. Weinberger (1995, 2003, 2007); Armony et al (1998); Apergis-Schoute et al (2014); Morris et al (1998, 2001).

54. Maren et al (2001); Goosens and Maren (2002); Herry et al (2008); Li and Rainnie (2014); Wolff et al (2014).

55. Pascoe and Kapp (1985); Wilensky et al (1999, 2000, 2006); Paré et al (2004); Duvarci et al (2011); Haubensak et al (2010); Ciocchi et al (2010); Li et al (2013); Duvarchi and Paré (2014); Pape and Paré (2010); Penzo et al (2014).

56. Quirk et al (1996); Maren et al (2001).

57. Morgan et al (1993); Morgan and LeDoux (1995); Quirk and Mueller (2008); Quirk et al (2006); Milad and Quirk (2012); Sotres-Bayon et al (2004, 2006); Likhtik et al (2005); Duvarci and Paré (2014).

58. Sotres-Bayon et al (2004); LeDoux (1996, 2002); Morgan et al (1993); Morgan and LeDoux (1995); Quirk and Mueller (2008); Quirk et al (2006); Milad and Quirk (2012); Sotres-Bayon et al (2004, 2006); Likhtik et al (2005); Duvarci and Paré (2014).

59. LeDoux (1996, 2002); Quirk and Mueller (2008); Quirk et al (2006); Milad and Quirk (2012); Sotres-Bayon et al (2004, 2006); Likhtik et al (2005); Duvarci and Paré (2014); Paré and Duvarci (2012); VanElzakker et al (2014); Gilmartin et al (2014); Gorman et al (1989); Davidson (2002); Bishop (2007); Shin and Liberzon (2010); Mathew et al (2008).

60. LeDoux (2013).

61. Phillips and LeDoux (1992, 1994); Kim and Fanselow (1992); Ji and Maren (2007); Maren (2005); Maren and Fanselow (1997); Sanders et al (2003).

62. Frankland et al (1998).

63. LaBar et al (1995); Bechara et al (1995).

64. LaBar et al (1998); Buchel et al (1998).

65. Morris et al (1998, 1999).

66. LaBar and Phelps (2005); Lonsdorf et al (2014); Marschner et al (2008); Chun and Phelps (1999); Huff et al (2011).

67. Schiller et al (2008); Phelps et al (2004); Hartley et al (2011); Kim et al (2011); Quirk and Beer (2006); Milad et al (2007); Delgado et al (2004, 2006, 2008); Schiller and Delgado (2010).

68. Olsson and Phelps (2004).

69. Ostroff et al (2010).

70. Structural plasticity: Lamprecht and LeDoux (2004); Ostroff et al (2010); Bourne and Harris (2012); Bailey and Kandel (2008); Martin (2004).

71. Kandel (1999, 2006).

72. LeDoux (1996, 2002); Cain et al (2010); Choi et al (2010).

73. Whether avoidance conditioning meets the criteria used to decide whether a response is truly instrumental (goal directed) has been questioned (Bolles 1970, 1972). We are currently working on experiments to resolve this long-standing controversy.

74. In this survey, I emphasize findings from signaled active avoidance in rodents that build upon the Pavlovian conditioning results we have acquired. Other tasks have been studied. Particularly noteworthy is the work by Michael Gabriel, who accumulated much information about the brain mechanisms of avoidance learning. But Gabriel was more concerned with avoidance as a window into learning than as an emotion regulation response. In addition, due to procedural differences and an emphasis on different brain functions, his findings are difficult to evaluate in relation to the work we have done. For a summary of Gabriel's work, see: Gabriel (1990); Gabriel and Orona (1982); Hart et al (1997).

75. LeDoux (2002).

76. Amorapanth et al (2000).

77. Cain et al (2010); Cain and LeDoux (2007, 2008); Choi et al (2010); Moscarello and LeDoux (2012); Campese et al (2013, 2014); Lázaro-Muñoz et al (2010); LeDoux et al (2010); McCue et al (2014); Martinez et al (2013); Galatzer-Levy et al (2014).

78. Choi et al (2010); Moscarello and LeDoux (2013).

79. LeDoux and Gorman (2001).

80. Choi et al (2010); Moscarello and LeDoux (2013).

81. Choi et al (2010); Moscarello and LeDoux (2013).

82. Moscarello and LeDoux (2013).

83. Wendler et al (2014); Lichtenberg et al (2014); Ramirez et al (2015).

84. Delgado et al (2009); Aupperle and Paulus (2010); Schiller and Delgado (2010); Schlund et al (2010, 2011, 2013); Schlund and Cataldo (2010).
85. Cain and LeDoux (2007).
86. Amorapanth et al (2000); Campese, Cain and LeDoux (unpublished data).
87. Everitt and Robbins (2005); Everitt et al (1989); Cardinal et al (2002).
88. Grace et al 2007; Grace and Sesack (2010); Gato and Grace (2008).
89. Berridge (2009); Berridge and Kringelbach (2013); Pecina et al (2006); Castro and Berridge (2014).
90. Lázaro-Muñoz et al (2010).
91. Fernando et al (2013); Morrison and Salzman (2010); Savage and Ramos (2009); Balleine and Killcross (2006); Rolls (2005); Holland and Gallagher (2004); Petrovich and Gallagher (2003); Cardinal et al (2002); Everitt et al (1999).
92. Everitt and Robbins (2005, 2013); Everitt et al (2008); Smith and Graybiel (2014); Devan et al (2011); Balleine and O'Doherty (2010); Packard (2009); Balleine (2005); Wickens et al (2007).
93. Wendler et al (2014).
94. Barlow (2002); Borkovec et al (2004); Foa and Kozak (1986).
95. See LeDoux and Gorman (2001) and LeDoux (2013).
96. Corbit and Balleine (2005); Holmes et al (2010); Holland (2004); Rescorla (1994).
97. For a summary, see Holmes et al (2010).
98. Talmi et al (2008); Bray et al (2008); Prevost et al (2012); Lewis et al (2013); Nadler et al (2011); Talmi et al (2008).
99. Campese et al (2013, 2014); McCue et al (2014).
100. Morrison and Salzman (2010).
101. Levy and Glimcher (2012); Rolls (2014).
102. Glimcher (2009); Paulus and Yu (2012); Kishida et al (2010); Rangel et al (2008); Bach and Dolan (2012); Bach et al (2011); Toelch et al (2013); Yoshida et al (2013); Clark et al (2008).
103. Alheid and Heimer (1988).
104. Davis et al (1997); Tye et al (2011); Adhikari (2014); Waddell et al (2006).
105. Somerville et al (2010, 2013); Grupe et al (2013).
106. Sink et al (2011, 2013); Davis et al (2010); Walker and Davis (2008); Liu and Liang (2009); Liu et al (2009); Liang et al (2001); Graeff (1994); Waddell et al (2006); Sajdyk (2008).
107. Davis et al (1997, 2010); Davis (2006); Walker and Davis (2008).
108. For summaries, see Whalen (1998); McDonald (1998); Cullinan et al (1993); Alheid et al (1998); Alheid and Heimer (1988); Davis et al (2010); Stamatakis et al (2014); Dong and Swanson (2004a, 2004b, 2006a, 2006b); Dong et al (2001).
109. Poulos et al (2010).
110. Whalen (1998); McDonald (1998); Cullinan et al (1993); Alheid et al (1998); Alheid and Heimer (1988); Davis et al (2010); Stamatakis et al (2014); Dong and Swanson (2004a, 2004b, 2006a, 2006b); Dong et al (2001); Pitkanen et al (1997); Pitkanen (2000).
111. O'Keefe and Nadel (1978); Moser et al (2014); Moser and Moser (2008); Kubie and Muller (1991); Muller et al (1987); Hartley et al (2014); Burgess and O'Keefe (2011); O'Keefe et al (1998); McNaughton et al (1996); Terrazas et al (2005).
112. Gray (1982); Gray and McNaughton (1996, 2000).
113. Anthony et al (2014).
114. Sparks and LeDoux (2000); Treit et al (1990).
115. Johansen (2013).
116. Jennings et al (2013); Kim, S.Y. et al (2013).
117. Risold and Swanson (1996); Swanson (1987); Groenewegen et al (1996, 1997, 1999); Grace et al (2007); Alheid and Heimer (1998); Amaral et al (1992); Pitaken et al (1997); Swanson (1983).

118. Mathew et al (2008); Charney (2003); Patel et al (2012); Vermetten and Bremner (2002); Southwick et al (2007); Yehuda and LeDoux (2007); Shin and Liberzon (2010); Rauch et al (2003, 2006); Dillon et al (2014); Tuescher et al (2011); Protopopescu et al (2005); Grupe and Nitschke (2013); Pitman et al (2001, 2012); Shin et al (2006).
119. Grupe and Nitschke (2013).
120. Nesse and Klaas (1994).
121. Grupe and Nitschke (2013).
122. Beck and Emery (1985); Barlow (2002).
123. Butler and Mathews (1983); Foa et al (1996); Mathews et al (1989); Bar-Haim et al (2007, 2010); Bishop (2007); Beck and Clark (1997); Eysenck et al (2007); Fox (1994); McTeague et al (2011); Bradley et al (1999); Buckley et al (2002); Öhman et al (2001); Mogg and Bradley (1998); Mineka et al (2012).
124. Rosen and Schulkin (1998); Sakai et al (2005); Semple et al (2000); Chung et al (2006); Furmark et al (2002); Atkin and Wager (2007); Nitschke et al (2009); Lorberbaum et al (2004); Guyer et al (2008).
125. Weinberger (1995); Kapp et al (1992); Davis and Whalen (2001).
126. Kalisch and Gerlicher (2014); Berggren and Derakshan (2013); Cisler and Koster (2010); Etkin et al (2011); Erk et al (2006); Vuilleumier (2002).
127. Etkin et al (2004); Schiller et al (2008); Hartley and Phelps (2010); Milad and Quirk (2012).
128. Lissek et al (2005, 2009); Woody and Rachman (1994); Grillon et al (2008, 2009); Jovanovic et al (2010, 2012); Waters et al (2009); Jovanovic and Norrholm (2011); Corcoran and Quirk, 2007; Maren et al (2013).
129. Gorman et al (2000).
130. Morgan et al (1993); LeDoux (1996, 2002); Morgan et al (1993); Sotres-Bayon et al (2004); Hartley and Phelps (2010); Quirk and Mueller (2008); Kolb (1990); Rolls (1992); Frysztak and Neafsey (1991); Markowska and Lukaszewska (1980); Goldin et al (2008); Delgado et al (2008); Hermann et al (2014); Grace and Rosenkranz (2002); Salomons et al (2014).
131. Jacobsen (1936); Mark and Ervin (1970); Teuber (1964); Nauta (1971); Myers (1972); Stuss and Benson (1986); Damasio et al (1990); Fuster (1989); Morgan et al (1993); LeDoux (1996, 2002); Milad and Quirk (2012); Sehlmeyer et al (2011); Rauch et al (2006); Likhtik et al (2005); Gilboa et al (2004); Barad (2005); Urry et al (2006); Milad et al (2014); Graham and Milad (2011).
132. Beck and Emery (1985); Barlow (2002); Borkovec et al (2004); Foa and Kozak (1986); Lovibond et al (2009).
133. The examples above are from Grupe and Nitschke (2013).
134. Aupperle and Paulus (2010); Shackman et al (2011).
135. Straube et al (2006); Hauner et al (2012); de Carvalho et al (2010); Schienle et al (2007); Klumpp et al (2013, 2014).
136. Yook et al (2010); Dupuy and Ladouceur (2008); Carleton (2012); Reuther et al (2013); Whiting et al (2014); McEvoy and Mahoney (2012); Mahoney and McEvoy (2012).
137. Somerville et al (2010); Bechtholt et al (2008); Grillon et al (2006); Baas et al (2002); Straube et al (2007); Alvarez et al (2011); Adhikari (2014).
138. Butler and Mathews (1983, 1987); Foa et al (1996); Gilboa-Schechtman et al (2000); Borkovec et al (1999); Stöber (1997); Mitte (2007); Volz et al (2003); Knutson et al (2005); Preuschoff et al (2008); Padoa-Schioppa and Assad (2006); Peters and Büchel (2010); Plassmann et al (2010); Rangel and Hare (2010); Schoenbaum et al (2011); Wallis (2012); Gottlich et al (2014).
139. Shackman et al (2011).
140. Summarized by Grupe and Nitschke (2013).

CHAPTER 5: HAVE WE INHERITED EMOTIONAL STATES OF MIND FROM OUR ANIMAL ANCESTORS?

1. Tinbergen (1951), pp. 4–5.
2. Darwin (1872).
3. Darwin (1859).
4. Duchenne (1862).
5. According to Schott (2013), Duchenne identified subtle inaccuracies in how the ancient sculptors viewed the role of facial muscles in emotional expressions.
6. Plutchick (1980), p. 3.
7. Keller (1973), p. 49.
8. Oliver Sacks (2014) wrote a piece in the *New York Review of Books* in which he cites Darwin's (1881) book on the habits of vegetable molds and worms.
9. Darwin, quoted in Knoll (1997), p. 15.
10. Romanes (1883).
11. Romanes (1882). Quoted in Keller (1973), p. 49.
12. Keller (1973), p. 49.
13. See Kennedy (1992); Mitchell et al (1996).
14. Knoll (1997).
15. Heyes (1994, 1995, 2008).
16. Morgan (1890–1891).
17. Paraphrased from Keller (1973), p. 51.
18. Keller (1973), p. 40.
19. Keller (1973), p. 51.
20. This paragraph is based on Boring (1950).
21. Wundt (1874).
22. James (1890).
23. James (1884).
24. Thorndike (1898).
25. Donahoe, J.W. (1999).
26. Summarized in Keller (1973).
27. Watson (1913, 1919, 1925).
28. Skinner (1938).
29. Watson (1925).
30. Skinner (1938, 1974).
31. Tolman (1932, 1935); Hull (1943).
32. Mowrer (1939, 1940, 1960); Mowrer and Lamoreaux (1942, 1946); Miller (1948); Dollard and Miller (1950).
33. McAllister and McAllister (1971); Masterson and Crawford (1982); Bolles and Fanselow (1980).
34. Skinner (1938).
35. Hess (1962), p. 57.
36. Sheffield and Roby (1950); Cofer (1972).
37. Berridge (1996); Berridge and Winkielman (2003); Castro and Berridge (2014); Winkielman et al (2005).
38. Morgan (1943); Stellar (1954); Konorski (1948, 1967); Hebb (1955); Bindra (1969); Rescorla and Solomon (1967).
39. LeDoux (2012, 2014).
40. LeDoux (2012, 2014).
41. For a discussion of "innateness," see Chapter 3.
42. Tomkins (1962, 1963).

43. Izard (1971, 1992, 2007).
44. Ekman (1977, 1984, 1999); Ekman and Friesen (1975).
45. Panksepp (1980, 1998); Johnson-Laird and Oatley (1992).
46. Ekman (1980, 1984, 1992a, 1992b, 1993).
47. Aoki et al (2014); Sabatinelli et al (2011).
48. http://www.nytimes.com/2009/02/15/weekinreview/15marsh.html?partner=rss&emc=rss&pagewanted=all&_r=0.
49. http://www.fastcompany.com/1800709/human-lie-detector-paul-ekman-decodes-faces-depression-terrorism-and-joy.
50. *Lie to Me,* on Fox. http://www.imdb.com/title/tt1235099/; http://www.theguardian.com/lifeandstyle/2009/may/12/psychology-lying-microexpressions-paul-ekman.
51. For a review of basic emotions in psychology, see Tracy and Randles (2011).
52. Scarantino (2009) and Prinz (2004) survey the philosophy of basic emotions.
53. Ortony and Turner (1990); Barrett (2006); Barrett et al (2007); LeDoux (2012).
54. Leys (2012).
55. Russell (1994).
56. Scherer and Ellgring (2007).
57. Rachel Adelson, "Detecting Deception," http://www.apa.org/monitor/julaug04/detecting.aspx. Retrieved Nov. 21, 2014.
58. Barrett et al (2006, 2007).
59. Barrett et al (2006, 2007); Russell (2009).
60. Mandler and Kessen (1959).
61. Discussed in LeDoux (2012).
62. Ekman (2003).
63. Panksepp (1998, 2005, 2012).
64. Damasio and Carvalho (2013).
65. Mineka and Ohman (2002); Ohman and Mineka (2001).
66. Cosmides and Tooby (1999, 2013); Tooby and Cosmides (2008).
67. One exception is Robert Levenson.
68. Panksepp (1998, 2005, 2012); Panksepp and Panksepp (2013); Vandekerckhove and Panksepp (2009, 2011).
69. Panksepp (1998), p. 234.
70. In addition to Panksepp's papers, his views are also represented in the writings of his colleague Douglass Watt (Watt, 2005), and in a "declaration" on animal consciousness (Low et al, 2012), which Panksepp contributed to but which was signed by a number of participants at a conference at the University of Cambridge in 2012. Note that the Low et al (2012) document originally appeared at this link: http://fcmconferenceorg/. Churchill College, University of Cambridge. It was removed from the Web, but a search on Dec. 24, 2014, revealed that it had reappeared at: http://fcmconference.org/img/CambridgeDeclaration OnConsciousness.pdf.
71. Panksepp (1998).
72. Panksepp (1998), p. 122.
73. Panksepp (1998), p. 26.
74. Panksepp (1998), p. 208.
75. Panksepp (1998), p. 213.
76. Panksepp uses capital letters for names of emotion command systems (e.g., FEAR). For simplicity, I have used lowercase letters for these command system names.
77. Sternson (2013); Lin et al (2011).
78. Panksepp (1998, 2011); Panksepp and Panksepp (2013).

79. Heath (1954, 1963, 1972); Heath and Mickle (1960).
80. Panksepp (1998), p. 213.
81. Vandekerckhove and Panksepp (2009, 2011).
82. Panksepp (1998).
83. Panksepp (1998), p. 214.
84. Heath (1954, 1963, 1972); Heath and Mickle (1960).
85. Crichton (1972).
86. Percy (1971).
87. For a summary of the controversy, see Baumeister (2000, 2006). He pointed out that because the main objective of the project was to conduct research, not to treat the patients, a key issue was whether the patients had given consent to participate. Although the standards of informed consent have risen considerably over the years, Baumeister concluded that even by the weaker standards of the time, proper consent had not been obtained.
88. See, for example, Gloor et al (1982); Halgren (1981); Halgren et al (1978); Lanteaume et al (2007); Nashold et al (1969); Sem-Jacobson (1968).
89. Baumeister (2006).
90. Berridge and Kringelbach (2008, 2011).
91. Panksepp (1998), p. 214.
92. Halgren (1981); Halgren et al (1978).
93. Halgren (1981); Halgren et al (1978).
94. Berrios and Markova (2013).
95. Lindsley (1951); Aston-Jones et al (1986, 1991, 2000); Saper (1987).
96. Festinger (1957); Schachter and Singer (1962).
97. Festinger (1957); Schachter and Singer (1962); Nisbett and Wilson (1977); Wilson (2002); Kelley (1967).
98. Schachter and Singer (1962).
99. Hooper and Teresi (1991), pp. 152–61.
100. LeDoux et al (1977); Gazzaniga and LeDoux (1978).
101. Heath (1964), p. 78.
102. James (1884, 1890).
103. Cannon (1927, 1929, 1931).
104. For summaries of the history and current status of research on the specificity of body feedback in emotion, see Friedman (2010), Critchley et al (2001, 2004), and Nicotra et al (2006). While this work shows there is some specificity, it is less convincing about the necessity of such feedback in feelings.
105. Tomkins (1962, 1963); Izard (1971, 1992, 2007).
106. Whissell (1985); Buck (1980).
107. Damasio (1994).
108. Damasio (1994, 1999); Damasio et al (2013); Damasio and Carvalho (2013).
109. Damasio (1994); Damasio and Carvalho (2013).
110. Damasio (1996).
111. Damasio (1999); Damasio and Carvalho (2013); Craig (2002, 2003, 2009).
112. McEwen and Lasley (2002); Sapolsky (1996).
113. McGaugh (2003).
114. Damasio (1994).
115. Damasio et al (2000).
116. Damasio et al (2013); Damasio and Carvalho (2013); Philippi et al (2012).
117. Craig (2002, 2003).
118. Craig (2009).

119. Gu et al (2013); Morris (2002); Critchley (2005, 2009); Jones et al (2010); Medford and Critchley (2010); Singer et al (2004); Singer (2006); Singer (2007).
120. Damasio et al (2013); Damasio and Carvalho (2013).
121. Philippi et al (2012).
122. Laureys and Schiff (2012).
123. Damasio et al (2000).
124. Mobbs et al (2007).
125. Damasio and Carvalho (2013).
126. Damasio and Carvalho (2013).
127. Damasio and Carvalho (2013).
128. Dickinson (2008).
129. Balleine and Dickinson (1991); Balleine et al (1995); Balleine (2005); Balleine and Dickinson (1998).
130. Gould and Lewontin (1979).
131. Dickinson (2008).
132. Summarized in Keller (1973).
133. Panksepp (1998), p. 38.
134. Everitt and Robbins (2005); Dickinson (2008); Castro and Berridge (2014); Winkielman and Berridge (2004).
135. Huber et al (2011); Baxter and Byrne (2006); Brembs (2003).
136. Johansen et al (2014).
137. Whitten et al (2011).
138. Berridge (1996); Berridge and Winkielman (2003); Castro and Berridge (2014).
139. Winkielman et al (2005).
140. Berridge and Winkielman (2003); Cabanac (1996).
141. Schultz (1997, 2002); Baudonnat et al (2013); Schultz (2013); Doll et al (2012); Berridge (2007); Dalley and Everitt (2009).
142. "Reward Lasers." *The Connectome,* http://theconnecto.me/2012/03/reward-lasers/. Retrieved October 30, 2014. "By hooking rats up to a tiny fiber-optic cable and firing lasers directly into their brains, a team led by Garret D. Stuber at the University of North Carolina at Chapel Hill School of Medicine were able to isolate specific neurochemical shifts that cause rats to feel pleasure or anxiety—and switch between them at will."
143. http://www.dailymail.co.uk/sciencetech/article-2347921/Why-love-chocolate-The-sweet -treat-releases-feel-good-chemical-dopamine-brains-causing-pupils-dilate.html; http://www .news-medical.net/health/Dopamine-Functions.aspx. Retrieved Dec. 23, 2014.
144. Everitt and Robbins (2005); Castro and Berridge (2014); Winkielman and Berridge (2004).
145. Huber et al (2011); Baxter and Byrne (2006); Brembs (2003); Bendesky et al (2011); Bendesky and Bargmann (2011); Lebetsky et al (2009); Bargmann (2006, 2012); Hawkins et al (2006); Kandel (2011); Byrne et al (1993); Glanzman (2010); Martin (2002, 2004).
146. This is admittedly a more far-fetched likelihood than that dopamine produces pleasure in an animal, but no less difficult to prove.
147. Berridge (2007).
148. van Zessen et al (2012).
149. "Reward Lasers." *The Connectome.* http://theconnecto.me/2012/03/reward-lasers/. Retrieved October 30, 2014.
150. McCarthy et al (2010); Baker et al (2004a, 2004b); Wiers and Stacy (2006).
151. Lamb et al (1991).
152. Fischman (1989); Fischman and Foltin (1992).
153. Koob (2013).
154. McCarthy et al (2010); Baker et al (2004a, 2004b); Wiers and Stacy (2006).

155. Summary of the positive effects of hypnosis on pain by the American Psychological Association. http://www.apa.org/research/action/hypnosis.aspx. Retrieved Nov. 17, 2014; Spiegel (2007); Butler et al (2005).

156. Hoeft et al (2012).

157. Fernandez and Turk (1992); Price and Harkins (1992); Rainville et al (1992).

158. The dependence of pleasure and pain on dedicated sensory processing systems makes these states different from emotions, such as fear, anger, joy, love, empathy, and so on. Both can be consciously felt in brains that can be conscious, but that does not mean that pleasure and pain are the same as conventional emotions.

159. Barrett (2006); Barrett et al (2007); Russell (2009).

160. LeDoux (2012, 2014).

CHAPTER 6: LET'S GET PHYSICAL: THE CONSCIOUSNESS PROBLEM

1. Maugham (1949).

2. For a discussion of creature versus mental state consciousness, see Piccinini (2007); Rosenthal (2002).

3. This is what I mean by creature consciousness and mental state consciousness, which may not be defined exactly the same way by others who use the terms.

4. For various views of animal consciousness, see Panksepp (1998, 2005, 2011); Dixon (2001); Edelman and Seth (2009); Bekoff (2007); Griffin (1985); Heyes (2008); Shea and Heyes (2010); Weiskrantz (1995); Masson and McCarthy (1996); Dickinson (2008); Grandin (2005); Singer (2005); Jane Goodall's Introduction in Hatkoff (2009), p. 13; LeDoux (2008, 2012, 2014, 2015).

5. Gross (2013); Gallup (1991); Hampton (2001); Griffin (1985); Burghardt (1985, 2004); Jane Goodall's Introduction in Hatkoff (2009), p. 13; Interview with primatologist Frans de Waal, http://www.wonderlance.com/february2011_scientech_fransdewaal.html, retrieved Nov. 5, 2014; Goodall (2013), "Should Apes Have Legal Rights?" *The Week*, http://theweek.com/article/index/247763/should-apes-have-legal-rights, retrieved Nov. 5, 2014.

6. Panksepp (1998).

7. Clayton and Dickinson (1998).

8. Panksepp (1998, 2011); Damasio (1994, 1999, 2010), Vandekerckhove and Panksepp (2009, 2011).

9. Edelman and Seth (2009).

10. http://www.plantconsciousness.com/. retrieved Nov. 5, 2014; http://forums.philosophyforums.com/threads/are-cells-conscious-52606.html, retrieved Nov. 5, 2014.

11. Tononi (2005); Chalmers (2013).

12. Dennett (1991); Churchland PM (1984, 1988a, 1988b); Churchland PS (1986, 2013); Graziano (2013); Lamme (2006).

13. Descartes (1637, 1644).

14. Descartes (1637); Shugg (1968); Rosenfield (1941); Haldane and Ross (1911).

15. Boring (1950).

16. Watson (1913, 1919, 1925).

17. Strachey (1966–74).

18. Gardner (1987).

19. Neisser (1967).

20. Lashley (1950).

21. Bargh (1997); Bargh and Ferguson (2000); Bargh and Morsella (2008); Wilson (2002); Wilson and Dunn (2004); Jacoby (1991); Kihlstrom (1987); Ohman (1988, 2002); Ohman and Soares (1991); Ohman and Mineka (2001); Ohman et al (2000); Mineka and Ohman (2002); Phelps (2006).

22. Freud (1915).
23. Dehaene et al (2006).
24. Frith et al (1999); Naccache and Dehaene (2007); Weiskrantz (1997); Dehaene et al (2003); Dehaene and Changeux (2004); Sergent and Reis (2007); Koch and Tsuchiya (2007).
25. Descartes (1637); Shugg (1968).
26. Dennett (1991).
27. Dennett (1991); Jackendoff (2007); Wittgenstein (1958); Alanen (2003).
28. Lazarus and McCleary (1951).
29. Shimojo (2014); Kouider and Dehaene (2007); Macknik (2006).
30. One issue that has come up in these various studies is the extent to which subjects deny awareness (because of masking or brain damage) or have degraded awareness rather than no awareness of the stimulus. Alternative measures have been suggested to get at the issue of how unaware people are when they claim to be unaware. For example, in one approach subjects are asked to rate their confidence in their judgments about whether they saw or didn't see stimuli that are presented in such a way that the potential for awareness is systematically varied by the experimenter. While the ability to be somewhat confident that a stimulus was seen or not can indicate whether some modicum of awareness was available, this is not the same as a full-blown conscious experience. The argument that degraded consciousness can influence choices is not very convincing because it leads to a circular process; if it influenced choice, the subjects must have been aware of it. For arguments that what appears to be unconscious registration of stimuli is really weakly conscious perception, see Szczepanowski and Pessoa (2007); Mitchell and Greening (2012). For arguments in favor of nonconscious registration, see Merikle et al (2001); Kouider and Dehaene (2007). While consciousness certainly exists in gradations rather than in an all-or-none state, degraded consciousness (e.g., "I think that there might have been some kind of fruit on the screen") is not the same as a full-blown conscious experience (e.g., "I saw a red apple with a stem and a worm hole"). The qualia experienced are completely different.
31. One way is the use of *bistable images*. In this approach, when a different image is shown to each eye, subjects can only be consciously aware of one at a time. This also allows the effects of the "unseen" image to be evaluated (Maier et al, 2012). Another approach involves continuous flash suppression (Yang et al, 2014; Sterzer et al, 2014). In most research approaches considered so far, consciousness has been treated as an all-or-none phenomenon. Some researchers have recently raised the question of whether we should think of consciousness along a continuum. In this view, we should not ask whether a stimulus was "seen" or "unseen," but instead should ask how confident the person is in having seen or not seen something (Sahraie et al, 1998; Tunney, 2005). This is still a verbal report and thus a subjective measure. Others argue that by placing wagers about having seen something, the degree of awareness can be estimated more directly (Persaud et al, 2007; Seth et al, 2008). But problems with this approach have also been pointed out (Overgaard et al, 2010).
32. Romanes (1882, 1883); Jane Goodall's Introduction in Hatkoff (2009), p. 13. For an explanation of why complex behavior is not proof of consciousness, see Smith et al (2012); Fleming et al (2012); Wynne (2004); Harley (1999).
33. The perils of this approach can be illustrated by results we considered in earlier chapters. When people undergo Pavlovian threat conditioning, they are often aware of the CS and US and the relation between these. But this kind of knowledge is not what underlies the ability of the CS to elicit the conditioned response (see Chapter 8).
34. Amsterdam (1972).
35. Povinelli et al (1997); Reiss and Marino (2001); Uchino and Watanabe (2014); Plotnik et al (2006); Gallup (1991); Keenan et al (2003).
36. Heyes (1994, 1995, 2008).

37. Heyes (2008) emphasizes the difference between studies that use tests of alternative hypotheses to distinguish conscious from nonconscious states (Hampton, 2001) and studies that start from the assumption that the animals are conscious and that then try to determine how this capacity is affected by some manipulation (Cowey and Stoerig, 1995; Leopold and Logothetis, 1996). These issues are also discussed by Smith et al (2012).
38. Weiskrantz (1977); Weiskrantz (1997), p. 75.
39. Cowey and Stoerig (1992), pp. 11–37.
40. Weiskrantz (1997), p. 75.
41. Smith et al (2012); Hampton (2009); Shea and Heyes (2010); Smith (2009).
42. Metcalfe and Shimamura (1994); Flavell (1979); Kornell (2009); Terrace and Metcalfe (2004).
43. Sahraie et al (1998); Tunney (2005).
44. Persaud et al (2007).
45. Persaud et al (2007).
46. Seth (2008); Overgaard et al (2010).
47. Smith et al (2012).
48. Crystal (2014); Heyes (2008); Fleming et al (2012); Wynne (2004); Harley (1999).
49. Smith et al (2012).
50. Tulving (2001, 2005).
51. Tulving (2005).
52. Tononi and Koch (2015).
53. Frith et al (1999).
54. Dennett (1991).
55. Edelman (1989); Jackendoff (2007); Wittgenstein (1958); Alanen (2003); Carruthers (1996, 2002); Macphail (1998, 2000); Bridgeman (1992); Chafe (1996); Fireman et al (2003); Lecours (1998); Ricciardelli (1993); Searle (2002); Sekhar (1948); Stamenov (1997); Subitzky (2003); Clark (1998); Bloom (2000); Rosenthal (1990b).
56. Rolls (2008).
57. Preuss (1995, 2001); Wise (2008).
58. Semendeferi et al (2011); Barbey et al (2012); Gazzaniga (2008); Preuss (2001); Wise (2008); Bendarik (2011); Falk (1990).
59. Gazzaniga and LeDoux (1978); Gazzaniga (2008).
60. But they are deprived of the ability to use language to learn in social situations and can suffer from this. Oliver Sacks (1989) has described some of the consequences.
61. LeDoux (2008). Others have similar ideas. See Jackendoff (1987, 2007); Dennett (1991).
62. Wittgenstein (1958), p. 223.
63. http://www.theconsciousnesscollective.com/. Retrieved Nov. 6, 2014.
64. http://www.nytimes.com/2012/12/10/nyregion/jamming-about-the-mind-at-qualia-fest .html?_r=0. Retrieved Nov. 6, 2014.
65. Nagel (1974).
66. Chalmers (1996). Chalmers was at the University of California at Santa Cruz when he wrote this book.
67. Chalmers (1996); Block (2007).
68. Chalmers said this to me in an email on February 19, 2015.
69. Edelman (2004); Block (2007); Papineau (2002); Dennett (1991); Rosenthal (1990a, 1993, 2005); Humphrey (2006).
70. For additional theories and discussions that review consciousness more broadly, see Seth et al (2008); Searle (2000); Seth (2009); Flanagan (2003); Hobson (2009); Edelman (2001, 2004); Hameroff and Penrose (2014); Tononi (2012); Metzinger (2008); Hurley (2008); O'Regan and Noë (2001); Papineau (2008); Humphrey (2006); Noe (2012); Greenfield (1995).

71. Johnson-Laird (1988, 1993); Dennett (1991); Norman and Shallice (1980); Shallice (1988); Baddeley (2000, 2001); Gardiner (2001); Schacter (1989, 1998); Schacter et al (1998); Frith et al (1999); Frith and Dolan (1996); Frith (1992, 2008); Courtney et al (1998).

72. Hassin et al (2009); Kintsch et al (1999); Cowan (1999); O'Reilly et al (1999); Ellis (2005); Ercetin and Alptekin (2013).

73. Rosenthal (2005; 2012); Armstrong (1979); Carruthers (1996, 2002, 2009, 2014); Lycan (1986, 1995).

74. Rosenthal (2005, 2012).

75. Rosenthal (2005, 2012).

76. Sekida (1985), p. 110.

77. Heyes (2008).

78. Cleeremans (2008, 2011).

79. Dennett (1991).

80. Gazzaniga (1988, 1998, 2008, 2012).

81. Weiskrantz (1997); Dehaene and Changeux (2004).

82. Weiskrantz (1997), p. 167.

83. Baars (1988, 2005); Baars et al (2013); Baars and Franklin (2007); Cho et al (1997).

84. Dehaene and Changeux (2004, 2011); Dehaene et al (1998, 2003); Dehaene and Naccache (2001).

85. Murray Shanahan and Bernard Baars; comment in Block (2007).

86. Brentano (1874/1924); Metzinger (2003); Burge (2006); Block (2007).

87. Block (2007), p. 485.

88. Block (1990, 1992, 1995a, 1995b, 2002, 2007).

89. Block (1990, 1992); he has since proposed calling phenomenal consciousness simply phenomenology.

90. Block (2007).

91. While recent studies show that preattentive sensory memory can involve representations that are influenced in a manner similar to conscious perceptual experiences (Vandenbroucke et al, 2012), this does not demonstrate that preattentive sensory processing or memory is consciously experienced.

92. Naccache and Dehaene (2007).

93. Desimone (1996); Miller and Desimone (1996); Miller et al (1996).

94. Zeman (2009).

95. Putnam (1960).

96. Fodor (1975).

97. Crick and Koch (1990, 1995, 2003); Koch (2004).

98. Livingston (2008); Purves and Lotto (2003).

99. Critchley (1953).

100. Humphrey (1970, 1974); Cowey and Stoerig (1995); Stoerig and Cowey (2007).

101. Weiskrantz (1997).

102. Milner and Goodale (2006).

103. Weizkrantz (1997).

104. Ungerleider and Mishkin (1982); Milner, D.A. and Goodale, M. (2006).

105. Frith et al (1999); Rees and Frith (2007); Lau and Passingham (2006); Dehaene and Naccache (2001); Dehaene et al (2003).

106. Meyer (2011).

107. Persaud et al (2011); Lau and Passingham (2006).

108. Vuilleumier et al (2008); Del Cul et al (2009); Pascual-Leone and Walsh (2001).

109. Weiskrantz (1997); Wheeler et al (1997); Courtney et al (1998); Knight and Grabowecky (2000); Maia and Cleeremans (2005); Bor and Seth (2007); Mazoyer et al (2001).

110. Edelman (1987).
111. Geschwind (1965a, 1965b); Jones and Powell (1970); Mesulam et al (1977); Damasio (1989).
112. Fuster (1985, 1991, 2006); Fuster and Bressler (2012); Goldman-Rakic (1995, 1996); Levy and Goldman-Rakic (2000).
113. Crick and Koch (1990, 1995, 2003); Koch (2004).
114. Recent studies by Koch show that changes in functional connectivity between visual prefrontal networks during visual perception support the importance of long-range connections in consciousness (Imamoglu et al, 2012).
115. Meyer (2011).
116. However, Koch does suggest that it might be possible for the visual cortex to create a simple kind of phenomenal consciousness without cognitive access if the prefrontal cortex is damaged (Christof Koch and Naotsugu Tsuchiya, comment in Block, 2007). He also now emphasizes dissociations between attention and consciousness.
117. Edelman (1987, 1989, 1993).
118. Gaillard et al (2009).
119. Lamme (2006); van Gaal and Lamme (2012).
120. Tononi (2005, 2012).
121. Rosenthal (2012).
122. Dehaene et al (2006).
123. Block (2005).
124. See Block (2007).
125. Block (2007).
126. Rees et al (2000, 2002); Driver and Vuilleumier (2001).
127. Berger and Posner (2000); Mesulam (1999); Critchley (1953).
128. They also suggest that certain thalamic areas as well as the claustrum may also contribute.
129. See responses to Block (2007).
130. Lau and Brown. http://consciousnessonline.com/2012/02/17/empty-thoughts-an-explanatory-problem-for-higher-order-theories-of-consciousness/. Retrieved Jan. 20, 2015.
131. Davies, M., http://www.mkdavies.net/Martin_Davies/Mind_files/Ischia1.pdf. Retrieved Jan. 20, 2015.
132. Another Oxford philosopher, Nicholas Shea, has taken a more positive view (Shea, 2012). He attempts to show that phenomenal consciousness is a natural kind and proposes ways to test that idea.
133. Zeman (2009).
134. Papineau (2008).
135. Merker (2007).
136. Dickinson (2008).
137. Merker (2007).
138. Renier et al (2014); Sadato (2006); Neville and Bavelier (2002); Sur et al (1999).
139. Lennenberg (1967); Basser (1962); Vanlancker-Sidtis (2004).
140. Vandekerckhove and Panksepp (2009).
141. Vandekerckhove and Panksepp (2011).
142. Dickinson (2008); Smith et al (2012); Fleming et al (2012); Wynne (2004); Harley (1999); Weiskrantz (1997).
143. Bor and Seth (2012); Prinz (2012); Baars (1988, 2005); Johnson-Laird (1988, 1993); Frith et al (1999); Frith and Dolan (1996); Frith (1992, 2008); Schacter (1989, 1998); Schacter et al (1998); Dehaene et al (2003); Dehaene and Changeux (2004); Naccache and Dehaene (2007).
144. Carretie (2014); Han and Marois (2014); Ansorge et al (2011); Jonides and Yantis (1988); Abrams and Christ (2003); Ohman and Mineka (2001); Vuilleumier and Driver (2007).
145. Prinz (2012); Bor and Seth (2012).

146. Cohen et al (2012).
147. van Boxtel et al (2010); Cohen et al (2012); Hassin et al (2009); Soto et al (2011).
148. Tsuchiya and Koch (2009); Ansorge et al (2011); Kiefer (2012).
149. van Gaal and Lamme (2012); Thakral (2011).
150. Behrmann and Plaut (2013).
151. Goldman-Rakic (1987, 1995, 1999); Fuster (1989, 2000, 2003); Curtis (2006); Miller and Cohen (2001); Bor and Seth (2012).
152. Faw (2003); Goel and Vartanian (2005); Barde and Thompson-Schill (2002); Muller et al (2002); D'Esposito et al (1999); Duncan and Owen (2000).
153. Rolls et al (2003); Rolls (2005); Kringelbach (2008); Damasio (1994, 1999); Faw (2003); Damasio (1994, 1999); Medford and Critchley (2010); Posner and Rothbart (1998); Mayr (2004); Vogt et al (1992); Devinsky et al (1995); Shenhav et al (2013); Carter et al (1999); Oakley (1999); Reinders et al (2003); Ochsner et al (2004); Medford and Critchley (2010); Hasson et al (2007); Crick and Koch (2005); Craig (2002, 2003, 2009, 2010); Bechara et al (2000); Clark et al (2008); Damasio et al (2013); Philippi et al (2012); Damasio and Carvalho (2013); Hinson et al (2002); Critchley et al (2004); Critchley (2005); Smith and Alloway (2010); Thomson (2014); Stevens (2005).
154. For example, see Phillipi et al (2012).
155. Cotterill (2001); O'Keefe (1985); Gray (2004); Kandel (2006).
156. van Gaal and Lamme (2012).
157. Demertzi et al (2013).
158. Demertzi et al (2011).
159. Crick and Koch (2003).

CHAPTER 7: IT'S PERSONAL: HOW MEMORY AFFECTS CONSCIOUSNESS

1. Butler (1917).
2. This quote is attributed to Aldous Huxley but the source could not be verified. *Aldous Huxley Quotes. Quotes.net.* Retrieved February 17, 2014, from http://www.quotes.net/quote/52460.
3. Tulving (1972, 1983, 2002, 2005).
4. Schacter (1985); Squire (1987, 1992).
5. Tulving (1989); Schacter (1985); Squire (1987, 1992).
6. Tulving (2002, 2005); Suddendorf and Corbalis (2010).
7. Tulving (2002, 2005); Suddendorf and Corbalis (2010).
8. LeDoux (2002).
9. Tulving (1983); Greenberg and Verfaellie (2010); Simons et al (2002).
10. Scoville and Milner (1957); Milner (1962, 1965, 1967).
11. Suzuki and Amaral (2004).
12. Indeed, studies show that while the hippocampus is necessary for conscious memory, the hippocampus engages in nonconscious processing of so-called hippocampal-dependent explicit memories (Hannula and Greene, 2012).
13. Wheeler et al (1997); Buckner and Koutstaal (1998); Garcia-Lazaro et al (2012); Lee et al (2000); Rugg et al (2002); Mayes and Montaldi (2001); Fletcher and Henson (2001); Yancey and Phelps (2001); Buckner et al (2000); Cabeza and Nyberg (2000).
14. Cabeza et al (2012); Schoo et al (2011); Hutchinson et al (2009).
15. Barrouillet et al (2004, 2007).
16. Vredeveldt et al (2011).
17. The distinction between active and inactive memory is borrowed from Lewis, who used it in a different context (Lewis, 1979).
18. Barbas (1992, 2000); Fuster (2008).

19. Vargha-Khadem et al (2001); de Haan et al (2006); Dickerson and Eichenbaum (2010); Mayes and Montaldi (2001).
20. Moscovitch et al (2005).
21. Strenziok et al (2013).
22. Bechara et al (1995); LaBar et al (1995).
23. Shimamura (1986); Wiggs and Martin (1998); Farah (1989); Hamann and Squire (1997).
24. Marcel (1983); Dehaene et al (1998, 2006); Naccache et al (2002); Greenwald et al (1996). For discussion of the limits of unconscious priming, see Abrams and Greenwald (2000); Merikle et al (1995).
25. Hamann and Squire (1997); Schacter (1997); Schacter and Buckner (1998).
26. Dell'Acqua and Grainger (1999).
27. Scoville and Milner (1957); Milner (1965); Corkin (1968); Squire (1987); Squire and Cohen (1984); Cohen and Squire (1980).
28. Squire (1987); Squire and Kandel (1999); LeDoux (1996).
29. Tulving (1972, 1983, 2002, 2005); Reber et al (1980); Seger (1994).
30. Marr (1971); Mizumori et al (1989); O'Reilly and McClelland (1994); Recce and Harris (1996); Willshaw and Buckingham (1990); Rolls (1996).
31. Liddell and Scott's *Lexicon.* http://www.perseus.tufts.edu/hopper/text?doc=Perseus%3Atext %3A1999.04.0058%3Aentry%3Dnoe%2Fw. Retrieved Nov. 7, 2014.
32. Tulving (2001, 2002, 2005); Gardiner (2001); Klein (2013); Metcalfe and Son (2012).
33. Conway (2005); Marsh and Roediger (2013).
34. Frith and Frith (2007).
35. Lewis (2013).
36. Gazzaniga (1988, 1998, 2008, 2012).
37. To more clearly distinguish "anoetic" from "autonoetic," I spell the former "a-noetic."
38. Tulving (1985); Ebbinghaus (1885/1964).
39. I contacted Tulving to clarify this ambiguity about whether he viewed a-noetic states as conscious or nonconscious states. From our email conversation, dated July 24, 2013, it seems that when Tulving used the term "a-noetic consciousness," he was not referring to mental state (phenomenal) consciousness. Instead, he was referring to states in which awareness is lacking but the organism is alive and able to process information and behave (something akin to creature consciousness). Tulving admitted to me that what he calls a-noetic consciousness would be called nonconscious states by most other scientists.
40. Vandekerckhove and Panksepp (2009, 2011).
41. Marcel (1983); Dehaene et al (1998, 2006); Naccache et al (2002); Greenwald et al (1996). For discussion of limits of unconscious priming, see Abrams and Greenwald (2000); Merikle et al (1995).
42. Shimamura (1986); Hamann and Squire (1997).
43. Taylor and Gray (2009); Gallistel (1989); Dickinson (2012); Clayton (2007); Premack (2007); Wasserman (1997); Mackintosh (1994); Clayton and Dickinson (1998).
44. Pahl et al (2013); Chittka and Jensen (2011); Srinivasan (2010); Webb (2012); Skorupski and Chittka (2006); Menzel and Giurfa (1999); Gould (1990); Giurfa (2013).
45. Tulving (2005).
46. Clayton and Dickinson (1998).
47. Menzel (2005).
48. Eichenbaum and Fortin (2005); Fortin et al (2004); Allen and Fortin (2013).
49. Clayton and Dickinson (1998).
50. Menzel (2009).
51. Clayton et al (2003); Suddendorf and Busby (2003); Suddendorf and Corbalis (2010).
52. Dickerson and Eichenbaum (2010).

53. McKenzie et al (2014).

54. Allen and Fortin (2013).

55. Eichenbaum (1992, 1994, 2002); Kesner (1995); Olton et al (1979); McNaughton (1998); Wilson and McNaughton (1994); McGaugh (2000).

56. Kesner and Churchwell (2011); Sullivan and Brake (2003); Thuault et al (2013).

57. Preuss (1995); Wise (2008).

58. Semendeferi et al (2011); Gazzaniga (2008).

59. Dere et al (2006); Menzel (2005); Belzung and Philippot (2007); Suddendorf and Butler (2013); Plotnik et al (2010); Salwiczek et al (2010); Suddendorf and Corbalis (2007, 2010); Suddendorf et al (2009).

60. Frith et al (1999); Naccache and Dehaene (2007); Weiskrantz (1997); Dehaene et al (2003); Dehaene and Changeux (2004); Claire Sergent and Geraint Rees, comment in Block (2007); Christof Koch and Naotsugu Tsuchiya, comment in Block (2007).

61. Panagiotaropoulos et al (2014, 2013); Safavi et al (2014).

62. Preuss (1995); Wise (2008); Semendeferi et al (2011); Gazzaniga (2008).

63. Morgan (1890–1891).

64. Boring (1950); Keller (1973); LeDoux (2014).

CHAPTER 8: FEELING IT: EMOTIONAL CONSCIOUSNESS

1. Fowles (1965).

2. Seligman (1971); Ohman and Mineka (2001). Prepared refers to the fact that some stimuli, because of our evolutionary history, are more likely to trigger responses than others.

3. Ohman and Mineka (2001); Ohman (2009); Whalen et al (1998); Whalen and Phelps (2009); Olsson and Phelps (2004); Esteves et al (1994).

4. For example, see: Lazarus and McCleary (1951); Ohman and Mineka (2001); Olsson and Phelps (2004); Lissek et al (2008); Alvarez et al (2008); Morris et al (1998, 1999); Critchley et al (2002, 2005); Williams et al (2006); Hamm et al (2003); Phelps (2005); Morris et al (1998, 1999); Whalen et al (1998); Etkin et al (2004); de Gelder et al (2005); Hariri et al (2002); Das et al (2005); Williams et al (2006); Luo et al (2009); Mitchell et al (2008); Vuilleumier (2005).

5. Lazarus and McCleary (1951); Ohman and Mineka (2001); Olsson and Phelps (2004); Lissek et al (2008); Alvarez et al (2008); Morris et al (1998, 1999); Critchley et al (2002, 2005); Williams et al (2006).

6. As discussed in Chapter 3, it's hard to experimentally distinguish when we are truly conscious in their decision-making as opposed to consciously rationalizing decisions after the fact, but it seems likely that some of our decisions are made consciously.

7. Morris et al (1998, 1999); Whalen et al (1998); Etkin et al (2004); de Gelder et al (2005); Hariri et al (2002); Das et al (2005); Williams et al (2006); Luo et al (2009); Mitchell et al (2008).

8. Vuilleumier and Schwartz (2001); Vuilleumier and Driver (2007); Anderson and Phelps (2001); Vuilleumier (2005); Hadj-Bouziane et al (2012).

9. Maratos, E.J. Dolan, R.J. et al, *Neuropsychologia* (2001) 39:910–20.

10. Ohman (2009); Ohman and Mineka (2001); Buchel and Dolan (2000); Dolan and Vuilleumier (2003); LaBar et al (1998); Anderson and Phelps (2001); Olsson and Phelps (2004); Raio et al (2008); Phelps (2006); Vuilleumier (2005); Morris et al (1998); Pasley et al (2004); Whalen et al (1998); Liddell et al (2005); Öhman (2002); Brooks et al (2012); Liddell et al (2005); Williams et al (2006); Zald (2003); Luo et al (2009); Mitchell et al (2008).

11. Morris et al (2001); Tamietto and de Gelder (2010); Vuilleumier and Schwartz (2001); Vuilleumier et al (2002); Van den Stock et al (2011); Ward et al (2005); de Gelder et al (1999).

12. LaBar et al (1995); Bechara et al (1995).

13. Bechara et al (1995).

14. Shanks and Dickinson (1990); Shanks and Lovibond (2002); Lovibond et al (2011); Mitchell et al (2009).
15. Schultz and Helmstetter (2010); Asli and Falaten (2012).
16. Knight et al (2009).
17. Knight et al (2009).
18. Bechara et al (1995).
19. Anderson and Phelps (2001); Vuilleumier (2005); Hadj-Bouziane et al (2012).
20. Gross and Canteras (2012).
21. Everitt and Robbins (1999); Cardinal et al (2002); Holland and Gallagher (1999, 2004); Balleine and Killcross (2006); Balleine et al (2003); Robbins et al (2008).
22. Jones and Mishkin (1972); Mishkin and Aggleton (1981); Van Hoesen and Pandya (1975).
23. LeDoux et al (1984); Romanski and LeDoux (1992); LeDoux (1996).
24. LeDoux (1996).
25. den Hulk et al (2003); Heerebout and Phaf (2010).
26. Morris et al (1999); Luo et al (2010).
27. Morris et al (2001); Morris et al (2001); Tamietto and de Gelder (2010); Vuilleumier and Schwartz (2001); Vuilleumier et al (2002); Van den Stock et al (2011); Ward et al (2005); de Gelder et al (1999).
28. Pourtois et al (2010); Pourtois et al (2013).
29. LeDoux (2008); Vuilleumier (2005); Pourtois et al (2013); Pessoa and Adolphs (2010).
30. Pessoa and Ungerleider (2004); Pessoa et al (2002).
31. Pessoa (2008, 2013); Pessoa and Adolphs (2010); Pessoa et al (2002).
32. See Mitchell and Greening (2012); Pessoa (2008, 2013); Pessoa and Adolphs (2010); Pessoa et al (2002); Pessoa and Ungerleider (2004).
33. Pessoa (2013); Pessoa and Adolphs (2010).
34. Repa et al (2001).
35. Luo et al (2010); Pourtois et al (2010).
36. Repa et al (2001); Josselyn (2010); Han et al (2007, 2009); Reijmers et al (2007); Garner et al (2012).
37. For a discussion of attentional amplification of processing in visual cortex, see the discussion of Koch and Crick's global workspace theories of consciousness in Chapter 6.
38. Mitchell and Greening (2012).
39. Mitchell and Greening (2012).
40. LeDoux (2008); Vuilleumier (2005); Pourtois et al (2013); Pessoa and Adolphs (2010).
41. Van den Bussche et al (2009a, 2009b); Kinoshita et al (2008); Kouider and Dehaene (2007); Abrams and Grinspan (2007); Gaillard et al (2006); Abrams et al (2002); Lin and He (2009); Yang et al (2014); Kang et al (2011).
42. See Mitchell and Greening (2012).
43. Raio et al (2012).
44. Bargh (1997); Bargh and Chartrand (1999); Bargh and Morsella (2009); Wilson (2002); Wilson and Dunn (2004); Greenwald and Banaji (1995); Phelps et al (2000); Devos and Banaji (2003); Debner and Jacoby (1994); Kihlstrom (1984, 1987, 1990); Kihlstrom et al (1992).
45. Bargh (1997).
46. Kubota et al (2012); Phelps et al (2000); Olsson et al (2005); Stanley et al (2011).
47. LeDoux (1996, 2002).
48. Bebko et al (2014); Silvers et al (2014); Gruber et al (2014); Blechert et al (2012); Ochsner et al (2002); Shurick et al (2012).
49. Bebko et al (2014).
50. Amaral et al (1992); Barbas (1992, 2002).

51. Buhle et al (2013).
52. Delgado et al (2008).
53. Simon (1967).
54. Armony et al (1995, 1997a, 1997b).
55. Armony et al (1997).
56. Eysenck et al (2007).
57. Anderson and Phelps (2001); Mitchell and Greening (2012); Williams et al (2006); Vuilleumier (2005); Hadj-Bouziane et al (2012); Ohman et al (2001a, 2001b); Anderson and Phelps (2001); Schmidt et al (2014); Kappenman et al (2014); Lin et al (2009); Ohman (2005); Mohanty and Sussman (2013); Vuilleumier and Driver (2007); Mineka and Ohman (2002); Ohman and Mineka (2001); Fox et al (2000); Vuilleumier and Schwartz (2001); Raymond et al (1992); Fox (2002).
58. Kapp et al (1992); Lang and Davis (2006); Davis and Whalen (2001); Holland and Gallagher (1999); Mohanty and Sussman (2013); Vuilleumier and Driver (2007); Mineka and Ohman (2002); Ohman and Mineka (2001); Fox et al (2000); Anderson and Phelps (2001); Bar et al (2006).
59. Raymond et al (1992).
60. Anderson and Phelps (2001).
61. Anderson and Phelps (2001).
62. Price et al (1987).
63. Amaral et al (1992, 2003).
64. Phelps et al (2006).
65. Mitchell and Greening (2012); Williams et al (2006).
66. Mitchell and Greening (2012).
67. Kapp et al (1992); Lang and Davis (2006); Morris et al (1997, 1998b); Hurlemann et al (2007); Aston-Jones et al (1991); Woodward et al (1991); Davis and Whalen (2001); Sara (1989, 2009); Sara et al (1994); Foote et al (1980, 1983).
68. Lindsley (1951).
69. Saper (1987).
70. McCormick (1989); McCormick and Bal (1994); Woodward et al (1991); Edeline (2012); Aston-Jones et al (1991); Aton (2013); Levy and Farrow (2001); Gordon et al (1988); Singer (1986); Morrison et al (1982); Arnsten (2011); Ramos and Arnsten (2007); Dalmaz et al (1993); Johansen et al (2014); Tully and Bolshakov (2010); Bijak (1996); Harley (1991); Coull (1998); Kapp et al (1992); Lang and Davis (2006); Hurlemann et al (2007); Davis and Whalen (2001).
71. Bentley et al (2003).
72. Foote et al (1983); Waterhouse and Woodward (1980); Hasselmo et al (1997).
73. Kapp et al (1992); Lang and Davis (2006); Davis and Whalen (2001); Weinberger (1995); Sears et al (2013).
74. Holland and Gallagher (1999); Gallagher and Holland (1994); Lee et al (2010).
75. Sears et al (2013).
76. Grupe and Nitschke (2013); Hayes et al (2012); Barlow (2002); Matthews and Wells (2000); Mathews et al (1989); McNally (1995); MacLeod and Hagen (1992); Lang et al (1990).
77. Miller and Cohen (2001); Botvinick et al (2001); Golkar et al (2012); Beer et al (2006); Shallice and Burgess (1996); Amstadter (2008); Gross (2002).
78. Vallesi et al (2009).
79. Gangestad and Snyder (2000); Riggio and Friedman (1982); Gyurak et al (2011).
80. Goleman (2005).
81. Wood et al (2013); Berridge and Arnsten (2013); Hart et al (2012).
82. Schachter and Singer (1962).
83. Posner et al (2005); Russell and Barrett (1999); Russell (2003); Kuppens et al (2013).

84. Schachter and Singer (1962); Wilson and Dunn (2004).

85. Forsyth and Eifert (1996).

86. Lewis (2013).

87. Forsyth and Eifert (1996).

88. Piaget (1971).

89. Posner et al (2005); Russell (2003, 2009); Izard (2007).

90. Barrett (2006, 2009a, 2012); Barrett et al (2007); Lindquist and Barrett (2008); Wilson-Mendenhall et al (2011, 2013); Russell (2003, 2009); Russell and Barrett (1999); Barrett and Russell (2015).

91. Kron et al (2010).

92. Although I had written about feelings in terms of ingredients for a while (LeDoux, 1996), I first used the soup analogy in LeDoux (2014). Lisa Barrett pointed out that she also proposed a cooking analogy (Barrett, 2009b).

93. Levi-Strauss (1962).

94. Prendergast and Forrest (1998), p. 169.

95. LeDoux (1996, 2002, 2008, 2012, 2014, 2015a, 2015b).

96. Barrett (2006, 2009a, 2012); Barrett et al (2007); Lindquist and Barrett (2008); Wilson-Mendenhall et al (2011, 2013); Russell (2003, 2009); Russell and Barrett (1999); Barrett and Russell (2015).

97. Barrett (2006, 2009a, 2012); Barrett et al (2007); Lindquist and Barrett (2008); Wilson-Mendenhall et al (2011, 2013); Russell (2003, 2009); Russell and Barrett (1999); Barrett and Russell (2015).

98. Noe (2012).

99. LeDoux (1996, 2002, 2008, 2012, 2014, 2015a, 2015b).

100. Lewis (2013).

101. Marks (1987).

102. Whorf (1956); Sapir (1921).

103. Prinz (2013); Zhu et al (2007); Hedden et al (2008); Bowerman and Levinson (2001); Gentner and Goldin-Meadow (2003); Kitayama and Markus (1994); Wierzbicka (1994); Russell (1991).

104. Adams et al (2010); Chiao et al (2008).

105. Forsyth and Eifert (1996); Staats and Eifert (1990).

106. LeDoux (1996), p. 302; LeDoux (2002).

107. Epstein (2013).

108. LeDoux (1996, 2002, 2008, 2012, 2014, 2015a, 2015b).

CHAPTER 9: FORTY MILLION ANXIOUS BRAINS

1. Hitchens (2010), p. 367.

2. Menand (2014), p. 64.

3. Kierkegaard (1980).

4. Gazzaniga (2012); Wegner (2003); Wilson (2002).

5. Horwitz and Wakefield (2012).

6. McNally (2009), p. 42.

7. "An Interview with Peter Lang." Retrieved Dec. 31, 2014, from: https://www.sprweb.org/student/interviews/interviewlang.htm.

8. Lang (1968, 1978, 1979); Lang et al (1990); Lang and McTeague (2009).

9. This is explained in Skinner's (1957) book *Verbal Behavior*. The linguist Noam Chomsky was highly critical of Skinner's view of language, and this was one of the factors that stimulated the power shift in psychology from behaviorists to cognitivists.

10. Forsyth and Eifert (1996).

11. See Kozak and Miller (1982); Zinbarg (1998).
12. Lang (1968); see also Kozak and Miller (1982); Kozak et al (1988); Zinbarg (1998).
13. Rachman (2004).
14. Summarized in Rachman (2004).
15. Rachman (2004).
16. Frith et al (1999); Naccache and Dehaene (2007); Weiskrantz (1997); Dehaene et al (2003); Dehaene and Changeux (2004); Claire Sergent and Geraint Rees, comment in Block (2007); Christof Koch and Naotsugu Tsuchiya, comment in Block (2007).
17. Zinbarg (1998).
18. Wilhelm and Roth (2001); Clark (1999); Beck (1970).
19. Lang (1977, 1979); Lang et al (1990, 2009).
20. As an aside, there is more discordance between what Lang called overt behaviors, such as avoidance, and physiological responses than between innate behaviors, such as freezing, and physiological responses. This occurs because innate behaviors have built-in physiological response patterns, whereas learned behaviors, such as avoidance, do not; this is relevant to the discordance discussion because human studies almost always focus on learned rather than innate behavior.
21. LeDoux (2012, 2014).
22. Mandler and Kessen (1959).
23. See Griebel and Holmes (2013); Belzung and Lemoine (2011).
24. Valenstein (1999).
25. Valenstein (1999).
26. Stossel (2013).
27. Valenstein (1999).
28. Skolnick (2012).
29. John Ericson, "U.S. Doctors Prescribing More Xanax, Valium, and Other Sedatives than Ever Before." *Medical Daily*, Mar 9, 2014. http://www.medicaldaily.com/us-doctors-prescribing-more-xanax-valium-and-other-sedatives-ever-270844. Retrieved Nov. 12, 2014.
30. Carson et al (2004).
31. See Griebel and Holmes (2013); Kumar et al (2013); Belzung and Griebel (2001).
32. This paragraph is based on Griebel and Holmes (2013).
33. Young et al (1998); Insel (2010); Striepens et al (2011); Neumann and Landgraf (2012); Dębiec (2005); Cochran et al (2013).
34. Eckstein et al (2014).
35. MacDonald and Feifel (2014).
36. Dodhia et al (2014).
37. Lafenetre et al (2007); Lutz (2007); Riebe et al (2012).
38. Neumeister (2013); Vinod and Hungund (2005).
39. This discussion, except the first point below, is based on Belzung and Griebel (2001); Griebel and Holmes (2013); Belzung and Lemoine (2011).
40. Spielberger (1966).
41. Horikawa and Yagi (2012).
42. Griebel and Holmes (2013).
43. Bush et al (2007); Cowansage et al (2013).
44. Griebel and Holmes (2013).
45. Burghardt et al (2004, 2007, 2013).
46. McLean et al (2011).
47. This is a result of the fact that circulating hormones of the female estrous cycle adds a level of variability to research designs.

48. Gray (1982).
49. Kagan (1994); Kagan and Snidman (1999); Rothbart et al (2000).
50. See Kendler et al (1992a, 1992b, 1994, 1995); Hettema et al (2001); Eysenck and Eysenck (1985); Eley et al (2003).
51. Stephens et al (1990); Little (1990); Watson (1990).
52. Cowan et al (2000, 2002).
53. Hyman (2007).
54. Fisher and Hariri (2013); Hariri and Holmes (2006); Hariri and Weinberger (2003); Hariri et al (2006).
55. Lesch et al (1996).
56. Dincheva et al.
57. Friedman (2015).
58. Reik (2007); Miller (2010); Mehler (2008).
59. Hartley et al (2012); Bishop et al (2006); Hariri et al (2006); Fisher and Hariri (2013); Hartley and Casey (2013); Casey et al (2011); Frielingsdorf et al (2010); Kaminsky et al (2008); Nestler (2012); McGowan et al (2009); Szyf et al (2008).
60. Horwitz and Wakefield (2012).
61. Kendler (2013).
62. Insel et al (2010).
63. Galatzer-Levy (2013).
64. Galatzer-Levy (2014).
65. This paragraph is based on Hyman (2007); Insel et al (2010); Dillon et al (2014).
66. However, DSM-5 has made more of a tilt in this direction.
67. Hyman (2007).
68. Insel et al (2010); Morris and Cuthbert (2012).
69. Simpson (2012).
70. Some circuits, such as those involving working memory, attention, and other executive functions are better explored in humans and nonhuman primates, but many of the other processes, especially those that involve subcortical circuits, can be readily studied in rodents. Further, molecular mechanisms are in many instances highly conserved and can even be studied in invertebrates.
71. Rauch et al (2006); Bremner (2006); Bishop (2007); Liberzon and Sripada (2008); Koenigs and Grafman (2009); Shin and Liberzon (2010); Hughes and Shin (2011); Olmos-Serrano and Corbin (2011); Holzschneider and Mulert (2011); Blackford and Pine (2012); Fredrikson and Faria (2013); Fisher and Hariri (2013); Ipser et al (2013); Schulz et al (2013); Bruhl et al (2014).
72. Etkin et al (2013); Zalla and Sperduti (2013); Dillon et al (2014); Apkarian et al (2013); Stone (2013); Chiapponi et al (2013); Mazefsky et al (2013); Kennedy and Adolphs (2012); Townsend and Altshuler (2012); Mihov and Hurlemann (2012); Hamilton et al (2012); Kile et al (2009); Jellinger (2008); Horinek et al (2007); Olmos-Serrano and Corbin (2011); Amaral et al (2008).
73. Personal communication from Tom Insel in an email dated July 8, 2014.
74. Grupe and Nitschke (2013).
75. Tiihonen et al (1997); Brandt et al (1998).
76. Clark and Beck (2010); Clark et al (1997).
77. Grupe and Nitschke (2013).
78. This bears some similarity to the Loewenstein et al (2001) dual risk assessment model discussed in Chapter 3.
79. Rachman (2004); Barlow (2002); Clark (1997); Salkoviskis (1996).
80. Festinger (1957); Schachter and Singer (1962); Heider (1958); Abelson (1983).

81. Gazzaniga and LeDoux (1978); Gazzaniga (2008, 2012).
82. Barrett (2013); Barrett and Russell (2014); Russell (2003); Clore and Ortony (2013).
83. Bouton et al (2001).
84. James (1980), vol. 1, pp. 291–92.

CHAPTER 10: CHANGING THE ANXIOUS BRAIN

1. Picoult (2011).
2. I am grateful to Stephan Hofmann of Boston University for guiding my research into cognitive therapy.
3. Please note that I am not qualified to offer therapeutic advice to individuals. If you are in need of help and find these ideas interesting and potentially useful, please speak with a professional to help you decide whether this material may be of value to your situation.
4. http://www.apa.org/topics/therapy/psychotherapy-approaches.aspx.
5. Freud (1917); Etchegoyen (2005).
6. Shedler (2010); McKay (2011); Sundberg (2001).
7. Greening (2006); Kramer et al (2009).
8. Wolpe (1969); Eysenck (1960); O'Leary and Wilson (1975); Yates (1970); Marks (1987); O'Donohue et al (2003); Lindsley et al (1953); O'Donohue (2001); Stampfl and Levis (1967); Bandura (1969); Ferster and Skinner (1957).
9. Beck (1970, 1976); Ellis (1957, 1980); Ellis and MacLaren (2005); Clark and Beck (2010); Beck (2014); Hofmann and Smits (2008); Leahy (2004).
10. Eifert and Forsyth (2005); Hayes et al (2006).
11. Khoury et al (2013); Evans et al (2008); Chiesa and Serretti (2011); Yook et al (2008); Goyal et al (2014); Chugh-Gupta et al (2013); Epstein (1995, 2008, 2013).
12. Hayes et al (2006).
13. Hammond (2010); Armfield and Heaton (2013); Golden (2012).
14. Shapiro (1999); McGuire et al (2014); Rathschlag and Memmert (2014); Nazari et al (2011); Lu (2010).
15. Faw (2003); Osaka (2007); Rolls et al (2003); Rolls (2005); Kringelbach (2008); Damasio (1994, 1999); Medford and Critchley (2010); Posner and Rothbart (1998); Mayr (2004); Vogt et al (1992); Devinsky et al (1995); Shenhav et al (2013); Carter et al (1999); Oakley (1999); Reinders et al (2003); Ochsner et al (2004); Hasson et al (2007); Crick and Koch (2005); Craig (2002, 2003, 2009, 2010); Bechara et al (2000); Clark et al (2008); Damasio et al (2013); Philippi et al (2012); Damasio and Carvalho (2013); Hinson et al (2002); Critchley et al (2004); Critchley (2005); Smith and Alloway (2010); Thomson (2014); Stevens (2005); Miller and Cohen (2001); Posner (1992, 1994); Posner and Dehaene (1994); Badgaiyan and Posner (1998); Bush et al (2000).
16. Hofmann (2008).
17. Forsyth and Eifert (1996).
18. Eysenck et al (2007); Borkovec et al (1998).
19. Hofmann (2008); Ramnero (2012); Powers et al (2010); Feske and Chambless (1995); Foa et al (1999); Ost et al (2001).
20. Craske et al (2008); Bouton et al (2001); Mineka (1985); Eelen and Vervliet (2006); Foa (2011).
21. Hofmann (2008); Craske et al (2014).
22. Cited in Marks (1987), p. 458.
23. Foa et al (1999); Hofmann (2008); Ramnero (2012); Powers et al (2010); Feske and Chambless (1995); Abramowitz (1997); Ost et al (2001); Mitte (2005); Rubin et al (2009); Hoyer and Beesdo-Baum (2012).
24. Craske et al (1992); Van der Heiden and ten Broecke (2009); Borkovec et al (1998); Neudeck and Wittchen (2012).

25. Ramnero (2012).
26. Mowrer (1947); Dollard and Miller (1950).
27. Wolpe (1958, 1969); Lindsley et al (1953); O'Donohue (2001); Stampfl and Levis (1967); Bandura (1969); Ferster and Skinner (1957).
28. Mowrer (1947); Dollard and Miller (1950); Miller (1948); Mowrer (1950, 1951).
29. Ricard and Lauterbach (2007); Hofmann (2008); Dymond and Roche (2009).
30. Wolpe (1958).
31. Abramowitz et al (2010); Foa et al (2007).
32. Meyer and Gelder (1963); Ramnero (2012).
33. Polin (1959); Stampfl and Levis (1967); Boulougouris and Marks (1969).
34. Foa and Kozak (1985).
35. Agras et al (1968); Barlow (2002).
36. Bandura (1977); Rachman (1977).
37. Rothbaum et al (2006); Gerardi et al (2008).
38. Hofmann (2008); Marks (1987).
39. Spence (1950); Rescorla and Wagner (1972); Bolles (1972); O'Keefe and Nadel (1974); Mackintosh (1994); Dickinson (1981).
40. Agras et al (1968).
41. Early cognitive approaches included rational emotive therapy (Ellis, 1957, 1980), cognitive restructuring (Goldfried et al, 1974), and cognitive-behavior therapy (Beck, 1970).
42. Levis (1999).
43. Beck (1970, 1976).
44. Levis (1999).
45. Beck (1970, 1976).
46. Beck (1970, 1976); Beck et al (2005); Beck and Haight (2014).
47. Ellis (1957, 1980); Ellis and MacLaren (2005).
48. Clark (1986).
49. Clark and Beck (2010).
50. Ehlers and Clark (2000); Ehlers et al (2005).
51. Hayes et al (1999); Eifert and Forsyth (2005); Hayes (2004).
52. Hofmann and Asmundson (2008).
53. Hofmann and Asmundson (2008).
54. Beck (1970, 1976); Beck et al (2005); Beck and Haight (2014).
55. Kubota et al (2012); Olsson et al (2005); Phelps et al (2000); Phelps (2001).
56. Marks (1987).
57. See Hofmann (2008); Feske and Chambless (1995).
58. Hofmann (2008).
59. Hofmann (2008); Craske (2008, 2014); Seligman and Johnston (1973); Bolles (1978); Rescorla and Wagner (1972); Rescorla (1988); Dykman (1965); Bouton et al (2001); Kirsch et al (2004); Dickinson (1981, 2012); Gallistel (1989); Bouton (1993, 2000, 2002); Holland and Bouton (1999); Pearce and Bouton (2001); Pickens and Holland (2004); Holland (1993, 2008); Balsam and Gallistel (2009); Gallistel and Gibbon (2000).
60. Hofmann (2008); Craske (2008, 2014).
61. Myers and Davis (2007); Bouton (1993, 2014).
62. Rescorla and Wagner (1972); Holland (1993, 2008); Pickens and Holland (2004); Bouton (1993, 2000, 2002); Holland and Bouton (1999); Pearce and Bouton (2001).
63. Rescorla and Wagner (1972).
64. Dickinson (2012); Roesch et al (2012); Goosens (2011); van der Meer and Redish (2010); Delgado et al (2008a); Schultz and Dickinson (2000); Schultz et al (1997).
65. Bouton (2005).

66. Morgan and LeDoux (1995, 1999); Morgan et al (1993, 2003); Quirk and Gehlert (2003); Milad et al (2006); Quirk and Beer (2006); Quirk et al (2006); Quirk and Mueller (2008); Milad and Quirk (2012); Myers and Davis (2002, 2007); Sotres-Bayon et al (2004, 2006); Sotres-Bayon and Quirk (2010); Walker and Davis (2002).

67. Phelps et al (2004); Delgado et al (2006, 2008); Rauch et al (2006); Hartley and Phelps (2010); Schiller et al (2013); Milad and Quirk (2012); Milad et al (2007); Linnman et al (2012).

68. Lang (1971); Rachman and Hodgson (1974).

69. Phelps et al (2004); Delgado et al (2008b); Schiller et al (2008, 2013); Hartley and Phelps (2010).

70. Ochsner and Gross (2005); Ochsner et al (2002). Both studies found medial and lateral PFC involvement, but evidence exists that the effect of top-down control involved connections from lateral PFC to semantic processing areas and that the effects of implicit regulation involved direct connections from medial PFC to the amygdala.

71. Foa and Kozak (1986); Salkovskis et al (2006); Foa and McNally (1996).

72. Foa and Kozak (1986); Foa (2011).

73. Lang (1977, 1979).

74. This summary of prolonged exposure is based on Foa (2011).

75. Myers and Davis (2007); Bouton (1993, 2014).

76. Rescorla and Wagner (1972).

77. Lang (1971); Rachman and Hodgson (1974).

78. McNally (2007); Dalgleish (2004); Brewin (2001).

79. Brewin (2001); Dalgleish (2004).

80. Siegel and Warren (2013); Siegel and Weinberger (2012).

81. Jacobs and Nadel (1985).

82. Barlow (2002); Durand and Barlow (2006); Hofmann (2011).

83. Borkovec et al (1998); Barlow (2002).

84. Eysenck et al (2007).

85. Borkovec et al (1998).

86. Borkovec et al (1998).

87. Eysenck et al (2007).

88. Newman and Borkovec (1995).

89. Hofmann et al (2013).

90. Barlow et al (2004).

91. Craske et al (2008, 2014).

92. Delgado et al (2008).

93. Schiller et al (2008, 2013); Schiller and Delgado (2010); Delgado et al (2008); Phelps et al (2004); Milad et al (2005); Milad et al (2007); Linnman et al (2012).

CHAPTER 11: THERAPY: LESSONS FROM THE LABORATORY

1. Pope (1803).

2. The President's Council on Bioethics. *Beyond Therapy: Biotechnology and the Pursuit of Happiness.* Washington, D.C., October 2003.

3. Nader et al (2000).

4. Blakeslee, Sandra (2000). "Brain Updating May Explain False Memories." *New York Times,* Sept. 19, 2000. http://www.nytimes.com/2000/09/19/health/brain-updating-machinery-may -explain-false-memories.html?module=Search&mabReward=relbias%3As%2C{%221% 22%3A%22RI%3A6%22}. Retrieved Nov. 14, 2014.

5. Cloitre, Marylene (2000). "Power to Erase False Memories." *New York Times,* Sept. 26, 2000. http://www.nytimes.com/2000/09/26/science/l-power-to-erase-memories-343382.html?

module=Search&mabReward=relbias%3Aw%2C{%221%22%3A%22RI%3A9%22}. Retrieved Nov. 14, 2014.

6. This idea is similar to the emotional processing theory developed by Peter Lang, Edna Foa, and Michael Kozak, as described in the last chapter.

7. Morgan et al (1993); Morgan and LeDoux (1995).

8. LeDoux et al (1989).

9. Milner (1963); Teuber (1972); Nauta (1971); Goldberg and Bilder (1987).

10. LeDoux (1996, 2002); Quirk and Mueller (2008); Quirk et al (2006); Milad and Quirk (2012); Sotres-Bayon et al (2004, 2006); Lithtik et al (2005); Duvarci and Paré (2014); Paré and Duvarci (2012); VanElzakker et al (2014); Gilmartin et al (2014); Gorman et al (1989); Davidson (2002); Bishop (2007); Shin and Liberzon (2010); Mathew et al (2008); Charney (2003); Casey et al (2011); Patel et al (2012); Vermetten and Bremner (2002); Southwick et al (2007); Yehuda and LeDoux (2007).

11. Milad and Quirk (2012); Quirk and Mueller (2008); Quirk et al (2006); Quirk and Gehlert (2003).

12. Morgan and LeDoux (1995); Sotres-Bayon and Quirk GJ (2010); Vidal-Gonzalez et al (2006).

13. Phillips and LeDoux (1992, 1994); Kim and Fanselow (1992); Frankland et al (1998).

14. Maren (2005); Ji and Maren (2007); Maren and Fanselow (1997); Sanders et al (2003).

15. LeDoux (2002); Johansen et al (2011); Paré et al (2004); Fanselow and Poulos (2005); Sah et al (2003, 2008); Marek et al (2013); Ehrlich et al (2009); Maren (2005); Maren and Quirk (2004); Pape and Paré (2010); Stork and Pape (2002); Duvarci and Paré (2014); Paré (2002).

16. The extinction circuitry is based on work summarized in Morgan et al (1993); Morgan and LeDoux (1995); Riebe et al (2012); Quirk et al (2010); Herry et al (2010); Ehrlich et al (2009); Paré et al (2004); Pape and Paré (2010); Paré and Duvarci (2012); Duvarci and Paré (2014); Maren et al (2013); Orsini and Maren (2012); Bouton et al (2006); Goode and Maren (2014); Rosenkranz et al (2003); Grace and Rosenkranz (2002); Ochsner et al (2004); Milad et al (2014); Graham and Milad (2011); Milad and Rauch (2007).

17. Macdonald (1985); Li et al (1996); Woodson et al (2000).

18. LeDoux (2002); Johansen et al (2011); *Bissière* et al (2003); Paré et al (2003); Ehrlich et al (2009); Tully et al (2007).

19. Pitkanen et al (1997); Paré and Smith (1993, 1993); Paré et al (1995); LeDoux (2002).

20. Paré and Duvarci (2012).

21. The exact pathways connecting the BA to the CeA are debated (for a discussion, see Amano et al, 2011). One circuit involves connections from BA to the intercalated cells, which then connect to the lateral CeA, whereas another pathway involves connections from BA (especially the accessory basal amygdala, also known as the basomedial amygdala), which then has connections to the medial division of the central nucleus. Both pathways likely contribute to extinction.

22. The distinction between threat and extinction neurons in BA was proposed by Herry et al (2010), though they called the threat neurons "fear" neurons.

23. Ciocchi et al (2010); Ehrlich et al (2009); Haubensak et al (2010).

24. Morgan et al (1995); Sotres-Bayon and Quirk (2010).

25. Quirk et al (2008, 2010); Paré et al (2004); Paré and Duvarci (2012); Rosenkranz et al (2003, 2006); Grace and Rosenkranz (2002).

26. Maren et al (2013).

27. Papini et al (2014); Fitzgerald et al (2014); Myskiw et al (2014); Rabinak and Pham (2014); Andero et al (2012); Bowers et al (2012); Lafenetre et al (2007); Dincheva et al (2014).

28. Thanks to Christopher Cain for this summary.

29. Bailey et al (1996); Dudai (1996).

30. Santini et al (2004); Lin et al (2003).
31. Tronson et al (2012).
32. Stevens (1994); Abel and Kandel (1998); Lee et al (2008); Alberini and Chen (2012); Josselyn et al (2004); Silva et al (1998); Yin and Tully (1996); Tully et al (2003); Josselyn (2010); Frankland et al (2004).
33. Lin et al (2003); Tronson et al (2012).
34. Johansen et al (2011, 2014).
35. Bouton and King (1983); Bouton and Nelson (1994); Carew and Rudy (1991); Bouton (2000).
36. Bouton (1988, 2000, 2005).
37. Bouton et al (2006); Holland and Bouton (1999); Maren et al (2013); Ji and Maren (2007); Lonsdorf et al (2014); Huff et al (2011); LaBar and Phelps (2005).
38. Goldstein and Kanfer (1979).
39. Craske et al (2014).
40. Hofmann et al (2013).
41. Pavlov (1927).
42. Baum (1988).
43. Brooks and Bouton (1993); Bouton et al (1993).
44. Silverstein (1967); James et al (1974).
45. Jacobs and Nadel (1985); Vervliet et al (2013); Rowe and Craske (1998); Bouton (1988).
46. Bouton (1993, 2002, 2004).
47. Bouton (1993, 2002, 2004).
48. Bouton et al (2006); Holland and Bouton (1999).
49. LaBar and Phelps (2005).
50. Rescorla and Heth (1975).
51. Myers and Davis (2007); Bouton (1993, 2014).
52. Jacobs and Nadel (1985).
53. Baker et al (2014); Holmes and Wellman (2009); Akirav and Maroun (2007); Miracle et al (2006); Izquierdo et al (2006); Deschaux et al (2013); Knox et al (2012); Raio et al (2014).
54. Radley et al (2006); Diorio et al (1993); Bhatnagar et al (1996); McEwen (2005).
55. Rodrigues et al (2009).
56. Clark (1988); McNally (1999); Rachman (1977).
57. Ost and Hugdahl (1983); Rimm et al (1977); Merckelbach et al (1989); Forsyth and Eifert (1996); Barlow (1988).
58. McEwen and Lasley (2002); Sapolsky (1998); McGaugh (2003); Rodrigues et al (2009); Cahill and McGaugh (1996); Roozendaal and McGaugh (2011); Roozendaal et al (2009); McEwen and Sapolsky (1995); Kim et al (2006); Zoladz and Diamond (2008); Shors (2006).
59. Reviewed in LeDoux (1996, 2002); McEwen and Lasley (2002); Rodrigues et al (2009); Roozendaal et al (2009).
60. Barlow (2002).
61. Raio et al (2012).
62. Bargh (1997); Bargh and Chartrand (1999); Bargh and Morsella (2008); Wilson (2002); Wilson and Dunn (2004); Greenwald and Banaji (1995); Phelps et al (2000); Devos and Banaji (2003); Debner and Jacoby (1994); Kihlstrom (1984, 1987, 1990); Kihlstrom et al (1992).
63. Groves and Thompson (1970); Kandel (1976).
64. Groves and Thompson (1970); Kandel (1976, 2001); Kandel and Schwartz (1982).
65. Hawkins et al (2006).
66. Foa and Kozak (1985, 1986); Foa (2011).
67. For a different view, see Craske (2014) and Vervliet et al (2013).
68. Craske et al (2008, 2014).

69. Rescorla and Wagner (1972).
70. Bouton et al (2004).
71. Rescorla (2000).
72. Rescorla (2006).
73. Craske et al (2014).
74. Yin et al (1994); Kramar et al (2012); Bello-Medina et al (2013); Sutton et al (2002); Rowe and Craske (1998); Chen et al (2012); Long and Fanselow (2012); Cain et al (2003); Martasian and Smith (1993); Martasian et al (1992).
75. Most studies find a benefit for spaced over massed extinction (Li and Westbrook, 2008; Urcelay et al, 2009; Long and Fanselow, 2012), but one study noted that an initial block of massed trials followed by spaced trials improved extinction (Cain et al, 2003).
76. Stevens (1994); Abel and Kandel (1998); Lee et al (2008); Alberini and Chen (2012); Josselyn et al (2004); Silva et al (1998); Yin and Tully (1996); Tully et al (2003); Josselyn (2010).
77. This is based on a study by Kogan et al (1997). Shenna Josselyn, an expert in CREB and memory (Josselyn, 2010), confirmed that roughly 60 minutes is required between sessions to allow additional learning to have access to CREB (based on email correspondence, August 23, 2014).
78. Santini et al (2004); Lin et al (2003).
79. Tronson et al (2012).
80. Kandel (1997, 2001, 2012).
81. Bailey et al (1996); Dudai (1996).
82. Tim Tully, who was a key researcher involved in discovering the role of CREB in memory, once told me he was considering setting up such a clinic.
83. Buzsaki (1991, 2011).
84. Kleim et al (2014).
85. Dardennes et al (2015).
86. Weisskopf and LeDoux (1999); Weisskopf et al (1999); Rodrigues et al (2001); Goosens and Maren (2003, 2004); Walker and Davis (2000, 2002).
87. Walker et al (2002); Ressler et al (2004); Davis et al (2006).
88. Hofmann et al (2012, 2014).
89. Fitzgerald et al (2014).
90. Barrett and Gonzalez-Lima (2004); Cai et al (2006); Yang et al (2006).
91. Soravia et al (2006); de Quervain et al (2011); Bentz et al (2010).
92. McEwen (2005); McEwen and Lasley (2002); Roozendaal et al (2009).
93. Sears et al (2013).
94. Flores et al (2014).
95. Johnson et al (2012); Mathew et al (2008).
96. Spyer and Gourine (2009); Urfy and Suarez (2014); Alheid and McCrimmon (2008); Wemmie (2011).
97. Wemmie (2011); Wemmie et al (2013).
98. Pidoplichko et al (2014); Shekhar et al (2003); Sajdyk and Shekhar (2000).
99. Esquivel et al (2010).
100. Wemmie (2011); Wemmie et al (2006).
101. Wemmie et al (2006); Sluka et al (2009).
102. Hofmann et al (2012); Neumeister (2013); Vinod and Hungund (2005); Riebe et al (2012); Lafenetre et al (2007); Papini et al (2014).
103. Cochran et al (2013); MacDonald and Feifel (2014); Kormos and Gaszner (2013); Kendrick et al (2014); Dodhia et al (2004); Insel (2010); Neumann and Landgraf (2012); Striepens et al (2011).

104. Benedict Carey. "LSD reconsidered for Therapy." *New York Times*, March 3, 2015. http://www.nytimes.com/2014/03/04/health/lsd-reconsidered-for-therapy.html?_r=0, retrieved on Feb. 21, 2014. Michael Pollan, "The Trip Treatment." *The New Yorker*, Feb. 9, 2015. Retrieved Feb. 21, 2015. Gasser, P., Kirchner, K., and Passie, T. "LSD-assisted psychotherapy for anxiety associated with a life-threatening disease: a qualitative study of acute and sustained subjective effects." *Journal of Psychopharmacol.* Jan. 29, 2015, (1):57–68.
105. Marin et al (2014).
106. Ressler and Mayberg (2007); Couto et al (2014); Lipsman et al (2013a, 2013b); Voon et al (2013); Heeramun-Aubeeluck and Lu (2013).
107. Rodriguez-Romaguera et al (2012); Whittle et al (2013); Do-Monte et al (2013).
108. Mantione et al (2014); Marin et al (2014).
109. Marin et al (2014).
110. Isserles et al (2013).
111. Pena et al (2012).
112. George et al (2008); Porges (2001).
113. Pena et al (2012).
114. Porges (2001).
115. Farah (2012); Farah et al (2004); Hariz et al (2013); Ragan et al (2013).
116. Ambasudhan et al (2014); Allen and Feigin (2014).
117. Mitra and Sapolsky (2010).
118. Cardinal et al (2002); Balleine and Killcross (2006).
119. Nehoff et al (2014); Toumey (2013); Jacob et al (2011).
120. Florczyk and Saha (2007).
121. Myers and Davis (2007); Bouton (1993, 2014).
122. Nader et al (2000).
123. Davis and Squire (1984); Martinez et al (1981); Agranoff et al (1966); Flexner and Flexner (1966); Barondes and Cohen (1967); Barondes (1970); Quartermain et al (1970); Dudai (2004).
124. Misanin et al (1968); Lewis (1979).
125. McGaugh (2004).
126. Sara (2000); Przybyslawski and Sara (1997).
127. Schafe et al (1999); Schafe and LeDoux (2000).
128. One idea that came about was that the drug just made extinction work faster and better. But reconsolidation didn't work like extinction. The memory seemed to be resistant to revival by spontaneous recovery, renewal, and reinstatement. It seemed much more persistent than extinction.
129. The many reviews of this topic include: Nader and Einarsson (2010); Wang et al (2009); Nader and Hardt (2009); Milton and Everitt (2010); Reichelt and Lee (2013); Tronson and Taylor (2007, 2013); Besnard et al (2012); Dudai (2006, 2012); Alberini and LeDoux (2013); Alberini (2013).
130. Kindt et al (2009, 2014); Bos et al (2014); Schwabe et al (2014); Chan and LaPaglia (2013); Lonergan et al (2013); Agren et al (2012); Hupbach et al (2007); Stickgold and Walker (2005).
131. Alberini (2005).
132. Alberini (2013).
133. Diaz-Mataix et al (2013).
134. Diaz-Mataix et al (2013).
135. Wang et al (2009).
136. Schiller et al (2010, 2013); Monfils et al (2009); Haubrich et al (2014); De Oliveira Alvares et al (2013); Diaz-Mataix et al (2013); Lee (2010); Hupbach et al (2008).

137. Hirst et al (2009).
138. Loftus (1996); Bonham and Gonzalez-Vallejo (2009).
139. Schacter (Dębiec 2001, 2012).
140. Johnson et al (2012); Kopelman (2010); Whitfield (2000); Loftus and Davis (2006); Laney and Loftus (2005); Loftus and Polage (1999); Stocks (1998).
141. Diaz-Mataix (2011); Dębiec et al (2010); Dębiec et al (2006).
142. Tronson and Taylor (2013); Milton and Everitt (2010).
143. Brunet et al (2011); Poundja et al (2012); Lonergan et al (2013); Brunet et al (2008); Kindt et al (2009).
144. Kindt (2014); Schiller and Phelps (2011); Lane et al (2014); Pitman et al (2015).
145. Dębiec and LeDoux (2006); Dębiec et al (2011). In these studies Dębiec either facilitated or inhibited receptors that bind norepinephrine (NE) in the lateral amygdala. Because NE receptors, via cAMP, modulate CREB-dependent protein synthesis, blocking these thus indirectly dampens protein synthesis, whereas facilitating these enhances protein synthesis.
146. Taubenfeld et al (2009); Pitman et at (2011).
147. Miller and Sweatt (2006); Alberini (2005); Alberini and LeDoux (2013).
148. Lattal and Wood (2013) suggest that certain molecular changes could take place during extinction that make it unobservable behaviorally but persistent in the brain, making it difficult to distinguish reconsolidation from "silent extinction."
149. Eisenberg et al (2003).
150. Sangha et al (2003); Pedreira and Maldonado (2003); Suzuki et al (2004).
151. Dudai and Eisenberg (2004).
152. Quirk and Mueller (2008).
153. Monfils et al (2009).
154. In a laboratory such as mine, where the researchers discuss findings openly, it's hard to pinpoint from whom the key idea emerges. It seems that in conversations involving Marie Monfils, Daniela Schiller, Chris Cain, and others, the idea underlying Monfils's project emerged.
155. Clem and Huganir (2010).
156. Schiller et al (2010).
157. Steinfurth et al (2014); Schiller et al (2013).
158. Kip et al (2014).
159. Xue et al (2012).
160. Baker et al (2013); Kindt and Soeter (2013).
161. Serrano et al (2005).
162. Pastalkova et al (2006).
163. Serrano et al (2008); Shema et al (2009); Shema et al (2007); von Kraus et al (2010).
164. Parts of this section are based on an Opinionator piece I contributed to the *New York Times* website on April 7, 2013. "For the Anxious, Avoidance Can Have an Upside." http://opinionator.blogs.nytimes.com/2013/04/07/for-the-anxious-avoidance-can-have-an-upside/?_php=true&_type=blogs&_r=0.
165. LeDoux and Gorman (2001).
166. Amorapanth et al (2000).
167. Although this procedure has been criticized (Church, 1964), it is still the best procedure for evaluating the effects of learning versus stimulus exposure.
168. van der Kolk (1994, 2006, 2014).
169. Bonanno and Burton (2013).
170. MacArthur Research Network description of coping strategies. http://www.macses.ucsf.edu/research/psychosocial/coping.php. Retrieved Jan. 26, 2015.

171. I wrote three articles for the Opinionator series on anxiety in *The New York Times*, the final one being "For the Anxious, Avoidance Can Have an Upside." *The New York Times*, April 7, 2013. See LeDoux (2013).

172. http://michaelroganphd.com/neuroscience-research/.

173. Dymond et al (2012); Dymond and Roche (2009).

174. Guz (1997); Haouzi et al (2006).

175. Spyer and Gourine (2009); Urfy and Suarez (2014); Alheid and McCrimmon (2008).

176. Urfy and Suarez (2014); Haouzi et al (2006); Mitchell and Berger (1975).

177. Porges (2001).

178. Porges (2001); Streeter et al (2012).

179. McGowan et al (2009); Johnson and Casey (2014); Casey et al (2010, 2011); Tottenham (2014); Perry and Sullivan (2014); Rincón-Cortés and Sullivan (2014); Sullivan and Holman (2010).

180. Eifert and Forsyth (2005); Hayes et al (2006).

181. Austin (1998).

182. Epstein (2013).

183. Austin (1998).

184. Austin (1998).

185. Davidson and Lutz (2008); Lutz et al (2007).

186. Davidson and Lutz (2008); Lutz et al (2007); Fox et al (2014); Zeidan et al (2014); Dickenson et al (2013); Davanger et al (2010); Jang et al (2011); Manna et al (2010).

187. Marchand (2014); Malinowski (2013); Chiesa et al (2013); Farb et al (2012); Rubia (2009); Lutz et al (2008); Deshmukh (2006).

188. Raichle and Snyder (2007); Gusnard et al (2001); Andrews-Hanna et al (2014); Barkhof et al (2014); Buckner (2013).

189. Malinowski (2013).

190. Anderson and Hanslmayr (2014); DePrince et al (2012); Anderson and Huddleston (2012); Whitmer and Gotlib (2013).

191. Malinowski (2013).

192. Epstein (1995); "Freud and Buddha" by Mark Epstein: http://spiritualprogressives.org/newsite/?p=651. Retrieved Feb. 8, 2015.

BIBLIOGRAPHY

Abel, T., and E. Kandel. "Positive and Negative Regulatory Mechanisms That Mediate Long-Term Memory Storage." *Brain Research. Brain Research Reviews* (Amsterdam) (1998) 26:360–78.

Abelson, R.P. "Whatever Became of Consistency Theory?" *Personality and Social Psychology Bulletin* (1983) 9:37–64.

Abrahams, V.C., S.M. Hilton, and A. Zbrozyna. "Active Muscle Vasodilatation Produced by Stimulation of the Brain Stem: Its Significance in the Defence Reaction." *Journal of Physiology* (1960) 154:491–513.

Abramowitz, J.S. "Effectiveness of Psychological and Pharmacological Treatments for Obsessive-Compulsive Disorder: "A Quantitative Review." *Journal of Consulting and Clinical Psychology* (1997) 65:44–52.

Abramowitz, J.S., B.J. Deacon, and S.P.H. Whiteside. *Exposure Therapy for Anxiety: Principles and Practice* (New York: Guilford Press, 2010).

Abrams, R.A., and S.E. Christ. "Motion Onset Captures Attention." *Psychological Science* (2003) 14:427–32.

Abrams, R.L., and A.G. Greenwald. "Parts Outweigh the Whole (Word) in Unconscious Analysis of Meaning." *Psychological Science* (2000) 11:118–24.

Abrams, R.L., and J. Grinspan. "Unconscious Semantic Priming in the Absence of Partial Awareness." *Consciousness and Cognition* (2007) 16:942–53; discussion 954–58.

Abrams, R.L., M.R. Klinger, and A.G. Greenwald. "Subliminal Words Activate Semantic Categories (Not Automated Motor Responses)." *Psychonomic Bulletin & Review* (2002) 9:100–6.

Adams, D.B. "Brain Mechanisms for Offense, Defense, and Submission." *Behavioral and Brain Sciences* (1979) 2:201–42.

Adams, R.B. Jr., et al. "Culture, Gaze and the Neural Processing of Fear Expressions." *Social Cognitive and Affective Neuroscience* (2010) 5:340–48.

Adhikari, A. "Distributed Circuits Underlying Anxiety." *Frontiers in Behavioral Neuroscience* (2014) 8:112.

Adolphs, R. "The Biology of Fear." *Current Biology* (2013) 23:R79–93.

Agranoff, B.W., R.E. Davis, and J.J. Brink. "Chemical Studies on Memory Fixation in Goldfish." *Brain Research* (1966) 1:303–9.

Agras, S., H. Leitenberg, and D.H. Barlow. "Social Reinforcement in the Modification of Agoraphobia." *Archives of General Psychiatry* (1968) 19:423–27.

Agren, T., et al. "Disruption of Reconsolidation Erases a Fear Memory Trace in the Human Amygdala." *Science* (2012) 337:1550–52.

Akirav, I., and M. Maroun. "The Role of the Medial Prefrontal Cortex-Amygdala Circuit in Stress Effects on the Extinction of Fear." *Neural Plasticity* (2007) 2007:308–73.

Alanen, L. *Descartes's Concept of Mind* (Cambridge, MA: Harvard University Press, 2003).

Alberini, C.M. "Mechanisms of Memory Stabilization: Are Consolidation and Reconsolidation Similar or Distinct Processes?" *Trends in Neurosciences* (2005) 28:51–56.

Alberini, C.M., ed. *Memory Consolidation* (New York: Elsevier, 2013).

Alberini, C.M., and D.Y. Chen. "Memory Enhancement: Consolidation, Reconsolidation and Insulin-like Growth Factor 2." *Trends in Neurosciences* (2012) 35:274–83.

Alberini, C.M., and J.E. LeDoux. "Memory Reconsolidation." *Current Biology* (2013) 23:R746–50.

Alheid, G.F., et al. "The Neuronal Organization of the Supracapsular Part of the Stria Terminalis in the Rat: The Dorsal Component of the Extended Amygdala." *Neuroscience* (1998) 84:967–96.

Alheid, G.F., and L. Heimer. "New Perspectives in Basal Forebrain Organization of Special Relevance for Neuropsychiatric Disorders: The Striatopallidal, Amygdaloid, and Corticopetal Components of Substantia Innominata." *Neuroscience* (1988) 27:1–39.

Alheid, G.F., and D.R. McCrimmon. "The Chemical Neuroanatomy of Breathing." *Respiratory Physiology & Neurobiology* (2008) 164:3–11.

Allen, P.J., and A. Feigin. "Gene-Based Therapies in Parkinson's Disease." *Neurotherapeutics: The Journal of the American Society for Experimental Neurotherapeutics* (2014) 11:60–67.

Allen, T.A., and N.J. Fortin. "The Evolution of Episodic Memory." *Proceedings of the National Academy of Sciences of the United States of America* (2013) 110(Suppl 2):10379–86.

Allman, J.M., et al. "The Anterior Cingulate Cortex. The Evolution of an Interface Between Emotion and Cognition." *Annals of the New York Academy of Sciences* (2001) 935:107–17.

Alvarez, R.P., et al. "Contextual Fear Conditioning in Humans: Cortical-Hippocampal and Amygdala Contributions." *Journal of Neuroscience* (2008) 28:6211–19.

Alvarez, R.P., et al. "Phasic and Sustained Fear in Humans Elicits Distinct Patterns of Brain Activity." *NeuroImage* (2011) 55:389–400.

Amano, T., S. Duvarci, D. Popa, and D. Paré. "The Fear Circuit Revisited: Contributions of the Basal Amygdala Nuclei to Conditioned Fear." *Journal of Neuroscience* (2011) 31: 1581–89.

Amaral, D.G., et al. "Topographic Organization of Projections from the Amygdala to the Visual Cortex in the Macaque Monkey." *Neuroscience* (2003) 118:1099–1120.

Amaral, D.G., et al. "Anatomical Organization of the Primate Amygdaloid Complex." In: *The Amygdala: Neurobiological Aspects of Emotion, Memory, and Mental Dysfunction*, ed. J.P. Aggleton (New York: Wiley-Liss, 1992), 1–66.

Amaral, D.G., C.M. Schumann, and C.W. Nordahl. "Neuroanatomy of Autism." *Trends in Neurosciences* (2008) 31:137–45.

Ambasudhan, R., et al. "Potential for Cell Therapy in Parkinson's Disease Using Genetically Programmed Human Embryonic Stem Cell-Derived Neural Progenitor Cells." *Journal of Comparative Neurology* (2014) 522:2845–56.

Amorapanth, P., J.E. LeDoux, and K. Nader. "Different Lateral Amygdala Outputs Mediate Reactions and Actions Elicited by a Fear-Arousing Stimulus." *Nature Neuroscience* (2000) 3:74–79.

Amorapanth, P., K. Nader, and J.E. LeDoux. "Lesions of Periaqueductal Gray Dissociate-Conditioned Freezing from Conditioned Suppression Behavior in Rats." *Learning & Memory* (1999) 6:491–99.

Amstadter, A. "Emotion Regulation and Anxiety Disorders." *Journal of Anxiety Disorders* (2008) 22:211–21.

Amsterdam, B. "Mirror Self-Image Reactions Before Age Two." *Developmental Psychobiology* (1972) 5:297–305.

Anagnostaras, S.G., G.D. Gale, and M.S. Fanselow. "Hippocampus and Contextual Fear Conditioning: Recent Controversies and Advances." *Hippocampus* (2001) 11:8–17.

Anders, S., et al. "When Seeing Outweighs Feeling: A Role for Prefrontal Cortex in Passive Control of Negative Affect in Blindsight." *Brain: A Journal of Neurology* (2009) 132:3021–31.

Anderson, A.K., and E.A. Phelps. "Lesions of the Human Amygdala Impair Enhanced Perception of Emotionally Salient Events." *Nature* (2001) 411:305–9.

Anderson, D.J., and R. Adolphs. "A Framework for Studying Emotions Across Species." *Cell* (2014) 157:187–200.

Anderson, M.C., and S. Hanslmayr. "Neural Mechanisms of Motivated Forgetting." *Trends in Cognitive Sciences* (2014) 18:279–92.

Anderson, M.C., and E. Huddleston. "Towards a Cognitive and Neurobiological Model of Motivated Forgetting." *Nebraska Symposium on Motivation* (2012) 58:53–120.

Andrews-Hanna, J.R., J. Smallwood, and R.N. Spreng. "The Default Network and Self-Generated Thought: Component Processes, Dynamic Control, and Clinical Relevance." *Annals of the New York Academy of Sciences* (2014) 1316:29–52.

Ansorge, U., G. Horstmann, and I. Scharlau. "Top-Down Contingent Feature-Specific Orienting with and Without Awareness of the Visual Input." *Advances in Cognitive Psychology/ University of Finance and Management in Warsaw* (2011) 7:108–19.

Anthony, T.E., N. Dee, A. Bernard, W. Lerchner, N. Heintz, and D.J. Anderson (2014). "Control of Stress-Induced Persistent Anxiety by an Extra-Amygdala Septohypothalamic Circuit." *Cell* 156:522–36.

Aoki, Y., S. Cortese, and M. Tansella. "Neural Bases of Atypical Emotional Face Processing in Autism: A Meta-Analysis of fMRI Studies." *The World Journal of Biological Psychiatry: The Official Journal of the World Federation of Societies of Biological Psychiatry* (2014) 1–10.

Apergis-Schoute, A.M., et al. "Extinction Resistant Changes in the Human Auditory Association Cortex Following Threat Learning." *Neurobiology of Learning and Memory* (2014) 113: 109–14.

Apkarian, A.V., et al. "Neural Mechanisms of Pain and Alcohol Dependence." *Pharmacology, Biochemistry, and Behavior* (2013) 112:34–41.

Ariëns Kappers, C.U., C.G. Huber, and E.C. Crosby. *The Comparative Anatomy of the Nervous System of Vertebrates, Including Man* (New York: Macmillan Company, 1936).

Armfield, J.M., and L.J. Heaton. "Management of Fear and Anxiety in the Dental Clinic: A Review." *Australian Dental Journal* (2013) 58:390–407; quiz 531.

Armony, J.L., and R.J. Dolan. "Modulation of Spatial Attention by Fear-Conditioned Stimuli: An Event-Related fMRI Study." *Neuropsychologia* (2002) 40:817–26.

Armony, J.L., G.J. Quirk, and J.E. LeDoux. "Differential Effects of Amygdala Lesions on Early and Late Plastic Components of Auditory Cortex Spike Trains during Fear Conditioning." *Journal of Neuroscience* (1998) 18:2592–2601.

Armony J.L., et al. "An Anatomically Constrained Neural Network Model of Fear Conditioning." *Behavioral Neuroscience* (1995) 109:246–57.

Armony, J.L., et al. "Computational Modeling of Emotion: Explorations Through the Anatomy and Physiology of Fear Conditioning." *Trends in Cognitive Sciences* (1997) 1:28–34.

Armony, J.L., et al. "Stimulus Generalization of Fear Responses: Effects of Auditory Cortex Lesions in a Computational Model and in Rats." *Cerebral Cortex* (1997) 7:157–65.

Armstrong, D.M. "Three Types of Consciousness." *CIBA Foundation Symposium* (1979) 235–53.

Arnold, M.B. *Emotion and Personality* (New York: Columbia University Press, 1960).

Arnsten, A.F. "Catecholamine Influences on Dorsolateral Prefrontal Cortical Networks." *Biological Psychiatry* (2011) 69:E89–99.

Asli, O., and M.A. Flaten. "In the Blink of an Eye: Investigating the Role of Awareness in Fear Responding by Measuring the Latency of Startle Potentiation." *Brain Sciences* (2012) 2:61–84.

Aston-Jones, G., C. Chiang, and T. Alexinsky. "Discharge of Noradrenergic Locus Coeruleus Neurons in Behaving Rats and Monkeys Suggests a Role in Vigilance." *Progress in Brain Research* (1991) 88:501–20.

Aston-Jones, G., and J.D. Cohen. "An Integrative Theory of Locus Coeruleus-Norepinephrine Function: Adaptive Gain and Optimal Performance." *Annual Review of Neuroscience* (2005) 28:403–50.

Aston-Jones, G., et al. "The Brain Nucleus Locus Coeruleus: Restricted Afferent Control of a Broad Efferent Network." *Science* (1986) 234:734–36.

Aston-Jones, G., J. Rajkowski, and J. Cohen. "Locus Coeruleus and Regulation of Behavioral Flexibility and Attention." *Progress in Brain Research* (2000) 126:165–82.

Atkin, A., and T.D. Wager. "Functional Neuroimaging of Anxiety: A Meta-Analysis of Emotional Processing in PTSD, Social Anxiety Disorder, and Specific Phobia." *The American Journal of Psychiatry* (2007) 164:1476–88.

Aton, S.J. "Set and Setting: How Behavioral State Regulates Sensory Function and Plasticity." *Neurobiology of Learning and Memory* (2013) 106:1–10.

Auden, W.H. *The Age of Anxiety: A Baroque Eclogue* (New York: Random House, 1947).

Auden, W.H. *The Age of Anxiety (Reissue)* (Princeton, NJ: Princeton University Press, 2011).

Aupperle, R.L., and M.P. Paulus. "Neural Systems Underlying Approach and Avoidance in Anxiety Disorders." *Dialogues in Clinical Neuroscience* (2010) 12:517–31.

Austin, J. *Zen and the Brain* (Cambridge, MA: MIT Press, 1998).

Baars, B.J. *A Cognitive Theory of Consciousness* (New York: Cambridge University Press, 1988).

Baars, B.J. "Global Workspace Theory of Consciousness: Toward a Cognitive Neuroscience of Human Experience." *Progress in Brain Research* (2005) 150:45–53.

Baars, B.J., and S. Franklin. "An Architectural Model of Conscious and Unconscious Brain Functions: Global Workspace Theory and IDA." *Neural Networks: The Official Journal of the International Neural Network Society* (2007) 20:955–61.

Baars, B.J., S. Franklin, and T.Z. Ramsoy "Global Workspace Dynamics: Cortical 'Binding and Propagation' Enables Conscious Contents." *Frontiers in Psychology* (2013) 4:200.

Baas, J.M., et al. "Benzodiazepines Have No Effect on Fear-Potentiated Startle in Humans." *Psychopharmacology (Berl)* (2002) 161:233–47.

Bach, D.R., and R.J. Dolan. "Knowing How Much You Don't Know: A Neural Organization of Uncertainty Estimates." *Nature Reviews Neuroscience* (2012) 13:572–86.

Bach, D.R., et al. "The Known Unknowns: Neural Representation of Second-Order Uncertainty, and Ambiguity." *Journal of Neuroscience* (2011) 31:4811–20.

Bacon, F. *Instauratio Magna. Novum Organum* (1620). London: John Brill, p. 68. Cited in Mandler and Kessen, *The Language of Psychology* (New York: John Wiley, 1964).

Baddeley, A. "Working Memory." *Science* (1992) 255:556–59.

Baddeley, A. "The Episodic Buffer: A New Component of Working Memory?" *Trends in Cognitive Sciences* (2000) 4:417–23.

Baddeley, A. "The Concept of Episodic Memory." *Philosophical Transactions of the Royal Society B: Biological Sciences* (2001) 356:1345–50.

Baddeley, A. "Working Memory and Language: An Overview." *Journal of Communication Disorders* (2003) 36:189–208.

Bailey, C.H., D. Bartsch, and E.R. Kandel. "Toward a Molecular Definition of Long-Term Memory Storage." *Proceedings of the National Academy of Sciences of the United States of America* (1996) 93:13445–52.

Bailey, C.H., and E.R. Kandel. "Synaptic Remodeling, Synaptic Growth and the Storage of Long-Term Memory in *Aplysia*." *Progress in Brain Research* (2008) 169:179–98.

Baker, K.D., et al. "A Window of Vulnerability: Impaired Fear Extinction in Adolescence." *Neurobiology of Learning and Memory* (2014) 113:90–100.

Baker, K.D., G.P. McNally, and R. Richardson. "Memory Retrieval Before or after Extinction Reduces Recovery of Fear in Adolescent Rats." *Learning & Memory* (2013) 20:467–73.

Baker, T.B., T.H. Brandon, and L. Chassin. "Motivational Influences on Cigarette Smoking." *Annual Review of Psychology* (2004) 55:463–91.

Baker, T.B., et al. "Addiction Motivation Reformulated: An Affective Processing Model of Negative Reinforcement." *Psychological Review* (2004) 111:33–51.

Balleine, B., and A. Dickinson. "Instrumental Performance Following Reinforcer Devaluation Depends upon Incentive Learning." *Quarterly Journal of Experimental Psychology B* (1991) 43:279–96.

Balleine, B., C. Gerner, and A. Dickinson. "Instrumental Outcome Devaluation Is Attenuated by the Anti-emetic Ondansetron." *Quarterly Journal of Experimental Psychology B* (1995) 48:235–51.

Balleine, B.W. "Neural Bases of Food-Seeking: Affect, Arousal and Reward in Corticostriato-limbic Circuits." *Physiology & Behavior* (2005) 86:717–30.

Balleine, B.W. "Sensation, Incentive Learning and the Motivational Control of Goal-Directed Action." In: *Neurobiology of Sensation and Reward*, ed. J.A. Gottfried (Boca Raton, FL: CRC Press, 2011), 287–310.

Balleine, B.W., and A. Dickinson. "Consciousness: The Interface Between Affect and Cognition." *Consciousness and Human Identity*, ed. J. Cornwall (Oxford, UK: Oxford University Press, 1998), 57–85.

Balleine, B.W., and A. Dickinson. "Goal-Directed Instrumental Action: Contingency and Incentive Learning and Their Cortical Substrates." *Neuropharmacology* (1998) 37:407–19.

Balleine, B.W., A.S. Killcross, and A. Dickinson. "The Effect of Lesions of the Basolateral Amygdala on Instrumental Conditioning." *Journal of Neuroscience* (2003) 23:666–75.

Balleine, B.W., and S. Killcross. "Parallel Incentive Processing: An Integrated View of Amygdala Function." *Trends in Neurosciences* (2006) 29:272–79.

Balleine, B.W., and J.P. O'Doherty. "Human and Rodent Homologies in Action Control: Corticostriatal Determinants of Goal-Directed and Habitual Action." *Neuropsychopharmacology* (2010) 35:48–69.

Balsam, P.D., and C.R. Gallistel. "Temporal Maps and Informativeness in Associative Learning." *Trends in Neurosciences* (2009) 32:73–78.

Bandelow, B., et al. Care WTFoMDiP, WFSBP Task Force on Anxiety Disorders OCD, PTSD. "Guidelines for the Pharmacological Treatment of Anxiety Disorders, Obsessive-Compulsive Disorder and Posttraumatic Stress Disorder in Primary Care." *International Journal of Psychiatry in Clinical Practice* (2012) 16:77–84.

Bandler, R., and P. Carrive. "Integrated Defence Reaction Elicited by Excitatory Amino Acid Microinjection in the Midbrain Periaqueductal Grey Region of the Unrestrained Cat." *Brain Research* (1988) 439:95–106.

Bandura, A. *Principles of Behavior Modification* (New York: Holt, 1969).

Bandura, A. *Social Learning Theory* (Englewood Cliffs, NJ: Prentice Hall, 1977).

Bar, M., et al. "Top-Down Facilitation of Visual Recognition." *Proceedings of the National Academy of Sciences of the United States of America* (2006) 103:449–54.

Barad, M. "Fear Extinction in Rodents: Basic Insight to Clinical Promise." *Current Opinion in Neurobiology* (2005) 15:710–15.

Barbas, H. "Architecture and Cortical Connections of the Prefrontal Cortex in the Rhesus Monkey." *Advances in Neurology* (1992) 57:91–115.

Barbas, H. "Connections Underlying the Synthesis of Cognition, Memory, and Emotion in Primate Prefrontal Cortices." *Brain Research Bulletin* (2000) 52:319–30.

Barbey, A.K., et al. "An Integrative Architecture for General Intelligence and Executive Function Revealed by Lesion Mapping." *Brain* (2012) 135:1154–64.

Bard, P. "A Diencephalic Mechanism for the Expression of Rage with Special Reference to the Sympathetic Nervous System." *American Journal of Physiology* (1928) 84:490–515.

Bard, P., and D.M. Rioch. "A Study of Four Cats Deprived of Neocortex and Additional Parts of the Forebrain." *Bulletin of the Johns Hopkins Hospital* (1937) 60:73–147.

Barde, L.H., and S.L. Thompson-Schill. "Models of Functional Organization of the Lateral Prefrontal Cortex in Verbal Working Memory: Evidence in Favor of the Process Model." *Journal of Cognitive Neuroscience* (2002) 14:1054–63.

Bargh, J.A. "The Automaticity of Everyday Life." In: *Advances in Social Cognition*, vol. 10, ed. R.S. Wyer (Mahwah, NJ: Erlbaum, 1997).

Bargh, J.A., and T.L. Chartrand. "The Unbearable Automaticity of Being." *American Psychologist* (1999) 54:462–79.

Bargh, J.A., and M.J. Ferguson. "Beyond Behaviorism: on the Automaticity of Higher Mental Processes." *Psychological Bulletin* (2000) 126:925–45.

Bargh, J.A., and E. Morsella. "The Unconscious Mind." *Perspectives on Psychological Science* (2008) 3:73–79.

Bargmann, C.I. "Comparative Chemosensation from Receptors to Ecology." *Nature* (2006) 444:295–301.

Bargmann, C.I. "Beyond the Connectome: How Neuromodulators Shape Neural Circuits." *BioEssays* (2012) 34:458–65.

Bar-Haim, Y., et al. "When Time Slows Down: The Influence of Threat on Time Perception in Anxiety." *Cognition and Emotion* (2010) 24:255–63.

Bar-Haim, Y., et al. "Threat-Related Attentional Bias in Anxious and Nonanxious Individuals: A Meta-Analytic Study." *Psychological Bulletin* (2007) 133:1–24.

Barkhof, F., S. Haller, and S.A. Rombouts. "Resting-State Functional MR Imaging: A New Window to the Brain." *Radiology* (2014) 272:29–49.

Barlow, D.H. *Anxiety and Its Disorders: The Nature and Treatment of Anxiety and Panic* (New York: Guilford, 1988).

Barlow, D.H. "Long-Term Outcome for Patients with Panic Disorder Treated with Cognitive-Behavioral Therapy." *The Journal of Clinical Psychiatry* (1990) 51(Suppl A):17–23.

Barlow, D.H. *Anxiety and Its Disorders: The Nature and Treatment of Anxiety and Panic* (New York: Guilford Press, 2002).

Barlow, D.H., L.B. Allen, and M.L. Choate. "Toward a Unified Treatment for Emotional Disorders." *Behavior Therapy* (2004) 35:205–30.

Barondes, S.H. "Cerebral Protein Synthesis Inhibitors Block Long-Term Memory." *International Review of Neurobiology* (1970) 12:177–205.

Barondes, S.H., and H.D. Cohen. "Comparative Effects of Cycloheximide and Puromycin on Cerebral Protein Synthesis and Consolidation of Memory in Mice." *Brain Research* (1967) 4:44–51.

Barres, B.A. "The Mystery and Magic of Glia: A Perspective on Their Roles in Health and Disease." *Neuron* (2008) 6:430–40.

Barrett, D., and F. Gonzalez-Lima. "Behavioral Effects of Metyrapone on Pavlovian Extinction." *Neuroscience Letters* (2004) 371:91–96.

Barrett, L.F. "Are Emotions Natural Kinds?" *Perspectives on Psychological Science* (2006) 1:28–58.

Barrett, L.F. "Solving the Emotion Paradox: Categorization and the Experience of Emotion." *Personality and Social Psychology Review* (2006) 10:20–46.

Barrett, L.F. "The Future of Psychology: Connecting Mind to Brain." *Perspectives on Psychological Science* (2009) 4:326–39.

Barrett, L.F. "Variety Is the Spice of Life: A Psychological Construction Approach to Understanding Variability in Emotion." *Cognition and Emotion* (2009) 23:1284–1306.

Barrett, L.F. "Emotions Are Real." *Emotion* (2012) 12:413–29.

Barrett, L.F. "Psychological Construction: The Darwinian Approach to the Science of Emotion." *Emotion Review* (2013) 5:379–89.

Barrett, L.F., et al. "Of Mice and Men: Natural Kinds of Emotions in the Mammalian Brain? A Response to Panksepp and Izard." *Perspectives on Psychological Science* (2007) 2:297–311.

Barrett, L.F., K.A. Lindquist, and M. Gendron. "Language as Context for the Perception of Emotion." *Trends in Cognitive Sciences* (2007) 11:327–32.

Barrett, L.F., and J.A. Russell, eds. *The Psychological Construction of Emotion* (New York: Guilford Press, 2014).

Barrouillet, P., S. Bernardin, and V. Camos. "Time Constraints and Resource Sharing in Adults' Working Memory Spans." *Journal of Experimental Psychology: General* (2004) 133:83–100.

Barrouillet, P., et al. "Time and Cognitive Load in Working Memory." *Journal of Experimental Psychology Learning, Memory, and Cognition* (2007) 33:570–85.

Barton, R.A., and C. Venditti. "Human Frontal Lobes Are Not Relatively Large." *Proceedings of the National Academy of Sciences of the United States of America* (2013) 110:9001–6.

Basser, L.S. "Hemiplegia of Early Onset and the Faculty of Speech with Special Reference to the Effects of Hemispherectomy." *Brain: A Journal of Neurology* (1962) 85:427–60.

Baudonnat, M., et al. "Heads for Learning, Tails for Memory: Reward, Reinforcement and a Role of Dopamine in Determining Behavioral Relevance Across Multiple Timescales." *Frontiers in Neuroscience* (2013) 7:175.

Baum, M. "Spontaneous Recovery from the Effects of Flooding (Exposure) in Animals." *Behaviour Research and Therapy* (1988) 26:185–86.

Baumeister, A.A. "The Tulane Electrical Brain Stimulation Program a Historical Case Study in Medical Ethics." *Journal of the History of the Neurosciences* (2000) 9:262–78.

Baumeister, A.A. "Serendipity and the Cerebral Localization of Pleasure." *Journal of the History of the Neurosciences* (2006) 15:92–98.

Beach, F.A. "Central Nervous Mechanisms Involved in the Reproductive Behavior of Vertebrates." *Psychological Bulletin* (1942) 39:200–26.

Beach, F.A. "The Descent of Instinct." *Psychological Review* (1955) 62:401–10.

Bebdarik, R. *The Human Condition* (New York: Springer, 2011).

Bebko, G.M., et al. "Attentional Deployment Is Not Necessary for Successful Emotion Regulation via Cognitive Reappraisal or Expressive Suppression." *Emotion* (2014) 14:504–12.

Bechara, A., H. Damasio, and A.R. Damasio. "Emotion, Decision Making and the Orbitofrontal Cortex." *Cerebral Cortex* (2000) 10:295–307.

Bechara, A., et al. "Deciding Advantageously Before Knowing the Advantageous Strategy." *Science* (1997) 275:1293–95.

Bechara, A., et al. "Double Dissociation of Conditioning and Declarative Knowledge Relative to the Amygdala and Hippocampus in Humans." *Science* (1995) 269:1115–18.

Bechtholt, A.J., R.J. Valentino, and I. Lucki. "Overlapping and Distinct Brain Regions Associated with the Anxiolytic Effects of Chlordiazepoxide and Chronic Fluoxetine." *Neuropsychopharmacology* (2008) 33:2117–30.

Beck, A.T. "Cognitive Therapy: Nature and Relation to Behavior Therapy." *Behavior Therapy* (1970) 1:184–200.

Beck, A.T. *Cognitive Therapy and the Emotional Disorders* (New York: International Universities Press, 1976).

Beck, A.T. "Cognitive Therapy. A 30-Year Retrospective." *The American Psychologist* (1991) 46: 368–75.

Beck, A.T., and D.A. Clark. "An Information Processing Model of Anxiety: Automatic and Strategic Processes." *Behaviour Research and Therapy* (1997) 35:49–58.

Beck, A.T., and G. Emer. *Anxiety Disorders and Phobias: A Cognitive Perspective* (New York: Basic Books, 1985).

Beck, A.T., G. Emery, and R.L Greenberg. *Anxiety Disorders and Phobias: A Cognitive Perspective* (New York: Basic Books, 2005).

Beck, A.T., and E.A. Haigh. "Advances in Cognitive Theory and Therapy: The Generic Cognitive Model." *Annual Review of Clinical Psychology* (2014) 10:1–24.

Beck, K.D., et al. "Vulnerability Factors in Anxiety: Strain and Sex Differences in the Use of Signals Associated with Non-Threat During the Acquisition and Extinction of Active-Avoidance Behavior." *Progress in Neuro-Psychopharmacology & Biological Psychiatry* (2011) 35:1659–70.

Beer, J.S., et al. "Orbitofrontal Cortex and Social Behavior: Integrating Self-Monitoring and Emotion-Cognition Interactions." *Journal of Cognitive Neuroscience* (2006) 18:871–79.

Behrmann, M., and D.C. Plaut. "Distributed Circuits, Not Circumscribed Centers, Mediate Visual Recognition." *Trends in Cognitive Science* (2013) 17:210–19.

Bekoff, M. *The Emotional Lives of Animals: A Leading Scientist Explores Animal Joy, Sorrow, and Empathy—And Why They Matter* (Novato, CA: New World Library, 2007).

Bell, A.M. "Approaching the Genomics of Risk-Taking Behavior." *Advances in Genetics* (2009) 68:83–104.

Bello-Medina, P.C., et al. "Differential Effects of Spaced vs. Massed Training in Long-Term Object-Identity and Object-Location Recognition Memory." *Behavioural Brain Research* (2013) 250:102–13.

Belzung, C., and G. Griebel. "Measuring Normal and Pathological Anxiety-Like Behaviour in Mice: A Review." *Behavioural Brain Research* (2001) 125:141–49.

Belzung, C., and M. Lemoine. "Criteria of Validity for Animal Models of Psychiatric Disorders: Focus on Anxiety Disorders and Depression." *Biology of Mood & Anxiety Disorders* (2011) 1:9.

Belzung, C., and P. Philippot. "Anxiety from a Phylogenetic Perspective: Is There a Qualitative Difference Between Human and Animal Anxiety?" *Neural Plasticity* (2007) 2007:59676.

Bem, D.J. "Self-Perception: An Alternative Interpretation to Cognitive Dissonance Phenomena." *Psychological Review* (1972) 74:183–200.

Bendesky, A., et al. "Catecholamine Receptor Polymorphisms Affect Decision-Making in *C. elegans.*" *Nature* (2011) 472:313–18.

Benison, S., and A.C. Barger. "Walter Bradford Cannon." In: *Dictionary of Scientific Biography*, vol. 15, ed. C.C. Gillispie (New York: Charles Scribner's Sons, 1978), 71–77.

Bentley, P., et al. "Cholinergic Enhancement Modulates Neural Correlates of Selective Attention and Emotional Processing." *NeuroImage* (2003) 20:58–70.

Bentz, D., et al. "Enhancing Exposure Therapy for Anxiety Disorders with Glucocorticoids: From Basic Mechanisms of Emotional Learning to Clinical Applications." *Journal of Anxiety Disorders* (2010) 24:223–30.

Beran, M.J., and J.D. Smith. "The Uncertainty Response in Animal-Metacognition Researchers." *The Journal of Comparative Psychology* (2014) 128:155–59.

Berg, H.C. "Bacterial Behaviour." *Nature* (1975) 254:389–92.

Berg, H.C. "Motile Behavior of Bacteria." *Physics Today* (2000) 53:24–29.

Berger, A., and M.I. Posner. "Pathologies of Brain Attentional Networks." *Neuroscience and Biobehavioral Reviews* (2000) 24:3–5.

Berger, S. "Conditioning Through Vicarious Instigation." *Psychological Review* (1962) 69:450–66.

Berger, T.W., B. Alger, and R.F. Thompson. "Neuronal Substrate of Classical Conditioning in the Hippocampus." *Science* (1976) 192:483–85.

Berggren, N., and N. Derakshan. "Attentional Control Deficits in Trait Anxiety: Why You See Them and Why You Don't." *Biological Psychology* (2013) 92:440–46.

Bernard, C. *An Introduction to the Study of Experimental Medicine* (New York: Dover Press, 1865/1957).

Berridge, C.W., and A.F. Arnsten. "Psychostimulants and Motivated Behavior: Arousal and Cognition." *Neuroscience and Biobehavioral Reviews* (2013) 37:1976–84.

Berridge, K.C. "Food Reward: Brain Substrates of Wanting and Liking." *Neuroscience and Biobehavioral Reviews* (1996) 20:1–25.

Berridge, K.C. "The Debate over Dopamine's Role in Reward: The Case for Incentive Salience." *Psychopharmacology (Berl)* (2007) 191:391–431.

Berridge, K.C., and M.L. Kringelbach. "Affective Neuroscience of Pleasure: Reward in Humans and Animals." *Psychopharmacology (Berl)* (2008) 199:457–80.

Berridge, K.C., and M.L. Kringelbach. "Building a Neuroscience of Pleasure and Well-Being." *Psychology of Well-Being* (2011) 1:1–3.

Berridge, K.C., and P. Winkielman. "What Is an Unconscious Emotion: The Case of Unconscious 'Liking.'" *Cognition and Emotion* (2003) 17:181–211.

Berrios, G.E., and I.S. Markova. "Is the Concept of 'Dimension' Applicable to Psychiatric Objects?" *World Psychiatry: Official Journal of the World Psychiatric Association* (2013) 12: 76–78.

Berthier, N.E., and J.W. Moore. "Disrupted Conditioned Inhibition of the Rabbit Nictitating Membrane Response Following Mesencephalic Lesions." *Physiology & Behavior* (1980) 25:667–73.

Besnard, A., J. Caboche, and S. Laroche. "Reconsolidation of Memory: A Decade of Debate." *Progress in Neurobiology* (2012) 99:61–80.

Bevilacqua, L., and D. Goldman. "Genetics of Impulsive Behaviour." *Philosophical Transactions of the Royal Society B: Biological Sciences* (2013) 368:20120380.

Bhatnagar, S., N. Shanks, and M.J. Meaney. "Plaque-Forming Cell Responses and Antibody Titers Following Injection of Sheep Red Blood Cells in Nonstressed, Acute, and/or Chronically Stressed Handled and Nonhandled Animals." *Developmental Psychobiology* (1996) 29: 171–81.

Bijak, M. "Monoamine Modulation of the Synaptic Inhibition in the Hippocampus." *Acta Neurobiologiae Experimentalis* (1996) 56:385–95.

Bindra, D. "Neuropsychological Interpretation of the Effects of Drive and Incentive-Motivation on General Activity and Instrumental Behavior." *Psychological Review* (1968) 75:1–22.

Bindra, D. "The Interrelated Mechanisms of Reinforcement and Motivation, and the Nature of Their Influence on Response." In: *Nebraska Symposium on Motivation*, W.J. Arnold and D. Levine, eds. (Lincoln: University of Nebraska Press, 1969), 1–33.

Bindra, D. "A Unified Interpretation of Emotion and Motivation." *Annals of the New York Academy of Sciences* (1969) 159:1071–83.

Bindra, D. "A Motivational View of Learning, Performance, and Behavior Modification." *Psychological Review* (1974) 81:199–213.

Bishop, S.J. "Neurocognitive Mechanisms of Anxiety: An Integrative Account." *Trends in Cognitive Sciences* (2007) 11:307–16.

Bishop, S.J., et al. "COMT Genotype Influences Prefrontal Response to Emotional Distraction." *Cognitive, Affective & Behavioral Neuroscience* (2006) 6:62–70.

Blackford, J.U., and D.S. Pine. "Neural Substrates of Childhood Anxiety Disorders: A Review of Neuroimaging Findings." *Child and Adolescent Psychiatric Clinics of North America* (2012) 21:501–25.

Blanchard, D.C., and R.J. Blanchard. "Innate and Conditioned Reactions to Threat in Rats with Amygdaloid Lesions." *Journal of Comparative and Physiological Psychology* (1972) 81:281–90.

Blanchard, D.C., and R.J. Blanchard. "Ethoexperimental Approaches to the Biology of Emotion." *Annual Review of Psychology* (1988) 39:43–68.

Blanchard, R.J., and D.C. Blanchard. "Crouching as an Index of Fear." *Journal of Comparative and Physiological Psychology* (1969) 67:370–75.

Blanchard, R.J., D.C. Blanchard, and K. Hor. "An Ethoexperimental Approach to the Study of Defense." In: *Ethoexperimental Approaches to the Study of Behavior*, vol. 48, eds. R.J. Blanchard et al. (Dordrecht, Netherlands: Kluwer Academic, 1989), 114–36.

Blanchard, R.J., K.K. Fukunaga, and D.C. Blanchard. "Environmental Control of Defensive Reactions to Footshock." *Bulletin of the Psychonomic Society* (1976) 8:129–30.

Blechert, J., et al. "See What You Think: Reappraisal Modulates Behavioral and Neural Responses to Social Stimuli." *Psychological Science* (2012) 23:346–53.

Blessing W.W. *The Lower Brainstem and Bodily Homeostasis.* (New York: Oxford University Press, 1997).

Bliss, T.V., and G.L. Collingridge. "A Synaptic Model of Memory: Long-Term Potentiation in the Hippocampus." *Nature* (1993) 361:31–39.

Block, N. "Consciousness and Accessibility." *Behavioral and Brain Sciences* (1990) 13:596–98.

Block, N. "Begging the Question Against Phenomenal Consciousness." *Behavioral and Brain Sciences* (1992) 15:205–6.

Block, N. "How Many Concepts of Consciousness?" *Behavioral and Brain Sciences* (1995) 18: 272–84.

Block, N. "On a Confusion About a Function of Consciousness." *Behavioral and Brain Sciences* (1995) 18:227–47.

Block, N. "Concepts of Consciousness." In: *Philosophy of Mind: Classical and Contemporary Readings*, ed. D. Chalmers (New York: Oxford University Press, 2002), 206–18.

Block, N. "Two Neural Correlates of Consciousness." *Trends in Cognitive Sciences* (2005) 9:46–52.

Block, N. "Consciousness, Accessibility, and the Mesh Between Psychology and Neuroscience." *Behavioral and Brain Sciences* (2007) 30:481–99; discussion 499–548.

Bloom, P. "Language and Thought: Does Grammar Make Us Smart?" *Current Biology* (2000) 10:R516–17.

Blumberg, M.S. "On the Origins of Complex Behaviors: From Innateness to Epigenesis" (keynote address). In: Conference Entitled "Hormonal Control of Circuits for Complex Behaviors," Janelia Farm Research Campus, Howard Hughes Medical Institute, Ashburn, Virginia, October 27–30, 2013.

Boardman, J., ed. *The Oxford History of Classical Art* (Oxford, UK: Oxford University Press, 1993).

Bolles, R.C. *Theory of Motivation* (New York: Harper and Row, 1967).

Bolles, R.C. "Species-Specific Defense Reactions and Avoidance Learning." *Psychological Review* (1970) 77:32–48.

Bolles, R.C. "The Avoidance Learning Problem." In: *The Psychology of Learning and Motivation*, vol. 6, ed. G.H. Bower (New York: Academic Press, 1972) 97–145.

Bolles, R.C. "The Role of Stimulus Learning in Defensive Behavior." In: *Cognitive Processes in Animal Behavior*, eds. S.H. Hulse et al. (Hillsdale, NJ: Erlbaum, 1978), 89–107.

Bolles, R.C., and A.C. Collier. "The Effect of Predictive Cues on Freezing in Rats." *Animal Learning & Behavior* (1976) 4:6–8.

Bolles, R.C., and M.S. Fanselow. "A Perceptual-Defensive-Recuperative Model of Fear and Pain." *Behavioral and Brain Sciences* (1980) 3:291–323.

Bonanno, G.A., and C.L. Burton. "Regulatory Flexibility: An Individual Differences Perspective on Coping and Emotion Regulation." *Perspectives on Psychological Science* (2013) 8:591–612.

Bonham, A.J., and C. Gonzalez-Vallejo. "Assessment of Calibration for Reconstructed Eye-Witness Memories." *Acta Psychologica* (2009) 131:34–52.

Bor, D., and A.K. Seth. "Consciousness and the Prefrontal Parietal Network: Insights from Attention, Working Memory, and Chunking." *Frontiers in Psychology* (2012) 3:63.

Boring, E.G. *A History of Experimental Psychology* (New York: Appleton-Century-Crofts, 1950).

Borkovec, T.D., O.M. Alcaine, and E. Behar. "Avoidance Theory of Worry and Generalized Anxiety Disorder." In: *Generalized Anxiety Disorders: Advances in Research and Practice*, eds. R.G. Heimberg et al. (New York: Guilford Press, 2004), 77–108.

Borkovec, T.D., H. Hazlett-Stevens, and M.L. Diaz. "The Role of Positive Beliefs About Worry in Generalized Anxiety Disorder and Its Treatment." *Clinical Psychology & Psychotherapy* (1999) 6:126–38.

Borkovec, T.D., W.J. Ray, and J. Stober. "Worry: A Cognitive Phenomenon Intimately Linked to Affective, Physiological, and Interpersonal Behavioral Processes." *Cognitive Therapy and Research* (1998) 22:561–76.

Bornemann, B., P. Winkielman, and E. van der Meer. "Can You Feel What You Do Not See? Using Internal Feedback to Detect Briefly Presented Emotional Stimuli." *International Journal of Psychophysiology: Official Journal of the International Organization of Psychophysiology* (2012) 85:116–24.

Bos, M.G., et al. "Stress Enhances Reconsolidation of Declarative Memory." *Psychoneuroendocrinology* (2014) 46:102–13.

Botvinick, M.M., et al. "Conflict Monitoring and Cognitive Control." *Psychological Review* (2001) 108:624–52.

Boulougouris, J.C., and I.M. Marks. "Implosion (Flooding)—A New Treatment for Phobias." *British Medical Journal* (1969) 2:721–23.

Bouret, S., et al. "Phasic Activation of Locus Ceruleus Neurons by the Central Nucleus of the Amygdala." *Journal of Neuroscience* (2003) 23:3491–97.

Bouret, S., and S.J. Sara. "Network Reset: A Simplified Overarching Theory of Locus Coeruleus Noradrenaline Function." *Trends in Neurosciences* (2005) 28:574–82.

Bourne, J.N., and K.M. Harris. "Nanoscale Analysis of Structural Synaptic Plasticity." *Current Opinion in Neurobiology* (2012) 22:372–82.

Bouton, M.E. "Context and Ambiguity in the Extinction of Emotional Learning: Implications for Exposure Therapy." *Behaviour Research and Therapy* (1988) 26:137–49.

Bouton, M.E. "Context, Time, and Memory Retrieval in the Interference Paradigms of Pavlovian Learning." *Psychological Bulletin* (1993) 114:80–99.

Bouton, M.E. "A Learning Theory Perspective on Lapse, Relapse, and the Maintenance of Behavior Change." *Health Psychology* (2000) 19:57–63.

Bouton, M.E. "Context, Ambiguity, and Unlearning: Sources of Relapse after Behavioral Extinction." *Biological Psychiatry* (2002) 52:976–86.

Bouton, M.E. "Context and Behavioral Processes in Extinction." *Learning & Memory* (2004) 11:485–94.

Bouton, M.E. "Behavior Systems and the Contextual Control of Anxiety, Fear, and Panic." In: *Emotion and Consciousness*, eds. L.F. Barrett et al. (New York: Guilford Press, 2005), 205–30.

Bouton, M.E. *Learning and Behavior: A Contemporary Synthesis* (Sunderland, MA: Sinauer Associates, 2007).

Bouton, M.E. "Why Behavior Change Is Difficult to Sustain." *Preventive Medicine* (2014) 68: 29–36.

Bouton, M.E., and R.C. Bolles. "Contextual Control of Extinction of Conditioned Fear." *Journal of Experimental Psychology: Animal Behavior Processes* (1979) 10:445–66.

Bouton, M.E., and R.C. Bolles. "Conditioned Fear Assessed by Freezing and by the Suppression of Three Different Baselines." *Animal Learning and Behavior* (1980) 8:429–34.

Bouton, M.E., and J.B. Nelson. "Context-Specificity of Target Versus Feature Inhibition in a Feature-Negative Discrimination." *Journal of Experimental Psychology: Animal Behavior Processes* (1994) 20:51–65.

Bouton, M.E., and D.A. King. "Contextual Control of the Extinction of Conditioned Fear: Tests for the Associative Value of the Context." *Journal of Experimental Psychology: Animal Behavior Processes* (1983) 9:248–65.

Bouton, M.E., S. Mineka, and D.H. Barlow. "A Modern Learning Theory Perspective on the Etiology of Panic Disorder." *Psychological Review* (2001) 108:4–32.

Bouton, M.E., et al. "Effects of Contextual Conditioning and Unconditional Stimulus Presentation on Performance in Appetitive Conditioning." *Quarterly Journal of Experimental Psychology* (1993) 46B:63–95.

Bouton, M.E., et al. "Contextual and Temporal Modulation of Extinction: Behavioral and Biological Mechanisms." *Biological Psychiatry* (2006) 60:352–60.

Bouton, M.E., A.M. Woods, and O. Pineno. "Occasional Reinforced Trials During Extinction Can Slow the Rate of Rapid Reacquisition." *Learning and Motivation* (2004) 35:371–90.

Bowerman, M., and S.C. Levinson, eds. *Language Acquisition and Conceptual Development* (Cambridge, UK: Cambridge University Press, 2001).

Bownds, M.D. *The Biology of Mind: Origins and Structures of Mind, Brain, and Consciousness.* (New York: John Wiley and Sons, 1999).

Boyden, E.S., et al. "Millisecond-Timescale, Genetically Targeted Optical Control of Neural Activity." *Nature Neuroscience* (2005) 8:1263–68.

Bracha, H.S., et al. "Does 'Fight or Flight' Need Updating?" *Psychosomatics* (2004) 45:448–49.

Bradley, B.P., et al. "Attentional Bias for Emotional Faces in Generalized Anxiety Disorder." *The British Journal of Clinical Psychology/The British Psychological Society* (1999) 38 (Pt 3): 267–78.

Brady, J.V., and H.F. Hunt. "An Experimental Approach to the Analysis of Emotional Behavior." *Journal of Psychology* (1955) 40:313–24.

Brain, P.F., et al., eds. *Fear and Defense* (London: Harwood Academic, 1990).

Brandt, C.A., J. Meller, L. Keweloh, K. Höschel, J. Staedt, D. Munz, and G. Stoppe. "Increased Benzodiazepine Receptor Density in the Prefrontal Cortex in Patients with Panic Disorder." *Journal of Neural Transmission* (1998) 105: 1325–33.

Bray, S., et al. "The Neural Mechanisms Underlying the Influence of Pavlovian Cues on Human Decision Making." *Journal of Neuroscience* (2008) 28:5861–66.

Bremner, J.D. "Traumatic Stress: Effects on the Brain." *Dialogues in Clinical Neuroscience* (2006) 8:445–61.

Brentano, F. *Psychologie vom empirischen Standpunkt* (Leipzig: Felix Meiner, 1874/1924).

Breuer, J., and S. Freud. *Studies on Hysteria* (New York: Hogarth Press, 1893–1895).

Breviglieri, C.P., et al. "Predation-Risk Effects of Predator Identity on the Foraging Behaviors of Frugivorous Bats." *Oecologia* (2013) 173:905–12.

Brewin, C.R. "A Cognitive Neuroscience Account of Posttraumatic Stress Disorder and Its Treatment." *Behaviour Research and Therapy* (2001) 39:373–93.

Bridgeman, B. "On the Evolution of Consciousness and Language." *PSYCOLOQUY* (1992) 3.

Brodal, A. *Neurological Anatomy* (New York: Oxford University Press, 1982).

Brooks, D.C., and M.E. Bouton. "A Retrieval Cue for Extinction Attenuates Spontaneous Recovery." *Journal of Experimental Psychology: Animal Behavior Processes* (1993) 19:77–89.

Brooks, S.J., et al. "Exposure to Subliminal Arousing Stimuli Induces Robust Activation in the Amygdala, Hippocampus, Anterior Cingulate, Insular Cortex and Primary Visual Cortex: A Systematic Meta-Analysis of fMRI Studies." *NeuroImage* (2012) 59:2962–73.

Brown, J.S., and I.E. Farber. "Emotions Conceptualized as Intervening Variables—With Suggestions Toward a Theory of Frustration." *Psychological Bulletin* (1951) 48:465–95.

Brown, T.H., E.W. Kairiss, and C.L. Keenan. "Hebbian Synapses: Biophysical Mechanisms and Algorithms." *Annual Review of Neuroscience* (1990) 13:475–511.

Brown, T.M., and E. Fee. "Walter Bradford Cannon—Pioneer Physiologist of Human Emotions." *American Journal of Public Health* (2002) 92:1594–95.

Bruhl, A.B., et al. "Neuroimaging in Social Anxiety Disorder: A Meta-Analytic Review Resulting in a New Neurofunctional Model." *Neuroscience and Biobehavioral Reviews* (2014) 47C:260–80.

Brunet, A., et al. "Does Reconsolidation Occur in Humans: A Reply." *Frontiers in Behavioral Neuroscience* (2011) 5:74.

Brunet, A., S.P. Orr, J. Tremblay, K. Robertson, K. Nader, and R.K. Pitman. "Effect of Post-Retrieval Propranolol on Psychophysiologic Responding During Subsequent Script-Driven Traumatic Imagery in Post-Traumatic Stress Disorder." *Journal of Psychiatric Research* (2008) 42:503–6.

Buchel, C., and R.J. Dolan. "Classical Fear Conditioning in Functional Neuroimaging." *Current Opinion in Neurobiology* (2000) 10:219–23.

Buchel, C., et al. "Brain Systems Mediating Aversive Conditioning: An Event-Related fMRI Study." *Neuron* (1998) 20:947–57.

Buck, R. "Nonverbal Behavior and the Theory of Emotion: The Facial Feedback Hypothesis." *Journal of Personality and Social Psychology* (1980) 38:811–24.

Buckley, T.C., E.B. Blanchard, and E.J. Hickling. "Automatic and Strategic Processing of Threat Stimuli: A Comparison Between PTSD, Panic Disorder, and Non-anxiety Controls." *Cognitive Therapy and Research* (2002) 26:97–115.

Buckner, R.L. "The Brain's Default Network: Origins and Implications for the Study of Psychosis." *Dialogues in Clinical Neuroscience* (2013) 15:351–58.

Buckner, R.L., and W. Koutstaal. "Functional Neuroimaging Studies of Encoding, Priming, and Explicit Memory Retrieval." *Proceedings of the National Academy of Sciences of the United States of America* (1998) 95:891–98.

Buckner, R.L., et al. "Cognitive Neuroscience of Episodic Memory Encoding." *Acta Psychologica* (2000) 105:127–39.

Buhle, J.T., et al. "Cognitive Reappraisal of Emotion: A Meta-Analysis of Human Neuroimaging Studies." *Cerebral Cortex* (2014) 24:2981–90.

Burge, T. "Reflections on Two Kinds of Consciousness." In: *Philosophical Essays, Vol II: Foundations of Mind*, ed. T. Burge (Oxford, UK: Oxford University Press, 2006), 392–419.

Burgess, N., and J. O'Keefe. "Models of Place and Grid Cell Firing and Theta Rhythmicity." *Current Opinion in Neurobiology* (2011) 21:734–44.

Burghardt, G.M. "Animal Awareness. Current Perceptions and Historical Perspective." *The American Psychologist* (1985) 40:905–19.

Burghardt, G.M. "Ground Rules for Dealing with Anthropomorphism." *Nature* (2004) 430:15.

Burghardt, N.S., et al. "Acute Selective Serotonin Reuptake Inhibitors Increase Conditioned Fear Expression: Blockade with a 5-HT(2C) Receptor Antagonist." *Biological Psychiatry* (2007) 62:1111–18.

Burghardt, N.S., et al. "Chronic Antidepressant Treatment Impairs the Acquisition of Fear Extinction." *Biological Psychiatry* (2013) 73:1078–86.

Burghardt, N.S., et al. "The Selective Serotonin Reuptake Inhibitor Citalopram Increases Fear After Acute Treatment but Reduces Fear with Chronic Treatment: A Comparison with Tianeptine." *Biological Psychiatry* (2004) 55:1171–78.

Bush, D.E., F. Sotres-Bayon, and J.E. LeDoux. "Individual Differences in Fear: Isolating Fear Reactivity and Fear Recovery Phenotypes." *Journal of Traumatic Stress* (2007) 20:413–22.

Butler, A.B., and W. Hodos. *Comparative Vertebrate Neuroanatomy: Evolution and Adaptation* (Hoboken, NJ: John Wiley & Sons, 2005).

Butler, G., and A. Mathews. "Cognitive Processes in Anxiety." *Advances in Behaviour Research and Therapy* (1983) 5:51–62.

Butler, G., and A. Mathews. "Anticipatory Anxiety and Risk Perception." *Cognitive Therapy and Research* (1987) 11:551–65.

Butler, L.D., et al. "Hypnosis Reduces Distress and Duration of an Invasive Medical Procedure for Children." *Pediatrics* (2005) 115:E77–85.

Buzsaki, G. "Network Properties of Memory Trace Formation in the Hippocampus." *Bollettino Della Societa Italiana Di Biologia Sperimentale* (1991) 67:817–35.

Buzsaki, G. *Rhythms of the Brain* (Oxford, UK: Oxford University Press, 2011).

Byrne, J.H., et al. "Roles of Second Messenger Pathways in Neuronal Plasticity and in Learning and Memory. Insights Gained from *Aplysia*." *Advances in Second Messenger and Phosphoprotein Research* (1993) 27:47–108.

Cabanac, M. "On the Origin of Consciousness, a Postulate and Its Corollary." *Neuroscience and Biobehavioral Reviews* (1996) 20:33–40.

Cabeza, R., E. Ciaramelli, and M. Moscovitch. "Cognitive Contributions of the Ventral Parietal Cortex: An Integrative Theoretical Account." *Trends in Cognitive Sciences* (2012) 16:338–52.

Cabeza, R., and L. Nyberg. "Neural Bases of Learning and Memory: Functional Neuroimaging Evidence." *Current Opinion in Neurology* (2000) 13:415–21.

Cahill, L., and J.L. McGaugh. "Modulation of Memory Storage." *Current Opinion in Neurobiology* (1996) 6:237–42.

Cai, W.H., et al. "Postreactivation Glucocorticoids Impair Recall of Established Fear Memory." *Journal of Neuroscience* (2006) 26:9560–66.

Cain, C.K., A.M. Blouin, and M. Barad. "Temporally Massed CS Presentations Generate More Fear Extinction Than Spaced Presentations." *Journal of Experimental Psychology: Animal Behavior Processes* (2003) 29:323–33.

Cain, C.K., J.S. Choi, and J.E. LeDoux. "Active Avoidance and Escape Learning." In: *Encyclopedia of Behavioral Neuroscience*, eds. G. Koob et al. (New York: Elsevier, 2010).

Cain, C.K., and J.E. LeDoux. "Escape from Fear: A Detailed Behavioral Analysis of Two Atypical Responses Reinforced by CS Termination." *Journal of Experimental Psychology: Animal Behavior Processes* (2007) 33:451–63.

Cain, C.K., and J.E. LeDoux. "Brain Mechanisms of Pavlovian and Instrumental Aversive Conditioning." In: *Handbook of Anxiety and Fear*, eds. R.J. Blanchard et al. (Waltham, MA: Academic Press, 2008), 103–24.

Campese, V., et al. "Development of an Aversive Pavlovian-to-Instrumental Transfer Task in Rat." *Frontiers in Behavioral Neuroscience* (2013) 7:176.

Campese, V.D., et al. "Lesions of Lateral or Central Amygdala Abolish Aversive Pavlovian-to-Instrumental Transfer in Rats." *Frontiers in Behavioral Neuroscience* (2014) 8:161.

Campos, A.C., et al. "Animal Models of Anxiety Disorders and Stress." *Revista Brasileira de Psiquiatria* (2013) 35(Suppl 2):S101–11.

Candland, D.K., et al. *Emotion* (Belmont, CA: Wadsworth, 1977).

Cannon, W.B. *Bodily Changes in Pain, Hunger, Fear, and Rage* (New York: Appleton, 1929).

Cannon, W.B. "Again the James-Lange and the Thalamic Theories of Emotion." *Psychological Review* (1931) 38:281–95.

Cannon, W.B. "The Role of Emotions in Disease." *Annals of Internal Medicine* (1936) 9:1453–65.

Cannon, W.B., and S.W. Britton. "Pseudoaffective Medulliadrenal Secretion." *American Journal of Physiology* (1925) 72:283–94.

Caporael, L.R., and C.M. Heyes. "Why Anthropomorphize? Folk Psychology and Other Stories." In: *Anthropomorphism, Anecdotes, and Animals,* eds. R.W. Mitchell et al. (Albany: SUNY Press, 1977) 59–73.

Cardinal, R.N., et al. "Emotion and Motivation: The Role of the Amygdala, Ventral Striatum, and Prefrontal Cortex." *Neuroscience and Biobehavioral Reviews* (2002) 26:321–52.

Carey B. "LSD Reconsidered for Therapy." *New York Times*, March 3, 2015. Retrieved on Feb. 21, 2014, from http://www.nytimes.com/2014/03/04/health/lsd-reconsidered-for-therapy.html?_r=0.

Carew, M.B., and J.W. Rudy. "Multiple Functions of Context During Conditioning: A Developmental Analysis." *Developmental Psychobiology* (1991) 24:191–209.

Carew, T.J., H.M. Pinsker, and E.R. Kandel. "Long-Term Habituation of a Defensive Withdrawal Reflex in *Aplysia*." *Science* (1972) 175:451–54.

Carew, T.J., E.T. Walters, and E.R. Kandel. "Associative Learning in *Aplysia*: Cellular Correlates Supporting a Conditioned Fear Hypothesis." *Science* (1981) 211:501–4.

Carleton, R.N. "The Intolerance of Uncertainty Construct in the Context of Anxiety Disorders: Theoretical and Practical Perspectives." *Expert Review of Neurotherapeutics* (2012) 12:937–47.

Carretie, L. "Exogenous (Automatic) Attention to Emotional Stimuli: A Review." *Cognitive, Affective & Behavioral Neuroscience* (2014) 14:1228–58.

Carruthers, P. *Language, Thought and Consciousness: An Essay in Philosophical Psychology* (Cambridge, UK: Cambridge University Press, 1996).

Carruthers, P. "The Cognitive Functions of Language." *Behavioral and Brain Sciences* (2002) 25:657–74; discussion 674–725.

Carruthers, P. "How We Know Our Own Minds: The Relationship Between Mindreading and Metacognition." *Behavioral and Brain Sciences* (2009) 32:121–138; discussion 138–82.

Carruthers, P. "Unconsciously Competing Goals Can Collaborate or Compromise as Well as Win or Lose." *Behavioral and Brain Sciences* (2014) 37:139–40.

Carson, W.H., H. Kitagawa, and C.B. Nemeroff. "Drug Development for Anxiety Disorders: New Roles for Atypical Antipsychotics." *Psychopharmacology Bulletin* (2004) 38(Suppl 1):38–45.

Carter, C.S., M.M. Botvinick, and J.D. Cohen. "The Contribution of the Anterior Cingulate Cortex to Executive Processes in Cognition." *Reviews in the Neurosciences* (1999) 10:49–57.

Casey, B.J., et al. "The Storm and Stress of Adolescence: Insights from Human Imaging and Mouse Genetics." *Developmental Psychobiology* (2010) 52:225–35.

Casey, B.J., et al. "Transitional and Translational Studies of Risk for Anxiety." *Depression and Anxiety* (2011) 28:18–28.

Casey B.J., R.M. Jones, and L.H. Somerville. "Braking and Accelerating of the Adolescent Brain." *Journal of Research on Adolescence.* (2011) 21:21–33.

Castro, D.C., and K.C. Berridge. "Advances in the Neurobiological Bases for Food 'Liking' Versus 'Wanting.'" *Physiology & Behavior* (2014) 136:22–30.

Chafe, W.L. "How Consciousness Shapes Language." *Pragmatics and Cognition* (1996) 4:35–54.

Chalmers, D. *The Conscious Mind* (New York: Oxford University Press, 1996).

Chalmers, D.J. "Panpsychism and Panprotopsychism." Amherst Lecture in Philosophy 2013. Also in (T. Alter and Y. Nagasawa, eds) *Russellian Monism* (New York: Oxford University Press, 2013); and in (G. Bruntrup and L. Jaskolla, eds.) *Panpsychism* (New York: Oxford University Press).

Chalmers, D.J. *Constructing the World* (New York: Oxford University Press, 2014).

Chamberlin, T.C. "The Method of Multiple Working Hypotheses." *Science* (1890) 15:92–96 (reprinted 1965, 148:754–59).

Chan, J.C., and J.A. LaPaglia. "Impairing Existing Declarative Memory in Humans by Disrupting Reconsolidation." *Proceedings of the National Academy of Sciences of the United States of America* (2013) 110:9309–13.

Charney, D.S. "Neuroanatomical Circuits Modulating Ear and Anxiety Behaviors." *Acta Psychiat Scand Suppl* (2003) 417:38–50.

Chen, C.C., et al. "Visualizing Long-Term Memory Formation in Two Neurons of the *Drosophila* Brain." *Science* (2012) 335:678–85.

Chiao, J.Y., et al. "Cultural Specificity in Amygdala Response to Fear Faces." *Journal of Cognitive Neuroscience* (2008) 20:2167–74.

Chiapponi, C., et al. "Age-Related Brain Trajectories in Schizophrenia: A Systematic Review of Structural MRI Studies." *Psychiatry Research* (2013) 214:83–93.

Chiesa, A., and A. Serretti. "Mindfulness Based Cognitive Therapy for Psychiatric Disorders: A Systematic Review and Meta-Analysis." *Psychiatry Research* (2011) 187:441–53.

Chiesa, A., A. Serretti, and J.C. Jakobsen. "Mindfulness: Top-Down or Bottom-Up Emotion Regulation Strategy?" *Clinical Psychology Review* (2013) 33:82–96.

Chittka, L., and K. Jensen. "Animal Cognition: Concepts from Apes to Bees." *Current Biology* (2011) 21:R116–19.

Chivers, D.P., G.E. Brown, and R.J.F. Smith. "The Evolution of Chemical Alarm Signals: Attracting Predators Benefits Alarm Signal Senders." *American Naturalists* (1996) 148: 649–59.

Cho, S.B., B.J. Baars, and J. Newman. "A Neural Global Workspace Model for Conscious Attention." *Neural Networks: The Official Journal of the International Neural Network Society* (1997) 10:1195–1206.

Choi, J.S., C.K. Cain, and J.E. LeDoux. "The Role of Amygdala Nuclei in the Expression of Auditory Signaled Two-Way Active Avoidance in Rats." *Learning & Memory* (2010) 17:139–47.

Chugh-Gupta, N., F.G. Baldassarre, and B.H. Vrkljan. "A Systematic Review of Yoga for State Anxiety: Considerations for Occupational Therapy." *Canadian Journal of Occupational Therapy* (2013) 80:150–70.

Chun, M.M., and E.A. Phelps. "Memory Deficits for Implicit Contextual Information in Amnesic Subjects with Hippocampal Damage." *Nature Neuroscience* (1999) 2:844–47.

Chung, Y.A., et al. "Alterations in Cerebral Perfusion in Posttraumatic Stress Disorder Patients Without Re-Exposure to Accident-Related Stimuli." *Clinical Neurophysiology: Official Journal of the International Federation of Clinical Neurophysiology* (2006) 117:637–42.

Church, R.M. "Systematic Effect of Random Error in the Yoked Control Design." *Psychological Bulletin* (1964) 62:122–31.

Church, R.M., et al. "Cardiac Responses to Shock in Curarized Dogs: Effects of Shock Intensity and Duration, Warning Signal, and Prior Experience with Shock." *Journal of Comparative and Physiological Psychology* (1966) 62:1–7.

Churchland, P. "Reduction and the Neurobiological Basis of Consciousness." In: *Consciousness in Contemporary Science*, eds. A. Marcel and E. Bisiach (Oxford, UK: Oxford University Press, 1988).

Churchland, P.M. *Matter and Consciousness* (Cambridge, MA: MIT Press, 1984).

Churchland, P.M. "Folk Psychology and the Explanation of Human Behavior." *Proceedings of the Aristotelian Society* (1988) 62:209–21.

Churchland, P.S. *Neurophilosophy: Toward a Unified Science of the Mind-Brain* (Cambridge, MA: MIT Press, 1986).

Churchland, P.S. *Touching a Nerve: The Self as Brain* (New York: W.W. Norton, 2013).

Ciocchi, S., et al. "Encoding of Conditioned Fear in Central Amygdala Inhibitory Circuits." *Nature* (2010) 468:277–82.

Cisler, J.M., and E.H. Koster. "Mechanisms of Attentional Biases Towards Threat in Anxiety Disorders: An Integrative Review." *Clinical Psychology Review* (2010) 30:203–16.

Clark, A. *Being There* (Cambridge, MA: MIT Press, 1998).

Clark, D.A., and A.T. Beck. *Cognitive Therapy of Anxiety Disorders* (New York: Guilford Press, 2010).

Clark, D.M. "A Cognitive Approach to Panic." *Behaviour Research and Therapy* (1986) 24:461–70.

Clark, D.M. "A Cognitive Model of Panic." In: *Panic: Psychological Perspective*, eds. S. Rachman and J.D. Maser (Hillsdale, NJ: Erlbaum, 1988), 71–89.

Clark, D.M. "Panic Disorder and Social Phobia." In: *Science and Practice of Cognitive Behaviour Therapy*, eds. D.M. Clark and C. Fairburn (Oxford, UK: Oxford University Press, 1997).

Clark, D.M. "Anxiety Disorders: Why They Persist and How to Treat Them." *Behaviour Research and Therapy* (1999) 37(Suppl 1):S5–27.

Clark, D.M., et al. "Misinterpretation of Body Sensations in Panic Disorder." *Journal of Consulting and Clinical Psychology* (1997) 65:203–13.

Clark, L., et al. "Differential Effects of Insular and Ventromedial Prefrontal Cortex Lesions on Risky Decision-Making." *Brain: A Journal of Neurology* (2008) 131:1311–22.

Clayton, N. "Animal Cognition: Crows Spontaneously Solve a Metatool Task." *Current Biology* (2007) 17:R894–95.

Clayton, N.S., T.J. Bussey, and A. Dickinson. "Can Animals Recall the Past and Plan for the Future?" *Nature Reviews Neuroscience* (2003) 4:685–91.

Clayton, N.S., and A. Dickinson. "Episodic-Like Memory During Cache Recovery by Scrub Jays." *Nature* (1998) 395:272–74.

Cleeremans, A. "Consciousness: The Radical Plasticity Thesis." *Progress in Brain Research* (2008) 168:19–33.

Cleeremans, A. "The Radical Plasticity Thesis: How the Brain Learns to Be Conscious." *Frontiers in Psychology* (2011) 2:86.

Clem, R.L., and R.L. Huganir. "Calcium-Permeable AMPA Receptor Dynamics Mediate Fear Memory Erasure." *Science* (2010) 330:1108–12.

Clement, Y., F. Calatayud, and C. Belzung. "Genetic Basis of Anxiety-Like Behaviour: A Critical Review." *Brain Research Bulletin* (2002) 57:57–71.

Clore, G. "Why Emotions Are Never Unconscious." In: *The Nature of Emotion: Fundamental Questions*, eds. P. Ekman and R.J. Davidson (New York: Oxford University Press, 1994), 285–90.

Clore, G., and T. Ketelaar. "Minding Our Emotions. On the Role of Automatic Unconscious Affect." In: *Advances in Social Cognition*, vol. 10, ed. R.S. Wyer (Mahwah, NJ: Erlbaum, 1997), 105–20.

Clore, G.L., and A. Ortony. "Psychological Construction in the OCC Model of Emotion." *Emotion Review* (2013) 5:335–43.

Cochran, D.M., et al. "The Role of Oxytocin in Psychiatric Disorders: A Review of Biological and Therapeutic Research Findings." *Harvard Review of Psychiatry* (2013) 21:219–47.

Cofer, C.N. *Motivation and Emotion* (Glenview, IL: Scott Foresman, 1972).

Cohen, D.H. "Involvement of the Avian Amygdalar Homologue (Archistriatum Posterior and Mediale) in Defensively Conditioned Heart Rate Change." *Journal of Comparative Neurology* (1975) 160:13–35.

Cohen, D.H. "Identification of Vertebrate Neurons Modified During Learning: Analysis of Sensory Pathways." In: *Primary Neural Substrates of Learning and Behavioral Change*, eds. D.L. Alkon and J. Farley (Cambridge, UK: Cambridge Press, 1984).

Cohen, M.A., et al. "The Attentional Requirements of Consciousness." *Trends in Cognitive Sciences* (2012) 16:411–17.

Cohen, N.J., and L. Squire. "Preserved Learning and Retention of Pattern-Analyzing Skill in Amnesia: Dissociation of Knowing How and Knowing That." *Science* (1980) 210:207–9.

Congdon, E., and T. Canli. "A Neurogenetic Approach to Impulsivity." *Journal of Personality* (2008) 76:1447–84.

Conway, M.A. "Memory and the Self." *Journal of Memory and Language* (2005) 53:594–628.

Coover, G.D., et al. "Corticosterone Responses, Hurdle-Jump Acquisition, and the Effects of Dexamethasone Using Classical Conditioning of Fear." *Hormones and Behavior* (1978) 11:279–94.

Corbit, L.H., and B.W. Balleine. "Double Dissociation of Basolateral and Central Amygdala Lesions on the General and Outcome-Specific Forms of Pavlovian-Instrumental Transfer." *Journal of Neuroscience* (2005) 25:962–70.

Corbit, L.H., and B.W. Balleine. "The General and Outcome-Specific Forms of Pavlovian-Instrumental Transfer Are Differentially Mediated by the Nucleus Accumbens Core and Shell." *Journal of Neuroscience* (2011) 31:11786–94.

Corbit, L.H., and P.H. Janak. "Ethanol-Associated Cues Produce General Pavlovian-Instrumental Transfer." *Alcoholism: Clinical and Experimental Research* (2007) 31:766–74.

Corbit, L.H., and P.H. Janak. "Inactivation of the Lateral but Not Medial Dorsal Striatum Eliminates the Excitatory Impact of Pavlovian Stimuli on Instrumental Responding." *Journal of Neuroscience* (2007) 27:13977–81.

Corbit, L.H., P.H. Janak, and B.W. Balleine. "General and Outcome-Specific Forms of Pavlovian-Instrumental Transfer: The Effect of Shifts in Motivational State and Inactivation of the Ventral Tegmental Area." *European Journal of Neuroscience* (2007) 26:3141–49.

Corcoran, K.A., and G.J. Quirk. "Recalling Safety: Cooperative Functions of the Ventromedial Prefrontal Cortex and the Hippocampus in Extinction." *CNS Spectrums* (2007) 12:200–6.

Corkin, S. "Acquisition of Motor Skill After Bilateral Medial Temporal Lobe Excision." *Neuropsychologia* (1968) 6:255–65.

Cosmides, L., and J. Tooby. "Evolutionary Psychology." In: *Encyclopedia of Cognitive Science* (Cambridge, MA: MIT Press, 1999), 295–97.

Cosmides, L., and J. Tooby. "Evolutionary Psychology: New Perspectives on Cognition and Motivation." *Annual Review of Psychology* (2013) 64:201–29.

Cotterill, R.M. "Cooperation of the Basal Ganglia, Cerebellum, Sensory Cerebrum and Hippocampus: Possible Implications for Cognition, Consciousness, Intelligence and Creativity." *Progress in Neurobiology* (2001) 64:1–33.

Coull, J.T. "Neural Correlates of Attention and Arousal: Insights from Electrophysiology, Functional Neuroimaging and Psychopharmacology." *Progress in Neurobiology* (1998) 55:343–61.

Courtney, S.M., et al. "The Role of Prefrontal Cortex in Working Memory: Examining the Contents of Consciousness." *Philosophical Transactions of the Royal Society B: Biological Sciences* (1998) 353:1819–28.

Couto, M.I., et al. "Depression and Anxiety Following Deep Brain Stimulation in Parkinson's Disease: Systematic Review and Meta-Analysis." *Acta Médica Portuguesa* (2014) 27:372–82.

Cowan, N. "An Embedded-Processes Model of Working Memory." In: *Models of Working Memory: Mechanisms of Active Maintenance and Executive Control*, eds. A. Miyake and P. Shah (New York: Cambridge University Press, 1999), 62–101.

Cowan, W.M., D.H. Harter, and E.R. Kandel. "The Emergence of Modern Neuroscience: Some Implications for Neurology and Psychiatry." *Annual Review of Neuroscience* (2000) 23:343–91.

Cowan, W.M., K.L. Kopnisky, and S.E. Hyman. "The Human Genome Project and Its Impact on Psychiatry." *Annual Review of Neuroscience* (2002) 25:1–50.

Cowansage, K.K., et al. "Basal Variability in CREB Phosphorylation Predicts Trait-Like Differences in Amygdala-Dependent Memory." *Proceedings of the National Academy of Sciences of the United States of America* (2013) 110:16645–50.

Cowey, A., and P. Stoerig. "Reflections on Blindsight." In: *The Neuropsychology of Consciousness*, eds. D. Milner and M. Rugg (London: Academic Press, 1992), 11–37.

Cowey, A., and P. Stoerig. "Blindsight in Monkeys." *Nature* (1995) 373:247–49.

Craig, A.D. "How Do You Feel? Interoception: The Sense of the Physiological Condition of the Body." *Nature Reviews Neuroscience* (2002) 3:655–66.

Craig, A.D. "Interoception: The Sense of the Physiological Condition of the Body." *Current Opinion in Neurobiology* (2003) 13:500–5.

Craig, A.D. "How Do You Feel—Now? The Anterior Insula and Human Awareness." *Nature Reviews Neuroscience* (2009) 10:59–70.

Craig, A.D. "The Sentient Self." *Brain Structure & Function* (2010) 214:563–77.

Craske, M.G., D.H. Barlow, and T.A. O'Leary. *Mastery of Your Anxiety and Worry* (Boulder, CO: Graywind Publications, 1992).

Craske, M.G., et al. "Optimizing Inhibitory Learning During Exposure Therapy." *Behaviour Research and Therapy* (2008) 46:5–27.

Craske, M.G., et al. "Maximizing Exposure Therapy: An Inhibitory Learning Approach." *Behaviour Research and Therapy* (2014) 58:10–23.

Cravo, S.L., S.F. Morrison, and D.J. Reis. "Differentiation of Two Cardiovascular Regions Within Caudal Ventrolateral Medulla." *American Journal of Physiology* (1991) 261:R985–94.

Crawley, J.N., and R. Paylor. "A Proposed Test Battery and Constellations of Specific Behavioral Paradigms to Investigate the Behavioral Phenotypes of Transgenic and Knockout Mice." *Hormones and Behavior* (1997) 31:197–211.

Crichton, M. *The Terminal Man* (New York: Knopf, 1972).

Crick, F., and C. Koch. "Toward a Neurobiological Theory of Consciousness." *Seminars in the Neurosciences* (1990) 2:263–75.

Crick, F., and C. Koch. "Are We Aware of Neural Activity in Primary Visual Cortex?" *Nature* (1995) 375:121–23.

Crick, F., and C. Koch "A Framework for Consciousness." *Nature Neuroscience* (2003) 6:119–26.

Crick, F.C., and C. Koch. "What Is the Function of the Claustrum?" *Philosophical Transactions of the Royal Society B: Biological Sciences* (2005) 360:1271–79.

Critchley, H.D. "Neural Mechanisms of Autonomic, Affective, and Cognitive Integration." *Journal of Comparative Neurology* (2005) 493:154–66.

Critchley, H.D. "Psychophysiology of Neural, Cognitive and Affective Integration: fMRI and Autonomic Indicants." *International Journal of Psychophysiology: Official Journal of the International Organization of Psychophysiology* (2009) 73:88–94.

Critchley, H.D., C.J. Mathias, and R.J. Dolan. "Neuroanatomical Basis for First- and Second-Order Representations of Bodily States." *Nature Neuroscience* (2001) 4:207–12.

Critchley, H.D., C.J. Mathias, and R.J. Dolan. "Fear Conditioning in Humans: The Influence of Awareness and Autonomic Arousal on Functional Neuroanatomy." *Neuron* (2002) 33:653–63.

Critchley, H.D., et al. "Activity in the Human Brain Predicting Differential Heart Rate Responses to Emotional Facial Expressions." *NeuroImage* (2005) 24:751–62.

Critchley, H.D., et al. "Neural Systems Supporting Interoceptive Awareness." *Nature Neuroscience* (2004) 7:189–95.

Critchley, M. *The Parietal Lobes* (London: Edward Arnold, 1953).

Crystal, J.D. "Where Is the Skepticism in Animal Metacognition?" *The Journal of Comparative Psychology* (2014) 128:152–54; discussion 160–162.

Cullinan, W.E., J.P. Herman, and S.J. Watson. "Ventral Subicular Interaction with the Hypothalamic Paraventricular Nucleus: Evidence for a Relay in the Bed Nucleus of the Stria Terminalis." *Journal of Comparative Neurology* (1993) 332:1–20.

Curtis, C.E. "Prefrontal and Parietal Contributions to Spatial Working Memory." *Neuroscience* (2006) 139:173–80.

da Costa Gomez, T.M., and M.M. Behbehani. "An Electrophysiological Characterization of the Projection from the Central Nucleus of the Amygdala to the Periaqueductal Gray of the Rat: The Role of Opioid Receptors." *Brain Research* (1995) 689:21–31.

Dalgleish, T. "Cognitive Approaches to Posttraumatic Stress Disorder: The Evolution of Multirepresentational Theorizing." *Psychological Bulletin* (2004) 130:228–60.

Dalley, J.W., and B.J. Everitt. "Dopamine Receptors in the Learning, Memory and Drug Reward Circuitry." *Seminars in Cell & Developmental Biology* (2009) 20:403–10.

Dalmaz, C., I.B. Introini-Colliso, and J.L. McGaugh. "Noradrenergic and Cholinergic Interactions in the Amygdala and the Modulation of Memory Storage." *Behavioural Brain Research* (1993) 58:167–74.

Daly, H.B. "Disruptive Effects of Scopolamine on Fear Conditioning and on Instrumental Escape Learning." *Journal of Comparative and Physiological Psychology* (1968) 66:579–83.

Damasio, A. *Descartes' Error: Emotion, Reason, and the Human Brain* (New York: Gosset/Putnam, 1994).

Damasio, A. *Self Comes to Mind: Constructing the Conscious Brain* (New York: Pantheon Books, 2010).

Damasio, A., and G.B. Carvalho. "The Nature of Feelings: Evolutionary and Neurobiological Origins." *Nature Reviews Neuroscience* (2013) 14:143–52.

Damasio, A., H. Damasio, and D. Tranel. "Persistence of Feelings and Sentience After Bilateral Damage of the Insula." *Cerebral Cortex* (2013) 4:833–46.

Damasio, A.R. "The Brain Binds Entities and Events by Multiregional Activation from Convergence Zones." *Neural Computation* (1989) 1:123–32.

Damasio, A.R. "The Somatic Marker Hypothesis and the Possible Functions of the Prefrontal Cortex." *Philosophical Transactions of the Royal Society B: Biological Sciences* (1996) 351:1413–20.

Damasio, A.R. *The Feeling of What Happens: Body and Emotion in the Making of Consciousness* (New York: Harcourt Brace, 1999).

Damasio, A.R., et al. "Subcortical and Cortical Brain Activity during the Feeling of Self-Generated Emotions." *Nature Neuroscience* (2000) 3:1049–56.

Damasio, A.R., D. Tranel, and H. Damasio. "Individuals with Sociopathic Behavior Caused by Frontal Damage Fail to Respond Autonomically to Social Stimuli." *Behavioral Brain Research* (1990) 41:91–94.

Danielsen, E.H., D.J. Magnuson, and T.S. Gray. "The Central Amygdaloid Nucleus Innervation of the Dorsal Vagal Complex in Rat: a Phaseolus Vulgaris Leucoagglutinin Lectin Anterograde Tracing Study." *Brain Research Bulletin* (1989) 22:705–15.

Dardennes, R., N. Alanbar, A. Docteur, S.M. Divac, and C. Mirabel-Sarron. "Letter to the Editor: Simply Avoiding Reactivating Fear Memory After Exposure Therapy May Help to Consolidate Fear Extinction Memory." *Psychological Medicine* (2015) 45:887.

Darwin, C. *The Origin of Species by Means of Natural Selection: Or, the Preservation of Favored Races in the Struggle for Life* (New York: Collier, 1859).

Darwin, C. *The Expression of the Emotions in Man and Animals* (London: Fontana Press, 1872).

Darwin C. *The Formation of Vegetable Mould Through the Action of Worms: With Observations on Their Habits* (London: John Murray 1881).

Das, P., et al. "Pathways for Fear Perception: Modulation of Amygdala Activity by Thalamo-Cortical Systems." *NeuroImage* (2005) 26:141–48.

Davanger, S., et al. "Meditation-Specific Prefrontal Cortical Activation During Acem Meditation: An fMRI Study." *Perceptual and Motor Skills* (2010) 111:291–306.

Davidson, R.J. "Anxiety and Affective Style: Role of Prefrontal Cortex and Amygdala." *Biological Psychiatry* (2002) 51:68–80.

Davidson, R.J., and A. Lutz. "Buddha's Brain: Neuroplasticity and Meditation." *IEEE Signal Processing Magazine* (2008) 25:176–74.

Davis, H.P., and L.R. Squire. "Protein Synthesis and Memory: A Review." *Psychological Bulletin* (1984) 96:518–59.

Davis, M. "The Role of the Amygdala in Conditioned Fear." In: *The Amygdala: Neurobiological Aspects of Emotion, Memory, and Mental Dysfunction*, ed. J.P. Aggleton (New York: Wiley-Liss, 1992), 255–306.

Davis, M. "Neural Systems Involved in Fear and Anxiety Measured with Fear-Potentiated Startle." *The American Psychologist* (2006) 61:741–56.

Davis, M., et al. "Effects of D-cycloserine on Extinction: Translation from Preclinical to Clinical Work." *Biological Psychiatry* (2006) 60:369–75.

Davis, M., D.L. Walker, and Y. Lee. "Amygdala and Bed Nucleus of the Stria Terminalis: Differential Roles in Fear and Anxiety Measured with the Acoustic Startle Reflex." *Philosophical Transactions of the Royal Society B: Biological Sciences* (1997) 352:1675–87.

Davis, M., et al. "Phasic vs. Sustained Fear in Rats and Humans: Role of the Extended Amygdala in Fear vs. Anxiety." *Neuropsychopharmacology* (2010) 35:105–35.

Davis, M., and P.J. Whalen. "The Amygdala: Vigilance and Emotion." *Molecular Psychiatry* (2001) 6:13–34.

Dawkins, R., and J.R. Krebs. "Arms Races Between and Within Species." *Proceedings of the Royal Society of London Series B, Containing Papers of a Biological Character Royal Society* (1979) 205:489–511.

Dayan, P., and N.D. Daw. "Decision Theory, Reinforcement Learning, and the Brain." *Cognitive, Affective & Behavioral Neuroscience* (2008) 8:429–53.

de Carvalho, M.R., M. Rozenthal, and A.E. Nardi. "The Fear Circuitry in Panic Disorder and Its Modulation by Cognitive-Behaviour Therapy Interventions." *The World Journal of Biological Psychiatry: The Official Journal of the World Federation of Societies of Biological Psychiatry* (2010) 11:188–98.

de Gelder, B., J.S. Morris, and R.J. Dolan. "Unconscious Fear Influences Emotional Awareness of Faces and Voices." *Proceedings of the National Academy of Sciences of the United States of America* (2005) 102:18682–87.

de Gelder, B., et al. "Non-Conscious Recognition of Affect in the Absence of Striate Cortex." *Neuroreport* (1999) 10:3759–63.

de Haan, M., et al. "Human Memory Development and Its Dysfunction After Early Hippocampal Injury." *Trends in Neurosciences* (2006) 29:374–81.

De Oca, B.M., et al. "Distinct Regions of the Periaqueductal Gray Are Involved in the Acquisition and Expression of Defensive Responses." *Journal of Neuroscience* (1998) 18:3426–32.

De Oliveira Alvares, L., et al. "Reactivation Enables Memory Updating, Precision-Keeping and Strengthening: Exploring the Possible Biological Roles of Reconsolidation." *Neuroscience* (2013) 244:42–48.

De Quervain, D.J., A. Aerni, G. Schelling, and B. Roozendaal. "Glucocorticoids and the Regulation of Memory in Health and Disease. *Frontiers in Neuroendocrinology* (2009) 30: 358–70.

de Quervain, D.J., et al. "Glucocorticoids Enhance Extinction-Based Psychotherapy." *Proceedings of the National Academy of Sciences of the United States of America* (2011) 108:6621–25.

Dębiec, J. "Peptides of Love and Fear: Vasopressin and Oxytocin Modulate the Integration of Information in the Amygdala." *BioEssays* (2005) 27:869–73.

Dębiec, J., D.E. Bush, and J.E. LeDoux. "Noradrenergic Enhancement of Reconsolidation in the Amygdala Impairs Extinction of Conditioned Fear in Rats—A Possible Mechanism for the Persistence of Traumatic Memories in PTSD." *Depression and Anxiety* (2011) 28: 186–93.

Dębiec, J., et al. "The Amygdala Encodes Specific Sensory Features of an Aversive Reinforcer." *Nature Neuroscience* (2010) 13:536–37.

Dębiec, J., V. Doyere, K. Nader, and J.E. LeDoux. "Directly Reactivated, but Not Indirectly Reactivated, Memories Undergo Reconsolidation in the Amygdala." *Proceedings of the National Academy of Sciences of the United States of America* (2006) 103:3428–33.

Dębiec, J., and J.E. LeDoux. "Noradrenergic Signaling in the Amygdala Contributes to the Reconsolidation of Fear Memory: Treatment Implications for PTSD. *Annals of the New York Academy of Sciences* (2006) 1071:521–24.

Debner, J.A., and L.L. Jacoby. "Unconscious Perception: Attention, Awareness, and Control." *Journal of Experimental Psychology Learning, Memory, and Cognition* (1994) 20:304–17.

Decety, J. "[Naturalizing Empathy]." *L'Encephale* (2002) 28:9–20.

Dehaene, S., and J.-P. Changeux. "Neural Mechanisms for Access to Consciousness." In: *The Cognitive Neurosciences 3rd Edition,* ed. M.S. Gazzaniga (Cambridge, MA: MIT Press, 2004), 1145–58.

Dehaene, S., and J.-P. Changeux. "Experimental and Theoretical Approaches to Conscious Processing." *Neuron* (2011) 70:200–27.

Dehaene, S., et al. "Conscious, Preconscious, and Subliminal Processing: A Testable Taxonomy." *Trends in Cognitive Sciences* (2006) 10:204–11.

Dehaene, S., M. Kerszberg, and J.-P. Changeux. "A Neuronal Model of a Global Workspace in Effortful Cognitive Tasks." *Proceedings of the National Academy of Sciences of the United States of America* (1998) 95:14529–34.

Dehaene, S., and L. Naccache. "Towards a Cognitive Neuroscience of Consciousness: Basic Evidence and a Workspace Framework." *Cognition* (2001) 79:1–37.

Dehaene, S., et al. "Imaging Conscious Semantic Priming." *Nature* (1998) 395:597–600.

Dehaene, S., C. Sergent, and J.-P. Changeux. "A Neuronal Network Model Linking Subjective Reports and Objective Physiological Data during Conscious Perception." *Proceedings of the National Academy of Sciences of the United States of America* (2003) 100:8520–25.

Deisseroth, K. "Optogenetics and Psychiatry: Applications, Challenges, and Opportunities." *Biological Psychiatry* (2012) 71:1030–32.

Del Cul, A., et al. "Causal Role of Prefrontal Cortex in the Threshold for Access to Consciousness." *Brain: a Journal of Neurology* (2009) 132:2531–40.

Delgado, M.R., and K.C. Dickerson. "Reward-Related Learning via Multiple Memory Systems." *Biological Psychiatry* (2012) 72:134–41.

Delgado, M.R., et al. "Avoiding Negative Outcomes: Tracking the Mechanisms of Avoidance Learning in Humans During Fear Conditioning." *Frontiers in Behavioral Neuroscience* (2009) 3:33.

Delgado, M.R., et al. "The Role of the Striatum in Aversive Learning and Aversive Prediction Errors." *Philosophical Transactions of the Royal Society B: Biological Sciences* (2008) 363: 3787–3800.

Delgado, M.R., et al. "Neural Circuitry Underlying the Regulation of Conditioned Fear and Its Relation to Extinction." *Neuron* (2008) 59:829–38.

Delgado, M.R., A. Olsson, and E.A. Phelps. "Extending Animal Models of Fear Conditioning to Humans." *Biological Psychology* (2006) 73:39–48.

Delgado, M.R., et al. "Emotion Regulation of Conditioned Fear: The Contributions of Reappraisal." Paper presented at the 11th Annual Meeting of the Cognitive Neuroscience Society San Francisco (2004).

Dell'Acqua, R., and J. Grainger. "Unconscious Semantic Priming from Pictures." *Cognition* (1999) 73:B1–B15.

Demertzi, A., et al. "Hypnotic Modulation of Resting State fMRI Default Mode and Extrinsic Network Connectivity." *Progress in Brain Research* (2011) 193:309–22.

Demertzi, A., A. Soddu, and S. Laureys. "Consciousness Supporting Networks." *Current Opinion in Neurobiology* (2013) 23:239–44.

Demertzis, K.H., and M.G. Kraske. "Cognitive-Behavioral Therapy for Anxiety Disorders in Primary Care." *Primary Psychiatry* (2005). Retrieved Dec. 19, 2014. http://primarypsychiatry .com/cognitive-behavioral-therapy-for-anxiety-disorders-in-primary-care.

den Dulk, P., B.T. Heerebout, and R.H. Phaf. "A Computational Study into the Evolution of Dual-Route Dynamics for Affective Processing." *Journal of Cognitive Neuroscience* (2003) 15:194–208.

Dennett, D.C. *Consciousness Explained* (Boston: Little, Brown and Company, 1991).

DePrince, A.P., et al. "Motivated Forgetting and Misremembering: Perspectives from Betrayal Trauma Theory." *Nebraska Symposium on Motivation* (2012) 58:193–242.

Dere, E., et al. "The Case for Episodic Memory in Animals." *Neuroscience and Biobehavioral Reviews* (2006) 30:1206–24.

Descartes, R. *Discourse on the Method* (Indianapolis: Hackett, 2007).

Descartes, R. *Principia Philosophiae* (Ghent University: Apud Ludovicum Elzevirium, 1644).

Deschaux, O., et al. "Post-Extinction Fluoxetine Treatment Prevents Stress-Induced Reemergence of Extinguished Fear." *Psychopharmacology (Berl)* (2013) 225:209–16.

Deshmukh, V.D. "Neuroscience of Meditation." *The Scientific World Journal* (2006) 6:2239–53.

Desiderato, O. "Generalization of Acquired Fear as a Function of CS Intensity and Number of Acquisition Trials." *Journal of Experimental Psychology* (1964) 67:41–47.

Desimone, R. "Neural Mechanisms for Visual Memory and Their Role in Attention." *Proceedings of the National Academy of Sciences of the United States of America* (1996) 93:13494–99.

D'Esposito, M., et al. "Maintenance Versus Manipulation of Information Held in Working Memory: An Event-Related fMRI Study." *Brain and Cognition* (1999) 41:66–86.

Devan, B.D., N.S. Hong, and R.J. McDonald. "Parallel Associative Processing in the Dorsal Striatum: Segregation of Stimulus-Response and Cognitive Control Subregions." *Neurobiology of Learning and Memory* (2011) 96:95–120.

Devinsky, O., M.J. Morrell, and B.A. Vogt. "Contributions of Anterior Cingulate Cortex to Behaviour." *Brain: A Journal of Neurology* (1995) 118:279–306.

Devos, T., and M.R. Banaji. "Implicit Self and Identity." *Annals of the New York Academy of Sciences* (2003) 1001:177–211.

Diaz-Mataix, L., et al. "Sensory-Specific Associations Stored in the Lateral Amygdala Allow for Selective Alteration of Fear Memories." *Journal of Neuroscience* (2011) 31:9538–43.

Diaz-Mataix, L., et al. "Detection of a Temporal Error Triggers Reconsolidation of Amygdala-Dependent Memories." *Current Biology* (2013) 23:467–72.

Dickenson, J., et al. "Neural Correlates of Focused Attention During a Brief Mindfulness Induction." *Social Cognitive and Affective Neuroscience* (2013) 8:40–47.

Dickerson, B.C., and H. Eichenbaum. "The Episodic Memory System: Neurocircuitry and Disorders." *Neuropsychopharmacology* (2010) 35:86–104.

Dickinson, A. "Conditioning and Associative Learning." *British Medical Bulletin* (1981) 37: 165–68.

Dickinson, A. "Why a Rat Is Not a Beast Machine." In: *Frontiers of Consciousness*, eds. L. Weiskrantz and M. Davies (Oxford, UK: Oxford University Press, 2008), 275–88.

Dickinson, A. "Associative Learning and Animal Cognition." *Philosophical Transactions of the Royal Society B: Biological Sciences* (2012) 367:2733–42.

Dickinson, E. *Emily Dickinson—Selected Poems* (New York: St. Martin's Press, 1992).

Dickson, B.J. "Wired for Sex: The Neurobiology of *Drosophila* Mating Decisions." *Science* (2008) 322:904–9.

Dielenberg, R.A., P. Carrive, and I.S. McGregor. "The Cardiovascular and Behavioral Response to Cat Odor in Rats: Unconditioned and Conditioned Effects." *Brain Research* (2001) 897: 228–37.

Dillon, D.G., et al. "Peril and Pleasure: An RDoC-Inspired Examination of Threat Responses and Reward Processing in Anxiety and Depression." *Depression and Anxiety* (2014) 31:233–49.

Dincheva, I., et al. "FAAH Genetic Variation Enhances Fronto-Amygdala Function in Mouse and Human." *Nature Communications* (2015) 6:6395.

Dincheva, I., S.S. Pattwell, L. Tessarollo, K.G. Bath, and F.S. Lee. "BDNF Modulates Contextual Fear Learning During Adolescence." *Developmental Neuroscience* (2014) 36:269–76.

Dinsmoor, J.A. "Variable-Interval Escape from Stimuli Accompanied by Shocks." *Journal of the Experimental Analysis of Behavior* (1962) 5:41–47.

Diorio, D., V. Viau, and M.J. Meaney. "The Role of the Medial Prefrontal Cortex (Cingulate Gyrus) in the Regulation of Hypothalamic-Pituitary-Adrenal Responses to Stress." *Journal of Neuroscience* (1993) 13:3839–47.

Dityatev, A.E., and V.Y. Bolshakov. "Amygdala, Long-Term Potentiation, and Fear Conditioning." *Neuroscientist* (2005) 11:75–88.

Dixon, B.A. "Animal Emotion." *Ethics and the Environment* (2001) 6:22–30.

Dodhia, S., et al. "Modulation of Resting-State Amygdala-Frontal Functional Connectivity by Oxytocin in Generalized Social Anxiety Disorder." *Neuropsychopharmacology* (2014) 39:2061–69.

Dolan, R.J., and P. Dayan. "Goals and Habits in the Brain." *Neuron* (2013) 80:312–25.

Dolan, R.J., and P. Vuilleumier. "Amygdala Automaticity in Emotional Processing." *Annals of the New York Academy of Sciences* (2003) 985:348–55.

Dolcos F., A.D. Iordan, and S. Dolcos (2011) "Neural Correlates of Emotion-Cognition Interactions: A Review of Evidence from Brain Imaging Investigations." *Journal of Cognitive Psychology* (Hove) 2011 Sep; 23(6):669-94.

Doll, B.B., D.A. Simon, and N.D. Daw. "The Ubiquity of Model-Based Reinforcement Learning." *Current Opinion in Neurobiology* (2012) 22:1075–81.

Dollard, J., and N.E. Miller. *Personality and Psychotherapy: An Analysis in Terms of Learning, Thinking, and Culture* (New York: McGraw-Hill, 1950).

Do-Monte, F.H., et al. "Deep Brain Stimulation of the Ventral Striatum Increases BDNF in the Fear Extinction Circuit." *Frontiers in Behavioral Neuroscience* (2013) 7:102.

Donahoe, J.W., and Edward L. Thorndike. "The Selectionist Connection." *Journal of the Experimental Analysis of Behavior* (1999) 72:451–54.

Dong, H.W., G.D. Petrovich, and L.W. Swanson. "Topography of Projections from Amygdala to Bed Nuclei of the Stria Terminalis." *Brain Research. Brain Research Reviews* (Amsterdam) (2001) 38:192–246.

Dong, H.W., and L.W. Swanson. "Organization of Axonal Projections from the Anterolateral Area of the Bed Nuclei of the Stria Terminalis." *Journal of Comparative Neurology* (2004) 468:277–98.

Dong, H.W., and L.W. Swanson. "Projections from Bed Nuclei of the Stria Terminalis, Posterior Division: Implications for Cerebral Hemisphere Regulation of Defensive and Reproductive Behaviors." *Journal of Comparative Neurology* (2004) 471:396–433.

Dong, H.W., and L.W. Swanson. "Projections from Bed Nuclei of the Stria Terminalis, Anteromedial Area: Cerebral Hemisphere Integration of Neuroendocrine, Autonomic, and Behavioral Aspects of Energy Balance." *Journal of Comparative Neurology* (2006) 494:142–78.

Dong, H.W., and L.W. Swanson. "Projections from Bed Nuclei of the Stria Terminalis, Dorsomedial Nucleus: Implications for Cerebral Hemisphere Integration of Neuroendocrine, Autonomic, and Drinking Responses." *Journal of Comparative Neurology* (2006) 494:75–107.

Driver, J., and P. Vuilleumier. "Perceptual Awareness and Its Loss in Unilateral Neglect and Extinction." *Cognition* (2001) 79:39–88.

Duchenne, G.-B. *Mécanisme de la physionomie humaine ou Analyse électrophysiologique de l'expression des passions applicable à la pratique des arts plastiques* (Paris: Jules Renouard, 1862).

Dudai, Y. "Consolidation: Fragility on the Road to the Engram." *Neuron* (1996) 17:367–70.

Dudai, Y. "The Neurobiology of Consolidations, or, How Stable Is the Engram?" *Annual Review of Psychology* (2004) 55:51–86.

Dudai Y. "Reconsolidation: The Advantage of Being Refocused." *Current Opinion in Neurobiology* (2006) 16:174–78.

Dudai, Y. "The Restless Engram: Consolidations Never End." *Annual Review of Neuroscience* (2012) 35:227–47.

Dudai, Y., and M. Eisenberg. "Rites of Passage of the Engram: Reconsolidation and the Lingering Consolidation Hypothesis." *Neuron* (2004) 44:93–100.

Duncan, J., and A.M. Owen. "Common Regions of the Human Frontal Lobe Recruited by Diverse Cognitive Demands." *Trends in Neurosciences* (2000) 23:475–83.

Dunsmoor, J.E., et al. "Aversive Learning Modulates Cortical Representations of Object Categories." *Cerebral Cortex* (2014) 24:2859–72.

Dupuy, J.B., and R. Ladouceur. "Cognitive Processes of Generalized Anxiety Disorder in Comorbid Generalized Anxiety Disorder and Major Depressive Disorder." *Journal of Anxiety Disorders* (2008) 22:505–14.

Durand, V.M., and D.H. Barlow. *Essentials of Abnormal Psychology* (Independence, KY: Cengage Learning, 2006).

Duvarci, S., and D. Paré. "Amygdala Microcircuits Controlling Learned Fear." *Neuron* (2014) 82:966–80.

Duvarci, S., D. Popa, and D. Paré. "Central Amygdala Activity During Fear Conditioning." *Journal of Neuroscience* (2011) 31:289–94.

Dykman, R.A. "Toward a Theory of Classical Conditioning: Cognitive, Emotional, and Motor Components of the Conditional Reflex." *Progress in Experimental Personality Research* (1965) 2:229–317.

Dymond, S., et al. "Safe from Harm: Learned, Instructed, and Symbolic Generalization Pathways of Human Threat-Avoidance." *PLoS One* (2012) 7:E47539.

Dymond, S., and B. Roche. "A Contemporary Behavior Analysis of Anxiety and Avoidance." *The Behavior Analyst* (2009) 32:7–27.

Ebbinghaus, H. *On Memory* (New York: Dover, 1885/1964).

Eckstein, M., et al. "Oxytocin Facilitates the Extinction of Conditioned Fear in Humans." *Biological Psychiatry* (2014) (in press).

Edeline, J.M. "Beyond Traditional Approaches to Understanding the Functional Role of Neuromodulators in Sensory Cortices." *Frontiers in Behavioral Neuroscience* (2012) 6: article 45. Published online July 30, 2012. doi: 10.3389/fnbeh.2012.00045 PMCID: PMC3407859.

Edelman, D.B., and A.K. Seth. "Animal Consciousness: A Synthetic Approach." *Trends in Neurosciences* (2009) 32:476–84.

Edelman, G. *Bright Air, Brilliant Fire: On the Matter of Mind* (New York: Basic Books, 1993).

Edelman, G. "Consciousness: The Remembered Present." *Annals of the New York Academy of Sciences* (2001) 929:111–22.

Edelman, G.M. *Neural Darwinism* (New York: Basic Books, 1987).

Edelman, G.M. *The Remembered Present* (New York: Basic Books, 1989).

Edelman, G.M. *Wider Than the Sky: The Phenomenal Gift of Consciousness* (New Haven, CT: Yale University Press, 2004).

Edinger, L. *Vorlesungen uber den Bau der nervosen Zentralorgane* (Leipzig: Vogel, 1908).

Edmunds, M. *Defence in Animals: A Survey of Anti-Predator Defences* (New York: Longman, 1974).

Eelen, P., and B. Vervliet. "Fear Conditioning and Clinical Implications: What Can We Learn from the Past?" In: *Fear and Learning: From Basic Processes to Clinical Implications*, eds. M.G. Craske, et al. (Washington, DC: American Psychological Association, 2006), 197–215.

Ehlers, A., and D.M. Clark. "A Cognitive Model of Posttraumatic Stress Disorder." *Behaviour Research and Therapy* (2000) 38:319–45.

Ehlers, A., et al. "Cognitive Therapy for Post-Traumatic Stress Disorder: Development and Evaluation." *Behaviour Research and Therapy* (2005) 43:413–31.

Ehrlich, I., et al. "Amygdala Inhibitory Circuits and the Control of Fear Memory." *Neuron* (2009) 62:757–71.

Eichenbaum, H. "The Hippocampal System and Declarative Memory in Animals." *Journal of Cognitive Neuroscience* (1992) 4:217–31.

Eichenbaum, H. "The Hippocampal System and Declarative Memory in Humans and Animals: Experimental Analysis and Historical Origins." In: *Memory System*, eds. D.L. Schacter and E. Tulving (Cambridge, MA: MIT Press, 1994), 147–201.

Eichenbaum, H. *The Cognitive Neuroscience of Memory* (New York: Oxford University Press, 2002).

Eichenbaum, H., and N.J. Fortin. "Bridging the Gap Between Brain and Behavior: Cognitive and Neural Mechanisms of Episodic Memory." *Journal of the Experimental Analysis of Behavior* (2005) 84:619–29.

Eifert, G.H., and J.P. Forsyth. *Acceptance and Commitment Therapy for Anxiety Disorders: A Practitioner's Treatment Guide to Using Mindfulness, Acceptance, and Values-Based Behavior Change Strategies* (Oakland, CA: New Harbinger Publications, 2005).

Eisenberg, M., et al. "Stability of Retrieved Memory: Inverse Correlation with Trace Dominance." *Science* (2003) 301:1102–4.

Ekman, P. "Universals and Cultural Differences in Facial Expressions of Emotions." In: *Nebraska Symposium on Motivation 1971*, ed. J. Cole (Lincoln: University of Nebraska Press, 1972), 207–83.

Ekman, P. "Biological and Cultural Contributions to Body and Facial Movement." In: *The Anthropology of the Body*, ed. J. Blacking (London: Academic Press, 1977), 39–84.

Ekman, P. "Biological and Cultural Contributions to Body and Facial Movement in the Expression of Emotions." In: *Explaining Emotions*, ed. A.O. Rorty (Berkeley: University of California Press, 1980).

Ekman, P. "Expression and Nature of Emotion." In: *Approaches to Emotion*, eds. K. Scherer and P. Ekman (Hillsdale, NJ: Erlbaum, 1984), 319–43.

Ekman, P. "Are There Basic Emotions?" *Psychological Review* (1992) 99:550–53.

Ekman, P. "An Argument for Basic Emotions." *Cognition and Emotion* (1992) 6:169–200.

Ekman, P. "Facial Expressions of Emotion: New Findings, New Questions." *Psychological Science* (1992) 3:34–38.

Ekman, P. "Facial Expression and Emotion." *American Psychologist* (1993) 48.

Ekman, P. "Basic Emotions." In: *Handbook of Cognition and Emotion*, eds. T. Dalgleish and M. Power (Chichester, UK: John Wiley and Sons, 1999), 45–60.

Ekman, P. *Emotions Revealed: Recognizing Faces and Feelings to Improve Communication and Emotional Life* (New York: Times Books, 2003).

Ekman, P., and W.V. Friesen. *Unmasking the Face* (Englewood, NJ: Prentice-Hall, 1975).

El-Amamy, H., and P.C. Holland. "Dissociable Effects of Disconnecting Amygdala Central Nucleus from the Ventral Tegmental Area or Substantia Nigra on Learned Orienting and Incentive Motivation." *European Journal of Neuroscience* (2007) 25:1557–67.

Eley, T.C., et al. "A Twin Study of Anxiety-Related Behaviours in Pre-School Children." *Journal of Child Psychology and Psychiatry, and Allied Disciplines* (2003) 44:945–60.

Eliasson, S., et al. "Activation of Sympathetic Vasodilator Nerves to the Skeletal Muscles in the Cat by Hypothalamic Stimulation." *Acta Physiologica Scandinavica* (1951) 23:333–51.

Ellis, A. "Rational Psychotherapy and Individual Psychology." *Journal of Individual Psychology* (1957) 13:38–44.

Ellis, A. "Rational-Emotive Therapy and Cognitive Behavior Therapy: Similarities and Differences." *Cognitive Therapy and Research* (1980) 4:325–40.

Ellis, A., and C. MacLaren. *Rational Emotive Behavior Therapy: a Therapist's Guide* (San Luis Obispo, CA: Impact Publishers, 2005).

Ellis, N. "At the Interface: Dynamic Interactions of Explicit and Implicit Language Knowledge." *Studies in Second Language Acquisition* (2005) 27:305–52.

Elman, J., et al. *Rethinking Innateness* (Cambridge, MA: MIT Press, 1997).

Emerson, R.W. *Society and Solitude* (Boston: Fields, Osgood & Co, 1870).

Emes, R.D., and S.G. Grant. "Evolution of Synapse Complexity and Diversity." *Annual Review of Neuroscience* (2012) 35:111–31.

Epstein, M. *Psychotherapy Without the Self: A Buddhist Perspective* (New Haven, CT: Yale University Press, 2009.)

Epstein, M. *Thoughts Without a Thinker: Psychotherapy from a Buddhist Perspective* (New York: Basic Books, 2013).

Epstein, M. *The Trauma of Everyday Life* (New York: Penguin Press, 2013).

Epstein, S. "The Nature of Anxiety with Emphasis upon Its Relationship to Expectancy." In: *Anxiety: Current Trends in Theory and Research*, ed. C.D. Speilberger (New York: Academic Press, 1972), 292–338.

Ercetin, G., and C.E.M. Alptekin. "The Explicit/Implicit Knowledge Distinction and Working Memory: Implications for Second Language Reading Comprehension." *Applied Psycholinguistics* (2013) 34:727–53.

Eriksson, S., R. Hurme, and M. Rhen. "Low-Temperature Sensors in Bacteria." *Philosophical Transactions of the Royal Society B: Biological Sciences* (2002) 357:887–93.

Erk, S., B. Abler, and H. Walter. "Cognitive Modulation of Emotion Anticipation." *European Journal of Neuroscience* (2006) 24:1227–36.

Erlich, J.C., D.E. Bush, and J.E. LeDoux. "The Role of the Lateral Amygdala in the Retrieval and Maintenance of Fear-Memories Formed by Repeated Probabilistic Reinforcement." *Frontiers in Behavioral Neuroscience* (2012) 6:16.

Esmoris-Arranz, F.J., J.L. Pardo-Vazquez, and G.A. Vazquez-Garcia. "Differential Effects of Forward or Simultaneous Conditioned Stimulus-Unconditioned Stimulus Intervals on the Defensive Behavior System of the Norway Rat (*Rattus norvegicus*)." *Journal of Experimental Psychology: Animal Behavior Processes* (2003) 29:334–40.

España, R.A., and T.E. Scammell. "Sleep Neurobiology from a Clinical Perspective." *Sleep* (2011) 34:845–58.

Esquivel, G., et al. "Acids in the Brain: A Factor in Panic?" *Journal of Psychopharmacology* (2010) 24:639–47.

Esteves, F., et al. "Nonconscious Associative Learning: Pavlovian Conditioning of Skin Conductance Responses to Masked Fear-Relevant Facial Stimuli." *Psychophysiology* (1994) 31:375–85.

Etchegoyen, R.H. *The Fundamentals of Psychoanalytic Technique* (New York: Karnac Books, 2005).

Etkin, A., T. Egner, and R. Kalisch. "Emotional Processing in Anterior Cingulate and Medial Prefrontal Cortex." *Trends in Cognitive Sciences* (2011) 15:85–93.

Etkin, A., A. Gyurak, and R. O'Hara. "A Neurobiological Approach to the Cognitive Deficits of Psychiatric Disorders." *Dialogues in Clinical Neuroscience* (2013) 15:419–29.

Etkin, A., et al. "Individual Differences in Trait Anxiety Predict the Response of the Basolateral Amygdala to Unconsciously Processed Fearful Faces." *Neuron* (2004) 44:1043–55.

Evans, J.S. "Dual-Processing Accounts of Reasoning, Judgment, and Social Cognition." *Annual Review of Psychology* (2008) 59:255–78.

Evans, J.S. *Thinking Twice: Two Minds in One Brain* (Oxford, UK: Oxford University Press, 2010).

Evans, J.S. "Rationality and the Illusion of Choice." *Frontiers in Psychology* (2014) 5:104.

Evans, S., et al. "Mindfulness-Based Cognitive Therapy for Generalized Anxiety Disorder." *Journal of Anxiety Disorders* (2008) 22:716–21.

Everitt, B., and T. Robbins. "Motivation and Reward." In: *Fundamental Neuroscience*, eds. M.J. Zigmond, et al. (San Diego: Academic Press, 1999).

Everitt, B.J., et al. "Review: Neural Mechanisms Underlying the Vulnerability to Develop Compulsive Drug-Seeking Habits and Addiction." *Philosophical Transactions of the Royal Society B: Biological Sciences* (2008) 363:3125–35.

Everitt, B.J., M. Cador, and T.W. Robbins. "Interactions Between the Amygdala and Ventral Striatum in Stimulus-Reward Associations: Studies Using a Second-Order Schedule of Sexual Reinforcement." *Neuroscience* (1989) 30:63–75.

Everitt, B.J., A. Dickinson, and T.W. Robbins. "The Neuropsychological Basis of Addictive Behaviour." *Brain Research. Brain Research Reviews* (Amsterdam) (2001) 36:129–38.

Everitt, B.J., et al. "Associative Processes in Addiction and Reward: The Role of Amygdala-Ventral Striatal Subsystems." *Annals of the New York Academy of Sciences* (1999) 877:412–38.

Everitt, B.J., and T.W. Robbins. "Neural Systems of Reinforcement for Drug Addiction: From Actions to Habits to Compulsion." *Nature Neuroscience* (2005) 8:1481–89.

Everitt, B.J., and T.W. Robbins. "From the Ventral to the Dorsal Striatum: Devolving Views of Their Roles in Drug Addiction." *Neuroscience and Biobehavioral Reviews* (2013) 37:1946–54.

Ewbank, M.P., E. Fox, and A.J. Calder. "The Interaction Between Gaze and Facial Expression in the Amygdala and Extended Amygdala Is Modulated by Anxiety." *Frontiers in Human Neuroscience* (2010) 4:56.

Eysenck, H.J. *Behaviour Therapy and the Neuroses* (London: Pergamon Press, 1960).

Eysenck, H.J. "Behavior Therapy." In: *Theoretical Foundations of Behavior Therapy*, eds. H.J. Eysenck and I. Martin (New York: Plenum, 1987), 3–36.

Eysenck, H.J. "Anxiety and the Natural History of Neurosis." In *Stress and Anxiety* (vol. 1), eds. C.D. Spielberger and I.G. Sarason (New York: Wiley, 1995), 51–94.

Eysenck, H.J., and M.W. Eysenck. *Personality and Individual Differences* (New York: Plenum, 1985).

Eysenck, M.W., et al. "Anxiety and Cognitive Performance: Attentional Control Theory." *Emotion* (2007) 7:336–53.

Falk, D. "Brain Evolution in *Homo*: The 'Radiator' Theory." *Behavioral and Brain Sciences* (1990) 13:333–44.

Fanselow, M.S. "Associative vs. Topographical Accounts of the Immediate Shock-Freezing Deficits in Rats: Implications for the Response Selection Rules Governing Species-Specific Defensive Reactions." *Learning and Motivation* (1986) 17:16–39.

Fanselow, M.S. "The Adaptive Function of Conditioned Defensive Behavior: An Ecological Approach to Pavlovian Stimulus-Substitution Theory." In: *Ethoexperimental Approaches to the Study of Behavior*, eds. R.J. Blanchard, et al. (Dordrecht, the Netherlands: Kluwer, 1989), 151–66.

Fanselow, M.S. "Contextual Fear, Gestalt Memories, and the Hippocampus." *Behavioural Brain Research* (2000) 110:73–81.

Fanselow, M.S., et al. "Ventral and Dorsolateral Regions of the Midbrain Periaqueductal Gray (PAG) Control Different Stages of Defensive Behavior: Dorsolateral, PAG Lesions Enhance the Defensive Freezing Produced by Massed and Immediate Shock." *Aggressive Behavior* (1995) 21:63–77.

Fanselow, M.S., and L.S. Lester. "A Functional Behavioristic Approach to Aversively Motivated Behavior: Predatory Imminence as a Determinant of the Topography of Defensive Behavior." In: *Evolution and Learning*, eds. R.C. Bolles and M.D. Beecher (Hillsdale, NJ: Erlbaum, 1988), 185–211.

Fanselow, M.S., and A.M. Poulos. "The Neuroscience of Mammalian Associative Learning." *Annual Review of Psychology* (2005) 56:207–34.

Farah, M.J. "Semantic and Perceptual Priming: How Similar Are the Underlying Mechanisms?" *Journal of Experimental Psychology: Human Perception and Performance* (1989) 15:188–94.

Farah, M.J. "Neuroethics: The Ethical, Legal, and Societal Impact of Neuroscience." *Annual Review of Psychology* (2012) 63:571–91.

Farah, M.J., et al. "Neurocognitive Enhancement: What Can We Do and What Should We Do?" *Nature Reviews Neuroscience* (2004) 5:421–25.

Farb, N.A., A.K. Anderson, and Z.V. Segal. "The Mindful Brain and Emotion Regulation in Mood Disorders." *Canadian Journal of Psychiatry / Revue Canadienne de Psychiatrie* (2012) 57:70–77.

Faw, B. "Pre-Frontal Executive Committee for Perception, Working Memory, Attention, Long-Term Memory, Motor Control, and Thinking: A Tutorial Review." *Consciousness and Cognition* (2003) 12:83–139.

Feinstein, J.S., et al. "Fear and Panic in Humans with Bilateral Amygdala Damage." *Nature Neuroscience* (2013) 16:270–72.

Fernandez, E., and D.C. Turk. "Sensory and Affective Components of Pain: Separation and Synthesis." *Psychological Bulletin* (1992) 112:205–17.

Fernandez de Molina, A., and R.W. Hunsperger. "Central Representation of Affective Reactions in Forebrain and Brain Stem: Electrical Stimulation of Amygdala, Stria Terminalis, and Adjacent Structures." *Journal of Physiology* (1959) 145:251–65.

Fernandez de Molina, A., and R.W. Hunsperger. "Organization of the Subcortical System Governing Defense and Flight Reactions in the Cat." *Journal of Physiology* (1962) 160:200–13.

Fernando, A.B., J.E. Murray, and A.L. Milton. "The Amygdala: Securing Pleasure and Avoiding Pain." *Frontiers in Behavioral Neuroscience* (2013) 7:190.

Fernando, C.T., et al. "Molecular Circuits for Associative Learning in Single-Celled Organisms." *Journal of the Royal Society Interface* (2009) 6:463–69.

Ferrier, D. *The Functions of the Brain* (New York: G. P. Putnam's Sons, 1886).

Feske, U., and D.L. Chambless. "Cognitive Behavioral Versus Exposure Only Treatment for Social Phobia: A Meta-Analysis." *Behavior Therapy* (1995) 26:695–720.

Festinger, L. *A Theory of Cognitive Dissonance* (Evanston, IL: Row Peterson, 1957).

Festinger, L. "Cognitive Dissonance." *Scientific American* (1962) 207:93–102.

File, S.E. "The Interplay of Learning and Anxiety in the Elevated Plus-Maze." *Behavioural Brain Research* (1993) 58:199–202.

File, S.E. "Animal Models of Different Anxiety States." *Advances in Biochemical Psychopharmacology* (1995) 48:93–113.

File, S.E. "Factors Controlling Measures of Anxiety and Responses to Novelty in the Mouse." *Behavioural Brain Research* (2001) 125:151–57.

File, S.E., et al. "Animal Tests of Anxiety." *Current Protocols in Neuroscience* (2004) Chapter 8: Unit 8 3.

File, S.E., and P. Seth. "A Review of 25 Years of the Social Interaction Test." *European Journal of Pharmacology* (2003) 463:35–53.

Fireman, G.D., T.E. McVay, and O.J. Flanagan, eds. *Narrative and Consciousness: Literature, Psychology and the Brain* (Oxford, UK: Oxford University Press, 2003).

Fischman, M.W. "Relationship Between Self-Reported Drug Effects and Their Reinforcing Effects: Studies with Stimulant Drugs." *NIDA Research Monograph* (1989) 92:211–30.

Fischman, M.W., and R.W. Foltin. "Self-Administration of Cocaine by Humans: A Laboratory Perspective." *CIBA Foundation Symposium* (1992) 166:165–73; discussion 173–80.

Fisher, P.M., and A.R. Hariri. "Identifying Serotonergic Mechanisms Underlying the Corticolimbic Response to Threat in Humans." *Philosophical Transactions of the Royal Society B: Biological Sciences* (2013) 368:20120192.

Fitzgerald, P.J., J.R. Seemann, and S. Maren. "Can Fear Extinction Be Enhanced? A Review of Pharmacological and Behavioral Findings." *Brain Research Bulletin* (2014) 105:46–60.

Flanagan, O. *The Problem of the Soul: Two Visions of Mind and How to Reconcile Them* (New York: Basic Books, 2003).

Flavell, J.H. "Metacognition and Cognitive Monitoring: A New Area of Cognitive-Developmental Inquiry." *The American Psychologist* (1979) 34:906–11.

Fleming, D. "Walter Bradford Cannon." In: *Dictionary of American Biography Supplement 3*, ed. W.T. James (New York: Charles Scribner's Sons, 1973), 133–37.

Fleming, S.M., R.J. Dolan, and C.D. Frith. "Metacognition: Computation, Biology and Function." *Philosophical Transactions of the Royal Society B: Biological Sciences* (2012) 367:1280–86.

Fletcher, G.J.O. "Two Uses of Folk Psychology: Implications for *Psychological Science*." *Philosophical Psychology* (1995) 8:221–38.

Fletcher, P.C., and R.N. Henson. "Frontal Lobes and Human Memory: Insights from Functional Neuroimaging." *Brain: A Journal of Neurology* (2001) 124:849–81.

Flexner, L.B., and J.B. Flexner. "Effect of Acetoxycycloheximide and of an Acetoxycycloheximide-Puromycin Mixture on Cerebral Protein Synthesis and Memory in Mice." *Proceedings of the National Academy of Sciences of the United States of America* (1966) 55:369–74.

Florczyk, S.J., and S. Saha. "Ethical Issues in Nanotechnology." *Journal of Long-Term Effects of Medical Implants* (2007) 17:271–80.

Flores A., et al. "The Hypocretin/Orexin System Mediates the Extinction of Fear Memories." *Neuropsychopharmacology* (2014) 39:2732–41.

Flower, T.P., M. Gribble, and A.R. Ridley. "Deception by Flexible Alarm Mimicry in an African Bird." *Science* (2014) 344:513–16.

Flynn, J.P. "The Neural Basis of Aggression in Cats." In: *Biology and Behavior: Neurophysiology and Emotion,* ed. D.C. Glass (New York: Rockefeller University Press and Russell Sage Foundation, 1967), 40–60.

Foa, E.B. "Prolonged Exposure Therapy: Past, Present, and Future." *Depression and Anxiety* (2011) 28:1043–47.

Foa, E.B., et al. "A Comparison of Exposure Therapy, Stress Inoculation Training, and Their Combination for Reducing Posttraumatic Stress Disorder in Female Assault Victims." *Journal of Consulting and Clinical Psychology* (1999) 67:194–200.

Foa, E.B., et al. "Cognitive Biases in Generalized Social Phobia." *Journal of Abnormal Psychology* (1996) 105:433–39.

Foa, E.B., E.A. Hembree, and B.O. Rothbaum. *Prolonged Exposure Therapy for PTSD: Emotional Processing of Traumatic Experiences Therapist Guide* (Oxford, UK: Oxford University Press, 2007).

Foa, E.B., and M.J. Kozak. "Treatment of Anxiety Disorders: Implications for Psychopathology." In: *Anxiety and the Anxiety Disorders*, eds. A.H. Tuma and J.D. Maser (Hillsdale, NJ: Erlbaum, 1985), 421–52.

Foa, E.B., and M.J. Kozak. "Emotional Processing of Fear: Exposure to Corrective Information." *Psychological Bulletin* (1986) 99:20–35.

Foa, E.B., and R. McNally. "Mechanics of Change in Exposure Therapy." In: *Current Controversies in the Anxiety Disorders*, ed. R.M. Rapee (New York: Guilford, 1996), 329–43.

Fodor, J. *The Language of Thought* (Cambridge, MA: Harvard University Press, 1975).

Foote, S.L., G. Aston-Jones, and F.E. Bloom. "Impulse Activity of Locus Coeruleus Neurons in Awake Rats and Monkeys Is a Function of Sensory Stimulation and Arousal." *Proceedings of the National Academy of Sciences of the United States of America* (1980) 77:3033–37.

Foote, S.L., F.E. Bloom, and G. Aston-Jones. "Nucleus Locus Ceruleus: New Evidence of Anatomical and Physiological Specificity." *Physiological Reviews* (1983) 63:844–914.

Forsyth, J.P., and G.H. Eifert. "The Language of Feeling and the Feeling of Anxiety: Contributions of the Behaviorisms Toward Understanding the Function-Altering Effects of Language." *Psychological Record* (1996) 46.

Fortin, N.J., S.P. Wright, and H. Eichenbaum. "Recollection-Like Memory Retrieval in Rats Is Dependent on the Hippocampus." *Nature* (2004) 431:188–91.

Fossat, P., et al. "Comparative Behavior. Anxiety-Like Behavior in Crayfish Is Controlled by Serotonin." *Science* (2014) 344:1293–97.

Fowles, J. *The Magus* (New York: Little, Brown and Company, 1965).

Fox, E. "Attentional Bias in Anxiety: A Defective Inhibition Hypothesis." *Cognition and Emotion* (1994) 8:165–96.

Fox, E. "Processing Emotional Facial Expressions: The Role of Anxiety and Awareness." *Cognitive, Affective & Behavioral Neuroscience* (2002) 2:52–63.

Fox, E., et al. "Facial Expressions of Emotion: Are Angry Faces Detected More Efficiently?" *Cognition and Emotion* (2000) 14:61–92.

Fox, K.C., et al. "Is Meditation Associated with Altered Brain Structure? A Systematic Review and Meta-Analysis of Morphometric Neuroimaging in Meditation Practitioners." *Neuroscience and Biobehavioral Reviews* (2014) 43:48–73.

Frankland, P.W., et al. "The Dorsal Hippocampus Is Essential for Context Discrimination but Not for Contextual Conditioning." *Behavioral Neuroscience* (1998) 112:863–74.

Frankland, P.W., et al. "Consolidation of CS and US Representations in Associative Fear Conditioning." *Hippocampus* (2004) 14:557–69.

Fredrikson, M., and V. Faria. "Neuroimaging in Anxiety Disorders." *Modern Trends in Pharmacopsychiatry* (2013) 29:47–66.

Freeman, D., and J. Freeman. *Anxiety: A Very Short Introduction* (Oxford, UK: Oxford University Press, 2012).

Freire, R.C., G. Perna, and A.E. Nardi. "Panic Disorder Respiratory Subtype: Psychopathology, Laboratory Challenge Tests, and Response to Treatment." *Harvard Review of Psychiatry* (2010) 18:220–29.

Freud, S. "The Unconscious." In: *The Standard Edition of the Complete Psychological Works of Sigmund Freud*, vol. 14, ed. J. Strachey (London: The Hogarth Press, 1915), 161–215.

Freud, S. *Introductory Lectures on Psychoanalysis* (Vienna: H. Heller, 1917).

Freud, S. *Beyond the Pleasure Principle* (New York: Bantam Books, 1959).

Friedman, B.H. "Feelings and the Body: The Jamesian Perspective on Autonomic Specificity of Emotion." *Biological Psychology* (2010) 84:383–93.

Friedman, R.A. "The Feel-Good Gene." In: "Sunday Review." *The New York Times* (New York: The New York Times Company, 2015).

Frielingsdorf, H., et al. "Variant Brain-Derived Neurotrophic Factor Val66Met Endophenotypes: Implications for Posttraumatic Stress Disorder." *Annals of the New York Academy of Sciences* (2010) 1208:150–57.

Frith, C., and R. Dolan. "The Role of the Prefrontal Cortex in Higher Cognitive Functions." *Brain Research. Cognitive Brain Research* (1996) 5:175–81.

Frith, C., R. Perry, and E. Lumer. "The Neural Correlates of Conscious Experience: An Experimental Framework." *Trends in Cognitive Sciences* (1999) 3:105–14.

Frith, C.D. "Consciousness, Information Processing and the Brain." *Journal of Psychopharmacology* (1992) 6:436–40.

Frith, C.D. "The Social Functions of Consciousness." In: *Frontiers of Consciousness: Chichele Lectures*, eds. L. Weiskrantz and M. Davies (Oxford, UK: Oxford University Press, 2008), 225–44.

Frith, C.D., and U. Frith. "Social Cognition in Humans." *Current Biology* (2007) 17:R724–32.

Frohardt, R.J., F.A. Guarraci, and M.E. Bouton. "The Effects of Neurotoxic Hippocampal Lesions on Two Effects of Context after Fear Extinction." *Behavioral Neuroscience* (2000) 114:227–40.

Frysztak, R.J., and E.J. Neafsey. "The Effect of Medial Frontal Cortex Lesions on Respiration, 'Freezing,' and Ultrasonic Vocalizations during Conditioned Emotional Responses in Rats." *Cerebral Cortex* (1991) 1:418–25.

Furmark, T., et al. "Common Changes in Cerebral Blood Flow in Patients with Social Phobia Treated with Citalopram or Cognitive-Behavioral Therapy." *Archives of General Psychiatry* (2002) 59:425–33.

Fuster, J. *The Prefrontal Cortex* (New York: Academic Press, 2008).

Fuster, J.M. "The Prefrontal Cortex, Mediator of Cross-Temporal Contingencies." *Human Neurobiology* (1985) 4:169–79.

Fuster, J.M. *The Prefrontal Cortex* (New York: Raven, 1989).

Fuster, J.M. "The Prefrontal Cortex and Its Relation to Behavior." *Progress in Brain Research* (1991) 87:201–11.

Fuster, J.M. "Prefrontal Neurons in Networks of Executive Memory." *Brain Research Bulletin* (2000) 52:331–36.

Fuster, J.M. *Cortex and Mind: Unifying Cognition* (Oxford, UK: Oxford University Press, 2003).

Fuster, J.M. "The Cognit: A Network Model of Cortical Representation." *International Journal of Psychophysiology: Official Journal of the International Organization of Psychophysiology* (2006) 60:125–32.

Fuster, J.M., and S.L. Bressler. "Cognit Activation: A Mechanism Enabling Temporal Integration in Working Memory." *Trends in Cognitive Sciences* (2012) 16:207–18.

Gabriel, M. "Functions of Anterior and Posterior Cingulate Cortex During Avoidance Learning in Rabbits." *Progress in Brain Research* (1990) 85:467–82.

Gabriel, M., and E. Orona. "Parallel and Serial Processes of the Prefrontal and Cingulate Cortical Systems During Behavioral Learning." *Brain Research Bulletin* (1982) 8:781–85.

Gaillard, R., et al. "Converging Intracranial Markers of Conscious Access." *PLoS Biology* (2009) 7:E61.

Gaillard, R., et al. "Nonconscious Semantic Processing of Emotional Words Modulates Conscious Access." *Proceedings of the National Academy of Sciences of the United States of America* (2006) 103:7524–29.

Galatzer-Levy, I.R. "Empirical Characterization of Heterogeneous Posttraumatic Stress Responses Is Necessary to Improve the Science of Posttraumatic Stress." *The Journal of Clinical Psychiatry* (2014) 75:E950–52.

Galatzer-Levy, I.R., and R.A. Bryant. "636,120 Ways to Have Posttraumatic Stress Disorder." *Perspectives in Psychological Science* (2013) 50:161–80.

Galatzer-Levy, I.R., et al. "Heterogeneity in Signaled Active Avoidance Learning: Substantive and Methodological Relevance of Diversity in Instrumental Defensive Responses to Threat Cues." *Frontiers in Systems Neuroscience* (2014) 8:179.

Galea, S., A. Nandi, and D. Vlahov. "The Epidemiology of Post-Traumatic Stress Disorder After Disasters." *Epidemiologic Reviews* (2005) 27:78–91.

Gallagher, M., and P.C. Holland. "The Amygdala Complex: Multiple Roles in Associative Learning and Attention." *Proceedings of the National Academy of Sciences of the United States of America* (1994) 91:11771–76.

Galliot, B. "Hydra, a Fruitful Model System for 270 Years." *International Journal of Developmental Biology* (2012) 56:411–23.

Gallistel, C.R. "Animal Cognition: The Representation of Space, Time and Number." *Annual Review of Psychology* (1989) 40:155–89.

Gallistel, C.R., Gibbon J. "Time, Rate, and Conditioning." *Psychological Review* (2000) 107: 289–344.

Gallistel, R. *The Organization of Action: A New Synthesis* (Hillsdale, NJ: Erlbaum, 1980).

Gallup, G. "Toward a Comparative Psychology of Self-Awareness: Species Limitations and Cognitive Consequences." In: *The Self: Interdisciplinary Approaches*, eds. J. Strauss and G.R. Goethals (New York: Springer, 1991).

Gangestad, S.W., and M. Snyder. "Self-Monitoring: Appraisal and Reappraisal." *Psychological Bulletin* (2000) 126:530–55.

Garcia-Lazaro, H.G., et al. "Neuroanatomy of Episodic and Semantic Memory in Humans: A Brief Review of Neuroimaging Studies." *Neurology India* (2012) 60:613–17.

Gardiner, J.M. "Episodic Memory and Autonoetic Consciousness: A First-Person Approach." *Philosophical Transactions of the Royal Society B: Biological Sciences* (2001) 356:1351–61.

Gardner, H. *The Mind's New Science: A History of the Cognitive Revolution* (New York: Basic Books, 1987).

Garner, A.R., et al. "Generation of a Synthetic Memory Trace." *Science* (2012) 335:1513–16.

Garrido, M.I., et al. "Functional Evidence for a Dual Route to Amygdala." *Current Biology* (2012) 22:129–34.

Garrity, P.A., et al. "Running Hot and Cold: Behavioral Strategies, Neural Circuits, and the Molecular Machinery for Thermotaxis in *C. elegans* and *Drosophila*." *Genes & Development* (2010) 24:2365–82.

Gasser P., K. Kirchner, and T. Passie. "LSD-Assisted Psychotherapy for Anxiety Associated with a Life-Threatening Disease: A Qualitative Study of Acute and Sustained Subjective Effects." *Journal of Psychopharmacology* (2015) 29:57–68.

Gazzaniga, M.S. *The Bisected Brain* (New York: Appleton-Century-Crofts, 1970).

Gazzaniga, M.S. *Mind Matters* (Cambridge, MA: MIT Press, 1988).

Gazzaniga, M.S. *The Mind's Past* (Berkeley: University of California Press, 1998).

Gazzaniga, M.S. "The Split Brain Revisited." *Scientific American* (1998) 279:50–55.

Gazzaniga, M.S. *Human: The Science Behind What Makes Us Unique* (New York: Ecco, 2008).

Gazzaniga, M.S. *Who's in Charge?: Free Will and the Science of the Brain* (New York: Ecco, 2012).

Gazzaniga, M.S., and J.E. LeDoux. *The Integrated Mind* (New York: Plenum, 1978).

Gentner, D., and S. Goldin-Meadow, eds. *Language in Mind: Advances in the Study of Language and Thought* (Cambridge, MA: MIT Press, 2003).

Genud-Gabai, R., O. Klavir, and R. Paz. "Safety Signals in the Primate Amygdala." *Journal of Neuroscience* (2013) 33:17986–94.

George, M.S., et al. "A Pilot Study of Vagus Nerve Stimulation (VNS) for Treatment-Resistant Anxiety Disorders." *Brain Stimulation* (2008) 1:112–21.

Gerardi, M., et al. "Virtual Reality Exposure Therapy Using a Virtual Iraq: Case Report." *Journal of Traumatic Stress* (2008) 21:209–13.

Geschwind, N. "The Disconnexion Syndromes in Animals and Man." Part I. *Brain: A Journal of Neurology* (1965) 88:237–94.

Geschwind, N. "The Disconnexion Syndromes in Animals and Man." Part II. *Brain: A Journal of Neurology* (1965) 88:585–644.

Gibson, R.W., and J.A. Pickett. "Wild Potato Repels Aphids by Release of Aphid Alarm Pheromone." *Nature* (1983) 302:608–9.

Gilboa, A., et al. "Functional Connectivity of the Prefrontal Cortex and the Amygdala in Post-traumatic Stress Disorder." *Biological Psychiatry* (2004) 55:263–72.

Gilboa-Schechtman, E., M.E. Franklin, and E.B. Foa. "Anticipated Reactions to Social Events: Differences Among Individuals with Generalized Social Phobia, Obsessive Compulsive Disorder, and Nonanxious Controls." *Cognitive Therapy and Research* (2000) 24:731–46.

Gilmartin, M.R., N.L. Balderston, and F.J. Helmstetter. "Prefrontal Cortical Regulation of Fear Learning." *Trends in Neurosciences* (2014) 37:455–64.

Giske, J., et al. "Effects of the Emotion System on Adaptive Behavior." *American Naturalist* (2013) 182:689–703.

Giurfa, M. "Cognition with Few Neurons: Higher-Order Learning in Insects." *Trends in Neurosciences* (2013) 36:285–94.

Glanzman, D.L. "Common Mechanisms of Synaptic Plasticity in Vertebrates and Invertebrates." *Current Biology* (2010) 20:R31–36.

Glimcher, P.W. *Decisions, Uncertainty, and the Brain: The Science of Neuroeconomics* (Cambridge, MA: MIT Press, 2003).

Glimcher, P.W. *Neuroeconomics Decision Making and the Brain* (San Diego: Academic Press, 2009).

Gloor, P., et al. "The Role of the Limbic System in Experiential Phenomena of Temporal Lobe Epilepsy." *Annals of Neurology* (1982) 12:129–44.

Gluck, M.A., E. Mercado, and C.E. Myers. *Learning and Memory: From Brain to Behavior* (New York: Worth Publishers, 2007).

Goddard, G. "Functions of the Amygdala." *Psychological Review* (1964) 62:89–109.

Godsil, B.P., and M.S. Fanselow. "Motivation." In: *Handbook of Psychology*, vol. 4, eds. A.F. Healy and R.W. Proctor (Hoboken, NJ: John Wiley & Sons, 2013), 32–60.

Goel, V., and O. Vartanian. "Dissociating the Roles of Right Ventral Lateral and Dorsal Lateral Prefrontal Cortex in Generation and Maintenance of Hypotheses in Set-Shift Problems." *Cerebral Cortex* (2005) 15:1170–77.

Goldberg, E., and R.M. Bilder Jr. "The Frontal Lobes and Hierarchical Organization of Cognitive Control." In: *The Frontal Lobes Revisited,* ed. E. Perecman (New York: IRBN Press, 1987), 159–87.

Golden, W.L. "Cognitive Hypnotherapy for Anxiety Disorders." *The American Journal of Clinical Hypnosis* (2012) 54:263–74.

Goldfried, M.R., E.T. Decenteceo, and L. Weinberg. "Systematic Rational Restructuring as a Self-Control Technique." *Behavior Therapy* (1974) 5:247–54.

Goldin, P.R., et al. "The Neural Bases of Emotion Regulation: Reappraisal and Suppression of Negative Emotion." *Biological Psychiatry* (2008) 63:577–86.

Goldman-Rakic, P.S. "Circuitry of Primate Prefrontal Cortex and Regulation of Behavior by Representational Memory." In: *Handbook of Physiology Section 1: The Nervous System Vol V, Higher Functions of the Brain,* ed. F. Plum (Bethesda, MD: American Physiological Society, 1987), 373–418.

Goldman-Rakic, P.S. "Architecture of the Prefrontal Cortex and the Central Executive." *Annals of the New York Academy of Sciences* (1995) 769:71–83.

Goldman-Rakic, P.S. "The Prefrontal Landscape: Implications of Functional Architecture for Understanding Human Mentation and the Central Executive." *Philosophical Transactions of the Royal Society B: Biological Sciences* (1996) 351:1445–53.

Goldman-Rakic, P.S. "Working Memory, Neural Basis." In: *MIT Encyclopedia of Cognitive Sciences*, eds. R.A. Wilson and F.C. Keil (Cambridge, MA: MIT Press, 1999).

Goldstein, A.P., and F.H. Kanfer, eds. *Maximizing Treatment Gains: Transfer Enhancement in Psychotherapy* (New York: Academic Press, 1979).

Goldstein, M.L. "Acquired Drive Strength as a Joint Function of Shock Intensity and Number of Acquisition Trials." *Journal of Experimental Psychology* (1960) 60:349–58.

Goleman, D. *Emotional Intelligence: Why It Can Matter More Than IQ* (New York: Bantam Books, 2005).

Golkar, A., et al. "Distinct Contributions of the Dorsolateral Prefrontal and Orbitofrontal Cortex during Emotion Regulation." *PLoS One* (2012) 7:E48107.

Goltz, F. "Der Hund ohne Grosshirn." *Pfluegers Archiv für die gesammte Physiologie des Menschen und der Tiere* (1892) 51:570–614.

Goode, T.D., and S. Maren. "Animal Models of Fear Relapse." *ILAR Journal / National Research Council, Institute of Laboratory Animal Resources* (2014) 55:246–58.

Goosens, K.A. "Hippocampal Regulation of Aversive Memories." *Current Opinion in Neurobiology* (2011) 21:460–66.

Goosens, K.A., and S. Maren. "Long-Term Potentiation as a Substrate for Memory: Evidence from Studies of Amygdaloid Plasticity and Pavlovian Fear Conditioning." *Hippocampus* (2002) 12:592–99.

Goosens, K.A., and S. Maren. "Pretraining NMDA Receptor Blockade in the Basolateral Complex, but Not the Central Nucleus, of the Amygdala Prevents Savings of Conditional Fear." *Behavioral Neuroscience* (2003) 117:738–50.

Goosens, K.A., and S. Maren. "NMDA Receptors Are Essential for the Acquisition, but Not Expression, of Conditional Fear and Associative Spike Firing in the Lateral Amygdala." *European Journal of Neuroscience* (2004) 20:537–48.

Gordon, B., E.E. Allen, and P.Q. Trombley. "The Role of Norepinephrine in Plasticity of Visual Cortex." *Progress in Neurobiology* (1988) 30:171–91.

Gorman, J.M., et al. "Neuroanatomical Hypothesis of Panic Disorder, Revised." *The American Journal of Psychiatry* (2000) 157:493–505.

Gorman, J.M., et al. "A Neuroanatomical Hypothesis for Panic Disorder." *The American Journal of Psychiatry* (1989) 146:148–61.

Gorwood, P., et al. "Genetics of Dopamine Receptors and Drug Addiction." *Human Genetics* (2012) 131:803–22.

Gottlich, M., et al. "Decreased Limbic and Increased Fronto-Parietal Connectivity in Unmedicated Patients with Obsessive-Compulsive Disorder." *Human Brain Mapping* (2014) 35: 5617–32.

Gould, J.L. "Honey Bee Cognition." *Cognition* (1990) 37:83–103.

Gould, S.J., and R.C. Lewontin. "The Spandrels of San Marco and the Panglossian Paradigm: A Critique of the Adaptationist Programme." *Proceedings of the Royal Society of London Series B, Containing Papers of a Biological Character Royal Society* (1979) 205:581–98.

Goyal, M., et al. "Meditation Programs for Psychological Stress and Well-Being: A Systematic Review and Meta-Analysis." *JAMA Internal Medicine* (2014) 174:357–68.

Grace, A.A., and J.A. Rosenkranz. "Regulation of Conditioned Responses of Basolateral Amygdala Neurons." *Physiology & Behavior* (2002) 77:489–93.

Graeff, F.G. "Neuroanatomy and Neurotransmitter Regulation of Defensive Behaviors and Related Emotions in Mammals." *Brazilian Journal of Medical and Biological Research = Revista Brasileira de Pesquisas Medicas e Biologicas / Sociedade Brasileira de Biofisica* [et al] (1994) 27:811–29.

Graham, B.M., and M.R. Milad. "The Study of Fear Extinction: Implications for Anxiety Disorders." *The American Journal of Psychiatry* (2011) 168:1255–65.

Grandin, T. *Animals in Translation* (New York: Mariner Books, 2005).

Grant, R., et al. "The Release of Catechols from the Adrenal Medulla on Activation of the Sympathetic Vasodilator Nerves to the Skeletal Muscles in the Cat by Hypothalamic Stimulation." *Acta Physiologica Scandinavica* (1958) 43:135–54.

Gray, J.A. *The Neuropsychology of Anxiety* (New York: Oxford University Press, 1982).

Gray, J.A. *The Psychology of Fear and Stress* (New York: Cambridge University Press, 1987).

Gray, J.A. *Consciousness: Creeping Up on the Hard Problem* (Oxford, UK: Oxford University Press, 2004).

Gray, J.A., and N. McNaughton. "The Neuropsychology of Anxiety: Reprise." *Nebraska Symposium on Motivation* (1996) 43:61–134.

Gray, J.A., and N. McNaughton. *The Neuropsychology of Anxiety*, 2nd ed (Oxford, UK: Oxford University Press, 2000).

Gray, T.S., and E.W. Bingaman. "The Amygdala: Corticotropin-Releasing Factor, Steroids, and Stress." *Critical Reviews in Neurobiology* (1996) 10:155–68.

Gray, T.S., M.E. Carney, and D.J. Magnuson. "Direct Projections from the Central Amygdaloid Nucleus to the Hypothalamic Paraventricular Nucleus: Possible Role in Stress-Induced Adrenocorticotropin Release." *Neuroendocrinology* (1989) 50:433–46.

Gray, T.S., et al. "Ibotenic Acid Lesions in the Bed Nucleus of the Stria Terminalis Attenuate Conditioned Stress Induced Increases in Prolactin, A.C.TH, and Corticosterone." *Neuroendocrinology* (1993) 57:517–24.

Graziano, M.S.A. *Consciousness and the Social Brain* (Oxford, UK: Oxford University Press, 2013).

Graziano, M.S.A. "Are We Really Conscious?" In: Sunday Review. *The New York Times* (New York: The New York Times Company, 2014).

Greenberg, D.L., and M. Verfaellie. "Interdependence of Episodic and Semantic Memory: Evidence from Neuropsychology." *Journal of the International Neuropsychological Society* (2010) 16:748–53.

Greene, J., and J. Cohen. "For the Law, Neuroscience Changes Nothing and Everything." *Philosophical Transactions of the Royal Society B: Biological Sciences* (2004) 359:1775–85.

Greenfield, S. *Journey to the Centers of the Mind: Toward a Science of Consciousness* (San Francisco: W. H. Freeman, 1995).

Greening, T. "Five Basic Postulates of Humanistic Psychology." *Journal of Humanistic Psychology* (2006) 46:239.

Greenwald, A.G., and M.R. Banaji. "Implicit Social Cognition: Attitudes, Self-Esteem, and Stereotypes." *Psychological Review* (1995) 102:4–27.

Greenwald, A.G., S.C. Draine, and R.L. Abrams. "Three Cognitive Markers of Unconscious Semantic Activation." *Science* (1996) 273:1699–1702.

Griebel, G., and A. Holmes. "50 Years of Hurdles and Hope in Anxiolytic Drug Discovery." *Nature Reviews Drug Discovery* (2013) 12:667–87.

Griffin, D.R. "Animal Consciousness." *Neuroscience and Biobehavioral Reviews* (1985) 9:615–22.

Griffiths, P.E. *What Emotions Really Are: The Problem of Psychological Categories* (Chicago: University of Chicago Press, 1997).

Griffiths, P.E. "Is Emotion a Natural Kind?" In: *Thinking About Feeling: Contemporary Philosophers on Emotions*, ed. R.C. Solomon (Oxford, UK: Oxford University Press, 2004), 233–49.

Grillon, C. "Models and Mechanisms of Anxiety: Evidence from Startle Studies." *Psychopharmacology (Berl)* (2008) 199:421–37.

Grillon, C., et al. "The Benzodiazepine Alprazolam Dissociates Contextual Fear from Cued Fear in Humans as Assessed by Fear-Potentiated Startle." *Biological Psychiatry* (2006) 60:760–66.

Grillon, C., et al. "Increased Anxiety During Anticipation of Unpredictable but Not Predictable Aversive Stimuli as a Psychophysiologic Marker of Panic Disorder." *American Journal of Psychiatry* (2008) 165:898–904.

Grillon, C., et al. "Increased Anxiety During Anticipation of Unpredictable Aversive Stimuli in Posttraumatic Stress Disorder but Not in Generalized Anxiety Disorder." *Biological Psychiatry* (2009) 66:47–53.

Groenewegen, H.J., C.I. Wright, and A.V. Beijer. "The Nucleus Accumbens: Gateway for Limbic Structures to Reach the Motor System?" *Progress in Brain Research* (1996) 107:485–511.

Groenewegen, H.J., C.I. Wright, A.V. Beijer, and P. Voorn. "Convergence and Segregation of Ventral Striatal Inputs and Outputs." *Annals of the New York Academy of Sciences* (1999) 877:49–63.

Groenewegen, H.J., C.I. Wright, and H.B. Uylings. "The Anatomical Relationships of the Prefrontal Cortex with Limbic Structures and the Basal Ganglia." *Journal of Psychopharmacology* (1997) 11:99–106.

Gross, C.T., and N.S. Canteras. "The Many Paths to Fear." *Nature Reviews Neuroscience* (2012) 13:651–58.

Gross, J.J. "Emotion Regulation: Affective, Cognitive, and Social Consequences." *Psychophysiology* (2002) 39:281–91.

Gross, M. "Elements of Consciousness in Animals." *Current Biology* (2013) 23:R981–83.

Groves, P.M., R. De Marco, and R.F. Thompson. "Habituation and Sensitization of Spinal Interneuron Activity in Acute Spinal Cat." *Brain Research* (1969) 14:521–25.

Groves, P.M., and R.F. Thompson. "Habituation: a Dual-Process Theory." *Psychological Review* (1970) 77:419–50.

Gruber, J., A.C. Hay, and J.J. Gross. "Rethinking Emotion: Cognitive Reappraisal Is an Effective Positive and Negative Emotion Regulation Strategy in Bipolar Disorder." *Emotion* (2014) 14: 388–96.

Grupe, D.W., and J.B. Nitschke. "Uncertainty and Anticipation in Anxiety: An Integrated Neurobiological and Psychological Perspective." *Nature Reviews Neuroscience* (2013) 14: 488–501.

Grupe, D.W., D.J. Oathes, and J.B. Nitschke. "Dissecting the Anticipation of Aversion Reveals Dissociable Neural Networks." *Cerebral Cortex* (2013) 23:1874–83.

Gu X., et al. "Anterior Insular Cortex and Emotional Awareness." *Journal of Comparative Neurology* (2013) 521:3371–88.

Gusnard, D.A., and M.E. Raichle. "Searching for a Baseline: Functional Imaging and the Resting Human Brain." *Nature Reviews Neuroscience* (2001) 2:685–94.

Guyer, A.E., et al. "Amygdala and Ventrolateral Prefrontal Cortex Function During Anticipated Peer Evaluation in Pediatric Social Anxiety." *Archives of General Psychiatry* (2008) 65:1303–12.

Guz, A. "Brain, Breathing and Breathlessness." *Respiration Physiology* (1997) 109:197–204.

Gyurak, A., J.J. Gross, and A. Etkin. "Explicit and Implicit Emotion Regulation: A Dual-Process Framework." *Cognition & Emotion* (2011) 25:400–12.

Hadj-Bouziane, F., et al. "Amygdala Lesions Disrupt Modulation of Functional MRI Activity Evoked by Facial Expression in the Monkey Inferior Temporal Cortex." *Proceedings of the National Academy of Sciences of the United States of America* (2012) 109:E3640–48.

Haldane, E.S., and G.R.T. Ross. *The Philosophical Works of Descartes* (Cambridge, UK: Cambridge University Press, 1911).

Halgren, E. "The Amygdala Contribution to Emotion and Memory: Current Studies in Humans." In: *The Amygdaloid Complex*, ed. Y. Ben-Ari (Amsterdam: Elsevier, 1981), 395–408.

Halgren, E., et al. "Mental Phenomena Evoked by Electrical Stimulation of the Human Hippocampal Formation and Amygdala." *Brain: A Journal of Neurology* (1978) 101:83–117.

Hall, J., et al. "Involvement of the Central Nucleus of the Amygdala and Nucleus Accumbens Core in Mediating Pavlovian Influences on Instrumental Behaviour." *European Journal of Neuroscience* (2001) 13:1984–92.

Hamann, S.B., and L.R. Squire. "Intact Priming for Novel Perceptual Representations in Amnesia." *Journal of Cognitive Neuroscience* (1997) 9:699–713.

Hameroff, S., and R. Penrose. "Consciousness in the Universe: A Review of the 'Orch OR' Theory." *Physics of Life Reviews* (2014) 11:39–78.

Hamilton, J.P., et al. "Functional Neuroimaging of Major Depressive Disorder: A Meta-Analysis and New Integration of Base Line Activation and Neural Response Data." *The American Journal of Psychiatry* (2012) 169:693–703.

Hamm, A.O., et al. "Affective Blindsight: Intact Fear Conditioning to a Visual Cue in a Cortically Blind Patient." *Brain: A Journal of Neurology* (2003) 126:267–75.

Hammond, D.C. "Hypnosis in the Treatment of Anxiety- and Stress-Related Disorders." *Expert Review of Neurotherapeutics* (2010) 10:263–73.

Hampton, R.R. "Rhesus Monkeys Know When They Remember." *Proceedings of the National Academy of Sciences of the United States of America* (2001) 98:5359–62.

Hampton, R.R. "Multiple Demonstrations of Metacognition in Nonhumans: Converging Evidence or Multiple Mechanisms?" *Comparative Cognition & Behavior Reviews* (2009) 4:17–28.

Han, J.H., et al. "Neuronal Competition and Selection during Memory Formation." *Science* (2007) 316:457–60.

Han, J.H., et al. "Selective Erasure of a Fear Memory." *Science* (2009) 323:1492–96.

Han, S.W., and R. Marois. "The Effects of Stimulus-Driven Competition and Task Set on Involuntary Attention." *Journal of Vision* (2014) 14.

Hannula, D.E., and A.J. Greene. "The Hippocampus Reevaluated in Unconscious Learning and Memory: At a Tipping Point?" *Frontiers in Human Neuroscience* (2012) 6:80.

Haouzi, P., B. Chenuel, and G. Barroche. "Interactions Between Volitional and Automatic Breathing during Respiratory Apraxia." *Respiratory Physiology & Neurobiology* (2006) 152:169–75.

Hariri, A.R., E.M. Drabant, and D.R. Weinberger. "Imaging Genetics: Perspectives from Studies of Genetically Driven Variation in Serotonin Function and Corticolimbic Affective Processing." *Biological Psychiatry* (2006) 59:888–97.

Hariri, A.R., and A. Holmes. "Genetics of Emotional Regulation: The Role of the Serotonin Transporter in Neural Function." *Trends in Cognitive Sciences* (2006) 10:182–91.

Hariri A.R., et al. "The Amygdala Response to Emotional Stimuli: A Comparison of Faces and Scenes." *NeuroImage* (2002) 17:317–23.

Hariri, A.R., and D.R. Weinberger. "Functional Neuroimaging of Genetic Variation in Serotonergic Neurotransmission." *Genes, Brain, and Behavior* (2003) 2:341–49.

Hariz, M., P. Blomstedt, and L. Zrinzo. "Future of Brain Stimulation: New Targets, New Indications, New Technology." *Movement Disorders: Official Journal of the Movement Disorder Society* (2013) 28:1784–92.

Harley, C. "Noradrenergic and Locus Coeruleus Modulation of the Perforant Path-Evoked Potential in Rat Dentate Gyrus Supports a Role for the Locus Coeruleus in Attentional and Memorial Processes." *Progress in Brain Research* (1991) 88:307–21.

Harley, H.E. "Consciousness in Dolphins? A Review of Recent Evidence." *Journal of Comparative Physiology A, Neuroethology, Sensory, Neural, and Behavioral Physiology* (2013) 199:565–82.

Harshey, R.M. "Bees Aren't the Only Ones: Swarming in Gram-Negative Bacteria." *Molecular Microbiology* (1994) 13:389–94.

Hart, C.L., et al. "Is Cognitive Functioning Impaired in Methamphetamine Users? A Critical Review." *Neuropsychopharmacology* (2012) 37:586–608.

Hart, M., A. Poremba, and M. Gabriel. "The Nomadic Engram: Overtraining Eliminates the Impairment of Discriminative Avoidance Behavior Produced by Limbic Thalamic Lesions." *Behavioural Brain Research* (1997) 82:169–77.

Hartley, C.A., and B.J. Casey. "Risk for Anxiety and Implications for Treatment: Developmental, Environmental, and Genetic Factors Governing Fear Regulation." *Annals of the New York Academy of Sciences* (2013) 1304:1–13.

Hartley, C.A., B. Fischl, and E.A. Phelps. "Brain Structure Correlates of Individual Differences in the Acquisition and Inhibition of Conditioned Fear." *Cerebral Cortex* (2011) 21:1954–62.

Hartley, C.A., et al. "Serotonin Transporter Polyadenylation Polymorphism Modulates the Retention of Fear Extinction Memory." *Proceedings of the National Academy of Sciences of the United States of America* (2012) 109:5493–98.

Hartley, C.A., and E.A. Phelps. "Changing Fear: The Neurocircuitry of Emotion Regulation." *Neuropsychopharmacology* (2010) 35:136–46.

Hartley, C.A., and E.A. Phelps. "Anxiety and Decision-Making." *Biological Psychiatry* (2012) 72:113–18.

Hartley, T., et al. "Space in the Brain: How the Hippocampal Formation Supports Spatial Cognition." *Philosophical Transactions of the Royal Society B: Biological Sciences* (2014) 369:20120510.

Hasselmo, M.E., et al. "Noradrenergic Suppression of Synaptic Transmission May Influence Cortical Signal-to-Noise Ratio." *Journal of Neurophysiology* (1997) 77:3326–39.

Hassin, R.R., et al. "Implicit Working Memory." *Consciousness and Cognition* (2009) 18:665–78.

Hasson, U., et al. "Abstract Coding of Audiovisual Speech: Beyond Sensory Representation." *Neuron* (2007) 56:1116–26.

Hatkoff, A. *The Inner World of Farm Animals* (New York: Stewart, Tabori, and Chang, 2009).

Haubensak, W., et al. "Genetic Dissection of an Amygdala Microcircuit That Gates Conditioned Fear." *Nature* (2010) 468:270–76.

Haubrich, J., et al. "Reconsolidation Allows Fear Memory to Be Updated to a Less Aversive Level Through the Incorporation of Appetitive Information." *Neuropsychopharmacology* (2014) 40: 315–326.

Hauner, K.K., et al. "Exposure Therapy Triggers Lasting Reorganization of Neural Fear Processing." *Proceedings of the National Academy of Sciences of the United States of America* (2012) 109:9203–8.

Hawkins, R.D., et al. "A Cellular Mechanism of Classical Conditioning in *Aplysia*: Activity-Dependent Amplification of Presynaptic Facilitation." *Science* (1983) 219:400–5.

Hawkins, R.D., E.R. Kandel, and C.H. Bailey. "Molecular Mechanisms of Memory Storage in *Aplysia*." *The Biological Bulletin* (2006) 210:174–91.

Hayes, J.P., M.B. Vanelzakker, and L.M. Shin. "Emotion and Cognition Interactions in PTSD: A Review of Neurocognitive and Neuroimaging Studies." *Frontiers in Integrative Neuroscience* (2012) 6:89.

Hayes, S.C. "Acceptance and Commitment Therapy, Relational Frame Theory, and the Third Wave of Behavioral and Cognitive Therapies." *Behavior Therapy* (2004) 35:639–65.

Hayes, S.C., et al. "Acceptance and Commitment Therapy: Model, Processes and Outcomes." *Behaviour Research and Therapy* (2006) 44:1–25.

Hayes, S.C., K. Strosahl, and K.G. Wilson. *Acceptance and Commitment Therapy: An Experiential Approach to Behavior Change* (New York: Guilford Press, 1999).

Heath, R.G. *Studies in Schizophrenia: A Multidisciplinary Approach to Mind-Brain Relationships* (Cambridge, MA: Harvard University Press, 1954).

Heath, R.G. "Electrical Self-Stimulation of the Brain in Man." *The American Journal of Psychiatry* (1963) 120:571–77.

Heath, R.G. (Ed.) *The Role of Pleasure in Human Behavior* (New York: Harper and Row, 1964).

Heath, R.G. "Pleasure and Brain Activity in Man. Deep and Surface Electroencephalograms During Orgasm." *The Journal of Nervous and Mental Disease* (1972) 154:3–18.

Heath, R.G., and W.A. Mickle. "Evaluation of Seven Years' Experience with Depth Electrode Studies in Human Patients." In: *Electrical Studies on the Unanesthetized Brain*, eds. E.R. Ramey and D.S. O'Doherty (New York: Hoeber, 1960), 214–47.

Hebb, D.O. *The Organization of Behavior* (New York: John Wiley and Sons, 1949).

Hebb, D.O. "Drives and the CNS (Conceptual Nervous System)." *Psychological Review* (1955) 62:243–54.

Heberlein, A.S., and R. Adolphs. "Impaired Spontaneous Anthropomorphizing Despite Intact Perception and Social Knowledge." *Proceedings of the National Academy of Sciences of the United States of America* (2004) 101:7487–91.

Hedden, T., et al. "Cultural Influences on Neural Substrates of Attentional Control." *Psychological Science* (2008) 19:12–17.

Heeramun-Aubeeluck, A., and Z. Lu. "Neurosurgery for Mental Disorders: A Review." *African Journal of Psychiatry* (2013) 16:177–81.

Heerebout, B.T., and R.H. Phaf. "Emergent Oscillations in Evolutionary Simulations: Oscillating Networks Increase Switching Efficacy." *Journal of Cognitive Neuroscience* (2010) 22:807–23.

Heidegger, M. *Being and Time*, trans. by John MacQuarrie and Edward Robinson (London: SCM Press, 1962). (Germany: 1927).

Heider, F. *The Psychology of Interpersonal Relations* (New York: John Wiley & Sons, 1958).

Heider, F., and M. Simmel. "An Experimental Study of Apparent Behavior." *American Journal of Psychology* (1944) 57:243–59.

Helmstetter, C., et al. "On the Bacterial Life Sequence." *Cold Spring Harbor Symposium on Quantitative Biology* (1968) 33:809–22.

Hennessey, T.M., W.B. Rucker, and C.G. McDiarmid. "Classical Conditioning in Paramecia." *Animal Learning and Behavior* (1979) 7:417–23.

Hermann, A., et al. "Brain Structural Basis of Cognitive Reappraisal and Expressive Suppression." *Social Cognitive and Affective Neuroscience* (2014) 9:1435–42.

Herrick, C.J. "The Functions of the Olfactory Parts of the Cerebral Cortex." *Proceedings of the National Academy of Sciences* (1933) 19:7–14.

Herrick, C.J. *The Brain of the Tiger Salamander* (Chicago: The University of Chicago Press, 1948).

Herry, C., et al. "Switching on and off Fear by Distinct Neuronal Circuits." *Nature* (2008) 454: 600–6.

Herry, C., et al. "Neuronal Circuits of Fear Extinction." *European Journal of Neuroscience* (2010) 31:599–612.

Hess, W.R. *Das Zwischenhirn. Syndrome, Lokalisationen, Funktionen.* (Basel: Schwabe, 1949).

Hess, W.R. *The Biology of Mind* (Chicago: University of Chicago Press, 1962).

Hess, W.R., and M. Brugger. "Das Subkortikale Zentrum der Affektiven Abwehrreaktion." *Helvetica Physiologica et Pharmacologica Acta* (1943) 1:35–52.

Hettema, J.M., et al. "The Genetic Covariation Between Fear Conditioning and Self-Report Fears." *Biological Psychiatry* (2008) 63:587–93.

Hettema, J.M., M.C. Neale, and K.S. Kendler. "A Review and Meta-Analysis of the Genetic Epidemiology of Anxiety Disorders." *The American Journal of Psychiatry* (2001) 158:1568–78.

Hettema, J.M., C.A. Prescott, and K.S. Kendler. "A Population-Based Twin Study of Generalized Anxiety Disorder in Men and Women." *The Journal of Nervous and Mental Disease* (2001) 189:413–20.

Heyes, C. "Beast Machines? Questions of Animal Consciousness." In: *Frontiers of Consciousness: Chichelle Lectures*, eds. L. Weiskrantz and M. Davies (Oxford, UK: Oxford University Press, 2008), 259–74.

Heyes, C.M. "Reflections on Self-Recognition in Primates." *Animal Behaviour* (1994) 47:909–19.

Heyes, C.M. "Self-Recognition in Primates: Further Reflections Create a Hall of Mirrors." *Animal Behaviour* (1995) 50:1533–42.

Higgins, G.A., and J.S. Schwaber. "Somatostatinergic Projections from the Central Nucleus of the Amygdala to the Vagal Nuclei." *Peptides* (1983) 4:657–62.

Hilton, S.M. "The Defense Reaction as a Paradigm for Cardiovascular Control." In: *Integrative Functions of the Autonomic Nervous System*, eds. C.M. Brooks, et al. (Tokyo: University of Tokyo Press, 1979), 443–49.

Hilton, S.M. "The Defence-Arousal System and Its Relevance for Circulatory and Respiratory Control." *The Journal of Experimental Biology* (1982) 100:159–74.

Hilton, S.M., and A.W. Zbrozyna. "Amydaloid Region for Defense Reactions and Its Efferent Pathway to the Brainstem." *Journal of Physiology* (1963) 165:160–73.

Hinson, J.M., T.L. Jameson, and P. Whitney. "Somatic Markers, Working Memory, and Decision Making." *Cognitive, Affective & Behavioral Neuroscience* (2002) 2:341–53.

Hirst, W., J. LeDoux, and S. Stein. "Constraints on the Processing of Indirect Speech Acts: Evidence from Aphasiology." *Brain and Language* (1984) 23:26–33.

Hirst, W., et al. "Long-Term Memory for the Terrorist Attack of September 11: Flashbulb Memories, Event Memories, and the Factors That Influence Their Retention." *Journal of Experimental Psychology: General* (2009) 138:161–76.

Hitchens, C. *Hitch-22: A Memoir* (London: Atlantic Books, 2010).

Hobson, A. "The Neurobiology of Consciousness: Lucid Dreaming Wakes Up." *International Journal of Dream Research* (2009) 2:41–44.

Hoebel, B.G. "Hypothalamic Self-Stimulation and Stimulation Escape in Relation to Feeding and Mating." *Federation Proceedings* (1979) 38:2454–61.

Hoeft, F., et al. "Functional Brain Basis of Hypnotizability." *Archives of General Psychiatry* (2012) 69:1064–72.

Hofmann, S.G. "Cognitive Processes During Fear Acquisition and Extinction in Animals and Humans: Implications for Exposure Therapy of Anxiety Disorders." *Clinical Psychological Review* (2008) 28:199–210.

Hofmann, S.G. *An Introduction to Modern CBT: Psychological Solutions to Mental Health Problems* (New York: Wiley-Blackwell, 2011).

Hofmann, S.G., and G.J. Asmundson. "Acceptance and Mindfulness-Based Therapy: New Wave or Old Hat?" *Clinical Psychological Review* (2008) 28:1–16.

Hofmann, S.G., G.J. Asmundson, and A.T. Beck. "The Science of Cognitive Therapy." *Behavior Therapy* (2013) 44:199–212.

Hofmann, S.G., K.K. Ellard, and G.J. Siegle. "Neurobiological Correlates of Cognitions in Fear and Anxiety: A Cognitive-Neurobiological Information-Processing Model." *Cognition & Emotion* (2012) 26:282–99.

Hofmann, S.G., A. Fang, and C.A. Gutner. "Cognitive Enhancers for the Treatment of Anxiety Disorders." *Restorative Neurology and Neuroscience* (2014) 32:183–95.

Hofmann, S.G., C.A. Gutner, and A. Asnaani. "Cognitive Enhancers in Exposure Therapy for Anxiety and Related Disorders." In: *Exposure Therapy: Rethinking the Model—Refining the Method*, eds. P. Neudeck and H.-U. Wittchen (New York: Springer, 2012), 89–110.

Hofmann, S.G., and J.A. Smits. "Cognitive-Behavioral Therapy for Adult Anxiety Disorders: A Meta-Analysis of Randomized Placebo-Controlled Trials." *The Journal of Clinical Psychiatry* (2008) 69:621–32.

Holland, P.C. "Cognitive Aspects of Classical Conditioning." *Current Opinion in Neurobiology* (1993) 3:230–36.

Holland, P.C. "Relations Between Pavlovian-Instrumental Transfer and Reinforcer Devaluation." *Journal of Experimental Psychology: Animal Behavior Processes* (2004) 30:104–17.

Holland, P.C. "Cognitive Versus Stimulus-Response Theories of Learning." *Learning & Behavior* (2008) 36:227–41.

Holland, P.C., and M.E. Bouton. "Hippocampus and Context in Classical Conditioning." *Current Opinion in Neurobiology* (1999) 9:195–202.

Holland, P.C., and M. Gallagher. "Amygdala Circuitry in Attentional and Representational Processes." *Trends in Cognitive Sciences* (1999) 3:65–73.

Holland, P.C., and M. Gallagher. "Amygdala-Frontal Interactions and Reward Expectancy." *Current Opinion in Neurobiology* (2004) 14:148–55.

Holmes, A., and C.L. Wellman. "Stress-Induced Prefrontal Reorganization and Executive Dysfunction in Rodents." *Neuroscience and Biobehavioral Reviews* (2009) 33:773–83.

Holmes, N.M., A.R. Marchand, and E. Coutureau. "Pavlovian to Instrumental Transfer: A Neurobehavioural Perspective." *Neuroscience and Biobehavioral Reviews* (2010) 34:1277–95.

Holzschneider, K., and C. Mulert. "Neuroimaging in Anxiety Disorders." *Dialogues in Clinical Neuroscience* (2011) 13:453–61.

Homayoun, H., and B. Moghaddam. "Differential Representation of Pavlovian-Instrumental Transfer by Prefrontal Cortex Subregions and Striatum." *European Journal of Neuroscience* (2009) 29:1461–76.

Hooper, J., and D. Teresi. *The Three-Pound Universe* (New York: G. P. Putnam, 1991).

Hopkins, D.A., and D. Holstege. "Amygdaloid Projections to the Mesencephalon, Pons, and Medulla Oblongata in the Cat." *Experimental Brain Research* (1978) 32:529–47.

Horikawa, M., and A. Yagi. "The Relationships Among Trait Anxiety, State Anxiety and the Goal Performance of Penalty Shoot-Out by University Soccer Players." *PLoS One* (2012) 7:E35727.

Horinek, D., A. Varjassyova, and J. Hort. "Magnetic Resonance Analysis of Amygdalar Volume in Alzheimer's Disease." *Current Opinion in Psychiatry* (2007) 20:273–77.

Horwitz, A.V., and J.C. Wakefield. *All We Have to Fear: Psychiatry's Transformation of Natural Anxieties into Mental Disorders* (New York: Oxford University Press, 2012).

Hoyer, J., and K. Beesdo-Baum. "Prolonged Imaginal Exposure Based on Worry Scenarios." In: *Exposure Therapy: Rethinking the Model—Refining the Method*, eds. P. Neudeck and H.-U. Wittchen (New York: Springer, 2012), 245–60.

Hubbard, D.T., et al. "Development of Defensive Behavior and Conditioning to Cat Odor in the Rat." *Physiology & Behavior* (2004) 80:525–30.

Huff, N.C., et al. "Revealing Context-Specific Conditioned Fear Memories with Full Immersion Virtual Reality." *Frontiers in Behavioral Neuroscience* (2011) 5:75.

Hughes, K.C., and L.M. Shin. "Functional Neuroimaging Studies of Post-Traumatic Stress Disorder." *Expert Review of Neurotherapeutics* (2011) 11:275–85.

Hull, C.L. *Principles of Behavior* (New York: Appleton-Century-Crofts, 1943).

Hull, C.L. *A Behavior System: An Introduction to Behavior Theory Concerning the Individual Organism* (New Haven, CT: Yale University Press, 1952).

Humphrey, N. *Seeing Red: A Study in Consciousness* (Cambridge, MA: Harvard University Press, 2006).

Humphrey, N.K. "What the Frog's Eye Tells the Monkey's Brain." *Brain, Behavior and Evolution* (1970) 3:324–37.

Humphrey, N.K. "Vision in a Monkey Without Striate Cortex: A Case Study." *Perception* (1974) 3:241–55.

Hunsperger, R.W. "Affektreaktionen auf elektrische Reizung im Hirnstamm der Katze." *Helvetica Physiologica et Pharmacologica Acta* (1956) 14:70–92.

Hunt, H.F., and J.V. Brady. "Some Effects of Electro-Convulsive Shock on a Conditioned Emotional Response ('Anxiety')." *Journal of Comparative and Physiological Psychology* (1951) 44:88–98.

Hupbach, A., et al. "Reconsolidation of Episodic Memories: A Subtle Reminder Triggers Integration of New Information." *Learning & Memory* (2007) 14:47–53.

Hupbach, A., et al. "The Dynamics of Memory: Context-Dependent Updating." *Learning & Memory* (2008) 15:574–79.

Hurlemann, R., et al. "Emotion-Induced Retrograde Amnesia Varies as a Function of Noradrenergic-Glucocorticoid Activity." *Psychopharmacology (Berl)* (2007) 194:261–69.

Hurley, S. "The Shared Circuits Model (SCM): How Control, Mirroring, and Simulation Can Enable Imitation, Deliberation, and Mindreading." *Behavioral and Brain Sciences* (2008) 31:1–22; discussion 22–58.

Hutchinson, J.B., M.R. Uncapher, and A.D. Wagner. "Posterior Parietal Cortex and Episodic Retrieval: Convergent and Divergent Effects of Attention and Memory." *Learning & Memory* (2009) 16:343–56.

Hygge, S., and A. Öhman. "Modeling Processes in the Acquisition of Fears: Vicarious Electrodermal Conditioning to Fear-Relevant Stimuli." *Journal of Personality and Social Psychology* (1978) 36:271–79.

Hyman, S.E. "Can Neuroscience Be Integrated into the DSM-V?" *Nature Reviews Neuroscience* (2007) 8:725–32.

Imamoglu, F., et al. "Changes in Functional Connectivity Support Conscious Object Recognition." *NeuroImage* (2012) 63:1909–17.

Insel, T., et al. "Research Domain Criteria (RDoC): Toward a New Classification Framework for Research on Mental Disorders." *The American Journal of Psychiatry* (2010) 167:748–51.

Insel, T.R. "The Challenge of Translation in Social Neuroscience: A Review of Oxytocin, Vaso-pressin, and Affiliative Behavior." *Neuron* (2010) 65:768–79.

Ipser, J.C., L. Singh, and D.J. Stein. "Meta-Analysis of Functional Brain Imaging in Specific Phobia." *Psychiatry and Clinical Neurosciences* (2013) 67:311–22.

Isserles, M., et al. "Effectiveness of Deep Transcranial Magnetic Stimulation Combined with a Brief Exposure Procedure in Post-Traumatic Stress Disorder—A Pilot Study." *Brain Stimulation* (2013) 6:377–83.

Izard, C.E. *The Face of Emotion* (New York: Appleton-Century-Crofts, 1971).

Izard, C.E. "Basic Emotions, Relations Among Emotions, and Emotion-Cognition Relations." *Psychological Review* (1992) 99:561–65.

Izard, C.E. "Basic Emotions, Natural Kinds, Emotion Schemas, and a New Paradigm." *Perspectives on Psychological Science* (2007) 2:260–80.

Izquierdo A., C.L. Wellman, and A. Holmes. "Brief Uncontrollable Stress Causes Dendritic Retraction in Infralimbic Cortex and Resistance to Fear Extinction in Mice." *Journal of Neuroscience* (2006) 26:5733–38.

Izquierdo, I., et al. "The Connection Between the Hippocampal and the Striatal Memory Systems of the Brain: a Review of Recent Findings." *Neurotoxicity Research* (2006) 10:113–21.

Jackendoff, R. *Consciousness and the Computational Mind* (Cambridge, MA: Bradford Books/MIT Press, 1987).

Jackendoff, R. *Language, Consciousness, Culture: Essays on Mental Structure* (Cambridge, MA: MIT Press, 2007).

Jacob, T., et al. "A Nanotechnology-Based Delivery System: Nanobots. Novel Vehicles for Molecular Medicine." *The Journal of Cardiovascular Surgery* (2011) 52:159–67.

Jacobs, W.J., and L. Nadel. "Stress-Induced Recovery of Fears and Phobias." *Psychological Review* (1985) 92:512–31.

Jacobsen, C.F. "Studies of Cerebral Function in Primates. I. The Functions of the Frontal Associations Areas in Monkeys." *Comparative Psychology Monographs* (1936) 13:3–60.

Jacoby, L.L. "A Process Dissociation Framework: Separating Automatic from Intentional Uses of Memory." *Journal of Memory and Learning* (1991) 30:513–41.

James, J.P., K.R. Daniels, and B. Hanson. "Overhabituation and Spontaneous Recovery of the Galvanic Skin Response." *Journal of Experimental Psychology* (1974) 102:732–34.

James, W. "What Is an Emotion?" *Mind* (1884) 9:188–205.

James, W. *Principles of Psychology* (New York: Holt, 1890).

Jang, J.H., et al. "Increased Default Mode Network Connectivity Associated with Meditation." *Neuroscience Letters* (2011) 487:358–62.

Jarvis, E.D., et al. "Avian Brains and a New Understanding of Vertebrate Brain Evolution." *Nature Reviews Neuroscience* (2005) 6:151–59.

Jellinger, K.A. "Neuropathological Aspects of Alzheimer Disease, Parkinson Disease and Frontotemporal Dementia." *Neuro-Degenerative Diseases* (2008) 5:118–21.

Jennings, J.H., et al. "Distinct Extended Amygdala Circuits for Divergent Motivational States." *Nature* (2013) 496:224–28.

Ji, J., and S. Maren. "Hippocampal Involvement in Contextual Modulation of Fear Extinction." *Hippocampus* (2007) 17:749–58.

Johansen, J.P. "Neuroscience: Anxiety Is the Sum of Its Parts." *Nature* (2013) 496:174–75.

Johansen, J.P., et al. "Molecular Mechanisms of Fear Learning and Memory." *Cell* (2011) 147: 509–24.

Johansen, J.P., et al. "Hebbian and Neuromodulatory Mechanisms Interact to Trigger Associative Memory Formation." *Proceedings of the National Academy of Sciences USA.* (2014) 111: E5584–92.

Johansen, J.P., et al. "Optical Activation of Lateral Amygdala Pyramidal Cells Instructs Associative Fear Learning." *Proceedings of the National Academy of Sciences of the United States of America* (2010) 107:12692–97.

Johnson, D.C., and B.J. Casey. "Easy to Remember, Difficult to Forget: The Development of Fear Regulation." *Developmental Cognitive Neuroscience* (2014). 11:42–55.

Johnson, M.K., et al. "The Cognitive Neuroscience of True and False Memories." *Nebraska Symposium on Motivation* (2012) 58:15–52.

Johnson, P.L., L.M. Federici, and A. Shekhar. "Etiology, Triggers and Neurochemical Circuits Associated with Unexpected, Expected, and Laboratory-Induced Panic Attacks." *Neuroscience and Biobehavioral Reviews* (2014). 46:429–54.

Johnson, P.L., et al. "Orexin, Stress, and Anxiety/Panic States." *Progress in Brain Research* (2012) 198:133–61.

Johnson-Laird, P.N. *The Computer and the Mind: An Introduction to Cognitive Science* (Cambridge, MA: Harvard University Press, 1988).

Johnson-Laird, P.N. "A Computational Analysis of Consciousness." In: *Consciousness in Contemporary Science*, eds. A.J. Marcel and E. Bisiach (Oxford, UK: Oxford University Press, 1993), 357–68.

Johnson-Laird, P.N., and K. Oatley. "The Language of Emotions: An Analysis of a Semantic Field." *Cognition and Emotion* (1989) 3:81–123.

Johnson-Laird, P.N., and K. Oatley. "Basic Emotions, Rationality, and Folk Theory." *Cognition and Emotion* (1992) 6:201–23.

Jones, C.E., et al. "Social Transmission of Pavlovian Fear: Fear-Conditioning by-Proxy in Related Female Rats." *Animal Cognition* (2014) 17:827–34.

Jones, C.L., J. Ward, and H.D. Critchley. "The Neuropsychological Impact of Insular Cortex Lesions." *Journal of Neurology, Neurosurgery, and Psychiatry* (2010) 81:611–18.

Jones, E.G., and T.P.S. Powell. "An Anatomical Study of Converging Sensory Pathways Within the *Cerebral Cortex* of the Monkey." *Brain: A Journal of Neurology* (1970) 93:793–820.

Jones, O.D. "Law, Evolution and the Brain: Applications and Open Questions." *Philosophical Transactions of the Royal Society B: Biological Sciences* (2004) 359:1697–1707.

Jonides, J., and S. Yantis. "Uniqueness of Abrupt Visual Onset in Capturing Attention." *Perception & Psychophysics* (1988) 43:346–54.

Josselyn, S.A. "Continuing the Search for the Engram: Examining the Mechanism of Fear Memories." *Journal of Psychiatry & Neuroscience* (2010) 35:221–28.

Josselyn, S.A., S. Kida, and A.J. Silva. "Inducible Repression of CREB Function Disrupts Amygdala-Dependent Memory." *Neurobiology of Learning and Memory* (2004) 82:159–63.

Jovanovic, T., et al. "Impaired Safety Signal Learning May Be a Biomarker of PTSD." *Neuropharmacology* (2012) 62:695–704.

Jovanovic, T., and S.D. Norrholm. "Neural Mechanisms of Impaired Fear Inhibition in Posttraumatic Stress Disorder." *Frontiers in Behavioral Neuroscience* (2011) 5:44.

Jovanovic, T., et al. "Impaired Fear Inhibition Is a Biomarker of PTSD but Not Depression." *Depression and Anxiety* (2010) 27:244–51.

Kagan, J. *Galen's Prophecy: Temperament in Human Nature* (New York: Basic Books, 1994).

Kagan, J. (2003) "Understanding the Effects of Temperament, Anxiety, and Guilt. Panel: The Affect of Emotions: Laying the Groundwork in Childhood." Library of Congress/NIMH Decade of the Brain Project. Jan 3, 2003. http://www.loc.gov/loc/brain/emotion/kagan .html.

Kagan, J., and N. Snidman. "Early Childhood Predictors of Adult Anxiety Disorders." *Biological Psychiatry* (1999) 46:1536–41.

Kahneman, D. *Thinking, Fast and Slow* (New York: Farrar, Straus and Giroux, 2011).

Kahneman, D., P. Slovic, and A. Tversky. *Judgement Under Uncertainty: Heuristics and Biases* (Cambridge, UK: Cambridge University Press, 1982).

Kalisch, R., and A.M. Gerlicher. "Making a Mountain Out of a Molehill: On the Role of the Rostral Dorsal Anterior Cingulate and Dorsomedial Prefrontal Cortex in Conscious Threat Appraisal, Catastrophizing, and Worrying." *Neuroscience and Biobehavioral Reviews* (2014) 42:1–8.

Kalish, H.I. "Strength of Fear as a Function of the Number of Acquisition and Extinction Trials." *Journal of Experimental Psychology* (1954) 47:1–9.

Kaminsky, Z., et al. "Epigenetics of Personality Traits: An Illustrative Study of Identical Twins Discordant for Risk-Taking Behavior." *Twin Research and Human Genetics: The Official Journal of the International Society for Twin Studies* (2008) 11:1–11.

Kanazawa, S. "Common Misconceptions About Science VI: 'Negative Reinforcement.'" *Psychology Today*. Post published by Satoshi Kanazawa on Jan. 3, 2010.

Kandel, E.R. *Cellular Basis of Behavior: An Introduction to Behavioral Neurobiology* (San Francisco: W.H. Freeman and Company, 1976).

Kandel, E.R. "From Metapsychology to Molecular Biology: Explorations into the Nature of Anxiety." *The American Journal of Psychiatry* (1983) 140:1277–93.

Kandel, E.R. "Genes, Synapses, and Long-Term Memory." *Journal of Cellular Physiology* (1997) 173:124–25.

Kandel, E.R. "Biology and the Future of Psychoanalysis: A New Intellectual Framework for Psychiatry Revisited." *The American Journal of Psychiatry* (1999) 156:505–24.

Kandel, E.R. "The Molecular Biology of Memory Storage: A Dialog Between Genes and Synapses." *Bioscience Reports* (2001) 21:565–611.

Kandel, E.R. "The Molecular Biology of Memory Storage: A Dialogue Between Genes and Synapses." *Science* (2001) 294:1030–38.

Kandel, E.R. *In Search of Memory: The Emergence of a New Science of Mind* (New York: W.W. Norton, 2006).

Kandel, E.R. "The Molecular Biology of Memory: aAMP, PKA, CRE, CREB-1, CREB-2, and CPEB." *Molecular Brain* (2012) 5:14.

Kandel, E.R., et al. "Classical Conditioning and Sensitization Share Aspects of the Same Molecular Cascade in *Aplysia*." *Cold Spring Harbor Symposium on Quantitative Biology* (1983) 48(Pt 2):821–30.

Kandel, E.R., and J.H. Schwartz. "Molecular Biology of Learning: Modulation of Transmitter Release." *Science* (1982) 218:433–43.

Kandel, E.R., and W.A. Spencer. "Cellular Neurophysiological Approaches to the Study of Learning." *Physiological Reviews* (1968) 48:65–134.

Kang, M.S., R. Blake, and G.F. Woodman. "Semantic Analysis Does Not Occur in the Absence of Awareness Induced by Interocular Suppression." *Journal of Neuroscience* (2011) 31: 13535–45.

Kapp, B.S., et al. "Amygdala Central Nucleus Lesions: Effect on Heart Rate Conditioning in the Rabbit." *Physiology & Behavior* (1979) 23:1109–17.

Kapp, B.S., J.P. Pascoe, and M.A. Bixler. "The Amygdala: A Neuroanatomical Systems Approach to Its Contributions to Aversive Conditioning." In: *Neuropsychology of Memory*, eds. N. Buttlers and L.R. Squire (New York: Guilford, 1984), 473–88.

Kapp, B.S., et al. "Amygdaloid Contributions to Conditioned Arousal and Sensory Information Processing." In: *The Amygdala: Neurobiological Aspects of Emotion, Memory, and Mental Dysfunction*, ed. J.P. Aggleton (New York: Wiley-Liss, 1992), 229–54.

Kappenman, E.S., A. Macnamara, and G.H. Proudfit. "Electrocortical Evidence for Rapid Allocation of Attention to Threat in the Dot-Probe Task." Social Cognitive and Affective Neuroscience (2014). Published online Dec. 4, 2014: doi: 10.3389/fpsyg.2014.01368.

Karplus, J.P., and A. Kreidl. "Gehirn und Sympathicus. I. Zwischenhirn Basis und Halssympathicus." *Archiv f d ges Physiologie (Pflüger's)* (1909) 129:138–44.

Kazdin, A.E., and G.T. Wilson. *Evaluation of Behavior Therapy: Issues, Evidence and Research Strategies* (Cambridge, MA: Ballinger, 1978).

Keenan, J.P., G.G. Gallup, and D. Falk. *The Face in the Mirror: The Search for the Origins of Consciousness* (London: Ecco/HarperCollins, 2003).

Keller, F.S. *The Definition of Psychology* (New York: Appleton-Century-Crofts, 1973).

Kelley, H.H. "Attribution Theory in Social Psychology." *Nebraska Symposium on Motivation* (1967) 15:192–238.

Kelley, H.H. "Common-Sense Psychology and Scientific Psychology." *Annual Review of Psychology* (1992) 43:1–24.

Kelso, S.R., A.H. Ganong, and T.H. Brown. "Hebbian Synapses in Hippocampus." *Proceedings of the National Academy of Sciences USA* (1986) 83:5326–30.

Kendler, K.S. "Major Depression and Generalised Anxiety Disorder. Same Genes, (Partly) Different Environments—Revisited." *The British Journal of Psychiatry Supplement* (1996) 68–75.

Kendler, K.S. "All We Have to Fear: Psychiatry's Transformation of Natural Anxieties into Mental Disorders." *American Journal of Psychiatry* (2013) 170:124–25.

Kendler, K.S., et al. "The Impact of Environmental Experiences on Symptoms of Anxiety and Depression Across the Life Span." *Psychological Science* (2011) 22:1343–52.

Kendler, K.S., C.O. Gardner, and P. Lichtenstein. "A Developmental Twin Study of Symptoms of Anxiety and Depression: Evidence for Genetic Innovation and Attenuation." *Psychological Medicine* (2008) 38:1567–75.

Kendler, K.S., et al. "Specificity of Genetic and Environmental Risk Factors for Use and Abuse/Dependence of Cannabis, Cocaine, Hallucinogens, Sedatives, Stimulants, and Opiates in Male Twins." *The American Journal of Psychiatry* (2003) 160:687–95.

Kendler, K.S., et al. "Generalized Anxiety Disorder in Women. A Population-Based Twin Study." *Archives of General Psychiatry* (1992) 49:267–72.

Kendler, K.S., et al. "Major Depression and Generalized Anxiety Disorder. Same Genes, (Partly) Different Environments?" *Archives of General Psychiatry* (1992) 49:716–22.

Kendler, K.S., et al. "Clinical Characteristics of Familial Generalized Anxiety Disorder." *Anxiety* (1994) 1:186–91.

Kendler, K.S., et al. "The Structure of the Genetic and Environmental Risk Factors for Six Major Psychiatric Disorders in Women. Phobia, Generalized Anxiety Disorder, Panic Disorder, Bulimia, Major Depression, and Alcoholism." *Archives of General Psychiatry* (1995) 52:374–83.

Kennedy, D.P., and R. Adolphs. "The Social Brain in Psychiatric and Neurological Disorders." *Trends in Cognitive Sciences* (2012) 16:559–72.

Kennedy, J.S. *The New Anthropomorphism* (New York: Cambridge University Press, 1992).

Kent, N.D., M.K. Wagner, and D.R. Gannon. "Effects of Unconditioned Response Restriction on Subsequent Acquisition of a Habit Motivated by 'Fear.'" *Psychological Reports* (1960) 6:335–38.

Kesner, R.P. "Learning and Memory in Rats with an Emphasis on the Role of the Hippocampal Formation." In: *Neurobiology of Comparative Cognition*, eds. R.P. Kesner and D.S. Olton (Hillsdale, NJ: Erlbaum, 1990), 179–204.

Kesner, R.P., and J.C. Churchwell. "An Analysis of Rat Prefrontal Cortex in Mediating Executive Function." *Neurobiology of Learning and Memory* (2011) 96:417–31.

Kessler, R.C., et al. "Posttraumatic Stress Disorder in the National Comorbidity Survey." *Archives of General Psychiatry* (1995) 52:1048–60.

Khoury, B., et al. "Mindfulness-Based Therapy: A Comprehensive Meta-Analysis." *Clinical Psychological Review* (2013) 33:763–71.

Kiefer, M. "Executive Control over Unconscious Cognition: Attentional Sensitization of Unconscious Information Processing." *Frontiers in Human Neuroscience* (2012) 6:61.

Kierkegaard, S. *The Concept of Anxiety: A Simple Psychologically Orienting Deliberation on the Dogmatic Issue of Hereditary Sin* (Princeton, NJ: Princeton University Press, 1980).

Kihlstrom, J.F. "Conscious, Subconscious, Unconscious: A Cognitive Perspective." In: *The Unconscious Reconsidered*, eds. K.S. Bowers and D. Meichenbaum (New York: John Wiley & Sons, 1984), 149–211.

Kihlstrom, J.F. "The Cognitive Unconscious." *Science* (1987) 237:1445–52.

Kihlstrom, J.F. "The Psychological Unconscious." In: *Handbook of Personality: Theory and Research*, ed. L. Pervin (New York: Guilford, 1990), 445–64.

Kihlstrom, J.F., T.M. Barnhardt, and D.J. Tataryn. "Implicit Perception." In: *Perception Without Awareness: Cognitive, Clinical, and Social Perspectives,* eds. R.F. Bornstein and T.S. Pittman (New York: Guilford Press, 1992), 17–54.

Kihlstrom, J.F., T.M. Barnhardt, and D.J. Tatryn. "The Psychological Unconscious: Found, Lost, Regained." *The American Psychologist* (1992) 47:788–91.

Kile, S.J., et al. "Alzheimer Abnormalities of the Amygdala with Kluver-Bucy Syndrome Symptoms: An Amygdaloid Variant of Alzheimer Disease." *Archives of Neurology* (2009) 66:125–29.

Kim, E.J., et al. "Social Transmission of Fear in Rats: The Role of 22-kHz Ultrasonic Distress Vocalization." *PLoS One* (2010) 5:E15077.

Kim, J.J., and M.S. Fanselow. "Modality-Specific Retrograde Amnesia of Fear." *Science* (1992) 256:675–77.

Kim, J.J., R.A. Rison, and M.S. Fanselow. "Effects of Amygdala, Hippocampus, and Periaqueductal Gray Lesions on Short- and Long-Term Contextual Fear." *Behavioral Neuroscience* (1993) 107:1093–98.

Kim, J.J., E.Y. Song, and T.A. Kosten. "Stress Effects in the Hippocampus: Synaptic Plasticity and Memory." *Stress* (2006) 9:1–11.

Kim, M.J., et al. "The Structural and Functional Connectivity of the Amygdala: From Normal Emotion to Pathological Anxiety." *Behavioural Brain Research* (2011) 223:403–10.

Kim, S.Y., et al. "Diverging Neural Pathways Assemble a Behavioural State from Separable Features in Anxiety." *Nature* (2013) 496:219–23.

Kindt, M. "A Behavioural Neuroscience Perspective on the Aetiology and Treatment of Anxiety Disorders." *Behaviour Research and Therapy* (2014) 62:24–36.

Kindt, M., and M. Soeter. "Reconsolidation in a Human Fear Conditioning Study: a Test of Extinction as Updating Mechanism." *Biological Psychology* (2013) 92:43–50.

Kindt, M., M. Soeter, and D. Sevenster. "Disrupting Reconsolidation of Fear Memory in Humans by a Noradrenergic Beta-Blocker." *Journal of Visualized Experiments* (2014). http://www.jove .com/video/52151/disrupting-reconsolidation-fear-memory-humans-noradrenergic.

Kindt, M., M. Soeter, and B. Vervliet. "Beyond Extinction: Erasing Human Fear Responses and Preventing the Return of Fear." *Nature Neuroscience* (2009) 12:256–58.

Kinoshita, S., K.I. Forster, and M.C. Mozer. "Unconscious Cognition Isn't That Smart: Modulation of Masked Repetition Priming Effect in the Word Naming Task." *Cognition* (2008) 107:623–49.

Kintsch, W., et al. "Eight Questions and Some General Issues." In: *Models of Working Memory: Mechanisms of Active Maintenance and Executive Control,* eds. A. Miyake and P. Shah (New York: Cambridge University Press, 1999), 412–41.

Kip, K.E., A. Shuman, D.F. Hernandez, D.M. Diamond, and L. Rosenzweig. "Case Report and Theoretical Description of Accelerated Resolution Therapy (ART) for Military-Related Post-Traumatic Stress Disorder." *Military Medicine* (2014) 179: 31–37.

Kirsch, I., et al. "The Role of Cognition in Classical and Operant Conditioning." *Journal of Clinical Psychology* (2004) 60:369–92.

Kishida, K.T., B. King-Casas, and P.R. Montague. "Neuroeconomic Approaches to Mental Disorders." *Neuron* (2010) 67:543–54.

Kitayama, S., and H.R. Markus, eds. *Emotion and Culture: Empirical Studies of Mutual Influence* (Washington, DC: American Psychological Association, 1994).

Kleim B., F.H. Wilhelm, L. Temp, J. Margraf, B.K. Wiederhold, and B. Rasch. "Sleep Enhances Exposure Therapy." *Psychological Medicine* (2014) 44:1511–19.

Klein, D. "Anxiety Reconceptualized." In: *New Research and Changing Concepts,* eds. D. Klein and J. Rabkin (New York: Raven, 1981).

Klein, D.F. "Delineation of Two Drug-Responsive Anxiety Syndromes." *Psychopharmacologia* (1964) 5:397–408.

Klein, D.F. "False Suffocation Alarms, Spontaneous Panics, and Related Conditions. An Integrative Hypothesis." *Archives of General Psychiatry* (1993) 50:306–17.

Klein, D.F. "Historical Aspects of Anxiety." *Dialogues in Clinical Neuroscience* (2002) 4:295–304.

Klein, D.F., and M. Fink. "Psychiatric Reaction Patterns to Imipramine." *The American Journal of Psychiatry* (1962) 119:432–38.

Klein, S.B. "Making the Case That Episodic Recollection Is Attributable to Operations Occurring at Retrieval Rather Than to Content Stored in a Dedicated Subsystem of Long-Term Memory." *Frontiers in Behavioral Neuroscience* (2013) 7:3.

Klumpp, H., et al. "Neural Response During Attentional Control and Emotion Processing Predicts Improvement After Cognitive Behavioral Therapy in Generalized Social Anxiety Disorder." *Psychological Medicine* (2014) 44:3109–21.

Klumpp, H., D.A. Fitzgerald, and K.L. Phan. "Neural Predictors and Mechanisms of Cognitive Behavioral Therapy on Threat Processing in Social Anxiety Disorder." *Progress in Neuro-Psychopharmacology & Biological Psychiatry* (2013) 45:83–91.

Kluver, H., and P.C. Bucy. "'Psychic Blindness' and Other Symptoms Following Bilateral Temporal Lobectomy in Rhesus Monkeys." *American Journal of Physiology* (1937) 119:352–53.

Knight, D.C., H.T. Nguyen, and P.A. Bandettini. "The Role of the Human Amygdala in the Production of Conditioned Fear Responses." *NeuroImage* (2005) 26:1193–1200.

Knight, D.C., H.T. Nguyen, and P.A. Bandettini. "The Role of Awareness in Delay and Trace Fear Conditioning in Humans." *Cognitive, Affective, & Behavioral Neuroscience* (2006) 6: 157–62.

Knight, D.C., N.S. Waters, and P.A. Bandettini. "Neural Substrates of Explicit and Implicit Fear Memory." *NeuroImage* (2009) 45:208–14.

Knight, R.T., and M. Grabowecky. "Prefrontal Cortex, Time and Consciousness." In: *The New Cognitive Neurosciences*, ed. M.S. Gazzaniga (Cambridge, MA: MIT Press, 2000).

Knoll, E. "Dogs, Darwinism, and English Sensibilities." In: *Anthropomorphism, Anecdotes, and Animals*, eds. R.W. Mitchell et al. (Albany: State University of New York Press, 1997), 12–21.

Knox, D., et al. "Single Prolonged Stress Disrupts Retention of Extinguished Fear in Rats." *Learning & Memory* (2012) 19:43–49.

Knutson, B., et al. "Distributed Neural Representation of Expected Value." *Journal of Neuroscience* (2005) 25:4806–12.

Koch, C. *The Quest for Consciousness: A Neurobiological Approach* (Denver: Roberts and Co., 2004).

Koch, C., and N. Tsuchiya. "Phenomenology Without Conscious Access Is a Form of Consciousness Without Top-Down Attention." *Behavioral and Brain Sciences* (2007) 30:509–10.

Koenigs, M., and J. Grafman. "Posttraumatic Stress Disorder: The Role of Medial Prefrontal Cortex and Amygdala." *Neuroscientist* (2009) 15:540–48.

Kogan, J.H., et al. "Spaced Training Induces Normal Long-Term Memory in CREB Mutant Mice." *Current Biology* (1997) 7:1–11.

Kolb, B.J. "Prefrontal Cortex." In: *The Cerebral Cortex of the Rat*, eds. B.J. Kolb and R.C. Tees (Cambridge, MA: MIT Press, 1990), 437–58.

Konorski, J. *Conditioned Reflexes and Neuron Organization* (Cambridge, UK: Cambridge University Press, 1948).

Konorski, J. *Integrative Activity of the Brain* (Chicago: University of Chicago Press, 1967).

Koob, G.F. "Negative Reinforcement in Drug Addiction: The Darkness Within." *Current Opinion in Neurobiology* (2013) 23:559–63.

Kopelman, M.D. "Varieties of Confabulation and Delusion." *Cognitive Neuropsychiatry* (2010) 15:14–37.

Kormos, V., and B. Gaszner. "Role of Neuropeptides in Anxiety, Stress, and Depression: From Animals to Humans." *Neuropeptides* (2013) 47:401–19.

Kornell, N. "Metacognition in Humans and Animals." *Current Directions in Psychological Science* (2009) 18:11–15.

Kotter, R., and N. Meyer. "The Limbic System: A Review of Its Empirical Foundation." *Behavioural Brain Research* (1992) 52:105–27.

Kouider, S., and S. Dehaene. "Levels of Processing During Non-Conscious Perception: A Critical Review of Visual Masking." *Philosophical Transactions of the Royal Society B: Biological Sciences* (2007) 362:857–75.

Kozak, M.J., E.B. Foa, and G. Steketee. "Process and Outcome of Exposure Treatment with Obsessive-Compulsives: Psychophysiological Indicators of Emotional Processing." *Behavior Therapy* (1988) 19:157–69.

Kozak, M.J., and G.A. Miller. "Hypothetical Constructs Versus Intervening Variables: A Reappraisal of the Three-Systems Model of Anxiety Assessment." *Behavioral Assessment* (1982) 4:347–58.

Kramer, E.A., et al. "Synaptic Evidence for the Efficacy of Spaced Learning." *Proceedings of the National Academy of Sciences of the United States of America* (2012) 109:5121–26.

Kramer, G.P., D.A. Bernstein, and V. Phares. *Introduction to Clinical Psychology* (Upper Saddle River, NJ: Pearson Prentice Hall, 2009).

Kringelbach, M.L. *The Pleasure Center: Trust Your Animal Instincts* (Oxford, UK: Oxford University Press, 2008).

Kron, A., et al. "Feelings Don't Come Easy: Studies on the Effortful Nature of Feelings." *Journal of Experimental Psychology: General* (2010) 139:520–34.

Kubiak, A. *Stages of Terror: Terrorism, Ideology, and Coercion as Theatre History* (Bloomington: Indiana University Press, 1991).

Kubie, J.L., and R.U. Muller. "Multiple Representations in the Hippocampus." *Hippocampus* (1991) 1:240–42.

Kubota, J.T., M.R. Banaji, and E.A. Phelps. "The Neuroscience of Race." *Nature Neuroscience* (2012) 15:940–48.

Kumar, V., Z.A. Bhat, and D. Kumar. "Animal Models of Anxiety: A Comprehensive Review." *Journal of Pharmacological and Toxicological Methods* (2013) 68:175–83.

Kupfermann, I. "Feeding Behavior in *Aplysia*: A Simple System for the Study of Motivation." *Behavioral Biology* (1974) 10:1–26.

Kupfermann, I. "Neural Control of Feeding." *Current Opinion in Neurobiology* (1994) 4:869–76.

Kupfermann, I., et al. "Behavioral Switching of Biting and of Directed Head Turning in *Aplysia*: Explorations Using Neural Network Models." *Acta Biologica Hungarica* (1992) 43:315–28.

Kuppens, P., et al. "The Relation Between Valence and Arousal in Subjective Experience." *Psychological Bulletin* (2013) 139:917–40.

Kwapis, J.L., and M.A. Wood. "Epigenetic Mechanisms in Fear Conditioning: Implications for Treating Post-Traumatic Stress Disorder." *Trends in Neuroscience* (2014) 37:706–20.

LaBar, K.S., et al. "Human Amygdala Activation During Conditioned Fear Acquisition and Extinction: A Mixed-Trial fMRI Study." *Neuron* (1998) 20:937–45.

LaBar, K.S., et al. "Impaired Fear Conditioning Following Unilateral Temporal Lobectomy in Humans." *Journal of Neuroscience* (1995) 15:6846–55.

LaBar, K.S., and E.A. Phelps. "Reinstatement of Conditioned Fear in Humans Is Context Dependent and Impaired in Amnesia." *Behavioral Neuroscience* (2005) 119:677–86.

Lafenetre, P., F. Chaouloff, and G. Marsicano. "The Endocannabinoid System in the Processing of Anxiety and Fear and How CB1 Receptors May Modulate Fear Extinction." *Pharmacological Research: The Official Journal of the Italian Pharmacological Society* (2007) 56:367–81.

Lamb, R.J., et al. "The Reinforcing and Subjective Effects of Morphine in Post-Addicts: A Dose-Response Study." *Journal of Pharmacology and Experimental Therapeutics* (1991) 259:1165–73.

Lamme, V.A. "Towards a True Neural Stance on Consciousness." *Trends in Cognitive Sciences* (2006) 10:494–501.

Lamprecht, R., and J. LeDoux. "Structural Plasticity and Memory." *Nature Reviews Neuroscience* (2004) 5:45–54.

Lane, R.D., et al. "Memory Reconsolidation, Emotional Arousal and the Process of Change in Psychotherapy: New Insights from Brain Science." *Behavioral and Brain Sciences* (2014) 1–80.

Laney, C., and E.F. Loftus. "Traumatic Memories Are Not Necessarily Accurate Memories." *Canadian Journal of Psychiatry* (2005) 50:823–28.

Lang, P. "The Application of Psychophysiological Methods to the Study of Psychotherapy and Behaviour Modification." In: *Handbook of Psychotherapy and Behaviour Change*, eds. A. Bergin and S. Garfield (New York: John Wiley & Sons, 1971).

Lang, P. "A Bioinformational Theory of Emotional Imagery." *Psychophysiology* (1979) 16:495–512.

Lang, P.J. "Fear Reduction and Fear Behavior: Problems in Treating a Construct." In: *Research in Psychotherapy*, vol. 3, ed. J.M. Schlien (Washington, DC: American Psychological Association, 1968), 90–103.

Lang, P.J. "Imagery in Therapy: An Information Processing Analysis of Fear." *Behavior Therapy* (1977) 8:862–86.

Lang, P.J. "Anxiety: Toward a Psychophysiological Definition." In: *Psychiatric Diagnosis: Exploration of Biological Criteria*, eds. H.S. Akiskal and W.L. Webb (New York: Spectrum, 1978), 265–389.

Lang, P.J. "The Emotion Probe: Studies of Motivation and Attention." *American Psychologist* (1995) 5:372–85.

Lang, P.J., M.M. Bradley, and B.N. Cuthbert. "Emotion, Attention, and the Startle Reflex." *Psychological Review* (1990) 97:377–95.

Lang, P.J., M.M. Bradley, B.N. Cuthbert. "Emotion, Motivation, and Anxiety: Brain Mechanisms and Psychophysiology." *Biological Psychiatry* (1998) 44:1248–63.

Lang, P.J., and M. Davis. "Emotion, Motivation, and the Brain: Reflex Foundations in Animal and Human Research." *Progress in Brain Research* (2006) 156:3–29.

Lang, P.J., and L.M. McTeague. "The Anxiety Disorder Spectrum: Fear Imagery, Physiological Reactivity, and Differential Diagnosis." *Anxiety, Stress, and Coping* (2009) 22:5–25.

Langley, J.N. "The Autonomic Nervous System." *Brain* (1903) 26:1–26.

Langerhans, R.B. "Evolutionary Consequences of Predation: Avoidance, Escape, and Diversification." In: *Predation in Organisms*, ed. A.M.T. Elewa (Berlin and Heidelberg: Springer, 2007), 177–220.

Lanteaume, L., et al. "Emotion Induction After Direct Intracerebral Stimulations of Human Amygdala." *Cerebral Cortex* (2007) 17:1307–13.

Lashley, K. "The Problem of Serial Order in Behavior." In: *Cerebral Mechanisms in Behavior*, ed. L.A. Jeffers (New York: Wiley, 1950).

Lattal, K.M., and M.A. Wood. "Epigenetics and Persistent Memory: Implications for Reconsolidation and Silent Extinction Beyond the Zero." *Nature Neuroscience* (2013) 16:124–29.

Lau, H., and R. Brown. "The Emperor's New Phenomenology? The Empirical Case for Conscious Experience Without First-Order Representations." http://philpapers.org/archive/BROTEN.1.pdf

Lau, H.C., and R.E. Passingham. "Relative Blindsight in Normal Observers and the Neural Correlate of Visual Consciousness." *Proceedings of the National Academy of Sciences of the United States of America* (2006) 103:18763–68.

Lau, J.Y., et al. "Distinct Neural Signatures of Threat Learning in Adolescents and Adults." *Proceedings of the National Academy of Sciences of the United States of America* (2011) 108:4500–5.

Laureys, S., and N.D. Schiff. "Coma and Consciousness: Paradigms (Re)framed by Neuroimaging." *NeuroImage* (2012) 61:478–91.

Lázaro-Muñoz, G., J.E. LeDoux, and C.K. Cain. "Sidman Instrumental Avoidance Initially Depends on Lateral Amygdala and Basal Amygdala and Is Constrained by Central Amygdala-Mediated Pavlovian Processes." *Biological Psychiatry* (2010) 67:1120–27.

Lazarus, R., R. McCleary. "Autonomic Discrimination Without Awareness: A Study of Subception." *Psychological Review* (1951) 58:113–22.

Lazarus, R.S. "Cognition and Motivation in Emotion." *American Psychologist* (1991) 46:352–67.

Lazarus, R.S. *Emotion and Adaptation* (New York: Oxford University Press, 1991).

Leahy, R.L. *Contemporary Cognitive Therapy: Theory, Research, and Practice* (New York: Guilford Press, 2004).

Lebestky, T., et al. "Two Different Forms of Arousal in *Drosophila* Are Oppositely Regulated by the Dopamine D1 Receptor Ortholog DopR via Distinct Neural Circuits." *Neuron* (2009) 64:522–36.

Lecours, A.R. "Language Contrivance on Consciousness (and Vice Versa)." In: *Consciousness: At the Frontiers of Neuroscience*, eds. H. Jasper et al. (Philadelphia: Lippincott-Raven, 1998), 167–80.

LeDoux, J.E. "Cognition and Emotion: Processing Functions and Brain Systems." In: *Handbook of Cognitive Neuroscience*, ed. M.S. Gazzaniga (New York: Plenum Publishing, 1984), 357–68.

LeDoux, J.E. "Emotion." In: *Handbook of Physiology 1: The Nervous System Vol. V., Higher Functions of the Brain*, ed. F. Plum (Bethesda, MD: American Physiological Society, 1987), 419–59.

LeDoux, J.E. "Emotion and the Limbic System Concept." *Concepts in Neuroscience* (1991) 2:169–99.

LeDoux, J.E. "Emotion and the Amygdala." In: *The Amygdala: Neurobiological Aspects of Emotion, Memory, and Mental Dysfunction*, ed. J.P. Aggleton (New York: Wiley-Liss, Inc., 1992), 339–51.

LeDoux, J.E. "Emotion, Memory and the Brain." *Scientific American* (1994) 270:50–57.

LeDoux, J.E. *The Emotional Brain* (New York: Simon and Schuster, 1996).

LeDoux, J.E. "Emotion Circuits in the Brain." *Annual Review of Neuroscience* (2000) 23:155–84.

LeDoux, J.E. *Synaptic Self: How Our Brains Become Who We Are* (New York: Viking, 2002).

LeDoux, J.E. "The Amygdala." *Current Biology* (2007) 17:R868–74.

LeDoux, J.E. "Emotional Colouration of Consciousness: How Feelings Come About." In: *Frontiers of Consciousness: Chichele Lectures*, eds. L. Weiskrantz and M. Davies (Oxford, UK: Oxford University Press, 2008), 69–130.

LeDoux, J.E. "Rethinking the Emotional Brain." *Neuron* (2012) 73:653–76.

LeDoux, J.E. "Evolution of Human Emotion: A View Through Fear." *Progress in Brain Research* (2012) 195:431–42.

LeDoux, J.E. "For the Anxious, Avoidance Can Have an Upside." *The New York Times*, April 7, 2013. Retrieved on Dec. 29, 2014, from http://opinionator.blogs.nytimes.com/2013/04/07/for-the-anxious-avoidance-can-have-an-upside/?_r=0.

LeDoux, J.E. "The Slippery Slope of Fear." *Trends in Cognitive Sciences* (2013) 17:155–56.

LeDoux, J.E. "Coming to Terms with Fear." *Proceedings of the National Academy of Sciences of the United States of America* (2014) 111:2871–78.

LeDoux, J.E. "Afterword: Emotional Construction in the Brain." In: *The Psychological Construction of Emotion*, eds. L.F. Barrett and J.A. Russell (New York: Guilford Press, 2014), 459–63.

LeDoux, J.E. "Feelings: What Are They and How Does the Brain Make Them?" *Daedalus* (2015) 144.

LeDoux, J.E., C. Blum, and W. Hirst. "Inferential Processing of Context: Studies of Cognitively Impaired Persons." *Brain and Language* (1983) 19:216–24.

LeDoux, J.E., and J.M. Gorman. "A Call to Action: Overcoming Anxiety Through Active Coping." *American Journal of Psychiatry* (2001) 158:1953–55.

LeDoux, J.E., et al. "Different Projections of the Central Amygdaloid Nucleus Mediate Autonomic and Behavioral Correlates of Conditioned Fear." *Journal of Neuroscience* (1988) 8:2517–29.

LeDoux, J.E., and E.A. Phelps. "Emotional Networks in the Brain." In: *Handbook of Emotions*, eds. M. Lewis, et al. (New York: Guilford Press, 2008), 159–79.

LeDoux, J.E., L.M. Romanski, and A.E. Xagoraris. "Indelibility of Subcortical Emotional Memories." *Journal of Cognitive Neuroscience* (1989) 1:238–43.

LeDoux, J.E., A. Sakaguchi, and D.J. Reis. "Behaviorally Selective Cardiovascular Hyperreactivity in Spontaneously Hypertensive Rats. Evidence for Hypoemotionality and Enhanced Appetitive Motivation." *Hypertension* (1982) 4:853–63.

LeDoux, J.E., A. Sakaguchi, and D.J. Reis. "Subcortical Efferent Projections of the Medial Geniculate Nucleus Mediate Emotional Responses Conditioned to Acoustic Stimuli." *Journal of Neuroscience* (1984) 4:683–98.

LeDoux, J.E., D. Schiller, and C. Cain. "Emotional Reaction and Action: From Threat Processing to Goal-Directed Behavior." In: *The Cognitive Neurosciences*, ed. M.S. Gazzaniga (Cambridge, MA: MIT Press, 2009), 905–24.

LeDoux, J.E., D.H. Wilson, and M.S. Gazzaniga. "A Divided Mind: Observations on the Conscious Properties of the Separated Hemispheres." *Annals of Neurology* (1977) 2:417–21.

Lee, A.C., T.W. Robbins, and A.M. Owen. "Episodic Memory Meets Working Memory in the Frontal Lobe: Functional Neuroimaging Studies of Encoding and Retrieval." *Critical Reviews in Neurobiology* (2000) 14:165–97.

Lee, H.J., M. Gallagher, and P.C. Holland. "The Central Amygdala Projection to the Substantia Nigra Reflects Prediction Error Information in Appetitive Conditioning." *Learning & Memory* (2010) 17:531–38.

Lee, J.L. "Memory Reconsolidation Mediates the Updating of Hippocampal Memory Content." *Frontiers in Behavioral Neuroscience* (2010) 4:168.

Lee, S., S-J. Kim, O.B. Kwon, J.H. Lee, and J.H. Kim. "Inhibitory Networks of the Amygdala for Emotional Memory." *Frontiers in Neural Circuits* Aug. 1, 2013. doi: 10.3389/fncir.2013.00129.

Lee, Y.S., et al. "Transcriptional Regulation of Long-Term Memory in the Marine Snail *Aplysia*." *Molecular Brain* (2008) 1:3.

Lehrman, D. "A Critique of Konrad Lorenz's Theory of Instinctive Behavior." *Quarterly Review of Biology* (1953) 28:337–63.

Lehrman, D.S. "Problems Raised by Instinct Theories." In: *Instinct: An Enduring Problem in Psychology*, eds. R.C. Birney and R.C. Teevan (New York: D. Van Nostrand Company, Inc., 1961), 152–64.

Lenneberg, E. *Biological Foundations of Language* (New York: John Wiley & Sons, 1967).

Leopold, D.A., and N.K. Logothetis. "Activity Changes in Early Visual Cortex Reflect Monkeys' Percepts During Binocular Rivalry." *Nature* (1996) 379:549–53.

Lesch, K.-P., et al. "Association of Anxiety-Related Traits with a Polymorphism in the Serotonin Transporter Gene Regulatory Region." *Science* (1996) 274:1527–31.

Levenson, R.W. "Basic Emotion Questions." *Emotion Review* (2011) 3:379–86.

Levenson, R.W., J. Soto, and N. Pole. "Emotion, Biology, and Culture." In: *Handbook of Cultural Psychology*, eds. S. Kitayama and D. Cohen (New York: Guilford Press, 2007), 780–96.

Lévi-Strauss, C. *La Pensée sauvage* (Paris, 1962): English translation as *The Savage Mind* (Chicago: The University of Chicago Press, 1966).

Levis, D.J. "The Case for a Return to a Two-Factor Theory of Avoidance: The Failure of Non-Fear Interpretations." In: *Contemporary Learning Theories: Pavlovian Conditioning and the Status of Traditional Learning Theory*, eds. S.B. Klein and R.R. Mowrer (Hillsdale, NJ: Erlbaum, 1989), 227–77.

Levis, D.J. "The Negative Impact of the Cognitive Movement on the Continued Growth of the Behavior Therapy Movement: A Historical Perspective." *Genetic, Social, and General Psychology Monographs* (1999) 125:157–71.

Levy, D.J., and P.W. Glimcher. "The Root of All Value: A Neural Common Currency for Choice." *Current Opinion in Neurobiology* (2012) 22:1027–38.

Levy, F., and M. Farrow. "Working Memory in ADHD: Prefrontal/Parietal Connections." *Current Drug Targets* (2001) 2:347–52.

Levy, R., and P.S. Goldman-Rakic. "Segregation of Working Memory Functions Within the Dorsolateral Prefrontal Cortex." *Experimental Brain Research* (2000) 133:23–32.

Lewis, A. "The Ambiguous Word 'Anxiety.'" *International Journal of Psychiatry* (1970) 9:62–79.

Lewis, A.H., et al. "Avoidance-Based Human Pavlovian-to-Instrumental Transfer." *European Journal of Neuroscience* (2013) 38:3740–48.

Lewis, D.J. "Psychobiology of Active and Inactive Memory." *Psychological Bulletin* (1979) 86: 1054–83.

Lewis, M. *The Rise of Consciousness and the Development of Emotional Life* (New York: Guilford Press, 2013).

Lewontin, R.C. "In the Beginning Was the Word." *Science* (2001) 291:1263–64.

Ley, R. "The 'Suffocation Alarm' Theory of Panic Attacks: A Critical Commentary." *Journal of Behavior Therapy and Experimental Psychiatry* (1994) 25:269–73.

Leys, R. "How Did Fear Become a Scientific Object and What Kind of Object Is It?" In: *Fear: Across the Disciplines*, eds. J. Plamper and B. Lazier (Pittsburgh: University of Pittsburgh Press, 2012), 51–77.

Li, C., and D.G. Rainnie. "Bidirectional Regulation of Synaptic Plasticity in the Basolateral Amygdala Induced by the D1-Like Family of Dopamine Receptors and Group II Metabotropic Glutamate Receptors." *Journal of Physiology* (2014) 592:4329–51.

Li, H., et al. "Experience-Dependent Modification of a Central Amygdala Fear Circuit." *Nature Neuroscience* (2013) 16:332–39.

Li, S.H., and R.F. Westbrook. "Massed Extinction Trials Produce Better Short-Term but Worse Long-Term Loss of Context Conditioned Fear Responses Than Spaced Trials." *Journal of Experimental Psychology: Animal Behavior Processes* (2008) 34:336–51.

Liang, K.C., H.C. Chen, and D.Y. Chen. "Posttraining Infusion of Norepinephrine and Corticotropin Releasing Factor into the Bed Nucleus of the Stria Terminalis Enhanced Retention in an Inhibitory Avoidance Task." *The Chinese Journal of Physiology* (2001) 44:33–43.

Liberzon, I., and C.S. Sripada. "The Functional Neuroanatomy of PTSD: A Critical Review." *Progress in Brain Research* (2008) 167:151–69.

Lichtenberg, N.T., et al. "Nucleus Accumbens Core Lesions Enhance Two-Way Active Avoidance." *Neuroscience* (2014) 258:340–46.

Liddell, B.J., et al. "A Direct Brainstem-Amygdala-Cortical 'Alarm' System for Subliminal Signals of Fear." *NeuroImage* (2005) 24:235–43.

Likhtik, E., et al. "Prefrontal Control of the Amygdala." *Journal of Neuroscience* (2005) 25: 7429–37.

Lim, D., K. Sanderson, and G. Andrews. "Lost Productivity Among Full-Time Workers with Mental Disorders." *The Journal of Mental Health Policy and Economics* (2000) 3:139–46.

Lin, C.H., et al. "The Similarities and Diversities of Signal Pathways Leading to Consolidation of Conditioning and Consolidation of Extinction of Fear Memory." *Journal of Neuroscience* (2003) 23:8310–17.

Lin, D., et al. "Functional Identification of an Aggression Locus in the Mouse Hypothalamus." *Nature* (2011) 470:221–26.

Lin, J.Y., S.O. Murray, and G.M. Boynton. "Capture of Attention to Threatening Stimuli Without Perceptual Awareness." *Current Biology* (2009) 19:1118–22.

Lin, Z., and S. He. "Seeing the Invisible: The Scope and Limits of Unconscious Processing in Binocular Rivalry." *Progress in Neurobiology* (2009) 87:195–211.

Lindquist, K.A., and L.F. Barrett. "Constructing Emotion: The Experience of Fear as a Conceptual Act." *Psychological Science* (2008) 19:898–903.

Lindquist, K.A., et al. "Language and the Perception of Emotion." *Emotion* (2006) 6:125–38.

Lindsley, D.B. "Emotions." In: *Handbook of Experimental Psychology*, ed. S.S. Stevens (New York: Wiley, 1951), 473–516.

Lindsley, O.R., B.F. Skinner, and H.C. Solomon. *Study of Psychotic Behavior, Studies in Behavior Therapy* (Harvard Medical School, Department of Psychiatry, Metropolitan State Hospital, Waltham, MA, 1953).

Linnman, C., et al. "Resting Amygdala and Medial Prefrontal Metabolism Predicts Functional Activation of the Fear Extinction Circuit." *The American Journal of Psychiatry* (2012) 169: 415–23.

Lipsman, N., P. Giacobbe, and A.M. Lozano. "Deep Brain Stimulation in Obsessive-Compulsive Disorder: Neurocircuitry and Clinical Experience." *Handbook of Clinical Neurology* (2013) 116:245–50.

Lipsman, N., B. Woodside, and A.M. Lozano. "Evaluating the Potential of Deep Brain Stimulation for Treatment-Resistant Anorexia Nervosa." *Handbook of Clinical Neurology* (2013) 116:271–76.

Lissek, S., et al. "Elevated Fear Conditioning to Socially Relevant Unconditioned Stimuli in Social Anxiety Disorder." *The American Journal of Psychiatry* (2008) 165:124–32.

Lissek, S., et al. "Classical Fear Conditioning in the Anxiety Disorders: A Meta-Analysis." *Behaviour Research and Therapy* (2005) 43:1391–1424.

Lissek, S., et al. "Impaired Discriminative Fear-Conditioning Resulting from Elevated Fear Responding to Learned Safety Cues Among Individuals with Panic Disorder." *Behaviour Research and Therapy* (2009) 47:111–18.

Little, P.F. "Gene Mapping and the Human Genome Mapping Project." *Current Opinion in Cell Biology* (1990) 2:478–84.

Litvin, Y., D.C. Blanchard, and R.J. Blanchard "Rat 22kHz Ultrasonic Vocalizations as Alarm Cries." *Behavioural Brain Research* (2007) 182:166–72.

Liu, T.L., D.Y. Chen, and K.C. Liang. "Post-Training Infusion of Glutamate into the Bed Nucleus of the Stria Terminalis Enhanced Inhibitory Avoidance Memory: An Effect Involving Norepinephrine." *Neurobiology of Learning and Memory* (2009) 91:456–65.

Liu, T.L., and K.C. Liang. "Posttraining Infusion of Cholinergic Drugs into the Ventral Subiculum Modulated Memory in an Inhibitory Avoidance Task: Interaction with the Bed Nucleus of the Stria Terminalis." *Neurobiology of Learning and Memory* (2009) 91:235–42.

Liubashina, O., V. Bagaev, and S. Khotiantsev. "Amygdalofugal Modulation of the Vago-Vagal Gastric Motor Reflex in Rat." *Neuroscience Letters* (2002) 325:183–86.

Livingstone, M. *Vision and Art: The Biology of Seeing* (New York: Abrams, 2008).

Loewenstein, G.F., et al. "Risk as Feelings." *Psychological Bulletin* (2001) 127:267–86.

Loftus, E.F. *Eyewitness Testimony* (Cambridge, MA: Harvard University Press, 1996).

Loftus, E.F., and D. Davis. "Recovered Memories." *Annual Review of Clinical Psychology* (2006) 2:469–98.

Loftus, E.F., and D.C. Polage. "Repressed Memories. When Are They Real? How Are They False?" *The Psychiatric Clinics of North America* (1999) 22:61–70.

Lonergan, M.H., et al. "Propranolol's Effects on the Consolidation and Reconsolidation of Long-Term Emotional Memory in Healthy Participants: A Meta-Analysis." *Journal of Psychiatry & Neuroscience* (2013) 38:222–31.

Long, V.A., and M.S. Fanselow. "Stress-Enhanced Fear Learning in Rats Is Resistant to the Effects of Immediate Massed Extinction." *Stress* (2012) 15:627–36.

Lonsdorf, T.B., J. Haaker, and R. Kalisch. "Long-Term Expression of Human Contextual Fear and Extinction Memories Involves Amygdala, Hippocampus and Ventromedial Prefrontal Cortex: A Reinstatement Study in Two Independent Samples." *Social Cognitive and Affective Neuroscience* (2014). First published online in Feb. 3, 2014. doi: 10.1093/scan/nsu018.

Lorberbaum, J.P., et al. "Neural Correlates of Speech Anticipatory Anxiety in Generalized Social Phobia." *Neuroreport* (2004) 15:2701–5.

Lorenz, K.Z. "The Comparative Method in Studying Innate Behavior Patterns." *Symposia of the Society for Experimental Biology* (1950) 4:221–68.

Lovibond, P.F., et al. "Awareness Is Necessary for Differential Trace and Delay Eyeblink Conditioning in Humans." *Biological Psychology* (2011) 87:393–400.

Lovibond, P.F., et al. "Safety Behaviours Preserve Threat Beliefs: Protection from Extinction of Human Fear Conditioning by an Avoidance Response." *Behaviour Research and Therapy* (2009) 47:716–20.

Low, P. "Cambridge Declaration on Consciousness in Non-Human Animals." (Also by J. Panksepp, D. Reiss, D. Edelman, B. van Swinderen, and C. Koch). Originally retrieved on Sept. 26,

2013, from http://fcmconferenceorg/Churchill College, University of Cambridge. This link was subsequently removed. A search on Dec. 24, 2014, revealed that the document was again available through this link: http://fcmconference.org/img/cambridgedeclara tiononconsciousness.pdf.

Lu, D.P. "Using Alternating Bilateral Stimulation of Eye Movement Desensitization for Treatment of Fearful Patients." *General Dentistry* (2010) 58:E140–47.

Luchicchi, A., et al. "Illuminating the Role of Cholinergic Signaling in Circuits of Attention and Emotionally Salient Behaviors." *Frontiers in Synaptic Neuroscience* (2014) 6:24.

Luo, Q., et al. "Emotional Automaticity Is a Matter of Timing." *Journal of Neuroscience* (2010) 30:5825–29.

Luo, Q. et al. "Visual Awareness, Emotion, and Gamma Band Synchronization." *Cerebral Cortex* (2009) 19:1896–1904.

Luppi, P.H., et al. "Afferent Projections to the Rat Locus Coeruleus Demonstrated by Retrograde and Anterograde Tracing with Cholera-Toxin B Subunit and *Phaseolus vulgaris leucoagglutinin.*" *Neuroscience* (1995) 65:119–60.

Lutz, A., J.P. Dunne, and R.J. Davidson. "Meditation and the Neuroscience of Consciousness: An Introduction." In: *The Cambridge Handbook of Consciousness,* eds. P.D. Zelazo et al. (Cambridge, UK: Cambridge University Press, 2007), 499–552.

Lutz, A., et al. "Attention Regulation and Monitoring in Meditation." *Trends in Cognitive Sciences* (2008) 12:163–69.

Lutz, B. "The Endocannabinoid System and Extinction Learning." *Molecular Neurobiology* (2007) 36:92–101.

Lycan, W.G. *Consciousness* (Cambridge, MA: Bradford Books/MIT Press, 1986).

Lycan, W.G. *Consciousness and Experience* (Cambridge, MA: Bradford Books/MIT Press, 1995).

Maccorquodale, K., and P.E. Meehl. "On a Distinction Between Hypothetical Constructs and Intervening Variables." *Psychological Review* (1948) 55:95–107.

Macdonald, K., and D. Feifel. "Oxytocin's Role in Anxiety: A Critical Appraisal." *Brain Research* (2014) 580:22–56.

Macefield, V.G., C. James, and L.A. Henderson. "Identification of Sites of Sympathetic Outflow at Rest and During Emotional Arousal: Concurrent Recordings of Sympathetic Nerve Activity and fMRI of the Brain." *International Journal of Psychophysiology: Official Journal of the International Organization of Psychophysiology* (2013) 89:451–59.

Mackintosh, N.J., ed. *Animal Learning and Cognition* (San Diego: Academic Press, 1994).

Macknik, S.L. "Visual Masking Approaches to Visual Awareness." *Progress in Brain Research* (2006) 155:177–215.

MacLean, P.D. "Psychosomatic Disease and the 'Visceral Brain': Recent Developments Bearing on the Papez Theory of Emotion." *Psychosomatic Medicine* (1949) 11:338–53.

MacLean, P.D. "Some Psychiatric Implications of Physiological Studies on Frontotemporal Portion of Limbic System (Visceral Brain)." *Electroencephalography and Clinical Neurophysiology* (1952) 4:407–18.

MacLean, P.D. "The Triune Brain, Emotion and Scientific Bias." In: *The Neurosciences: Second Study Program,* ed. F.O. Schmitt (New York: Rockefeller University Press, 1970), 336–49.

MacLeod, C., and R. Hagan. "Individual Differences in the Selective Processing of Threatening Information, and Emotional Responses to a Stressful Life Event." *Behaviour Research and Therapy* (1992) 30:151–61.

Macnab, R.M., and D.E. Koshland Jr. "The Gradient-Sensing Mechanism in Bacterial Chemotaxis." *Proceedings of the National Academy of Sciences of the United States of America* (1972) 69:2509–12.

Macphail, E.M. *The Evolution of Consciousness* (Oxford, UK: Oxford University Press, 1998).

Macphail, E.M. "The Search for a Mental Rubicon." In: *The Evolution of Cognition,* eds. C. Heyes and L. Huber (Cambridge, MA: MIT Press, 2000), 253–71.

Magee, J.C., and D. Johnston. "A Synaptically Controlled, Associative Signal for Hebbian Plasticity in Hippocampal Neurons." *Science* (1997) 275:209–13.

Mahler, S.V., and K.C. Berridge. "What and When to 'Want'? Amygdala-Based Focusing of Incentive Salience upon Sugar and Sex." *Psychopharmacology (Berl)* (2012) 221:407–26.

Mahoney, A.E., and P.M. McEvoy. "A Transdiagnostic Examination of Intolerance of Uncertainty Across Anxiety and Depressive Disorders." *Cognitive Behaviour Therapy* (2012) 41:212–22.

Maia, T.V., and A. Cleeremans. "Consciousness: Converging Insights from Connectionist Modeling and Neuroscience." *Trends in Cognitive Sciences* (2005) 9:397–404.

Maier, A., et al. "Introduction to Research Topic—Binocular Rivalry: A Gateway to Studying Consciousness." *Frontiers in Human Neuroscience* (2012) 6:263.

Makari, G. "In the Arcadian Woods." *The New York Times,* April 16, 2012. Retrieved Nov. 30, 2014, from http://opinionator.blogs.nytimes.com/2012/04/16/in-the-arcadian-woods/?_r=0.

Malinowski, P. "Neural Mechanisms of Attentional Control in Mindfulness Meditation." *Frontiers in Neuroscience* (2013) 7:8.

Mandler, G., and W. Kessen. *The Language of Psychology* (New York: John Wiley & Sons, 1959).

Manna, A., et al. "Neural Correlates of Focused Attention and Cognitive Monitoring in Meditation." *Brain Research Bulletin* (2010) 82:46–56.

Mantione, M., et al. "Cognitive-Behavioural Therapy Augments the Effects of Deep Brain Stimulation in Obsessive-Compulsive Disorder." *Psychological Medicine* (2014) 44:3515–22.

Marcel, A.J. "Conscious and Unconscious Perception: Experiments on Visual Masking and Word Recognition." *Cognitive Psychology* (1983) 15:197–237.

Marchand, W.R. "Neural Mechanisms of Mindfulness and Meditation: Evidence from Neuroimaging Studies." *World Journal of Radiology* (2014) 6:471–79.

Marder, E. "Neuromodulation of Neuronal Circuits: Back to the Future." *Neuron* (2012) 76:1–11.

Marek, R., et al. "The Amygdala and Medial Prefrontal Cortex: Partners in the Fear Circuit." *Journal of Physiology* (2013) 591:2381–91.

Maren, S. "Synaptic Mechanisms of Associative Memory in the Amygdala." *Neuron* (2005) 47: 783–86.

Maren, S., and M.S. Fanselow. "The Amygdala and Fear Conditioning: Has the Nut Been Cracked?" *Neuron* (1996) 16:237–40.

Maren, S., and M.S. Fanselow. "Electrolytic Lesions of the Fimbria/Fornix, Dorsal Hippocampus, or Entorhinal Cortex Produce Anterograde Deficits in Contextual Fear Conditioning in Rats." *Neurobiology of Learning and Memory* (1997) 67:142–49.

Maren, S., K.L. Phan, and I. Liberzon. "The Contextual Brain: Implications for Fear Conditioning, Extinction and Psychopathology." *Nature Reviews Neuroscience* (2013) 14:417–28.

Maren, S., and G.J. Quirk. "Neuronal Signalling of Fear Memory." *Nature Reviews Neuroscience* (2004) 5:844–52.

Maren, S., S.A. Yap, and K.A. Goosens. "The Amygdala Is Essential for the Development of Neuronal Plasticity in the Medial Geniculate Nucleus During Auditory Fear Conditioning in Rats." *Journal of Neuroscience* (2001) 21:RC135.

Marewski, J.N., and G. Gigerenzer. "Heuristic Decision Making in Medicine." *Dialogues in Clinical Neuroscience* (2012) 14:77–89.

Marin, M.F., et al. "Device-Based Brain Stimulation to Augment Fear Extinction: Implications for PTSD Treatment and Beyond." *Depression and Anxiety* (2014) 31:269–78.

Mark, V.H., and F.R. Ervin. *Violence and the Brain* (New York: Harper & Row, 1970).

Markowska, A., and I. Lukaszewska. "Emotional Reactivity After Frontomedial Cortical, Neostriatal or Hippocampal Lesions in Rats." *Acta Neurobiologiae Experimentalis* (1980) 40: 881–93.

Marks, I. *Fears, Phobias, and Rituals: Panic, Anxiety and Their Disorders* (New York: Oxford University Press, 1987).

Marks, I., and A. Tobena. "Learning and Unlearning Fear: A Clinical and Evolutionary Perspective." *Neuroscience and Biobehavioral Reviews* (1990) 14:365–84.

Marr, D. "Simple Memory: A Theory for Archicortex." *Philosophical Transactions of the Royal Society B: Biological Sciences* (1971) 262:23–81.

Marschner, A., et al. "Dissociable Roles for the Hippocampus and the Amygdala in Human Cued Versus Context Fear Conditioning." *Journal of Neuroscience* (2008) 28:9030–36.

Marsh, E.J., and H.L. Roediger. "Episodic and Autobiographical Memory." In: *Handbook of Psychology: Volume 4, Experimental Psychology*, eds. A.F. Healy and R.W. Proctor (New York: John Wiley & Sons, 2013), 472–94.

Martasian, P.J., and N.F. Smith. "A Preliminary Resolution of the Retention of Distributed vs. Massed Response Prevention in Rats." *Psychological Reports* (1993) 72:1367–77.

Martasian, P.J., et al. "Retention of Massed vs. Distributed Response-Prevention Treatments in Rats and a Revised Training Procedure." *Psychological Reports* (1992) 70:339–55.

Martin, K.C. "Local Protein Synthesis During Axon Guidance and Synaptic Plasticity." *Current Opinion in Neurobiology* (2004) 14:305–10.

Martin, S.J., P.D. Grimwood, and R.G.M. Morris. "Synaptic Plasticity and Memory: An Evaluation of the Hypothesis." *Annual Review of Neuroscience* (2000) 23:649–711.

Martinez, J.L. Jr., R.A. Jensen, and J.L. McGaugh. "Attenuation of Experimentally-Induced Amnesia." *Progress in Neurobiology* (1981) 16:155–86.

Martinez, R.C., et al. "Active vs. Reactive Threat Responding Is Associated with Differential c-Fos Expression in Specific Regions of Amygdala and Prefrontal Cortex." *Learning & Memory* (2013) 20:446–52.

Marx, M.H. "Intervening Variable or Hypothetical Construct?" *Psychological Review* (1951) 58:235–47.

Masson, J.M., and S. McCarthy. *When Elephants Weep: The Emotional Lives of Animals* (New York: Delacorte, 1996).

Masterson, F.A., and M. Crawford. "The Defense Motivation System: A Theory of Avoidance Behavior." *Behavioral and Brain Sciences* (1982) 5:661–96.

Masuda, A., et al. "Multisensory Interaction Mediates the Social Transmission of Avoidance in Rats: Dissociation from Social Transmission of Fear." *Behavioural Brain Research* (2013) 252:334–38.

Mather, J. "Consciousness in Cephalopods?" *Journal of Cosmology* (2011) 14.

Mathew, S.J., R.B. Price, and D.S. Charney. "Recent Advances in the Neurobiology of Anxiety Disorders: Implications for Novel Therapeutics." *American Journal of Medical Genetics Part C, Seminars in Medical Genetics* (2008) 148C:89–98.

Mathews, A., et al. "Implicit and Explicit Memory Bias in Anxiety." *Journal of Abnormal Psychology* (1989) 98:236–40.

Mathews, A., A. Richards, and M. Eysenck. "Interpretation of Homophones Related to Threat in Anxiety States." *Journal of Abnormal Psychology* (1989) 98:31–34.

Matthews, G., and A. Wells. "Attention, Automaticity, and Affective Disorder." Behavior *Modification* (2000) 24:69–93.

Maugham, W.S. (1949) *A Writer's Notebook*, p. 78. (London: William Heinemann). Quoted by Alan Cowey. "TMS and Visual Awareness," Ch 27. In: *Oxford Handbook of Transcranial Stimulation*. eds. E. Wasserman, C. Epstein, and U. Ziemann (Oxford, UK: Oxford University Press, 2008).

May, R. *The Meaning of Anxiety* (New York: W.W. Norton, 1950).

Mayes, A.R., and D. Montaldi. "Exploring the Neural Bases of Episodic and Semantic Memory: The Role of Structural and Functional Neuroimaging." *Neuroscience and Biobehavioral Reviews* (2001) 25:555–73.

Mayr, U. "Conflict, Consciousness, and Control." *Trends in Cognitive Sciences* (2004) 8:145–48.

Mazefsky, C.A., et al. "The Role of Emotion Regulation in Autism Spectrum Disorder." *Journal of the American Academy of Child and Adolescent Psychiatry* (2013) 52:679–88.

Mazoyer, B., et al. "Cortical Networks for Working Memory and Executive Functions Sustain the Conscious Resting State in Man." *Brain Research Bulletin* (2001) 54:287–98.

McAllister, D.E., and W.R. McAllister. "Fear Theory and Aversively Motivated Behavior: Some Controversial Issues." In: *Fear, Avoidance, and Phobias: A Fundamental Analysis*, ed. M.R. Denny (Hillsdale, NJ: Erlbaum, 1991).

McAllister, D.E., et al. "Magnitude and Shift of Reward in Instrumental Aversive Learning in Rats." *Journal of Comparative and Physiological Psychology* (1972) 80:490–501.

McAllister, D.E., et al. "Escape-from-Fear Performance as Affected by Handling Method and an Additional CS-Shock Treatment." *Animal Learning and Behavior* (1980) 8:417–23.

McAllister, W.R., and D.E. McAllister. "Behavioral Measurement of Conditioned Fear." In: *Aversive Conditioning and Learning*, ed. F.R. Brush (New York: Academic Press, 1971), 105–79.

McCarthy, D.E., et al. "Negative Reinforcement: Possible Clinical Implications of an Integrative Model." In: *Substance Abuse and Emotion*, ed. J. Kassel (Washington, DC: American Psychological Association, 2010), 15–42.

McCormick, D.A. "Cholinergic and Noradrenergic Modulation of Thalamocortical Processing." *Trends in Neurosciences* (1989) 12:215–21.

McCormick, D.A., and T. Bal. "Sensory Gating Mechanisms of the Thalamus." *Current Opinion in Neurobiology* (1994) 4:550–56.

McCue, M.G., J.E. LeDoux, and C.K. Cain. "Medial Amygdala Lesions Selectively Block Aversive Pavlovian-Instrumental Transfer in Rats." *Frontiers in Behavioral Neuroscience* (2014) 8:329.

McDannald, M.A., et al. "Learning Theory: A Driving Force in Understanding Orbitofrontal Function." *Neurobiology of Learning and Memory* (2014) 108:22–27.

McDonald, A.J. "Cortical Pathways to the Mammalian Amygdala." *Progress in Neurobiology* (1998) 55:257–332.

McEvoy, P.M., and A.E. Mahoney. "To Be Sure, to Be Sure: Intolerance of Uncertainty Mediates Symptoms of Various Anxiety Disorders and Depression." *Behavior Therapy* (2012) 43: 533–45.

McEwen, B.S. "Glucocorticoids, Depression, and Mood Disorders: Structural Remodeling in the Brain." *Metabolism: Clinical and Experimental* (2005) 54:20–23.

McEwen, B.S., and E.N. Lasley. *The End of Stress as We Know It* (Washington, DC: Joseph Henry Press, 2002).

McEwen, B.S., and R.M. Sapolsky. "Stress and Cognitive Function." *Current Opinion in Neurobiology* (1995) 5:205–16.

McGaugh, J.L. "Memory—A Century of Consolidation." *Science* (2000) 287:248–51.

McGaugh, J.L. *Memory and Emotion: The Making of Lasting Memories* (London: The Orion Publishing Group, 2003).

McGaugh, J.L. "Memory Reconsolidation Hypothesis Revived but Restrained: Theoretical Comment on Biedenkapp and Rudy (2004)." *Behavioral Neuroscience* (2004) 118:1140–42.

McGowan, P.O., et al. "Epigenetic Regulation of the Glucocorticoid Receptor in Human Brain Associates with Childhood Abuse." *Nature Neuroscience* (2009) 12:342–48.

McGrath, P.T., et al. "Quantitative Mapping of a Digenic Behavioral Trait Implicates Globin Variation in *C. elegans* Sensory Behaviors." *Neuron* (2009) 61:692–99.

McGuire, T.M., C.W. Lee, and P.D. Drummond. "Potential of Eye Movement Desensitization and Reprocessing Therapy in the Treatment of Post-Traumatic Stress Disorder." *Psychology Research and Behavior Management* (2014) 7:273–83.

McKay, D. "Methods and Mechanisms in the Efficacy of Psychodynamic Psychotherapy." *The American Psychologist* (2011) 66:147–48; discussion 152–54.

McKenzie, S., et al. "Hippocampal Representation of Related and Opposing Memories Develop Within Distinct, Hierarchically Organized Neural Schemas." *Neuron* (2014) 83:202–15.

McLean, C.P., et al. "Gender Differences in Anxiety Disorders: Prevalence, Course of Illness, Comorbidity and Burden of Illness." *Journal of Psychiatric Research* (2011) 45:1027–35.

McNally, R. "Theoretical Approaches to Fear and Anxiety." In: *Anxiety Sensitivity: Theory, Research, and Treatment of the Fear of Anxiety*, ed. S. Taylor (Hillsdale, NJ: Erlbaum, 1999), 3–16.

McNally, R. "Anxiety." In: *Oxford Companion to Emotion and the Affective Sciences,* eds. D. Sander and Scherer (Oxford, UK: Oxford University Press, 2009).

McNally, R.J. *Panic Disorder: A Critical Analysis* (New York: Guilford Press, 1994).

McNally, R.J. "Automaticity and the Anxiety Disorders." *Behaviour Research and Therapy* (1995) 33:747–54.

McNally, R.J. "Mechanisms of Exposure Therapy: How Neuroscience Can Improve Psychological Treatments for Anxiety Disorders." *Clinical Psychology Review* (2007) 27:750–59.

McNaughton, B.L. "The Neurophysiology of Reminiscence." *Neurobiology of Learning and Memory* (1998) 70:252–67.

McNaughton, B.L., et al. "Deciphering the Hippocampal Polyglot: The Hippocampus as a Path Integration System." *The Journal of Experimental Biology* (1996) 199:173–85.

McNaughton, N. *Biology and Emotion* (Cambridge, UK: Cambridge University Press, 1989).

McNaughton, N., and P.J. Corr. "A Two-Dimensional Neuropsychology of Defense: Fear/Anxiety and Defensive Distance." *Neuroscience and Biobehavioral Reviews* (2004) 28:285–305.

McTeague, L.M., et al. "Social Vision: Sustained Perceptual Enhancement of Affective Facial Cues in Social Anxiety." *NeuroImage* (2011) 54:1615–24.

Medford, N., and H.D. Critchley. "Conjoint Activity of Anterior Insular and Anterior Cingulate Cortex: Awareness and Response." *Brain Structure & Function* (2010) 214:535–49.

Mehler, M.F. "Epigenetic Principles and Mechanisms Underlying Nervous System Functions in Health and Disease." *Progress in Neurobiology* (2008) 86:305–41.

Menand, L. "The Prisoner of Stress: What Does Anxiety Mean?" *The New Yorker* (New York: Condé Nast, 2014).

Menzel, E. "Progress in the Study of Chimpanzee Recall and Episodic Memory." In: *The Missing Link in Cognition,* eds. H. Terrace and J. Metcalfe (Oxford, UK: Oxford University Press, 2005), 188–224.

Menzel, R. "Serial Position Learning in Honeybees." *PLoS One* (2009) 4:E4694.

Menzel, R., and M. Giurfa. "Cognition by a Mini Brain." *Nature* (1999) 400:718–19.

Merckelbach, H., et al. "Conditioning Experiences and Phobias." *Behaviour Research and Therapy* (1989) 27:657–62.

Merikle, P.M., S. Joordens, and J.A. Stolz. "Measuring the Relative Magnitude of Unconscious Influences." *Consciousness and Cognition* (1995) 4:422–39.

Merikle, P.M., D. Smilek, and J.D. Eastwood. "Perception Without Awareness: Perspectives from Cognitive Psychology." *Cognition* (2001) 79:115–34.

Merker, B. "Consciousness Without a Cerebral Cortex: A Challenge for Neuroscience and Medicine." *Behavioral and Brain Sciences,* a discussion (2007) 30:63–81, 81–134.

Mesulam, M.M. "Spatial Attention and Neglect: Parietal, Frontal and Cingulate Contributions to the Mental Representation and Attentional Targeting of Salient Extrapersonal Events." *Philosophical Transactions of the Royal Society B: Biological Sciences* (1999) 354:1325–46.

Mesulam, M.M., and E.J. Mufson. "Insula of the Old World Monkey. I. Architectonics in the Insulo-Orbito-Temporal Component of the Paralimbic Brain." *Journal of Comparative Neurology* (1982) 212:1–22.

Metcalfe, J., and A.P. Shimamura. *Metacognition: Knowing About Knowing* (Cambridge, MA: Bradford Books, 1994).

Metcalfe, J., and L.K. Son. "Anoetic, Noetic and Autonoetic Metacognition." In: *The Foundations of Metacognition,* eds. M. Beran et al. (Oxford, UK: Oxford University Press, 2012).

Metzinger, T. *Being No One* (Cambridge, MA: MIT Press, 2003).

Metzinger, T. "Empirical Perspectives from the Self-Model Theory of Subjectivity: A Brief Summary with Examples." *Progress in Brain Research* (2008) 168:215–45.

Meuret, A.E., and S.G. Hofmann. "Anxiety Disorders in Adulthood." In: *Handbook of Neurodevelopmental and Genetic Disorders in Adults,* eds. S. Goldstein and C. Reynolds (New York: Guilford Press, 2005), 172–94.

Meyer, K. "Primary Sensory Cortices, Top-Down Projections and Conscious Experience." *Progress in Neurobiology* (2011) 94:408–17.

Meyer, V., and M.G. Gelder. "Behaviour Therapy and Phobic Disorders." *The British Journal of Psychiatry: The Journal of Mental Science* (1963) 109:19–28.

Mihov, Y., and R. Hurlemann. "Altered Amygdala Function in Nicotine Addiction: Insights from Human Neuroimaging Studies." *Neuropsychologia* (2012) 50:1719–29.

Milad, M.R., et al. "Thickness of Ventromedial Prefrontal Cortex in Humans Is Correlated with Extinction Memory." *Proceedings of the National Academy of Sciences of the United States of America* (2005) 102:10706–11.

Milad, M.R., and G.J. Quirk. "Fear Extinction as a Model for Translational Neuroscience: Ten Years of Progress." *Annual Review of Psychology* (2012) 63:129–51.

Milad, M.R., and S.L. Rauch. "The Role of the Orbitofrontal Cortex in Anxiety Disorders." *Annals of the New York Academy of Sciences* (2007) 1121:546–61.

Milad, M.R., et al. "Fear Extinction in Rats: Implications for Human Brain Imaging and Anxiety Disorders." *Biological Psychology* (2006) 73:61–71.

Milad, M.R., B.L. Rosenbaum, and N.M. Simon. "Neuroscience of Fear Extinction: Implications for Assessment and Treatment of Fear-Based and Anxiety Related Disorders." *Behaviour Research and Therapy* (2014) 62:17–23.

Milad, M.R., et al. "Recall of Fear Extinction in Humans Activates the Ventromedial Prefrontal Cortex and Hippocampus in Concert." *Biological Psychiatry* (2007) 62:446–54.

Millan, M.J. "The Neurobiology and Control of Anxious States." *Progress in Neurobiology* (2003) 70:83–244.

Millan, M.J., and M. Brocco. "The Vogel Conflict Test: Procedural Aspects, Gamma-Aminobutyric Acid, Glutamate and Monoamines." *European Journal of Pharmacology* (2003) 463:67–96.

Miller, C.A., and J.D. Sweatt. "Amnesia or Retrieval Deficit? Implications of a Molecular Approach to the Question of Reconsolidation." *Learning & Memory* (2006) 13:498–505.

Miller, E.K., and J.D. Cohen. "An Integrative Theory of Prefrontal Cortex Function." *Annual Review of Neuroscience* (2001) 24:167–202.

Miller, E.K., and R. Desimone. "Parallel Neuronal Mechanisms for Short-Term Memory." *Science* (1994) 263:520–22.

Miller, E.K., C.A. Erickson, and R. Desimone. "Neural Mechanisms of Visual Working Memory in Prefrontal Cortex of the Macaque." *The Journal of Neuroscience* (1996) 16:5154–67.

Miller, G. "Epigenetics. The Seductive Allure of Behavioral Epigenetics." *Science* (2010) 329: 24–27.

Miller, N.E. "An Experimental Investigation of Acquired Drives." *Psychological Bulletin* (1941) 38:534–35.

Miller, N.E. "Studies of Fear as an Acquirable Drive: I. Fear as Motivation and Fear Reduction as Reinforcement in the Learning of New Responses." *Journal of Experimental Psychology* (1948) 38:89–101.

Miller, N.E. "Learnable Drives and Rewards." In: *Handbook of Experimental Psychology,* ed. S.S. Stevens (New York: Wiley, 1951), 435–72.

Milner, B. "Les troubles de la memoire accompagnant des lesions hippocampiques bilaterales." In: *Physiologie de l'Hippocampe,* ed. P. Plassouant (Paris: Centre de la Recherche Scientifique, 1962).

Milner, B. "Effects of Different Brain Lesions on Card Sorting: The Role of the Frontal Lobes." *Archives of Neurology* (1963) 9:90–100.

Milner, B. "Memory Disturbances After Bilateral Hippocampal Lesions in Man." In: *Cognitive Processes and Brain,* eds. P.M. Milner and S.E. Glickman (Princeton, NJ: Van Nostrand, 1965).

Milner, B. "Brain Mechanisms Suggested by Studies of Temporal Lobes." In: *Brain Mechanisms Underlying Speech and Language,* ed. F.L. Darley (New York: Grune and Stratton, 1967).

Milner, D., and M. Goodale. *The Visual Brain in Action* (Oxford, UK: Oxford University Press, 2006).

Milton, A.L., and B.J. Everitt. "The Psychological and Neurochemical Mechanisms of Drug Memory Reconsolidation: Implications for the Treatment of Addiction." *European Journal of Neuroscience* (2010) 31:2308–19.

Mineka, S. "The Role of Fear in Theories of Avoidance Learning, Flooding, and Extinction." *Psychological Bulletin* (1979) 86:985–1010.

Mineka, S. "Animal Models of Anxiety-Based Disorders: Their Usefulness and Limitation." In: *Anxiety and Anxiety Disorders,* eds. A.H. Tuma and J.D. Maser (England: Lawrence Erlbaum Associates, 1985).

Mineka, S., and M. Cook. "Mechanisms Involved in the Observational Conditioning of Fear." *Journal of Experimental Psychology: General* (1993) 122:23–38.

Mineka, S., and A. Öhman. "Phobias and Preparedness: The Selective, Automatic, and Encapsulated Nature of Fear." *Biological Psychiatry* (2002) 52:927–37.

Mineka, S., E. Rafaeli, and I. Yovel. "Cognitive Biases in Emotional Disorders: Information Processing and Social-Cognitive Perspectives." In: *Handbook of Affective Sciences,* eds. R.J. Davidson, et al. (New York: Oxford University Press, 2012), 976–1009.

Minue, S., et al. "Identification of Factors Associated with Diagnostic Error in Primary Care." *BMC Family Practice* (2014) 15:92.

Miracle, A.D., et al. "Chronic Stress Impairs Recall of Extinction of Conditioned Fear." *Neurobiology of Learning and Memory* (2006) 85:213–18.

Misanin, J.R., R.R. Miller, and D.J. Lewis. "Retrograde Amnesia Produced by Electroconvulsive Shock After Reactivation of a Consolidated Memory Trace." *Science* (1968) 160:554–55.

Mitchell, C.J., J. De Houwer, and P.F. Lovibond. "The Propositional Nature of Human Associative Learning." *Behavioral and Brain Sciences* (2009) 32:183–98; discussion 198–246.

Mitchell, D.G., and S.G. Greening. "Conscious Perception of Emotional Stimuli: Brain Mechanisms." *Neuroscientist* (2012) 18:386–98.

Mitchell, D.G., et al. "The Interference of Operant Task Performance by Emotional Distracters: An Antagonistic Relationship Between the Amygdala and Frontoparietal Cortices." *NeuroImage* (2008) 40:859–68.

Mitchell, R.A., and A.J. Berger. "Neural Regulation of Respiration." *The American Review of Respiratory Disease* (1975) 111:206–24.

Mitchell, R.W., N.S. Thompson, and H.L. Miles, eds. *Anthropomorphism, Anecdotes, and Animals* (New York: SUNY Press, 1996).

Mitchell, S.H. "The Genetic Basis of Delay Discounting and Its Genetic Relationship to Alcohol Dependence." *Behavioural Processes* (2011) 87:10–17.

Mitra, R., and R.M. Sapolsky. "Gene Therapy in Rodent Amygdala Against Fear Disorders." *Expert Opinion on Biological Therapy* (2010) 10:1289–1303.

Mitte, K. "Meta-Analysis of Cognitive-Behavioral Treatments for Generalized Anxiety Disorder: A Comparison with Pharmacotherapy." *Psychological Bulletin* (2005) 131:785–95.

Mitte, K. "Anxiety and Risky Decision-Making: The Role of Subjective Probability and Subjective Costs of Negative Events." *Personality and Individual Differences* (2007) 43:243–53.

Mizumori, S.J., et al. "Preserved Spatial Coding in Hippocampal, C.A.1 Pyramidal Cells During Reversible Suppression of C.A.3c Output: Evidence for Pattern Completion in Hippocampus." *Journal of Neuroscience* (1989) 9:3915–28.

Mobbs, D., et al. "When Fear Is Near: Threat Imminence Elicits Prefrontal-Periaqueductal Gray Shifts in Humans." *Science* (2007) 317:1079–83.

Mogg, K., and B.P. Bradley. "A Cognitive-Motivational Analysis of Anxiety." *Behaviour Research and Therapy* (1998) 36:809–48.

Mohanty, A., and T.J. Sussman. "Top-Down Modulation of Attention by Emotion." *Frontiers in Human Neuroscience* (2013) 7:102.

Monfils, M.H., et al. "Extinction-Reconsolidation Boundaries: Key to Persistent Attenuation of Fear Memories." *Science* (2009) 324:951–55.

Montaigne, M. de. *Michel de Montaigne—The Complete Essays* (New York: Penguin Classics, 1993).

Morgan, C.L. *Animal Life and Intelligence* (Boston: Ginn & Company, 1890–1891).

Morgan, C.T. *Physiological Psychology* (New York: McGraw-Hill, 1943).

Morgan, C.T. "Physiological Mechanisms of Motivation." *Nebraska Symposium on Motivation* (1957) 5:1–43.

Morgan, M.A., and J.E. LeDoux. "Differential Contribution of Dorsal and Ventral Medial Prefrontal Cortex to the Acquisition and Extinction of Conditioned Fear in Rats." *Behavioral Neuroscience* (1995) 109:681–88.

Morgan, M.A., and J.E. LeDoux. "Contribution of Ventrolateral Prefrontal Cortex to the Acquisition and Extinction of Conditioned Fear in Rats." *Neurobiology of Learning and Memory* (1999) 72:244–51.

Morgan, M.A., L.M. Romanski, and J.E. LeDoux. "Extinction of Emotional Learning: Contribution of Medial Prefrontal Cortex." *Neuroscience Letters* (1993) 163:109–13.

Morgan, M.A., J. Schulkin, and J.E. LeDoux. "Ventral Medial Prefrontal Cortex and Emotional Perseveration: The Memory for Prior Extinction Training." *Behavioural Brain Research* (2003) 146:121–30.

Morris, J.S. "How Do You Feel?" *Trends in Cognitive Sciences* (2002) 6:317–19.

Morris, J.S., C. Buchel, and R.J. Dolan. "Parallel Neural Responses in Amygdala Subregions and Sensory Cortex During Implicit Fear Conditioning." *NeuroImage* (2001) 13:1044–52.

Morris, J.S., et al. "Differential Extrageniculostriate and Amygdala Responses to Presentation of Emotional Faces in a Cortically Blind Field." *Brain: A Journal of Neurology* (2001) 124: 1241–52.

Morris, J.S., K.J. Friston, and R.J. Dolan. "Neural Responses to Salient Visual Stimuli." *Proceedings of the Royal Society of London Series B, Containing Papers of a Biological Character Royal Society* (1997) 264:769–75.

Morris, J.S., K.J. Friston, and R.J. Dolan. "Experience-Dependent Modulation of Tonotopic Neural Responses in Human Auditory Cortex." *Proceedings Biological Sciences / the Royal Society* (1998) 265:649–57.

Morris, J.S., A. Öhman, and R.J. Dolan. "Conscious and Unconscious Emotional Learning in the Human Amygdala." *Nature* (1998) 393:467–70.

Morris, J.S., A. Öhman, and R.J. Dolan. "A Subcortical Pathway to the Right Amygdala Mediating 'Unseen' Fear." *Proceedings of the National Academy of Sciences of the United States of America* (1999) 96:1680–85.

Morris, S.E., and B.N. Cuthbert. "Research Domain Criteria: Cognitive Systems, Neural Circuits, and Dimensions of Behavior." *Dialogues in Clinical Neuroscience* (2012) 14:29–37.

Morrison, J.H., et al. "Noradrenergic and Serotonergic Fibers Innervate Complementary Layers in Monkey Primary Visual Cortex: An Immunohistochemical Study." *Proceedings of the National Academy of Sciences of the United States of America* (1982) 79:2401–5.

Morrison, S.E., and C.D. Salzman. "Re-Valuing the Amygdala." *Current Opinion in Neurobiology* (2010) 20:221–30.

Morrison, S.F., and D.J. Reis. "Responses of Sympathetic Preganglionic Neurons to Rostral Ventrolateral Medullary Stimulation." *American Journal of Physiology* (1991) 261:R1247–56.

Moruzzi, G., and H.W. Magoun. "Brain Stem Reticular Formation and Activation of the EEG. *Electroencephalography and Clinical Neurophysiology* (1949) 1:455–73.

Moscarello, J.M., and J.E. LeDoux. "Active Avoidance Learning Requires Prefrontal Suppression of Amygdala-Mediated Defensive Reactions." *Journal of Neuroscience* (2013) 33:3815–23.

Moscovitch, M., et al. "Functional Neuroanatomy of Remote Episodic, Semantic and Spatial Memory: A Unified Account Based on Multiple Trace Theory." *Journal of Anatomy* (2005) 207:35–66.

Moser, E.I., and M.B. Moser. "A Metric for Space." *Hippocampus* (2008) 18:1142–56.

Moser, E.I., et al. "Grid Cells and Cortical Representation." *Nature Reviews Neuroscience* (2014) 15:466–81.

Mougi, A. "Coevolution in a One Predator-Two Prey System." *PLoS One* (2010) 5:E13887.

Mowrer, O.H. "A Stimulus-Response Analysis of Anxiety and Its Role as a Reinforcing Agent." *Psychological Review* (1939) 46:553–65.

Mowrer, O.H. "Anxiety-Reduction and Learning." *Journal of Experimental Psychology* (1940) 27:497–516.

Mowrer, O.H. "On the Dual Nature of Learning: a Reinterpretation of 'Conditioning' and 'Problem Solving.'" *Harvard Educational Review* (1947) 17:102–48.

Mowrer, O.H. *Learning Theory and Personality Dynamics* (New York: The Ronald Press Co., 1950).

Mowrer, O.H. "Two-Factor Learning Theory: Summary and Comment." *Psychological Review* (1951) 58:350–54.

Mowrer, O.H. *Learning Theory and Behavior* (New York: Wiley, 1960).

Mowrer, O.H., and R.R. Lamoreaux. "Avoidance Conditioning and Signal Duration: A Study of Secondary Motivation and Reward." *Psychological Monographs* (1942) 54.

Mowrer, O.H., and R.R. Lamoreaux. "Fear as an Intervening Variable in Avoidance Conditioning." *Journal of Comparative Psychology* (1946) 39:29–50.

Moyer, K.E. *The Psychobiology of Aggression* (New York: Harper & Row, 1976).

Muller, N.G., L. Machado, and R.T. Knight. "Contributions of Subregions of the Prefrontal Cortex to Working Memory: Evidence from Brain Lesions in Humans." *Journal of Cognitive Neuroscience* (2002) 14:673–86.

Muller, R.U., J.L. Kubie, and J.B. Ranck Jr. "Spatial Firing Patterns of Hippocampal Complex-Spike Cells in a Fixed Environment." *Journal of Neuroscience* (1987) 7:1935–50.

Munk, H. "Weitere Mittheilungen zur Physiologie der Grosshirnrinde." *Verhandlungen der Physiologischen Gesellschaft zu Berlin* (1878) 162–78.

Myers, K.M., and M. Davis. "Behavioral and Neural Analysis of Extinction." *Neuron* (2002) 36:567–84.

Myers, K.M., and M. Davis. "Mechanisms of Fear Extinction." *Molecular Psychiatry* (2007) 12:120–50.

Myers, R.E. "Role of Prefrontal and Anterior Temporal Cortex in Social Behavior and Affect in Monkeys." *Acta Neurobiologiae Experimentalis* (1972) 32:567–79.

Naccache, L., E. Blandin, and S. Dehaene. "Unconscious Masked Priming Depends on Temporal Attention." *Psychological Science* (2002) 13:416–24.

Naccache, L., and S. Dehaene. "Reportability and Illusions of Phenomenality in the Light of the Global Neuronal Workspace Model." *Behavioral and Brain Sciences* (2007) 30:518–20.

Nader, K., and E.O. Einarsson. "Memory Reconsolidation: An Update." *Annals of the New York Academy of Sciences* (2010) 1191:27–41.

Nader, K., and O. Hardt. "A Single Standard for Memory: The Case for Reconsolidation." *Nature Reviews Neuroscience* (2009) 10:224–34.

Nader, K., and G.E. Schafe, and J.E. LeDoux. "Fear Memories Require Protein Synthesis in the Amygdala for Reconsolidation after Retrieval." *Nature* (2000) 406:722–26.

Nadel, L., and O. Hardt. "Update on Memory Systems and Processes." *Neuropsychopharmacology* (2011) 36: 251–73.

Nadim, F., and D. Bucher. "Neuromodulation of Neurons and Synapses." *Current Opinion in Neurobiology* (2014) 29C:48–56.

Nadler, N., M.R. Delgado, and A.R. Delamater. "Pavlovian to Instrumental Transfer of Control in a Human Learning Task." *Emotion* (2011) 11:1112–23.

Nagel, T. "What Is It Like to Be a Bat?" *Philosophical Review* (1974) 83:4435–50.

Nashold, B.S. Jr., W.P. Wilson, and D.G. Slaughter. "Sensations Evoked by Stimulation in the Midbrain of Man." *Journal of Neurosurgery* (1969) 30:14–24.

Nauta, W.J. "The Problem of the Frontal Lobe: A Reinterpretation." *Journal of Psychiatric Research* (1971) 8:167–87.

Nauta, W.J.H, and H.J. Karten. "A General Profile of the Vertebrate Brain, with Sidelights on the Ancestry of *Cerebral Cortex.*" In: *The Neurosciences: Second Study Program,* ed. F.O. Schmitt (New York: The Rockefeller University Press, 1970), 7–26.

Nazari, H., et al. "Comparison of Eye Movement Desensitization and Reprocessing with Citalopram in Treatment of Obsessive-Compulsive Disorder." *International Journal of Psychiatry in Clinical Practice* (2011) 15:270–74.

Nehoff, H., et al. "Nanomedicine for Drug Targeting: Strategies Beyond the Enhanced Permeability and Retention Effect." *International Journal of Nanomedicine* (2014) 9:2539–55.

Neisser, U. *Cognitive Psychology* (Englewood Cliffs, NJ: Prentice Hall, 1967).

Nemoda, Z., A. Szekely, and M. Sasvari-Szekely. "Psychopathological Aspects of Dopaminergic Gene Polymorphisms in Adolescence and Young Adulthood." *Neuroscience and Biobehavioral Reviews* (2011) 35:1665–86.

Nesse, R.M., and R. Klaas. "Risk Perception by Patients with Anxiety Disorders." *The Journal of Nervous and Mental Disease* (1994) 182:465–70.

Nestler, E.J. "Transcriptional Mechanisms of Drug Addiction." *Clinical Psychopharmacology and Neuroscience: The Official Scientific Journal of the Korean College of Neuropsychopharmacology* (2012) 10:136–43.

Neudeck, P., and H.-U. Wittchen. *Exposure Therapy: Rethinking the Model—Refining the Method* (New York: Springer, 2012).

Neumann, I.D., and R. Landgraf. "Balance of Brain Oxytocin and Vasopressin: Implications for Anxiety, Depression, and Social Behaviors." *Trends in Neurosciences* (2012) 35:649–59.

Neumeister, A. "The Endocannabinoid System Provides an Avenue for Evidence-Based Treatment Development for PTSD. *Depression and Anxiety* (2013) 30:93–96.

Neville, H., and D. Bavelier. "Human Brain Plasticity: Evidence from Sensory Deprivation and Altered Language Experience." *Progress in Brain Research* (2002) 138:177–88.

Newell, B.R., and D.R. Shanks. "Unconscious Influences on Decision Making: A Critical Review." *Behavioral and Brain Sciences* (2014) 37:1–19.

Newman, M.G., and T.D. Borkovec. "Cognitive-Behavioral Treatment of Generalized Anxiety Disorder." *Clinical Psychology* (1995) 48:5–7.

Nguyen, P.V. "CREB and the Enhancement of Long-Term Memory." *Trends in Neurosciences* (2001) 24:314.

Nicotra, A., et al. "Emotional and Autonomic Consequences of Spinal Cord Injury Explored Using Functional Brain Imaging." *Brain: a Journal of Neurology* (2006) 129:718–28.

Nisbett, R.E., and T.D. Wilson. "Telling More Than We Can Know: Verbal Reports on Mental Processes." *Psychological Review* (1977) 84:231–59.

Nitschke, J.B., et al. "Anticipatory Activation in the Amygdala and Anterior Cingulate in Generalized Anxiety Disorder and Prediction of Treatment Response." *The American Journal of Psychiatry* (2009) 166:302–10.

Noë, A. *Varieties of Presence* (Cambridge, MA: Harvard University Press, 2012).

Norman, D.A., and T. Shallice. "Attention to Action: Willed and Automatic Control of Behavior." In: *Consciousness and Self-Regulation*, eds. R.J. Davidson, et al. (New York: Plenum, 1980), 1–18.

Northcutt, R.G. "Changing Views of Brain Evolution." *Brain Research Bulletin* (2001) 55:663–74.

Oakley, D.A. "Hypnosis and Conversion Hysteria: A Unifying Model." *Cognitive Neuropsychiatry* (1999) 4:243–65.

Ochsner, K.N., et al. "Rethinking Feelings: An fMRI Study of the Cognitive Regulation of Emotion." *Journal of Cognitive Neuroscience* (2002) 14:1215–29.

Ochsner, K.N., and J.J. Gross. "The Cognitive Control of Emotion." *Trends in Cognitive Sciences* (2005) 9:242–49.

Ochsner, K.N., et al. "For Better or for Worse: Neural Systems Supporting the Cognitive Down- and Up-Regulation of Negative Emotion." *NeuroImage* (2004) 23:483–99.

O'Donohue, W.T. *A History of the Behavioral Therapies: Founders' Personal Histories* (Reno, NV: Context Press, 2001).

O'Donohue, W.T., et al. eds. *A History of the Behavioral Therapies: Founders' Personal Histories* (New York: Wiley, 2003).

Öhman, A. "Automaticity and the Amgydala: Nonconscious Responses to Emotional Faces." *Current Directions in Psychological Science* (2002) 11:62–66.

Öhman, A. "The Role of the Amygdala in Human Fear: Automatic Detection of Threat." *Psychoneuroendocrinology* (2005) 30:953–58.

Öhman, A. "Has Evolution Primed Humans to 'Beware the Beast'?" *Proceedings of the National Academy of Sciences of the United States of America* (2007) 104:16396–97.

Öhman, A. "Of Snakes and Faces: An Evolutionary Perspective on the Psychology of Fear." *Scandinavian Journal of Psychology* (2009) 50:543–52.

Öhman, A. "Nonconscious Control of Autonomic Responses: A Role for Pavlovian Conditioning?" *Biological Psychology* (1988) 27:113–35.

Öhman, A., Flykt A., Esteves F. "Emotion Drives Attention: Detecting the Snake in the Grass." *Journal of Experimental Psychology: General* (2001) 130:466–78.

Öhman, A., D. Lundqvist, and F. Esteves. "The Face in the Crowd Revisited: A Threat Advantage with Schematic Stimuli." *Journal of Personality and Social Psychology* (2001) 80:381–96.

Öhman, A., and S. Mineka. "Fears, Phobias, and Preparedness: Toward an Evolved Module of Fear and Fear Learning." *Psychological Review* (2001) 108:483–522.

O'Keefe, J. "Is Consciousness the Gateway to the Hippocampal Cognitive Map? A Speculative Essay on the Neural Basis of Mind." In: *Brain & Mind,* ed. D.A. Oakley (New York: Methuen & Co, 1985).

O'Keefe, J., et al. "Place Cells, Navigational Accuracy, and the Human Hippocampus." *Philosophical Transactions of the Royal Society B: Biological Sciences* (1998) 353:1333–40.

O'Keefe, J., and L. Nadel. *The Hippocampus as a Cognitive Map* (Oxford, UK: Clarendon Press, 1978).

Olds, J. "Pleasure Centers in the Brain." *Scientific American* (1956) 195:105–16.

Olds, J. "Self Stimulation of the Brain." *Science* (1958) 127:315–24.

Olds, J. *Drives and Reinforcement* (New York: Raven, 1977).

Olds, J., and P. Milner. "Positive Reinforcement Produced by Electrical Stimulation of Septal and Other Regions of the Brain." *Journal of Comparative and Physiological Psychology* (1954) 47:419–27.

O'Leary, K.D., and G.T. Wilson. *Behavior Therapy: Application and Outcome* (Englewood Cliffs, NJ: Prentice-Hall, 1975).

Olmos-Serrano, J.L., and J.G. Corbin. "Amygdala Regulation of Fear and Emotionality in Fragile X Syndrome." *Developmental Neuroscience* (2011) 33:365–78.

Olsson, A., et al. "The Role of Social Groups in the Persistence of Learned Fear." *Science* (2005) 309:785–87.

Olsson, A., K.I. Nearing, and E.A. Phelps. "Learning Fears by Observing Others: The Neural Systems of Social Fear Transmission." *Social Cognitive and Affective Neuroscience* (2007) 2:3–11.

Olsson, A., and E.A. Phelps. "Learned Fear of 'Unseen' Faces After Pavlovian, Observational, and Instructed Fear." *Psychological Science* (2004) 15:822–28.

Olsson, A., and E.A. Phelps. "Social Learning of Fear." *Nature Neuroscience* (2007) 10:1095–1102.

Olton, D., J.T. Becker, and G.E. Handleman. "Hippocampus, Space and Memory." *Behavioral and Brain Sciences* (1979) 2:313–65.

O'Regan, J.K., and A. Noe. "A Sensorimotor Account of Vision and Visual Consciousness." *Behavioral and Brain Sciences* (2001) 24:939–973; discussion 973–1031.

O'Reilly, R.C., T.S. Braver, and J.D. Cohen. "A Biologically Based Computational Model of Working Memory." In: *Models of Working Memory: Mechanisms of Active Maintenance and Executive Control,* eds. A. Miyake and P. Shah (New York: Cambridge University Press, 1999), 375–411.

O'Reilly, R.C., and J.L. McClelland. "Hippocampal Conjunctive Encoding, Storage, and Recall: Avoiding a Trade-off." *Hippocampus* (1994) 4:661–82.

Orsini, C.A., and S. Maren. "Neural and Cellular Mechanisms of Fear and Extinction Memory Formation." *Neuroscience and Biobehavioral Reviews* (2012) 36:1773–1802.

Ortony, A., and G.L. Clore. "Emotions, Moods, and Conscious Awareness." *Cognition and Emotion* (1989) 3:125–37.

Ortony, A., G.L. Clore, and A. Collins. *The Cognitive Structure of Emotions* (Cambridge, UK: Cambridge University Press, 1988).

Ortony, A., and T.J. Turner. "What's Basic About Basic Emotions?" *Psychological Review* (1990) 97:315–31.

Osaka, N. "[Active Consciousness and the Prefrontal Cortex: A Working-Memory Approach]. Shinrigaku Kenkyu: *The Japanese Journal of Psychology* (2007) 77:553–66.

Ost, L.G., and K. Hugdahl. "Acquisition of Agoraphobia, Mode of Onset and Anxiety Response Patterns." *Behaviour Research and Therapy* (1983) 21:623–31.

Ost, L.G., et al. "One-Session Treatment of Specific Phobias in Youths: A Randomized Clinical Trial." *Journal of Consulting and Clinical Psychology* (2001) 69:814–24.

Ostroff, L.E., et al. "Fear and Safety Learning Differentially Affect Synapse Size and Dendritic Translation in the Lateral Amygdala." *Proceedings of the National Academy of Sciences of the United States of America* (2010) 107:9418–23.

Overgaard, M., et al. "Optimizing Subjective Measures of Consciousness." *Consciousness and Cognition* (2010) 19:682–684; discussion 685–86.

Owen, A.M. "Detecting Consciousness: A Unique Role for Neuroimaging." *Annual Review of Psychology* (2013) 64:109–33.

Owen, A.M., and M.R. Coleman. "Functional MRI in Disorders of Consciousness: Advantages and Limitations." *Current Opinion in Neurology* (2007) 20:632–37.

Owen, A.M., et al. "Detecting Awareness in the Vegetative State." *Science* (2006) 313:1402.

Packard, M.G. "Anxiety, Cognition, and Habit: A Multiple Memory Systems Perspective." *Brain Research* (2009) 1293:121–28.

Padoa-Schioppa C., and J.A. Assad. "Neurons in the Orbitofrontal Cortex Encode Economic Value." *Nature* (2006) 441:223–26.

Pahl, M., A. Si, and S. Zhang. "Numerical Cognition in Bees and Other Insects." *Frontiers in Psychology* (2013) 4:162.

Panagiotaropoulos, T.I., V. Kapoor, and N.K. Logothetis. "Desynchronization and Rebound of Beta Oscillations During Conscious and Unconscious Local Neuronal Processing in the Macaque Lateral Prefrontal Cortex." *Frontiers in Psychology* (2013) 4:603.

Panagiotaropoulos, T.I., V. Kapoor, and N.K. Logothetis. "Subjective Visual Perception: From Local Processing to Emergent Phenomena of Brain Activity." *Philosophical Transactions of the Royal Society B: Biological Sciences* (2014) 369:20130534.

Panksepp, J. "Aggression Elicited by Electrical Stimulation of the Hypothalamus in Albino Rat." *Physiology & Behavior* (1971) 6:321–29.

Panksepp, J. "Hypothalamic Integration of Behavior: Rewards, Punishments, and Related Psychological Processes." In: *Handbook of the Hypothalamus Vol. 3, Behavioral Studies of the Hypothalamus*, eds. P.J. Morgane and J. Panksepp (New York: Marcel Dekker, 1980), 289–431.

Panksepp, J. "Toward a General Psychobiological Theory of Emotions." *Behavioral and Brain Sciences* (1982) 5:407–67.

Panksepp, J. *Affective Neuroscience* (New York: Oxford University Press, 1998).

Panksepp, J. "Emotions as Natural Kinds Within the Mammalian Brain." In: *Handbook of Emotions*, eds. M. Lewis and J.M. Haviland-Jones (New York: Guilford Press, 2000), 137–56.

Panksepp, J. "Affective Consciousness: Core Emotional Feelings in Animals and Humans." *Consciousness and Cognition* (2005) 14:30–80.

Panksepp, J. "Neurologizing the Psychology of Affects: How Appraisal-Based Constructivism and Basic Emotion Theory Can Coexist." *Perspectives on Psychological Science* (2007) 2:281–96.

Panksepp, J. "The Basic Emotional Circuits of Mammalian Brains: Do Animals Have Affective Lives?" *Neuroscience and Biobehavioral Reviews* (2011) 35:1791–1804.

Panksepp, J. "Cross-Species Affective Neuroscience Decoding of the Primal Affective Experiences of Humans and Related Animals." *PLoS One* (2011) 6:E21236.

Panksepp, J. *The Archaeology of Mind: Neuroevolutionary Origins of Human Emotion* (New York: W.W. Norton & Company, 2012).

Panksepp, J., et al. "The Psycho- and Neurobiology of Fear Systems in the Brain." In: *Fear, Avoidance, and Phobias,* ed. M.R. Denny (Hillsdale, NJ: Erlbaum, 1991), 7–59.

Pape, H.C., and D. Paré. "Plastic Synaptic Networks of the Amygdala for the Acquisition, Expression, and Extinction of Conditioned Fear." *Physiological Reviews* (2010) 90:419–63.

Papez, J.W. *Comparative Neurology* (New York: Thomas Y. Crowell, 1929).

Papineau, D. "Functionalism." In: *Routledge Encyclopedia of Philosophy,* ed. E. Craig (London: Routledge, 1998).

Papineau, D. *Thinking About Consciousness* (Oxford, UK: Oxford University Press, 2002).

Papineau, D. "Explanatory Gaps and Dualist Intuitions." In: *Frontiers of Consciousness: Chichele Lectures,* eds. L. Weiskrantz and M. Davies (Oxford, UK: Oxford University Press, 2008), 55–68.

Papini, S., et al. "Toward a Translational Approach to Targeting the Endocannabinoid System in Posttraumatic Stress Disorder: a Critical Review of Preclinical Research." *Biological Psychology* (2014) 104C:8–18.

Paré, D. "Mechanisms of Pavlovian Fear Conditioning: Has the Engram Been Located?" *Trends in Neurosciences* (2002) 25:436–437; discussion 437–38.

Paré, D., and D.R. Collins. "Neuronal Correlates of Fear in the Lateral Amygdala: Multiple Extracellular Recordings in Conscious Cats." *Journal of Neuroscience* (2000) 20:2701–10.

Paré, D., and S. Duvarci. "Amygdala Microcircuits Mediating Fear Expression and Extinction." *Current Opinion in Neurobiology* (2012) 22:717–23.

Paré, D., G.J. Quirk, and J.E. LeDoux. "New Vistas on Amygdala Networks in Conditioned Fear." *Journal of Neurophysiology* (2004) 92:1–9.

Paré D., Y. Smith, J.F. Paré. "Intra-Amygdaloid Projections of the Basolateral and Basomedial Nuclei in the Cat: Phaseolus Vulgaris-Leucoagglutinin Anterograde Tracing at the Light and Electron Microscopic Level." *Neuroscience* (1995) 69:567–83.

Paré, D., and Y. Smith. "The Intercalated Cell Masses Project to the Central and Medial Nuclei of the Amygdala in Cats." *Neuroscience* (1993) 57:1077–90.

Paré, D., and Y. Smith. "GABAergic Projection from the Intercalated Cell Masses of the Amygdala to the Basal Forebrain in Cats." *Journal of Comparative Neurology* (1994) 344:33–49.

Park, D.B., J.V. Dobson, and J.D. Losek. "All That Wheezes Is Not Asthma: Cognitive Bias in Pediatric Emergency Medical Decision Making." *Pediatric Emergency Care* (2014) 30:104–7.

Pascoe, J.P., and B.S. Kapp. "Electrophysiological Characteristics of Amygdaloid Central Nucleus Neurons During Pavlovian Fear Conditioning in the Rabbit." *Behavioural Brain Research* (1985) 16:117–33.

Pascual-Leone A., and V. Walsh. "Fast Backprojections from the Motion to the Primary Visual Area Necessary for Visual Awareness." *Science* (2001) 292:510–12.

Pasley, B.N., L.C. Mayes, and R.T. Schultz. "Subcortical Discrimination of Unperceived Objects During Binocular Rivalry." *Neuron* (2004) 42:163–72.

Pastalkova, E., et al. "Storage of Spatial Information by the Maintenance Mechanism of LTP." *Science* (2006) 313:1141–44.

Patel, R., R.N. Spreng, L.M. Shin, and T.A. Girard. "Neurocircuitry Models of Posttraumatic Stress Disorder and Beyond: A Meta-Analysis of Functional Neuroimaging Studies." *Neuroscience & Biobehavioral Reviews* 36: 2130–42.

Paulus, M.P., and A.J. Yu. "Emotion and Decision-Making: Affect-Driven Belief Systems in Anxiety and Depression." *Trends in Cognitive Sciences* (2012) 16:476–83.

Pavlov, I.P. *Conditioned Reflexes* (New York: Dover, 1927).

Pavlov, K.A., D.A. Chistiakov, and V.P. Chekhonin. "Genetic Determinants of Aggression and Impulsivity in Humans." *Journal of Applied Genetics* (2012) 53:61–82.

Pearce, J.M., and M.E. Bouton. "Theories of Associative Learning in Animals." *Annual Review of Psychology* (2001) 52:111–39.

Pedreira, M.E., and H. Maldonado. "Protein Synthesis Subserves Reconsolidation or Extinction Depending on Reminder Duration." *Neuron* (2003) 38:863–69.

Pena, D.F., N.D. Engineer, and C.K. McIntyre. "Rapid Remission of Conditioned Fear Expression with Extinction Training Paired with Vagus Nerve Stimulation." *Biological Psychiatry* (2013) 73:1071–77.

Penzo, M.A., V. Robert, and B. Li. "Fear Conditioning Potentiates Synaptic Transmission onto Long-Range Projection Neurons in the Lateral Subdivision of Central Amygdala." *The Journal of Neuroscience: The Official Journal of the Society for Neuroscience* (2014) 34:2432–37.

Percy, W. *Love in the Ruins* (New York: Farrar, Straus and Giroux, 1971).

Perry, R., and R.M. Sullivan. "Neurobiology of Attachment to an Abusive Caregiver: Short-Term Benefits and Long-Term Costs." *Developmental Psychobiology* (2014) 56:1626–34.

Persaud, N., et al. "Awareness-Related Activity in Prefrontal and Parietal Cortices in Blindsight Reflects More Than Superior Visual Performance." *NeuroImage* (2011) 58:605–11.

Persaud, N., P. McLeod, and A. Cowey. "Post-Decision Wagering Objectively Measures Awareness." *Nature Neuroscience* (2007) 10:257–61.

Pessoa, L. "On the Relationship Between Emotion and Cognition." *Nature Reviews Neuroscience* (2008) 9:148–58.

Pessoa, L. *The Cognitive-Emotional Brain: From Interactions to Integration* (Cambridge, MA: MIT Press, 2013).

Pessoa, L., and R. Adolphs. "Emotion Processing and the Amygdala: From a 'Low Road' to 'Many Roads' of Evaluating Biological Significance." *Nature Reviews Neuroscience* (2010) 11: 773–83.

Pessoa, L., S. Kastner, and L.G. Ungerleider. "Attentional Control of the Processing of Neural and Emotional Stimuli." *Brain Research. Cognitive Brain Research* (2002) 15:31–45.

Pessoa, L., et al. "Neural Processing of Emotional Faces Requires Attention." *Proceedings of the National Academy of Sciences of the United States of America* (2002) 99:11458–63.

Pessoa, L., and L.G. Ungerleider. "Neuroimaging Studies of Attention and the Processing of Emotion-Laden Stimuli." *Progress in Brain Research* (2004) 144:171–82.

Peters, J., and C. Buchel. "Neural Representations of Subjective Reward Value." *Behavioural Brain Research* (2010) 213:135–41.

Petrovich, G.D., and M. Gallagher. "Amygdala Subsystems and Control of Feeding Behavior by Learned Cues." *Annals of the New York Academy of Sciences* (2003) 985:251–62.

Phelps, E.A. "Faces and Races in the Brain." *Nature Neuroscience* (2001) 4:775–76.

Phelps, E.A. "Emotion and Cognition: Insights from Studies of the Human Amygdala." *Annual Review of Psychology* (2006) 57:27–53.

Phelps, E.A., et al. "Extinction Learning in Humans; Role of the Amygdala and vmPFC." *Neuron* (2004) 43:897–905.

Phelps, E.A., and J.E. LeDoux. "Contributions of the Amygdala to Emotion Processing: From Animal Models to Human Behavior." *Neuron* (2005) 48:175–87.

Phelps, E.A., S. Ling, and M. Carrasco. "Emotion Facilitates Perception and Potentiates the Perceptual Benefits of Attention." *Psychological Science* (2006) 17:292–99.

Phelps, E.A., et al. "Performance on Indirect Measures of Race Evaluation Predicts Amygdala Activation." *Journal of Cognitive Neuroscience* (2000) 12:729–38.

Philippi, C.L., et al. "Preserved Self-Awareness Following Extensive Bilateral Brain Damage to the Insula, Anterior Cingulate, and Medial Prefrontal Cortices." *PLoS One* (2012) 7:E38413.

Phillips, R.G., and J.E. LeDoux. "Differential Contribution of Amygdala and Hippocampus to Cued and Contextual Fear Conditioning." *Behavioral Neuroscience* (1992) 106:274–85.

Phillips, R.G., and J.E. LeDoux. "Lesions of the Dorsal Hippocampal Formation Interfere with Background but Not Foreground Contextual Fear Conditioning." *Learning & Memory* (1994) 1:34–44.

Piaget, J. *Biology and Knowledge* (Edinburgh: Edinburgh University Press, 1971).

Piccinini, G. "The Ontology of Creature Consciousness: A Challenge for Philosophy." *Behavioral and Brain Sciences* (2007) 30:103–4.

Pickens, C.L., and P.C. Holland. "Conditioning and Cognition." *Neuroscience and Biobehavioral Reviews* (2004) 28:651–61.

Picoult, J. *Sing You Home* (New York: Simon and Schuster, 2011), 322.

Pidoplichko, V.I., et al. "ASIC1a Activation Enhances Inhibition in the Basolateral Amygdala and Reduces Anxiety." *Journal of Neuroscience* (2014) 34:3130–41.

Pine, D.S., et al. "Methods for Developmental Studies of Fear Conditioning Circuitry." *Biological Psychiatry* (2001) 50:225–28.

Pinel, J.P.J., and D. Treit. "Burying as a Defensive Response in Rats." *Journal of Comparative and Physiological Psychology* (1978) 92:708–12.

Pinsker, H.M., et al. "Long-Term Sensitization of a Defensive Withdrawal Reflex in *Aplysia*." *Science* (1973) 182:1039–42.

Pirri, J.K., and M.J. Alkema. "The Neuroethology of *C. elegans* Escape." *Current Opinion in Neurobiology* (2012) 22:187–93.

Pitkänen, A. "Connectivity of the Rat Amygdaloid Complex." In: *The Amygdala: A Functional Analysis*, ed. J.P. Aggleton (Oxford, UK: Oxford University Press, 2000), 31–115.

Pitkänen, A., V. Savander, and J.E. LeDoux. "Organization of Intra-Amygdaloid Circuitries in the Rat: An Emerging Framework for Understanding Functions of the Amygdala." *Trends in Neurosciences* (1997) 20:517–23.

Pitman, R.K., M.R. Milad, S.A. Igoe, M.G. Vangel, S.P. Orr, A. Tsareva, K. Gamache, and K. Nader. "Systemic Mifepristone Blocks Reconsolidation of Cue-Conditioned Fear; Propranolol Prevents This Effect." *Behavioral Neuroscience* (2011) 125:632–38.

Pitman, R.K., S.P. Orr, and A.Y. Shalev. "Once Bitten, Twice Shy: Beyond the Conditioning Model of PTSD." *Biological Psychiatry* (1993) 33:145–46.

Pitman, R.K., A.M. Rasmusson, K.C. Koenen, L.M. Shin, S.P. Orr, M.W. Gilbertson, M.R. Milad, and I. Liberzon. "Biological Studies of Post-Traumatic Stress Disorder." *Nature Reviews Neuroscience* (2012) 13:769–87.

Pitman, R.K., L.M. Shin, and S.L. Rauch. "Investigating the Pathogenesis of Posttraumatic Stress Disorder with Neuroimaging." *Journal of Clinical Psychiatry Supplement* (2001) 17:47–54.

Plassmann, H., J.P. O'Doherty, and A. Rangel. "Appetitive and Aversive Goal Values Are Encoded in the Medial Orbitofrontal Cortex at the Time of Decision Making." *Journal of Neuroscience* (2010) 30:10799–808.

Plotnik, J.M., et al. "Self-Recognition in the Asian Elephant and Future Directions for Cognitive Research with Elephants in Zoological Settings." *Zoo Biology* (2010) 29:179–91.

Plotnik, J.M., F.B. de Waal, D. Reiss. "Self-Recognition in an Asian Elephant." *Proceedings of the National Academy of Sciences of the United States of America* (2006) 103:17053–57.

Plutchik, R. *Emotion: A Psychoevolutionary Synthesis* (New York: Harper & Row, 1980).

Pollan, M. "The Trip Treatment." *The New Yorker*, Feb. 9, 2015. Retrieved Feb. 21, 2015.

Polin, A.T. "The Effects of Flooding and Physical Suppression as Extinction Techniques on an Anxiety Motivated Avoidance Locomotor Response." *Journal of Psychology* (1959) 47:235–45.

Pope, A. *Eloisa to Abelard* (Zurich: Orell, Fusli, 1803).

Porges, S.W. "The Polyvagal Theory: Phylogenetic Substrates of a Social Nervous System." *International Journal of Psychophysiology: Official Journal of the International Organization of Psychophysiology* (2001) 42:123–46.

Posner, M.I., and M.K. Rothbart. "Attention, Self-Regulation and Consciousness." *Philosophical Transactions of the Royal Society B: Biological Sciences* (1998) 353:1915–27.

Poulos, A.M., et al. "Compensation in the Neural Circuitry of Fear Conditioning Awakens Learning Circuits in the Bed Nuclei of the Stria Terminalis." *Proceedings of the National Academy of Sciences of the United States of America* (2010) 107:14881–86.

Poundja, J., et al. "Trauma Reactivation Under the Influence of Propranolol: An Examination of Clinical Predictors." *European Journal of Psychotraumatology* (2012) 3.

Pourtois, G., A. Schettino, and P. Vuilleumier. "Brain Mechanisms for Emotional Influences on Perception and Attention: What Is Magic and What Is Not." *Biological Psychology* (2013) 92: 492–512.

Pourtois, G., et al. "Temporal Precedence of Emotion over Attention Modulations in the Lateral Amygdala: Intracranial, ERP Evidence from a Patient with Temporal Lobe Epilepsy." *Cognitive, Affective & Behavioral Neuroscience* (2010) 10:83–93.

Povinelli, D.J., et al. "Chimpanzees Recognize Themselves in Mirrors." *Animal Behavior* (1997) 53:1083–88.

Powers, M.B., et al. "Helping Exposure Succeed: Learning Theory Perspectives on Treatment Resistance and Relapse." In: *Avoiding Treatment Failures in the Anxiety Disorders*, eds. M.W. Otto and S.G. Hofmann (New York: Springer, 2010), 31–49.

Premack, D. "Human and Animal Cognition: Continuity and Discontinuity." *Proceedings of the National Academy of Sciences of the United States of America* (2007) 104:13861–67.

Prendergast, S., and S. Forrest. "'Shorties, Low-Lifers, Hardnuts and Kings': Boys, Emotions and Embodiment in School." In: *Emotions in Social Life: Critical Themes and Contemporary Issues*, G. Bendelow and S. J. Williams, eds (New York: Routledge, 1997), 155–72.

Preter, M., and D.F. Klein. "Panic, Suffocation False Alarms, Separation Anxiety and Endogenous Opioids." *Progress in Neuro-Psychopharmacology & Biological Psychiatry* (2008) 32: 603–12.

Preuschoff, K., S.R. Quartz, and P. Bossaerts. "Human Insula Activation Reflects Risk Prediction Errors as Well as Risk." *Journal of Neuroscience* (2008) 28:2745–52.

Preuss, T.M. "Do Rats Have Prefrontal Cortex? The Rose-Woolsey-Akert Program Reconsidered." *Journal of Cognitive Neuroscience* (1995) 7:1–24.

Preuss, T.M. "The Discovery of Cerebral Diversity: An Unwelcome Scientific Revolution." In: *Evolutionary Anatomy of Primate Cerebral Cortex*, eds. D. Falk and K.R. Gibson (Cambridge, UK: Cambridge University Press, 2001), 138–64.

Prevost, C., et al. "Neural Correlates of Specific and General Pavlovian-to-Instrumental Transfer Within Human Amygdalar Subregions: A High-Resolution fMRI Study." *Journal of Neuroscience* (2012) 32:8383–90.

Price, D.D., and S.W. Harkins. "The Affective-Motivational Dimension of Pain a Two-Stage Model." *APS Journal* (1992) 1:229–39.

Price, J.L., and D.G. Amaral. "An Autoradiographic Study of the Projections of the Central Nucleus of the Monkey Amygdala." *Journal of Neuroscience* (1981) 1:1242–59.

Price, J.L., F.T. Russchen, and D.G. Amaral. "The Limbic Region. II: The Amygdaloid Complex." In: *Handbook of Chemical Neuroanatomy Vol 5: Integrated Systems of the CNS, Pt 1*, eds. A. Bjorklund, et al. (Amsterdam: Elsevier, 1987), 279–388.

Prinz, J. "Which Emotions Are Basic?" In: *Emotion, Evolution and Rationality*, eds. P. Cruise and D. Evans (Oxford, UK: Oxford University Press, 2004), 69–87.

Prinz, J.J. *The Conscious Brain: How Attention Engenders Experience* (New York: Oxford University Press, 2012).

Prinz, J.J. *Beyond Human Nature: How Culture and Experience Shape Our Lives* (London: Penguin, 2013).

Protopopescu, X., et al. "Differential Time Courses and Specificity of Amygdala Activity in Posttraumatic Stress Disorder Subjects and Normal Control Subjects." *Biological Psychiatry* (2005) 57:464–73.

Przybyslawski, J., and S.J. Sara. "Reconsolidation of Memory After Its Reactivation." *Behavioural Brain Research* (1997) 84:241–46.

Purves, D., and R.B. Lotto. *Why We See What We Do: An Empirical Theory of Vision* (Sunderland, MA: Sinauer Associates, 2003).

Putnam, H. "Minds and Machines." In: *Dimensions of Mind*, ed. S. Hook (New York: Collier Books, 1960).

Qin, S., et al. "Amygdala Subregional Structure and Intrinsic Functional Connectivity Predicts Individual Differences in Anxiety During Early Childhood." *Biological Psychiatry* (2014) 75: 892–900.

Quartermain, D., B.S. McEwen, and E.C. Azmitia Jr. "Amnesia Produced by Electroconvulsive Shock or Cycloheximide: Conditions for Recovery." *Science* (1970) 169:683–86.

Quirk, G., et al. "Emotional Memory: A Search for Sites of Plasticity." *Cold Spring Harbor Symposium on Quantitative Biology* (1996) 61:247–57.

Quirk, G.J., J.L. Armony, and J.E. LeDoux. "Fear Conditioning Enhances Different Temporal Components of Tone-Evoked Spike Trains in Auditory Cortex and Lateral Amygdala." *Neuron* (1997) 19:613–24.

Quirk, G.J., and J.S. Beer. "Prefrontal Involvement in the Regulation of Emotion: Convergence of Rat and Human Studies." *Current Opinion in Neurobiology* (2006) 16:723–27.

Quirk, G.J., R. Garcia, and F. Gonzalez-Lima. "Prefrontal Mechanisms in Extinction of Conditioned Fear." *Biological Psychiatry* (2006) 60:337–43.

Quirk, G.J., and D.R. Gehlert. "Inhibition of the Amygdala: Key to Pathological States?" *Annals of the New York Academy of Sciences* (2003) 985:263–72.

Quirk, G.J., and D. Mueller. "Neural Mechanisms of Extinction Learning and Retrieval." *Neuropsychopharmacology* (2008) 33:56–72.

Quirk, G.J., et al. "Erasing Fear Memories with Extinction Training." *Journal of Neuroscience* (2010) 30:14993–97.

Quirk, G.J., C. Repa, and J.E. LeDoux. "Fear Conditioning Enhances Short-Latency Auditory Responses of Lateral Amygdala Neurons: Parallel Recordings in the Freely Behaving Rat." *Neuron* (1995) 15:1029–39.

Rachman, S. "Systematic Desensitization." *Psychological Bulletin* (1967) 67:93–103.

Rachman, S. "The Conditioning Theory of Fear-Acquisition: A Critical Examination." *Behaviour Research and Therapy* (1977) 15:375–87.

Rachman, S. *Fear and Courage* (New York: W.H. Freeman, 1990).

Rachman, S. *Anxiety* (Hove, East Sussex: Psychology Press, 1998).

Rachman, S. *Anxiety* (Hove, East Sussex: Psychology Press, 2004).

Rachman, S., and R. Hodgson. "I. Synchrony and Desynchrony in Fear and Avoidance." *Behaviour Research and Therapy* (1974) 12:311–18.

Radley, J.J., et al. "Repeated Stress Induces Dendritic Spine Loss in the Rat Medial Prefrontal Cortex." *Cerebral Cortex* (2006) 16:313–20.

Raes, A.K., et al. "Do CS-US Pairings Actually Matter? A Within-Subject Comparison of Instructed Fear Conditioning with and Without Actual CS-US Pairings." *PLoS One* (2014) 9:E84888.

Ragan, C.I., et al. "What Should We Do About Student Use of Cognitive Enhancers? An Analysis of Current Evidence." *Neuropharmacology* (2013) 64:588–95.

Raichle, M.E., and A.Z. Snyder. "A Default Mode of Brain Function: A Brief History of an Evolving Idea." *NeuroImage* (2007) 37:1083–90; discussion 1097–99.

Rainville, P., et al. "A Psychophysical Comparison of Sensory and Affective Responses to Four Modalities of Experimental Pain." *Somatosensory & Motor Research* (1992) 9:265–77.

Raio, C.M., et al. "Acute Stress Impairs the Retrieval of Extinction Memory in Humans." *Neurobiology of Learning and Memory* (2014) 112:212–21.

Raio, C.M., et al. "Nonconscious Fear Is Quickly Acquired but Swiftly Forgotten." *Current Biology* (2012) 22:R477–79.

Ramirez, F., et al. (2015) "Active avoidance requires a serial basal amygdala to nucleus accumbens shell circuit," J. Neurosicne (in press).

Ramnero, J. "Exposure Therapy for Anxiety Disorders: Is There Room for Cognitive Interventions?" In: *Exposure Therapy: Rethinking the Model—Refining the Method*, eds. P. Neudeck and H.-U. Wittchen (New York: Springer, 2012), 275–98.

Ramos, B.P., and A.F. Arnsten. "Adrenergic Pharmacology and Cognition: Focus on the Prefrontal Cortex." *Pharmacology & Therapeutics* (2007) 113:523–36.

Rangel, A., C. Camerer, and P.R. Montague. "A Framework for Studying the Neurobiology of Value-Based Decision Making." *Nature Reviews Neuroscience* (2008) 9:545–56.

Rangel, A., and T. Hare. "Neural Computations Associated with Goal-Directed Choice." *Current Opinion in Neurobiology* (2010) 20:262–70.

Ranson, S.W., and H.W. Magoun. "The Hypothalamus." *Ergebnis der Physiologie* (1939) 41:56–163.

Rao-Ruiz, P., et al. "Retrieval-Specific Endocytosis of GluA2-AMPARS Underlies Adaptive Reconsolidation of Contextual Fear." *Nature Neuroscience* (2011) 14:1302–8.

Rathschlag, M., and D. Memmert. "Reducing Anxiety and Enhancing Physical Performance by Using an Advanced Version of EMDR: A Pilot Study." *Brain and Behavior* (2014) 4:348–55.

Ratner, S.C. "Comparative Aspects of Hypnosis." In: *Handbook of Clinical and Experimental Hypnosis*, ed. J.E. Gordon (New York: Macmillan, 1967).

Ratner, S.C. "Animal's Defenses: Fighting in Predator-Prey Relations." In: *Nonverbal Communication of Aggression*, ed. P. Pliner, et al. (New York: Plenum, 1975).

Rauch, S.L., L.M. Shin, and E.A. Phelps. "Neurocircuitry Models of Posttraumatic Stress Disorder and Extinction: Human Neuroimaging Research—Past, Present, and Future." *Biological Psychiatry* (2006) 60:376–82.

Rauch, S.L., L.M. Shin, and C.I. Wright. "Neuroimaging Studies of Amygdala Function in Anxiety Disorders." *Annals of the New York Academy of Sciences* (2003) 985:389–410.

Raymond, J.E., K.L. Shapiro, and K.M. Arnell. "Temporary Suppression of Visual Processing in an RSVP Task: An Attentional Blink?" *Journal of Experimental Psychology: Human Perception and Performance* (1992) 18:849–60.

Reber, A.S., et al. "On the Relationship Between Implicit and Explicit Modes in the Learning of a Complex Rule Structure." *Journal of Experimental Psychology: Human Learning and Memory* (1980) 6:492–502.

Recce, M., and K.D. Harris. "Memory for Places: A Navigational Model in Support of Marr's Theory of Hippocampal Function." *Hippocampus* (1996) 6:735–48.

Redelmeier, D.A. "Improving Patient Care. The Cognitive Psychology of Missed Diagnoses." *Annals of Internal Medicine* (2005) 142:115–20.

Rees, G., and C. Frith. "Methodologies for Identifying the Neural Correlates of Consciousness." In: *A Companion to Consciousness*, eds. M. Velmans and S. Schneider (Oxford, UK: Blackwell, 2007).

Rees, G., et al. "Unconscious Activation of Visual Cortex in the Damaged Right Hemisphere of a Parietal Patient with Extinction." *Brain: A Journal of Neurology* (2000) 123(Pt 8):1624–33.

Rees, G., et al. "Neural Correlates of Conscious and Unconscious Vision in Parietal Extinction." *Neurocase* (2002) 8:387–93.

Reichelt, A.C., and J.L. Lee. "Memory Reconsolidation in Aversive and Appetitive Settings." *Frontiers in Behavioral Neuroscience* (2013) 7:118.

Reijmers, L.G., et al. "Localization of a Stable Neural Correlate of Associative Memory." *Science* (2007) 317:1230–33.

Reik, W. "Stability and Flexibility of Epigenetic Gene Regulation in Mammalian Development." *Nature* (2007) 447:425–32.

Reinders, A.A., et al. "One Brain, Two Selves." *NeuroImage* (2003) 20:2119–25.

Reiner, A. "An Explanation of Behavior." *Science* (1990) 250:303–5.

Reis, D.J., and J.E. LeDoux. "Some Central Neural Mechanisms Governing Resting and Behaviorally Coupled Control of Blood Pressure." *Circulation* (1987) 76:12–19.

Reiss, D., and L. Marino. "Mirror Self-Recognition in the Bottlenose Dolphin: A Case of Cognitive Convergence." *Proceedings of the National Academy of Sciences of the United States of America* (2001) 98:5937–42.

Renier, L., A.G. De Volder, and J.P. Rauschecker. "Cortical Plasticity and Preserved Function in Early Blindness." *Neuroscience and Biobehavioral Reviews* (2014) 41:53–63.

Repa, J.C., et al. "Two Different Lateral Amygdala Cell Populations Contribute to the Initiation and Storage of Memory." *Nature Neuroscience* (2001) 4:724–31.

Rescorla, R.A. "Behavioral Studies of Pavlovian Conditioning." *Annual Review of Neuroscience* (1988) 11:329–52.

Rescorla, R.A. "Transfer of Instrumental Control Mediated by a Devalued Outcome." *Animal Learning & Behavior* (1994) 22:27–33.

Rescorla, R.A. "Extinction Can Be Enhanced by a Concurrent Excitor." *Journal of Experimental Psychology: Animal Behavior Processes* (2000) 26:251–60.

Rescorla, R.A. "Deepened Extinction from Compound Stimulus Presentation." *Journal of Experimental Psychology: Animal Behavior Processes* (2006) 32:135–44.

Rescorla, R.A., and C.D. Heth. "Reinstatement of Fear to an Extinguished Conditioned Stimulus." *Journal of Experimental Psychology: Animal Behavior Processes* (1975) 104:88–96.

Rescorla, R.A., and R.L. Solomon. "Two Process Learning Theory: Relationships Between Pavlovian Conditioning and Instrumental Learning." *Psychological Review* (1967) 74:151–82.

Rescorla, R.A., and A.R. Wagner. "A Theory of Pavlovian Conditioning: Variations in the Effectiveness of Reinforcement and Nonreinforcement." In: *Classical Conditioning II: Current Research and Theory*, eds. A.A. Black and W.F. Prokasy (New York: Appleton-Century-Crofts, 1972), 64–99.

Ressler, K.J., and H.S. Mayberg. "Targeting Abnormal Neural Circuits in Mood and Anxiety Disorders: From the Laboratory to the Clinic." *Nature Neuroscience* (2007) 10:1116–24.

Ressler, K.J., et al. "Cognitive Enhancers as Adjuncts to Psychotherapy: Use of D-cycloserine in Phobic Individuals to Facilitate Extinction of Fear." *Archives of General Psychiatry* (2004) 61:1136–44.

Reuther, E.T., et al. "Intolerance of Uncertainty as a Mediator of the Relationship Between Perfectionism and Obsessive-Compulsive Symptom Severity." *Depression and Anxiety* (2013) 30:773–77.

Ricciardelli, L.A. "Two Components of Metalinguistic Awareness: Control of Linguistic Processing and Analysis of Linguistic Knowledge." *Applied Psycholinguistics* (1993) 14:349–67.

Richard, D.C.S., and D. Lauterbach. *Handbook of Exposure Therapy* (San Diego: Academic Press, 2007).

Riebe, C.J., et al. "Fear Relief—Toward a New Conceptual Frame Work and What Endocannabinoids Gotta Do with It." *Neuroscience* (2012) 204:159–85.

Riggio, R.E., and H.S. Friedman. "The Interrelationships of Self-Monitoring Factors, Personality Traits, and Nonverbal Social Skills." *Journal of Nonverbal Behavior* (1982) 7:33–45.

Rimm, D.C., et al. "An Exploratory Investigation of the Origin and Maintenance of Phobias." *Behaviour Research and Therapy* (1977) 15:231–38.

Rincón-Cortés, M., and R.M. Sullivan. "Early Life Trauma and Attachment: Immediate and Enduring Effects on Neurobehavioral and Stress Axis Development." *Frontiers in Endocrinology* (2014) 5:33.

Risold, P.Y., and L.W. Swanson. "Structural Evidence for Functional Domains in the Rat Hippocampus." *Science* (1996) 272:1484–86.

Risold, P.Y., and L.W. Swanson. "Connections of the Rat Lateral Septal Complex." *Brain Research Reviews* (1997) 24:115-195.

Robbins, T.W., K.D. Ersche, and B.J. Everitt. "Drug Addiction and the Memory Systems of the Brain." *Annals of the New York Academy of Sciences* (2008) 1141:1–21.

Robinson, T.E., and K.C. Berridge. "Review. The Incentive Sensitization Theory of Addiction: Some Current Issues." *Philosophical Transactions of the Royal Society B: Biological Sciences* (2008) 363:3137–46.

Rodrigues, S.M., J.E. LeDoux, and R.M. Sapolsky. "The Influence of Stress Hormones on Fear Circuitry." *Annual Review of Neuroscience* (2009) 32:289–313.

Rodrigues, S.M., G.E. Schafe, and J.E. LeDoux. "Intra-Amygdala Blockade of the NR2B Subunit of the NMDA Receptor Disrupts the Acquisition but Not the Expression of Fear Conditioning." *Journal of Neuroscience* (2001) 21:6889–96.

Rodrigues, S.M., G.E. Schafe, and J.E. LeDoux. "Molecular Mechanisms Underlying Emotional Learning and Memory in the Lateral Amygdala." *Neuron* (2004) 44:75–91.

Rodriguez-Romaguera, J., F.H. Do Monte, and G.J. Quirk. "Deep Brain Stimulation of the Ventral Striatum Enhances Extinction of Conditioned Fear." *Proceedings of the National Academy of Sciences of the United States of America* (2012) 109:8764–69.

Roesch, M.R., et al. "Surprise! Neural Correlates of Pearce-Hall and Rescorla-Wagner Coexist Within the Brain." *European Journal of Neuroscience* (2012) 35:1190–1200.

Rogan, M.T., et al. "Distinct Neural Signatures for Safety and Danger in the Amygdala and Striatum of the Mouse." *Neuron* (2005) 46:309–20.

Rogan, M.T., U.V. Staubli, and J.E. LeDoux. "Fear Conditioning Induces Associative Long-Term Potentiation in the Amygdala." *Nature* (1997) 390:604–7.

Rogan, M.T., et al. "Long-Term Potentiation in the Amygdala: Implications for Memory." In: *Neuronal Mechanisms of Memory Formation*, ed. C. Holscher (Cambridge, UK: Cambridge University Press, 2001), 58–76.

Rolls, E.T. "Neurophysiology and Functions of the Primate Amygdala." In: *The Amygdala: Neurobiological Aspects of Emotion, Memory, and Mental Dysfunction*, ed. J.P. Aggleton (New York: Wiley-Liss, 1992), 143–65.

Rolls, E.T. "A Theory of Hippocampal Function in Memory." *Hippocampus* (1996) 6:601–20.

Rolls, E.T. *Emotion Explained* (New York: Oxford University Press, 2005).

Rolls, E.T. "Emotion, Higher-Order Syntactic Thoughts, and Consciousness." In: *Frontiers of Consciousness: Chichele Lectures*, eds. L. Weiskrantz and M. Davies (Oxford, UK: Oxford University Press, 2008), 131–67.

Rolls, E.T. *Emotion and Decision-Making Explained* (Oxford, UK: Oxford University Press, 2014).

Rolls, E.T., M.L. Kringelbach, and I.E. De Araujo. "Different Representations of Pleasant and Unpleasant Odours in the Human Brain." *European Journal of Neuroscience* (2003) 18:695–703.

Romanes, G.J. *Animal Intelligence* (London: Kegan Paul, Trench & Co., 1882).

Romanes, G.J. *Mental Evolution in Animals* (London: Kegan Paul, Trench & Co., 1883).

Romanski, L.M., and J.E. LeDoux. "Equipotentiality of Thalamo-Amygdala and Thalamo-Cortico-Amygdala Circuits in Auditory Fear Conditioning." *Journal of Neuroscience* (1992) 12:4501–9.

Roozendaal, B., B.S. McEwen, and S. Chattarji. "Stress, Memory and the Amygdala." *Nature Reviews Neuroscience* (2009) 10:423–33.

Roozendaal, B., and J.L. McGaugh. "Memory Modulation." *Behavioral Neuroscience* (2011) 125:797–824.

Rorie, A.E., and W.T. Newsome. "A General Mechanism for Decision-Making in the Human Brain?" *Trends in Cognitive Sciences* (2005) 9:41–43.

Rosen, J.B. "The Neurobiology of Conditioned and Unconditioned Fear: A Neurobehavioral System Analysis of the Amygdala." *Behavioral and Cognitive Neuroscience Reviews* (2004) 3:23–41.

Rosen, J.B., and J. Schulkin. "From Normal Fear to Pathological Anxiety." *Psychological Review* (1998) 105:325–50.

Rosenblueth, A., and N. Wiener. Quoted in Lewontin, R.C. "In the Beginning Was the Word." (2001) *Science* 291:1264.

Rosenfield, L.C. *From Beast-Machine to Man-Machine: Animal Soul in French Letters from Descartes to la Mettrie* (New York: Octagon Books, 1941).

Rosenkranz, J.A., H. Moore, and A.A. Grace. "The Prefrontal Cortex Regulates Lateral Amygdala Neuronal Plasticity and Responses to Previously Conditioned Stimuli." *Journal of Neuroscience* (2003) 23:11054–64.

Rosenthal, D. "A Theory of Consciousness." In: *University of Bielefeld Mind and Brain Technical Report* 40. *Perspectives in Theoretical Psychology and Philosophy of Mind* (ZiF). (Bielefeld, Germany: University of Bielefeld, 1990).

Rosenthal, D. "Higher-Order Thoughts and the Appendage Theory of Consciousness." *Philosophical Psychology* (1993) 6:155–66.

Rosenthal, D. "Explaining Consciousness." In: *Philosophy of Mind: Classical and Contemporary Readings*, ed. D.J. Chalmers (Oxford, UK: Oxford University Press, 2002), 406–17.

Rosenthal, D. "Higher-Order Awareness, Misrepresentation and Function." *Philosophical Transactions of the Royal Society B: Biological Sciences* (2012) 367:1424–38.

Rosenthal, D.M. "Why Are Verbally Expressed Thoughts Conscious?" *Bielefeld Report* (1990). In: *University of Bielefeld Mind and Brain Technical Report* 40. *Perspectives in Theoretical Psychology and Philosophy of Mind* (ZiF). (Bielefeld, Germany: University of Bielefeld, 1990).

Rosenthal, D.M. *Consciousness and Mind* (Oxford: Oxford University Press, 2005).

Roth, W.T. "Physiological Markers for Anxiety: Panic Disorder and Phobias." *International Journal of Psychophysiology: Official Journal of the International Organization of Psychophysiology* (2005) 58:190–98.

Rothbart, M.K., S.A. Ahadi, and D.E. Evans. "Temperament and Personality: Origins and Outcomes." *Journal of Personality and Social Psychology* (2000) 78:122–35.

Rothbaum, B.O., et al. "Virtual Reality Exposure Therapy and Standard (In Vivo) Exposure Therapy in the Treatment of Fear of Flying." *Behavior Therapy* (2006) 37:80–90.

Rothfield, L., S. Justice, and J. Garcia-Lara. "Bacterial Cell Division." *Annual Review of Genetics* (1999) 33:423–48.

Rowe, M.K., and M.G. Craske. "Effects of an Expanding-Spaced vs. Massed Exposure Schedule on Fear Reduction and Return of Fear." *Behaviour Research and Therapy* (1998) 36:701–17.

Royer, S., and D. Paré. "Bidirectional Synaptic Plasticity in Intercalated Amygdala Neurons and the Extinction of Conditioned Fear Responses." *Neuroscience* (2002) 115:455–62.

Rubia, K. "The Neurobiology of Meditation and Its Clinical Effectiveness in Psychiatric Disorders." *Biological Psychology* (2009) 82:1–11.

Rubin, D.B., et al. "Dosed Versus Prolonged Exposure in the Treatment of Fear: An Experimental Evaluation and Review of Behavioral Mechanisms." *Journal of Anxiety Disorders* (2003) 23:806–12.

Rugg, M.D., L.J. Otten, and R.N. Henson. "The Neural Basis of Episodic Memory: Evidence from Functional Neuroimaging." *Philosophical Transactions of the Royal Society B: Biological Sciences* (2002) 357:1097–1110.

Russell, J.A. "Natural Language Concepts of Emotion." In: *Perspectives in Personality*, vol. 3, eds. R. Hogan et al. (London: Jessica Kingsley, 1991).

Russell, J.A. "Is There Universal Recognition of Emotion from Facial Expression? A Review of the Cross-Cultural Studies." *Psychological Bulletin* (1994) 115:102–41.

Russell, J.A. "Core Affect and the Psychological Construction of Emotion." *Psychological Review* (2003) 110:145–72.

Russell, J.A. "Emotion, Core Affect, and Psychological Construction." *Cognition and Emotion* (2009) 23:1259–83.

Russell, J.A. "From a Psychological Constructionist Perspective." In: *Categorical Versus Dimensional Models of Affect: A Seminar on the Theories of Panksepp and Russell*, eds. P. Zachar and R. Ellis (Amsterdam: John Benjamins, 2012).

Russell, J.A. "The Greater Constructionist Project for Emotion." In: *The Psychological Construction of Emotion*, eds. L.F. Barrett and J.A. Russell (New York: Guilford Press, 2014).

Russell, J.A., and L.F. Barrett. "Core Affect, Prototypical Emotional Episodes, and Other Things Called Emotion: Dissecting the Elephant." *Journal of Personality and Social Psychology* (1999) 76:805–19.

Ryugo, D.K., and N.M. Weinberger. "Differential Plasticity of Morphologically Distinct Neuron Populations in the Medial Geniculate Body of the Cat During Classical Conditioning." *Behavioral Biology* (1978) 22:275–301.

Sabatinelli, D., et al. "Emotional Perception: Meta-Analyses of Face and Natural Scene Processing." *NeuroImage* (2011) 54:2524–33.

Sacks, O. "The Mental Life of Plants and Worms." *The New York Review of Books*, Apr. 3, 2014.

Sacks. O. (1989) *Seeing Voices: A Journey into the World of the Deaf* (Oakland: University of California Press, 1989).

Sadato, N. "Cross-Modal Plasticity in the Blind Revealed by Functional Neuroimaging." *Supplements to Clinical Neurophysiology* (2006) 59:75–79.

Safavi, S., et al. "Is the Frontal Lobe Involved in Conscious Perception?" *Frontiers in Psychology* (2014) 5:1063.

Sah, P., et al. "The Amygdaloid Complex: Anatomy and Physiology." *Physiological Reviews* (2003) 83:803–34.

Sah, P., R.F. Westbrook, and A. Luthi. "Fear Conditioning and Long-Term Potentiation in the Amygdala: What Really Is the Connection?" *Annals of the New York Academy of Sciences* (2008) 1129:88–95.

Saha, S. "Role of the Central Nucleus of the Amygdala in the Control of Blood Pressure: Descending Pathways to Medullary Cardiovascular Nuclei." *Clinical and Experimental Pharmacology & Physiology* (2005) 32:450–56.

Sahraie, A., L. Weiskrantz, and J.L. Barbur. "Awareness and Confidence Ratings in Motion Perception without Geniculo-Striate Projection." *Behavioural Brain Research* (1998) 96: 71–77.

Sajdyk, T., et al. "Chronic Inhibition of GABA Synthesis in the Bed Nucleus of the Stria Terminalis Elicits Anxiety-Like Behavior." *Journal of Psychopharmacology* (2008) 22: 633–41.

Sajdyk, T.J., and A. Shekhar. "Sodium Lactate Elicits Anxiety in Rats After Repeated GABA Receptor Blockade in the Basolateral Amygdala." *European Journal of Pharmacology* (2000) 394:265–73.

Sakaguchi, A., J.E. LeDoux, and D.J. Reis. "Sympathetic Nerves and Adrenal Medulla: Contributions to Cardiovascular-Conditioned Emotional Responses in Spontaneously Hypertensive Rats." *Hypertension* (1983) 5:728–38.

Sakai, Y., et al. "Cerebral Glucose Metabolism Associated with a Fear Network in Panic Disorder." *Neuroreport* (2005) 16:927–31.

Salkovskis, P. "The Cognitive Approach to Anxiety: Threat Beliefs, Safety-Seeking Behaviours and the Special Case of Health Anxiety and Obsessions." In: *The Frontiers of Cognitive Therapy*, ed. P. Salkovskis (New York: Guilford Press, 1996), 48–74.

Salkovskis, P.M., et al. "Belief Disconfirmation Versus Habituation Approaches to Situational Exposure in Panic Disorder with Agoraphobia: A Pilot Study." *Behaviour Research and Therapy* (2006) 45:877–85.

Salomons, T.V., et al. "Neural Emotion Regulation Circuitry Underlying Anxiolytic Effects of Perceived Control over Pain." *Journal of Cognitive Neuroscience* (2014) 1–12.

Salwiczek, L.H., A. Watanabe, and N.S. Clayton. "Ten Years of Research into Avian Models of Episodic-Like Memory and Its Implications for Developmental and Comparative Cognition." *Behavioural Brain Research* (2010) 215:221–34.

Samuels, E.R., and E. Szabadi. "Functional Neuroanatomy of the Noradrenergic Locus Coeruleus: Its Roles in the Regulation of Arousal and Autonomic Function Part I: Principles of Functional Organisation." *Current Neuropharmacology* (2008) 6:235–53.

Sanders, M.J., B.J. Wiltgen, and M.S. Fanselow. "The Place of the Hippocampus in Fear Conditioning." *European Journal of Pharmacology* (2003) 463:217–23.

Sanderson, W.C., and D.H. Barlow. "Clients' Answers to Interviewer's Question, 'Do You Worry Excessively About Minor Things?' (From a Description of Patients Diagnosed with

DSM-III-R Generalized Anxiety Disorder)." *Journal of Nervous and Mental Disease* (1990) 178:590.

Sangha, S., A. Scheibenstock, and K. Lukowiak. "Reconsolidation of a Long-Term Memory in Lymnaea Requires New Protein and RNA Synthesis and the Soma of Right Pedal Dorsal 1." *Journal of Neuroscience* (2003) 23:8034–40.

Santini, E., et al. "Consolidation of Fear Extinction Requires Protein Synthesis in the Medial Prefrontal Cortex." *Journal of Neuroscience* (2004) 24:5704–10.

Saper, C.B. "Diffuse Cortical Projection Systems: Anatomical Organization and Role in Cortical Function." In: *Handbook of Physiology 1: The Nervous System Vol. V., Higher Functions of the Brain, Vol. V,* eds. V.B. Mountcastle et al. (Bethesda, MD: American Physiological Society, 1987), 169–210.

Saper, C.B., T.E. Scammell, and J. Lu. "Hypothalamic Regulation of Sleep and Circadian Rhythms." *Nature* (2005) 437:1257–63.

Sapir, E. *Language: An Introduction to the Study of Speech* (New York: Harcourt Brace, 1921).

Sapolsky, R.M. "Why Stress Is Bad for Your Brain." *Science* (1996) 273:749–50.

Sapolsky, R.M. *Why Zebras Don't Get Ulcers* (New York: Freeman, 1998).

Sara, S.J. "Noradrenergic-Cholinergic Interaction: Its Possible Role in Memory Dysfunction Associated with Senile Dementia." *Archives of Gerontology and Geriatrics Supplement* (1989) 1:99–108.

Sara, S.J. "Retrieval and Reconsolidation: Toward a Neurobiology of Remembering." *Learning & Memory* (2000) 7:73–84.

Sara, S.J. "The Locus Coeruleus and Noradrenergic Modulation of Cognition." *Nature Reviews Neuroscience* (2009) 10:211–23.

Sara, S.J., and S. Bouret. "Orienting and Reorienting: The Locus Coeruleus Mediates Cognition Through Arousal." *Neuron* (2012) 76:130–41.

Sara, S.J., A. Vankov, and A. Herve. "Locus Coeruleus-Evoked Responses in Behaving Rats: A Clue to the Role of Noradrenaline in Memory." *Brain Research Bulletin* (1994) 35:457–65.

Sarter, M.F., and H.J. Markowitsch. "Involvement of the Amygdala in Learning and Memory: A Critical Review, with Emphasis on Anatomical Relations." *Behavioral Neuroscience* (1985) 99:342–80.

Sartre, J.-P. *Being and Nothingness* (Paris: Gallimard, 1943).

Savage, L.M., and R.L. Ramos. "Reward Expectation Alters Learning and Memory: The Impact of the Amygdala on Appetitive-Driven Behaviors." *Behavioural Brain Research* (2009) 198: 1–12.

Scarantino, A. "Core Affect and Natural Affective Kinds." *Philosophy of Science* (2009) 76: 940–57.

Schachter, S., and J.E. Singer. "Cognitive, Social, and Physiological Determinants of Emotional State." *Psychological Review* (1962) 69:379–99.

Schacter, D. *The Seven Sins of Memory* (Boston: Houghton-Mifflin, 2001).

Schacter, D.L. "Multiple Forms of Memory in Humans and Animals." In: *Memory Systems of the Brain: Animal and Human Cognitive Processes,* eds. N.M. Weinberger, et al. (New York: Guilford, 1985), 351–79.

Schacter, D.L. "On the Relation Between Memory and Consciousness: Dissociable Interactions and Conscious Experience." In: *Varieties of Memory and Consciousness: Essays in Honour of Endel Tulving,* eds. H.L.I. Roediger and F.I.M. Craik (Hillsdale, NJ: Erlbaum, 1989), 355–89.

Schacter, D.L. "The Cognitive Neuroscience of Memory: Perspectives from Neuroimaging Research." *Philosophical Transactions of the Royal Society B: Biological Sciences* (1997) 352: 1689–95.

Schacter, D.L. "Memory and Awareness." *Science* (1998) 280:59–60.

Schacter, D.L. "Constructive Memory: Past and Future." *Dialogues in Clinical Neuroscience* (2012) 14:7–18.

Schacter, D.L., and R.L. Buckner. "Priming and the Brain." *Neuron* (1998) 20:185–95.

Schacter, D.L., R.L. Buckner, and W. Koutstaal. "Memory, Consciousness and Neuroimaging." *Philosophical Transactions of the Royal Society B: Biological Sciences* (1998) 353:1861–78.

Schafe, G.E., and J.E. LeDoux. "Memory Consolidation of Auditory Pavlovian Fear Conditioning Requires Protein Synthesis and Protein Kinase A in the Amygdala." *Journal of Neuroscience* (2000) 20:RC96.

Schafe, G.E., and J.E. LeDoux. "Neural and Molecular Mechanisms of Fear Memory." In: *Learning & Memory: A Comprehensive Reference: Molecular Mechanisms,* ed. J.D. Sweatt (New York: Academic Press, 2008).

Schafe, G.E., et al. "Memory Consolidation for Contextual and Auditory Fear Conditioning Is Dependent on Protein Synthesis, PKA, and MAP Kinase." *Learning & Memory* (1999) 6:97–110.

Scherer, K. "Emotions as Episodes of Subsystem Synchronization Driven by Nonlinear Appraisal Processes." In: *Emotion, Development, and Self-Organization: Dynamic Systems Approaches to Emotional Development,* eds. M. Lewis and I. Granic (New York: Cambridge University Press, 2000), 70–99.

Scherer, K.R. "Emotion as a Multicomponent Process: A Model and Some Cross-Cultural Data." *Review of Personality and Social Psychology* (1984) 5:37–63.

Scherer, K.R. "Neuroscience Findings Are Consistent with Appraisal Theories of Emotion; but Does the Brain 'Respect' Constructionism?" *Behavioral and Brain Sciences* (2012) 35:163–64.

Scherer, K.R., and H. Ellgring. "Are Facial Expressions of Emotion Produced by Categorical Affect Programs or Dynamically Driven by Appraisal?" *Emotion* (2007) 7:113–30.

Schienle, A., et al. "Symptom Provocation and Reduction in Patients Suffering from Spider Phobia: An fMRI Study on Exposure Therapy." *European Archives of Psychiatry and Clinical Neuroscience* (2007) 257:486–93.

Schiller, D., et al. "Evidence for Recovery of Fear Following Immediate Extinction in Rats and Humans." *Learning & Memory* (2008) 15:394–402.

Schiller, D., and M.R. Delgado. "Overlapping Neural Systems Mediating Extinction, Reversal and Regulation of Fear." *Trends in Cognitive Sciences* (2010) 14:268–76.

Schiller, D., et al. "Extinction During Reconsolidation of Threat Memory Diminishes Prefrontal Cortex Involvement." *Proceedings of the National Academy of Sciences of the United States of America* (2013) 110:20040–45.

Schiller, D., et al. "From Fear to Safety and Back: Reversal of Fear in the Human Brain." *Journal of Neuroscience* (2008) 28:11517–25.

Schiller, D., et al. "Preventing the Return of Fear in Humans Using Reconsolidation Update Mechanisms." *Nature* (2010) 463:49–53.

Schiller, D., and E.A. Phelps. "Does Reconsolidation Occur in Humans?" *Frontiers in Behavioral Neuroscience* (2011) 5:24.

Schlund, M.W., and M.R. Cataldo. "Amygdala Involvement in Human Avoidance, Escape and Approach Behavior." *NeuroImage* (2010) 53:769–76.

Schlund, M.W., et al. "Neuroimaging the Temporal Dynamics of Human Avoidance to Sustained Threat." *Behavioural Brain Research* (2013) 257:148–55.

Schlund, M.W., S. Magee, and C.D. Hudgins. "Human Avoidance and Approach Learning: Evidence for Overlapping Neural Systems and Experiential Avoidance Modulation of Avoidance Neurocircuitry." *Behavioural Brain Research* (2011) 225:437–48.

Schlund, M.W., et al. "Nothing to Fear? Neural Systems Supporting Avoidance Behavior in Healthy Youths." *NeuroImage* (2010) 52:710–19.

Schmidt, L.J., A.V. Belopolsky, and J. Theeuwes. "Attentional Capture by Signals of Threat." *Cognition & Emotion* (2014) 1–8.

Schneiderman, N., et al. "CNS Integration of Learned Cardiovascular Behavior." In: *Limbic and Autonomic Nervous System Research,* ed. L.V. Dicara (New York: Plenum, 1974), 277–309.

Schoenbaum, G., and M. Roesch. "Orbitofrontal Cortex, Associative Learning, and Expectancies." *Neuron* (2005) 47:633–36.

Schoenbaum, G., et al. "Does the Orbitofrontal Cortex Signal Value?" *Annals of the New York Academy of Sciences* (2011) 1239:87–99.

Schoo, L.A., et al. "The Posterior Parietal Paradox: Why Do Functional Magnetic Resonance Imaging and Lesion Studies on Episodic Memory Produce Conflicting Results?" *Journal of Neuropsychology* (2011) 5:15–38.

Schott, G. "Duchenne Superciliously 'Corrects' the Laocoön: Sculptural Considerations in the Mecanisme de la Physionomie Humaine." *Journal of Neurology, Neurosurgery, and Psychiatry* (2013) 84:10–13.

Schrader, G.A. *Existential Philosophers; Kierkegaard to Merleu-Ponty* (New York: McGraw-Hill, 1967).

Schultz, D.H., and F.J. Helmstetter. "Classical Conditioning of Autonomic Fear Responses Is Independent of Contingency Awareness." *Journal of Experimental Psychology: Animal Behavior Processes* (2010) 36:495–500.

Schultz, W. "Dopamine Neurons and Their Role in Reward Mechanisms." *Current Opinion in Neurobiology* (1997) 7:191–97.

Schultz, W. "Getting Formal with Dopamine and Reward." *Neuron* (2002) 36:241–63.

Schultz, W. "Updating Dopamine Reward Signals." *Current Opinion in Neurobiology* (2013) 23:229–38.

Schultz, W., P. Dayan, and P.R. Montague. "A Neural Substrate of Prediction and Reward." *Science* (1997) 275:1593–99.

Schultz, W., and A. Dickinson. "Neuronal Coding of Prediction Errors." *Annual Review of Neuroscience* (2000) 23:473–500.

Schulz, C., M. Mothes-Lasch, and T. Straube. "Automatic Neural Processing of Disorder-Related Stimuli in Social Anxiety Disorder: Faces and More." *Frontiers in Psychology* (2013) 4:282.

Schwabe, L., K. Nader, and J.C. Pruessner. "Reconsolidation of Human Memory: Brain Mechanisms and Clinical Relevance." *Biological Psychiatry* (2014) 76:274–80.

Schwaber, J.S., et al. "Amygdaloid and Basal Forebrain Direct Connections with the Nucleus of the Solitary Tract and the Dorsal Motor Nucleus." *Journal of Neuroscience* (1982) 2:1424–38.

Scoville, W.B., and B. Milner. "Loss of Recent Memory After Bilateral Hippocampal Lesions." *Journal of Neurology and Psychiatry* (1957) 20:11–21.

Searle, J.R. "Consciousness." *Annual Review of Neuroscience* (2000) 23:557–78.

Searle, J.R. *Consciousness and Language* (Cambridge, UK: Cambridge University Press, 2002).

Sears, R.M., et al. "Orexin/Hypocretin System Modulates Amygdala-Dependent Threat Learning Through the Locus Coeruleus." *Proceedings of the National Academy of Sciences of the United States of America* (2013) 110:20260–65.

Seger, C.A. "Implicit Learning." *Psychological Bulletin* (1994) 115:163–96.

Sehlmeyer, C., et al. "Neural Correlates of Trait Anxiety in Fear Extinction." *Psychological Medicine* (2011) 41:789–98.

Sekhar, A.C. "Language and Consciousness." *Indian Journal of Psychology* (1948) 23:79–84.

Sekida, K. *Zen Training: Methods and Philosophy* (New York: Weatherhill, 1985).

Seligman, M.E., and J.C. Johnston. "A Cognitive Theory of Avoidance Learning." In: *Contemporary Approaches to Conditioning and Learning*, eds. F.J. McGuigan and D.B. Lumsden (Oxford, UK: V. H. Winston & Sons, 1973), 69–110.

Seligman, M.E.P. "Phobias and Preparedness." *Behavior Therapy* (1971) 2:307–20.

Selye, H. *The Stress of Life* (New York: McGraw-Hill, 1956).

Semendeferi, K., et al. "Spatial Organization of Neurons in the Frontal Pole Sets Humans Apart from Great Apes." *Cerebral Cortex* (2011) 21:1485–97.

Sem-Jacobson, C.W. *Depth-Electroencephalographic Stimulation of the Human Brain and Behavior* (Springfield, IL: Charles C. Thomas, 1968).

Semple, W.E., et al. "Higher Brain Blood Flow at Amygdala and Lower Frontal Cortex Blood Flow in PTSD Patients with Comorbid Cocaine and Alcohol Abuse Compared with Normals." *Psychiatry* (2000) 63:65–74.

Sergent, C., and G. Rees. "Conscious Access Overflows Overt Report." *Behavioral and Brain Sciences* (2007) 30:523–24.

Serrano, P., et al. "PKMzeta Maintains Spatial, Instrumental, and Classically Conditioned Long-Term Memories." *PLoS Biology* (2008) 6:2698–2706.

Serrano, P., Y. Yao, and T.C. Sacktor. "Persistent Phosphorylation by Protein Kinase Mzeta Maintains Late-Phase Long-Term Potentiation." *Journal of Neuroscience* (2005) 25:1979–84.

Seth, A. "Explanatory Correlates of Consciousness: Theoretical and Computational Challenges." *Cognitive Computation* (2009) 1:50–63.

Seth, A.K. "Post-Decision Wagering Measures Metacognitive Content, Not Sensory Consciousness." *Consciousness and Cognition* (2008) 17:981–83.

Seth, A.K., et al. "Measuring Consciousness: Relating Behavioural and Neurophysiological Approaches." *Trends in Cognitive Sciences* (2008) 12:314–21.

Shackman, A.J., et al. "The Integration of Negative Affect, Pain and Cognitive Control in the Cingulate Cortex." *Nature Reviews Neuroscience* (2011) 12:154–67.

Shadlen, M.N., and R. Kiani. "Decision Making as a Window on Cognition." *Neuron* (2013) 80:791–806.

Shalev, A.Y., Y. Ragel-Fuchs, and R.K. Pitman. "Conditioned Fear and Psychological Trauma." *Biological Psychiatry* (1992) 31:863–65.

Shallice, T. "Information Processing Models of Consciousness." In: *Consciousness in Contemporary Science*, eds. A. Marcel and E. Bisiach (Oxford, UK: Oxford University Press, 1988), 305–33.

Shallice, T., and P. Burgess. "The Domain of Supervisory Processes and Temporal Organization of Behaviour." *Philosophical Transactions of the Royal Society B: Biological Sciences* (1996) 351:1405–11.

Shanks, D.R., and A. Dickinson. "Contingency Awareness in Evaluative Conditioning: A Comment on Baeyens, Eelen, and van den Bergh." *Cognition and Emotion* (1990) 4:19–30.

Shanks, D.R., and P.F. Lovibond. "Autonomic and Eyeblink Conditioning Are Closely Related to Contingency Awareness: Reply to Wiens and Öhman (2002) and Manns et al. (2002)." *Journal of Experimental Psychology: Animal Behavior Processes* (2002) 28:38–42.

Shapiro, F. "Eye Movement Desensitization and Reprocessing (EMDR) and the Anxiety Disorders: Clinical and Research Implications of an Integrated Psychotherapy Treatment." *Journal of Anxiety Disorders* (1999) 13:35–67.

Shea, N. "Methodological Encounters with the Phenomenal Kind." *Philosophy and Phenomenological Research* (2012) 84:307–44.

Shea, N., and C. Heyes. "Metamemory as Evidence of Animal Consciousness: The Type That Does the Trick." *Biology & Philosophy* (2010) 25:95–110.

Shedler, J. "The Efficacy of Psychodynamic Psychotherapy." *The American Psychologist* (2010) 65:98–109.

Sheffield, F.D., and T.B. Roby. "Reward Value of a Non-Nutritive Sweet Taste." *Journal of Comparative Physiology* and Psychology (1950) 43:471–81.

Shekhar, A., et al. "The Amygdala, Panic Disorder, and Cardiovascular Responses." *Annals of the New York Academy of Sciences* (2003) 985:308–25.

Shema, R., et al. "Boundary Conditions for the Maintenance of Memory by PKMzeta in Neocortex." *Learning & Memory* (2009) 16:122–28.

Shema, R., T.C. Sacktor, and Y. Dudai. "Rapid Erasure of Long-Term Memory Associations in the Cortex by an Inhibitor of PKM Zeta." *Science* (2007) 317:951–53.

Shenhav, A., M.M. Botvinick, and J.D. Cohen. "The Expected Value of Control: An Integrative Theory of Anterior Cingulate Cortex Function." *Neuron* (2013) 79:217–40.

Shimamura, A.P. "Priming Effects of Amnesia: Evidence for a Dissociable Memory Function." *The Quarterly Journal of Experimental Psychology A, Human Experimental Psychology* (1986) 38:619–44.

Shimojo, S. "Postdiction: Its Implications on Visual Awareness, Hindsight, and Sense of Agency." *Frontiers in Psychology* (2014) 5:196.

Shin, L.M., and I. Liberzon. "The Neurocircuitry of Fear, Stress, and Anxiety Disorders." *Neuropsychopharmacology* (2010) 35:169–91.

Shin, L.M., S.L. Rauch, and R.K. Pitman. "Amygdala, Medial Prefrontal Cortex, and Hippocampal Function in PTSD." *Annals of the New York Academy of Sciences* (2006) 1071:67–79.

Shors, T.J. "Stressful Experience and Learning Across the Lifespan." *Annual Review of Psychology* (2006) 57:55–85.

Shugg, W. "The Cartesian Beast-Machine in English Literature (1663–1750)." *Journal of the History of Ideas* (1968) 29:279–92.

Shurick, A.A., et al. "Durable Effects of Cognitive Restructuring on Conditioned Fear." *Emotion* (2012) 12:1393–97.

Siegel, A., and H. Edinger. "Neural Control of Aggression and Rage Behavior." In: *Handbook of the Hypothalamus, Vol. 3, Behavioral Studies of the Hypothalamus,* eds. P.J. Morgane and J. Panksepp (New York: Marcel Dekker, 1981), 203–40.

Siegel, P., and R. Warren. "Less Is Still More: Maintenance of the Very Brief Exposure Effect 1 Year Later." *Emotion* (2013) 13:338–44.

Siegel, P., and J. Weinberger. "Less Is More: The Effects of Very Brief Versus Clearly Visible Exposure." *Emotion* (2012) 12:394–402.

Silva, A.J., et al. "CREB and Memory." *Annual Review of Neuroscience* (1998) 21:127–48.

Silvers, J.A., et al. "Bad and Worse: Neural Systems Underlying Reappraisal of High- and Low-Intensity Negative Emotions." *Social Cognitive and Affective Neuroscience* (2014) 10:172–79.

Silverstein, A. "Unlearning, Spontaneous Recovery, and the Partial-Reinforcement Effect in Paired-Associate Learning." *Journal of Experimental Psychology* (1967) 73:15–21.

Simon, H.A. "Motivational and Emotional Controls of Cognition." *Psychological Review* (1967) 74:29–39.

Simons, J.S., K.S. Graham, and J.R. Hodges. "Perceptual and Semantic Contributions to Episodic Memory: Evidence from Semantic Dementia and Alzheimer's Disease." *Journal of Memory and Language* (2002) 47:197–213.

Simpson, H.B. "The RDoC Project: A New Paradigm for Investigating the Pathophysiology of Anxiety." *Depression and Anxiety* (2012) 29:251–52.

Singer, P. *In Defense of Animals: The Second Wave* (New York: Wiley-Blackwell, 2005).

Singer, T. "The Neuronal Basis and Ontogeny of Empathy and Mind Reading: Review of Literature and Implications for Future Research." *Neuroscience and Biobehavioral Reviews* (2006) 30:855–63.

Singer, T. "The Neuronal Basis of Empathy and Fairness." *Novartis Foundation Symposium* (2007) 278:20–30; discussion 30–40, 89–96, 216–21.

Singer, T., et al. "Empathy for Pain Involves the Affective but Not Sensory Components of Pain." *Science* (2004) 303:1157–62.

Singer, W. "The Brain as a Self-Organizing System." *European Archives of Psychiatry and Neurological Sciences* (1986) 236:4–9.

Sink, K.S., M. Davis, and D.L. Walker. "CGRP Antagonist Infused into the Bed Nucleus of the Stria Terminalis Impairs the Acquisition and Expression of Context but Not Discretely Cued Fear." *Learning & Memory* (2013) 20:730–39.

Sink, K.S., et al. "Calcitonin Gene-Related Peptide in the Bed Nucleus of the Stria Terminalis Produces an Anxiety-Like Pattern of Behavior and Increases Neural Activation in Anxiety-Related Structures." *Journal of Neuroscience* (2011) 31:1802–10.

Skinner, B.F. *The Behavior of Organisms: An Experimental Analysis* (New York: Appleton-Century-Crofts, 1938).

Skinner, B.F. "Are Theories of Learning Necessary?" *Psychological Review* (1950) 57:193–216.

Skinner, B.F. *Science and Human Behavior* (New York: Free Press, 1953).

Skinner, B.F. *About Behaviorism* (New York: Knopf, 1974).

Skolnick, P. "Anxioselective Anxiolytics: On a Quest for the Holy Grail." *Trends in Pharmacological Sciences* (2012) 33:611–20.

Skorupski, P., and L. Chittka. "Animal Cognition: An Insect's Sense of Time?" *Current Biology* (2006) 16:R851–53.

Sluka, K.A., O.C. Winter, and J.A. Wemmie. "Acid-Sensing Ion Channels: A New Target for Pain and CNS Diseases." *Current Opinion in Drug Discovery & Development* (2009) 12: 693–704.

Smith, C.A., and P.C. Ellsworth. "Patterns of Cognitive Appraisal in Emotion." *Journal of Personality and Social Psychology* (1985) 56:339–53.

Smith, D. "It's Still the 'Age of Anxiety.' Or Is It?" *The New York Times* (New York: The New York Times Company, 2012).

Smith, J.B., and K.D. Alloway. "Functional Specificity of Claustrum Connections in the Rat: Interhemispheric Communication Between Specific Parts of Motor Cortex." *Journal of Neuroscience* (2010) 30:16832–44.

Smith, J.D. "The Study of Animal Metacognition." *Trends in Cognitive Sciences* (2009) 13: 389–96.

Smith, J.D., J.J. Couchman, and M.J. Beran. "The Highs and Lows of Theoretical Interpretation in Animal-Metacognition Research." *Philosophical Transactions of the Royal Society B: Biological Sciences* (2012) 367:1297–1309.

Smith, K.S., and A.M. Graybiel. "Investigating Habits: Strategies, Technologies and Models." *Frontiers in Behavioral Neuroscience* (2014) 8:39.

Smith, O.A., et al. "Functional Analysis of Hypothalamic Control of the Cardiovascular Responses Accompanying Emotional Behavior." *Federation Proceedings* (1980) 39:2487–94.

Solomon, R.L. "The Opponent-Process Theory of Acquired Motivation: The Costs of Pleasure and the Benefits of Pain." *The American Psychologist* (1980) 35:691–712.

Somerville, L.H., et al. "Interactions Between Transient and Sustained Neural Signals Support the Generation and Regulation of Anxious Emotion." *Cerebral Cortex* (2013) 23:49–60.

Somerville, L.H., P.J. Whalen, and W.M. Kelley. "Human Bed Nucleus of the Stria Terminalis Indexes Hypervigilant Threat Monitoring." *Biological Psychiatry* (2010) 68:416–24.

Soravia, L.M., et al. "Glucocorticoids Reduce Phobic Fear in Humans." *Proceedings of the National Academy of Sciences of the United States of America* (2006) 103:5585–90.

Sotres-Bayon, F., D.E. Bush, and J.E. LeDoux. "Emotional Perseveration: An Update on Prefrontal-Amygdala Interactions in Fear Extinction." *Learning & Memory* (2004) 11:525–35.

Sotres-Bayon, F., C.K. Cain, and J.E. LeDoux. "Brain Mechanisms of Fear Extinction: Historical Perspectives on the Contribution of Prefrontal Cortex." *Biological Psychiatry* (2006) 60:329–36.

Sotres-Bayon, F., and G.J. Quirk. "Prefrontal Control of Fear: More Than Just Extinction." *Current Opinion in Neurobiology* (2010) 20:231–35.

Southwick, S.M., L.L. Davids, D.E. Aikins, A. Rasmusson, J. Barron, and C.A. Morgan. "Neurobiological Alterations Associated with PTSD." (2007) New York: Guilford Publications.

Spannuth, B.M., et al. "Investigation of a Central Nucleus of the Amygdala/Dorsal Raphe Nucleus Serotonergic Circuit Implicated in Fear-Potentiated Startle." *Neuroscience* (2011) 179:104–19.

Sparks, P.D., and J.E. LeDoux. "The Septal Complex as Seen Through the Context of Fear." In: *The Behavioral Neuroscience of the Septal Region,* Numan, R., ed, (New York: Springer-Verlag, 2000), 234–69.

Spence, K.W. "Cognitive Versus Stimulus-Response Theories of Learning." *Psychological Review* (1950) 57:159–72.

Spiegel, D. "The Mind Prepared: Hypnosis in Surgery." *Journal of the National Cancer Institute* (2007) 99:1280–81.

Spielberger, C.D., ed. *Anxiety and Behavior* (New York: Academic Press, 1966).

Spyer, K.M., and A.V. Gourine. "Chemosensory Pathways in the Brainstem Controlling Cardiorespiratory Activity." *Philosophical Transactions of the Royal Society B: Biological Sciences* (2009) 364:2603–10.

Squire, L. *Memory and Brain* (New York: Oxford, 1987).

Squire, L.R. "Declarative and Nondeclarative Memory: Multiple Brain Systems Supporting Learning and Memory." *Journal of Cognitive Neuroscience* (1992) 4:232–43.

Squire, L.R., and N.J. Cohen. "Human Memory and Amnesia." In: *Neurobiology of Learning and Memory,* eds. G. Lynch, et al. (New York: Guilford, 1984).

Squire, L.R., and E.R. Kandel. *Memory: From Mind to Molecules* (New York: Scientific American Library, 1999).

Srinivasan, M.V. "Honey Bees as a Model for Vision, Perception, and Cognition." *Annual Review of Entomology* (2010) 55:267–84.

Staats, A.W., and G.H. Eifert. "The Paradigmatic Behaviorism Theory of Emotions: Basis for Unification." *Clinical Psychology Review* (1990) 10:539–66.

Stamatakis, A.M., et al. "Amygdala and Bed Nucleus of the Stria Terminalis Circuitry: Implications for Addiction-Related Behaviors." *Neuropharmacology* (2014) 76 Pt B:320–28.

Stamenov, M.I., ed. *Language Structure, Discourse and the Access to Consciousness* (Philadelphia: John Benjamins Publishing Co., 1997).

Stampfl, T.G., and D.J. Levis. "Essentials of Implosive Therapy: A Learning-Theory-Based Psychodynamic Behavioral Therapy." *Journal of Abnormal Psychology* (1967) 72:496–503.

Stanley, D.A., et al. "Implicit Race Attitudes Predict Trustworthiness Judgments and Economic Trust Decisions." *Proceedings of the National Academy of Sciences of the United States of America* (2011) 108:7710–15.

Stein, D.J. "Panic Disorder: The Psychobiology of External Threat and Introceptive Distress." *CNS Spectrums* (2008) 13:26–30.

Stein, D.J., E. Hollander, and B.O. Rothbaum, eds. *Textbook of Anxiety Disorders* (Arlington, VA: American Psychiatric Publishing, 2009).

Steinfurth, E.C., et al. "Young and Old Pavlovian Fear Memories Can Be Modified with Extinction Training during Reconsolidation in Humans." *Learning & Memory* (2014) 21:338–41.

Stellar, E. "The Physiology of Motivation." *Psychological Review* (1954) 61:5–22.

Stephens, J.C., et al. "Mapping the Human Genome: Current Status." *Science* (1990) 250:237–44.

Sternson, S.M. "Hypothalamic Survival Circuits: Blueprints for Purposive Behaviors." *Neuron* (2013) 77:810–24.

Sterzer, P., et al. "Neural Processing of Visual Information Under Interocular Suppression: A Critical Review." *Frontiers in Psychology* (2014) 5:453.

Stevens, C.F. "CREB and Memory Consolidation." *Neuron* (1994) 13:769–70.

Stevens, C.F. "Consciousness: Crick and the Claustrum." *Nature* (2005) 435:1040–41.

Stickgold, R., and M.P. Walker "Memory Consolidation and Reconsolidation: What Is the Role of Sleep?" *Trends in Neurosciences* (2005) 28:408–15.

Stober, J. "Trait Anxiety and Pessimistic Appraisal of Risk and Chance." *Personality and Individual Differences* (1997) 22:465–76.

Stocks, J.T. "Recovered Memory Therapy: a Dubious Practice Technique." *Social Work* (1998) 43:423–36.

Stoerig, P., and A. Cowey. "Blindsight." *Current Biology* (2007) 17:R822–24.

Stone, M.H. "The Brain in Overdrive: A New Look at Borderline and Related Disorders." *Current Psychiatry Reports* (2013) 15:399.

Stork, O., and H.C. Pape. "Fear Memory and the Amygdala: Insights from a Molecular Perspective." *Cell and Tissue Research* (2002) 310:271–77.

Stossel, S. *My Age of Anxiety: Fear, Hope, Dread, and the Search for Peace of Mind* (New York: Knopf, 2013).

Strachey, J. *Standard Edition of the Complete Works of Sigmund Freud* (London: The Hogarth Press and the Institute of Psychoanalysis, 1966–74).

Straube, T., et al. "Effects of Cognitive-Behavioral Therapy on Brain Activation in Specific Phobia." *NeuroImage* (2006) 29:125–35.

Straube, T., H.J. Mentzel, and W.H. Miltner. "Waiting for Spiders: Brain Activation during Anticipatory Anxiety in Spider Phobics." *NeuroImage* (2007) 37:1427–36.

Streeter, C.C., et al. "Effects of Yoga on the Autonomic Nervous System, Gamma-Aminobutyric-Acid, and Allostasis in Epilepsy, Depression, and Post-traumatic Stress Disorder." *Medical Hypotheses* (2012) 78:571–79.

Strenziok, M., et al. "Differential Contributions of Dorso-Ventral and Rostro-Caudal Prefrontal White Matter Tracts to Cognitive Control in Healthy Older Adults." *PLoS One* (2013) 8:E81410.

Striedter, G.F. *Principles of Brain Evolution* (Sunderland: Sinauer Associates, 2005).

Striepens, N., et al. "Prosocial Effects of Oxytocin and Clinical Evidence for Its Therapeutic Potential." *Frontiers in Neuroendocrinology* (2011) 32:426–50.

Stuss, D.T., and D.F. Benson. *The Frontal Lobes* (New York: Raven Press, 1986).

Suarez, S.D., and G.G. Gallup Jr. "An Ethological Analysis of Open-Field Behavior in Rats and Mice." *Learning and Motivation* (1981) 12:342–63.

Subitzky, E. "I Am a Conscious Essay." *Journal of Consciousness Studies* (2003) 10:64–66.

Sudakov, S.K., et al. "Estimation of the Level of Anxiety in Rats: Differences in Results of Open-Field Test, Elevated Plus-Maze Test, and Vogel's Conflict Test." *Bulletin of Experimental Biology and Medicine* (2013) 155:295–97.

Suddendorf, T., D.R. Addis, and M.C. Corballis. "Mental Time Travel and the Shaping of the Human Mind." *Philosophical Transactions of the Royal Society B: Biological Sciences* (2009) 364:1317–24.

Suddendorf, T., and J. Busby. "Mental Time Travel in Animals?" *Trends in Cognitive Sciences* (2003) 7:391–96.

Suddendorf, T., and D.L. Butler. "The Nature of Visual Self-Recognition." *Trends in Cognitive Sciences* (2013) 17:121–27.

Suddendorf, T., and M.C. Corballis. "The Evolution of Foresight: What Is Mental Time Travel, and Is It Unique to Humans?" *Behavioral and Brain Sciences* (2007) 30:299–313; discussion 313–51.

Suddendorf, T., and M.C. Corballis. "Behavioural Evidence for Mental Time Travel in Nonhuman Animals." *Behavioural Brain Research* (2010) 215:292–98.

Sugrue, L.P., G.S. Corrado, and W.T. Newsome. "Choosing the Greater of Two Goods: Neural Currencies for Valuation and Decision Making." *Nature Reviews Neuroscience* (2005) 6:363–75.

Sullivan, G.M., et al. "Lesions in the Bed Nucleus of the Stria Terminalis Disrupt Corticosterone and Freezing Responses Elicited by a Contextual but Not a Specific Cue-Conditioned Fear Stimulus." *Neuroscience* (2004) 128:7–14.

Sullivan, R.M., and W.G. Brake. "What the Rodent Prefrontal Cortex Can Teach Us About Attention-Deficit/Hyperactivity Disorder: The Critical Role of Early Developmental Events on Prefrontal Function." *Behavioural Brain Research* (2003) 146:43–55.

Sullivan, R.M., and P.J. Holman. "Transitions in Sensitive Period Attachment Learning in Infancy: The Role of Corticosterone." *Neuroscience and Biobehavioral Reviews* (2010) 34:835–44.

Sundberg, N. *Clinical Psychology: Evolving Theory, Practice, and Research* (Englewood Cliffs: Prentice Hall, 2001).

Sur, M., A. Angelucci, and J. Sharma. "Rewiring Cortex: The Role of Patterned Activity in Development and Plasticity of Neocortical Circuits." *Journal of Neurobiology* (1999) 41:33–43.

Sutton, M.A., et al. "Interaction Between Amount and Pattern of Training in the Induction of Intermediate- and Long-Term Memory for Sensitization in *Aplysia*." *Learning & Memory* (2002) 9:29–40.

Suzuki, A., et al. "Memory Reconsolidation and Extinction Have Distinct Temporal and Biochemical Signatures." *Journal of Neuroscience* (2004) 24:4787–95.

Suzuki, W.A., and D.G. Amaral. "Functional Neuroanatomy of the Medial Temporal Lobe Memory System." *Cortex; a Journal Devoted to the Study of the Nervous System and Behavior* (2004) 40:220–22.

Swanson, L.W. "The Hippocampus and the Concept of the Limbic System." In: *Neurobiology of the Hippocampus*, ed. W. Seifert (London: Academic Press, 1983), 3–19.

Szczepanowski, R., and L. Pessoa. "Fear Perception: Can Objective and Subjective Awareness Measures Be Dissociated?" *Journal of Vision* (2007) 7:10.

Szyf, M., P. McGowan, and M.J. Meaney. "The Social Environment and the Epigenome." *Environmental and Molecular Mutagenesis* (2008) 49:46–60.

Takahashi, L.K., et al. "The Smell of Danger: A Behavioral and Neural Analysis of Predator Odor-Induced Fear." *Neuroscience and Biobehavioral Reviews* (2005) 29:1157–67.

Takeuchi, Y., et al. "Direct Amygdaloid Projections to the Dorsal Motor Nucleus of the Vagus Nerve: A Light and Electron Microscopic Study in the Rat." *Brain Research* (1983) 280:143–47.

Talmi, D., et al. "Human Pavlovian-Instrumental Transfer." *Journal of Neuroscience* (2008) 28: 360–68.

Tamietto, M., and B. de Gelder. "Neural Bases of the Non-Conscious Perception of Emotional Signals." *Nature Reviews Neuroscience* (2010) 11:697–709.

Taubenfeld, S.M., J.S. Riceberg, A.S. New, and C.M. Alberini. "Preclinical Assessment for Selectively Disrupting a Traumatic Memory via Postretrieval Inhibition of Glucocorticoid Receptors." *Biological Psychiatry* (2009) 65:249–57.

Tauber, A.I. *Freud, the Reluctant Philosopher* (Princeton, NJ: Princeton University Press, 2010).

Taylor, A.H., and R.D. Gray. "Animal Cognition: Aesop's Fable Flies from Fiction to Fact." *Current Biology* (2009) 19:R731–32.

Taylor, S., J.S. Abramowitz, and D. McKay. "Non-Adherence and Non-Response in the Treatment of Anxiety Disorders." *Journal of Anxiety Disorders* (2012) 26:583–89.

Tedesco, V., et al. "Extinction, Applied After Retrieval of Auditory Fear Memory, Selectively Increases Zinc-Finger Protein 268 and Phosphorylated Ribosomal Protein S6 Expression in Prefrontal Cortex and Lateral Amygdala." *Neurobiology of Learning and Memory* (2014) 115: 78–85.

Terrace, H., and J. Metcalfe. *The Missing Link in Cognition: Origins of Self-Reflective Consciousness* (New York: Oxford University Press, 2004).

Terrazas, A., et al. "Self-Motion and the Hippocampal Spatial Metric." *Journal of Neuroscience* (2005) 25:8085–96.

Teuber, H.L. "The Riddle of Frontal Lobe Function in Man." In: *The Frontal Granular Cortex and Behavior*, eds. J.M. Warren and K. Akert (New York: McGraw-Hill, 1964), 410–77.

Teuber, H.L. "Unity and Diversity of Frontal Lobe Functions." *Acta Neurobiologiae Experimentalis* (1972) 32:615–56.

Thakral, P.P. "The Neural Substrates Associated with Inattentional Blindness." *Consciousness and Cognition* (2011) 20:1768–75.

Thompson, R.F., T.W. Berger, and J. Madden 4th. "Cellular Processes of Learning and Memory in the Mammalian CNS." *Annual Review of Neuroscience* (1983) 6:447–91.

Thomson, H. "Consciousness On-Off Switch Discovered Deep in Brain." *New Scientist* (2014) magazine issue 2976. Posted July 2, 2014, by Helen Thomson.

Thorndike, E.L. "Animal Intelligence: An Experimental Study of the Associative Processes in Animals." *Psychological Monographs* (1898) 2:109.

Thorndike, E.L. *The Psychology of Learning* (New York: Teachers College, 1913).

Thuault, S.J., et al. "Prefrontal Cortex, H.C.N1 Channels Enable Intrinsic Persistent Neural Firing and Executive Memory Function." *Journal of Neuroscience* (2013) 33:13583–99.

Tiihonen, J., et al. "Cerebral Benzodiazepine Receptor Binding and Distribution in Generalized Anxiety Disorder: A Fractal Analysis." *Molecular Psychiatry* (1997) 2(6):463–71. DOI: 10.1098/rstb.2014.0167.

Tinbergen, N. *The Study of Instinct* (New York: Oxford University Press, 1951).

Toelch, U., D.R. Bach, and R.J. Dolan. "The Neural Underpinnings of an Optimal Exploitation of Social Information Under Uncertainty." *Social Cognitive and Affective Neuroscience* (2013) 9:1746–53.

Tolkien, J.R.R. *The Lord of the Rings* (London: Allen & Unwin, 1955).

Tolman, E.C. *Purposive Behavior in Animals and Men* (New York: Century, 1932).

Tolman, E.C. "Psychology vs. Immediate Experience." *Philosophy of Science* (1935) 2:356–80.

Tomkins, S.S. *Affect, Imagery, Consciousness* (New York: Springer, 1962).

Tomkins, S.S. *Affect, Imagery, Consciousness* (New York: Springer, 1963).

Tononi, G. "Consciousness, Information Integration, and the Brain." *Progress in Brain Research* (2005) 150:109–26.

Tononi, G. "Integrated Information Theory of Consciousness: An Updated Account." *Archives Italiennes De Biologie* (2012) 150:293–329.

Tononi, G., and C. Koch. "Consciousness: Here, There and Everywhere?" *Philosophical Transactions of the Royal Society B: Biological Sciences* (2015) 370: 20140167. http://dx.doi.org/10.1098/rstb.2014.0167.

Tooby, J., and L. Cosmides. "The Evolutionary Psychology of the Emotions and Their Relationship to Internal Regulatory Variables." In: *Handbook of Emotions*, eds. M.D. Lewis, et al. (New York: Guilford Press, 2008), 114–37.

Tottenham, N. "The Importance of Early Experiences for Neuro-Affective Development." *Current Topics in Behavioral Neurosciences* (2014) 16:109–29.

Toumey, C. "Nanobots Today." *Nature Nanotechnology* (2013) 8:475–76.

Townsend, J., and L.L. Altshuler. "Emotion Processing and Regulation in Bipolar Disorder: A Review." *Bipolar Disorders* (2012) 14:326–39.

Tracy, J.L., and D. Randles. "Four Models of Basic Emotions: A Review of Ekman and Cordaro, Izard, Levenson, and Panksepp and Watt." *Emotion Review* (2011) 3:397–405.

Treit, D., and C. Pesold. "Septal Lesions Inhibit Fear Reactions in Two Animal Models of Anxiolytic Drug Action." *Physiological Behavior* (1990) 47:365–71.

Tronson, N.C., and J.R. Taylor. "Molecular Mechanisms of Memory Reconsolidation." *Nature Reviews Neuroscience* (2007) 8:262–75.

Tronson, N.C., and J.R. Taylor. "Addiction: A Drug-Induced Disorder of Memory Reconsolidation." *Current Opinion in Neurobiology* (2013) 23:573–80.

Tronson, N.C., et al. "Distinctive Roles for Amygdalar CREB in Reconsolidation and Extinction of Fear Memory." *Learning & Memory* (2012) 19:178–81.

Tryon, W.W., and D. McKay. "Memory Modification as an Outcome Variable in Anxiety Disorder Treatment." *Journal of Anxiety Disorders* (2009) 23:546–56.

Tsuchiya, N., and C. Koch. "The Relationship Between Consciousness and Attention." In: *The Neurology of Consciousness*, eds. S. Laureys and G. Tononi (New York: Elsevier, 2009), 63–77.

Tuescher, O., et al. "Differential Activity of Subgenual Cingulate and Brainstem in Panic Disorder and PTSD." *Journal of Anxiety Disorders* (2011) 25:251–57.

Tully, K., and V.Y. Bolshakov. "Emotional Enhancement of Memory: How Norepinephrine Enables Synaptic Plasticity." *Molecular Brain* (2010) 3:15.

Tully, T., et al. "Targeting the CREB Pathway for Memory Enhancers." *Nature Reviews Drug Discovery* (2003) 2:267–77.

Tulving, E. "Episodic and Semantic Memory." In: *Organization of Memory*, eds. E. Tulving and W. Donaldson (New York: Academic Press, 1972), 382–403.

Tulving, E. *Elements of Episodic Memory* (New York: Oxford University Press, 1983).

Tulving, E. "Ebbinghaus's Memory: What Did He Learn and Remember?" *Journal of Experimental Psychology: Learning, Memory and Cognition* (1985) 11:485–90.

Tulving, E. "Memory: Performance, Knowledge, and Experience." *European Journal of Cognitive Psychology* (1989) 1:3–26.

Tulving, E. "Episodic Memory and Common Sense: How Far Apart?" *Philosophical Transactions of the Royal Society B: Biological Sciences* (2001) 356:1505–15.

Tulving, E. "The Origin of Autonoesis in Episodic Memory." In: *The Nature of Remembering: Essays in Honor of Robert G. Crowder*, eds. H.L. Roediger et al. (Washington, DC: American Psychological Association, 2001), 17–34.

Tulving, E. "Chronestesia: Conscious Awareness of Subjective Time." In: *Principles of Frontal Lobe Functions*, eds. D.T. Stuss and R.C. Knight (New York: Oxford University Press, 2002), 311–25.

Tulving, E. "Episodic Memory: From Mind to Brain." *Annual Review of Psychology* (2002) 53: 1–25.

Tulving, E. "Episodic Memory and Autonoesis: Uniquely Human?" In: *The Missing Link in Cognition*, eds. H.S. Terrace and J. Metcalfe (New York: Oxford University Press, 2005), 4–56.

Tunney, R.J. "Sources of Confidence Judgments in Implicit Cognition." *Psychonomic Bulletin & Review* (2005) 12:367–73.

Tversky, A., and D. Kahneman. "Judgment Under Uncertainty: Heuristics and Biases." *Science* (1974) 185:1124–31.

Tye, K.M., et al. "Amygdala Circuitry Mediating Reversible and Bidirectional Control of Anxiety." *Nature* (2011) 471:358–62.

Uchino, E., and S. Watanabe. "Self-Recognition in Pigeons Revisited." *Journal of the Experimental Analysis of Behavior* (2014) 102:327–34.

Ungerleider, L.G., and M. Mishkin. "Two Cortical Visual Systems." In: *Analysis of Visual Behavior*, eds. D.J. Ingle et al. (Cambridge, MA: MIT Press, 1982), 549–86.

Urcelay, G.P., D.S. Wheeler, and R.R. Miller. "Spacing Extinction Trials Alleviates Renewal and Spontaneous Recovery." *Learning & Behavior* (2009) 37:60–73.

Urfy, M.Z., and J.I. Suarez. "Breathing and the Nervous System." *Handbook of Clinical Neurology* (2014) 119:241–50.

Urry, H.L., et al. "Amygdala and Ventromedial Prefrontal Cortex Are Inversely Coupled During Regulation of Negative Affect and Predict the Diurnal Pattern of Cortisol Secretion Among Older Adults." *Journal of Neuroscience* (2006) 26:4415–25.

Uvnas, B. "Central Cardiovascular Control." In: *Handbook of Physiology: Neurophysiology*, vol II, eds. J. Field et al. (Washington, DC: American Physiological Society, 1960), 1131–62.

Valenstein, E. "Stability and Plasticity of Motivational Systems." In: *The Neurosciences: Second Study Program*, ed. F.O. Schmitt (New York: Rockefeller University Press, 1970), 207–17.

Valenstein, E. *Blaming the Brain* (New York: Free Press, 1999).

Vallesi, A., et al. "When Time Shapes Behavior: fMRI Evidence of Brain Correlates of Temporal Monitoring." *Journal of Cognitive Neuroscience* (2009) 21:1116–26.

Van Bockstaele, E.J., J. Chan, and V.M. Pickel. "Input from Central Nucleus of the Amygdala Efferents to Pericoerulear Dendrites, Some of Which Contain Tyrosine Hydroxylase Immunoreactivity." *Journal of Neuroscience Research* (1996) 45:289–302.

van Boxtel, J.J., N. Tsuchiya, and C. Koch. "Consciousness and Attention: On Sufficiency and Necessity." *Frontiers in Psychology* (2010) 1:217.

Van den Bussche, E., K. Notebaert, and B. Reynvoet. "Masked Primes Can Be Genuinely Semantically Processed: A Picture Prime Study." *Experimental Psychology* (2009) 56:295–300.

Van den Bussche, E., W. Van den Noortgate, and B. Reynvoet. "Mechanisms of Masked Priming: A Meta-Analysis." *Psychological Bulletin* (2009) 135:452–77.

Van den Stock, J., et al. "Cortico-Subcortical Visual, Somatosensory, and Motor Activations for Perceiving Dynamic Whole-Body Emotional Expressions with and Without Striate Cortex (V1)." *Proceedings of the National Academy of Sciences of the United States of America* (2011) 108:16188–93.

Van der Heiden, C., and E. ten Broecke. "The When, Why, and How of Worry Exposure." *Cognitive and Behavioral Practice* (2009) 16:386–93.

van der Kolk, B. *The Body Keeps the Score: Brain, Mind and Body in the Healing of Trauma* (New York: Viking Adult, 2014).

van der Kolk, B.A. "The Body Keeps the Score: Memory and the Evolving Psychobiology of Posttraumatic Stress." *Harvard Review of Psychiatry* (1994) 1:253–65.

van der Kolk, B.A. "Clinical Implications of Neuroscience Research in PTSD." *Annals of the New York Academy of Sciences* (2006) 1071:277–93.

van der Kooy, D., et al. "The Organization of Projections from the Cortex, Amygdala, and Hypothalamus to the Nucleus of the Solitary Tract in Rat." *Journal of Comparative Neurology* (1984) 224:1–24.

van der Meer, M.A., and A.D. Redish. "Expectancies in Decision Making, Reinforcement Learning, and Ventral Striatum." *Frontiers in Neuroscience* (2010) 4:6.

van Gaal, S., and V.A. Lamme. "Unconscious High-Level Information Processing: Implication for Neurobiological Theories of Consciousness." *Neuroscientist* (2012) 18:287–301.

Van Hoesen, G.W., and D.N. Pandya. "Some Connections of the Entorhinal (Area 28) and Perirhinal (Area 35) Cortices of the Rhesus Monkey." *Brain Research* (1975) 95:1–24.

van Zessen, R., et al. "Activation of VTA GABA Neurons Disrupts Reward Consumption." *Neuron* (2012) 73:1184–94.

Vandekerckhove, M., and J. Panksepp. "The Flow of Anoetic to Noetic and Autonoetic Consciousness: A Vision of Unknowing (Anoetic) and Knowing (Noetic) Consciousness in the Remembrance of Things Past and Imagined Futures." *Consciousness and Cognition* (2009) 18:1018–28.

Vandekerckhove, M., and J. Panksepp. "A Neurocognitive Theory of Higher Mental Emergence: From Anoetic Affective Experiences to Noetic Knowledge and Autonoetic Awareness." *Neuroscience and Biobehavioral Reviews* (2011) 35:2017–25.

Vandenbroucke, A.R., et al. "Non-Attended Representations Are Perceptual Rather Than Unconscious in Nature." *PLoS One* (2012) 7:E50042.

VanElzakker, M.B., et al. "From Pavlov to PTSD: The Extinction of Conditioned Fear in Rodents, Humans, and Anxiety Disorders." *Neurobiology of Learning and Memory* (2014) 113:3–18.

Vanlancker-Sidtis, D. "When Only the Right Hemisphere Is Left: Studies in Language and Communication." *Brain and Language* (2004) 91:199–211.

Vargha-Khadem, F., D.G. Gadian, and M. Mishkin. "Dissociations in Cognitive Memory: The Syndrome of Developmental Amnesia." *Philosophical Transactions of the Royal Society B: Biological Sciences* (2001) 356:1435–40.

Varley, R., and M. Siegal. "Evidence for Cognition Without Grammar from Causal Reasoning and 'Theory of Mind' in an Agrammatic Aphasic Patient." *Current Biology* (2000) 10:723–26.

Vaughan, E., and A.E. Fisher. "Male Sexual Behavior Induced by Intracranial Electrical Stimulation." *Science* (1962) 137:758–60.

Veening, J.G., L.W. Swanson, and P.E. Sawchenko. "The Organization of Projections from the Central Nucleus of the Amygdala to Brainstem Sites Involved in Central Autonomic Regulation: A Combined Retrograde Transport-Immunohistochemical Study." *Brain Research* (1984) 303:337–57.

Velmans, M. *Understanding Consciousness* (Philadelphia: Routledge, 2000).

Vermeij, G.J. *Evolution and Escalation: An Ecological History of Life* (Princeton, NJ: Princeton University Press, 1987).

Vervliet, B., M.G. Craske, and D. Hermans. "Fear Extinction and Relapse: State of the Art." *Annual Review of Clinical Psychology* (2013) 9:215–48.

Vermetten, E., and J.D. Bremner. "Circuits and Systems in Stress. II. Applications to Neurobiology and Treatment in Posttraumatic Stress Disorder." *Depression and Anxiety* (2002) 16: 14–38.

Vianna, M.R., A.S. Coitinho, and I. Izquierdo. "Role of the Hippocampus and Amygdala in the Extinction of Fear-Motivated Learning." *Current Neurovascular Research* (2004) 1:55–60.

Vickers, K., and R.J. McNally. "Respiratory Symptoms and Panic in the National Comorbidity Survey: A Test of Klein's Suffocation False Alarm Theory." *Behaviour Research and Therapy* (2005) 43:1011–18.

Vidal-Gonzalez, I., et al. "Microstimulation Reveals Opposing Influences of Prelimbic and Infralimbic Cortex on the Expression of Conditioned Fear." *Learning & Memory* (2006) 13: 728–33.

Vinod, K.Y., and B.L. Hungund. "Endocannabinoid Lipids and Mediated System: Implications for Alcoholism and Neuropsychiatric Disorders." *Life Sciences* (2005) 77:1569–83.

Vogt, B.A., D.M. Finch, and C.R. Olson. "Functional Heterogeneity in Cingulate Cortex: The Anterior Executive and Posterior Evaluative Regions." *Cerebral Cortex* (1992) 2:435–43.

Volkow, N.D., et al. "Overlapping Neuronal Circuits in Addiction and Obesity: Evidence of Systems Pathology." *Philosophical Transactions of the Royal Society B: Biological Sciences* (2008) 363:3191–200.

Volpe, B.T., J.E. LeDoux, and M.S. Gazzaniga. "Information Processing of Visual Stimuli in an 'Extinguished' Field." *Nature* (1979) 282:722–24.

Volz, K.G., R.I. Schubotz, and D.Y. Von Cramon. "Predicting Events of Varying Probability: Uncertainty Investigated by fMRI." *NeuroImage* (2003) 19:271–80.

von Kraus, L.M., T.C. Sacktor, and J.T. Francis. "Erasing Sensorimotor Memories via PKMzeta Inhibition." *PLoS One* (2010) 5:E11125.

Voon, V., N.A. Howell, and P. Krack. "Psychiatric Considerations in Deep Brain Stimulation for Parkinson's Disease." *Handbook of Clinical Neurology* (2013) 116:147–54.

Vredeveldt, A., G.J. Hitch, and A.D. Baddeley. "Eye Closure Helps Memory by Reducing Cognitive Load and Enhancing Visualisation." *Memory & Cognition* (2011) 39:1253–63.

Vuilleumier, P. "Perceived Gaze Direction in Faces and Spatial Attention: A Study in Patients with Parietal Damage and Unilateral Neglect." *Neuropsychologia* (2002) 40:1013–26.

Vuilleumier, P. "How Brains Beware: Neural Mechanisms of Emotional Attention." *Trends in Cognitive Sciences* (2005) 9:585–94.

Vuilleumier, P., et al. "Neural Response to Emotional Faces with and Without Awareness: Event-Related fMRI in a Parietal Patient with Visual Extinction and Spatial Neglect." *Neuropsychologia* (2002) 40:2156–66.

Vuilleumier, P., and J. Driver. "Modulation of Visual Processing by Attention and Emotion: Windows on Causal Interactions Between Human Brain Regions." *Philosophical Transactions of the Royal Society B: Biological Sciences* (2007) 362:837–55.

Vuilleumier, P., and S. Schwartz. "Beware and Be Aware: Capture of Spatial Attention by Fear-Related Stimuli in Neglect." *Neuroreport* (2001) 12:1119–22.

Vuilleumier, P., et al. "Abnormal Attentional Modulation of Retinotopic Cortex in Parietal Patients with Spatial Neglect." *Current Biology* (2008) 18:1525–29.

Waddell, J., R.W. Morris, and M.E. Bouton. "Effects of Bed Nucleus of the Stria Terminalis Lesions on Conditioned Anxiety: Aversive Conditioning with Long-Duration Conditional Stimuli and Reinstatement of Extinguished Fear." *Behavioral Neuroscience* (2006) 120: 324–36.

Wakefield, J.C. "Meaning and Melancholia: Why DSM Cannot (Entirely) Ignore the Patient's Intentional System." In: *Making Diagnosis Meaningful: Enhancing Evaluation and Treatment of Psychological Disorders*, ed. J.W. Barron (Washington, DC: American Psychological Association Press, 1998), 29–72.

Walasek, G., M. Wesierska, and K. Zielinski. "Conditioning of Fear and Conditioning of Safety in Rats." *Acta Neurobiologiae Experimentalis* (1995) 55:121–32.

Walker, D.L., and M. Davis. "Double Dissociation Between the Involvement of the Bed Nucleus of the Stria Terminalis and the Central Nucleus of the Amygdala in Startle Increases Produced by Conditioned Versus Unconditioned Fear." *The Journal of Neuroscience* (1997) 17:9375–83.

Walker, D.L., and M. Davis. "Involvement of NMDA Receptors Within the Amygdala in Short- Versus Long-Term Memory for Fear Conditioning as Assessed with Fear-Potentiated Startle." *Behavioral Neuroscience* (2000) 114:1019–33.

Walker, D.L., and M. Davis. "Light-Enhanced Startle: Further Pharmacological and Behavioral Characterization." *Psychopharmacology (Berl)* (2002) 159:304–10.

Walker, D.L., and M. Davis. "The Role of Amygdala Glutamate Receptors in Fear Learning, Fear-Potentiated Startle, and Extinction." *Pharmacology, Biochemistry, and Behavior* (2002) 71:379–92.

Walker, D.L., and M. Davis. "Role of the Extended Amygdala in Short-Duration Versus Sustained Fear: A Tribute to Dr. Lennart Heimer." *Brain Structure & Function* (2008) 213: 29–42.

Walker, D.L., et al. "Facilitation of Conditioned Fear Extinction by Systemic Administration or Intra-Amygdala Infusions of D-cycloserine as Assessed with Fear-Potentiated Startle in Rats." *Journal of Neuroscience* (2002) 22:2343–51.

Wallace, D.M., D.J. Magnuson, and T.S. Gray. "The Amygdalo-Brainstem Pathway: Selective Innervation of Dopaminergic, Noradrenergic and Adrenergic Cells in the Rat." *Neuroscience Letters* (1989) 97:252–58.

Wallis, J.D. "Cross-Species Studies of Orbitofrontal Cortex and Value-Based Decision-Making." *Nature Neuroscience* (2012) 15:13–19.

Walters, E.T., T.J. Carew, and E.R. Kandel. "Classical Conditioning in *Aplysia californica*." *Proceedings of the National Academy of Sciences of the United States of America* (1979) 76: 6675–79.

Wang, L., et al. "Hierarchical Chemosensory Regulation of Male-Male Social Interactions in *Drosophila*." *Nature Neuroscience* (2011) 14:757–62.

Wang, S.H., L. de Oliveira Alvares, and K. Nader. "Cellular and Systems Mechanisms of Memory Strength as a Constraint on Auditory Fear Reconsolidation." *Nature Neuroscience* (2009) 12:905–12.

Ward, R., S. Danziger, and S. Bamford. "Response to Visual Threat Following Damage to the Pulvinar." *Current Biology* (2005) 15:571–73.

Wasserman, E.A. "The Science of Animal Cognition: Past, Present, and Future." *Journal of Experimental Psychology: Animal Behavior Processes* (1997) 23:123–35.

Wassum, K.M., et al. "Differential Dependence of Pavlovian Incentive Motivation and Instrumental Incentive Learning Processes on Dopamine Signaling." *Learning & Memory* (2011) 18:475–83.

Waterhouse, B.D., and D.J. Woodward. "Interaction of Norepinephrine with Cerebrocortical Activity Evoked by Stimulation of Somatosensory Afferent Pathways in the Rat." *Experimental Neurology* (1980) 67:11–34.

Waters, A.M., J. Henry, and D.L. Neumann. "Aversive Pavlovian Conditioning in Childhood Anxiety Disorders: Impaired Response Inhibition and Resistance to Extinction." *Journal of Abnormal Psychology* (2009) 118:311–21.

Watson, J. "Behaviorism." In: *The Behavior of Organisms*, ed. B.F. Skinner (New York: Appleton-Century-Crofts, 1938).

Watson, J.B. "Psychology as the Behaviorist Views It." *Psychological Review* (1913) 20:158–77.

Watson, J.B. *Psychology from the Standpoint of a Behaviorist* (Philadelphia: Lippincott, 1919).

Watson, J.B. *Behaviorism* (New York: W.W. Norton, 1925).

Watson, J.D. "The Human Genome Project: Past, Present, and Future." *Science* (1990) 248:44–49.

Watt, D.F. "Panksepp's Common Sense View of Affective Neuroscience Is Not the Common-sense View in Large Areas of Neuroscience." *Consciousness and Cognition* (2005) 14:81–88.

Webb, B. "Cognition in Insects." *Philosophical Transactions of the Royal Society B: Biological Sciences* (2012) 367:2715–22.

Wegner, D. *The Illusion of Conscious Will* (Cambridge, MA: MIT Press, 2002).

Wegner, D.M. "The Mind's Best Trick: How We Experience Conscious Will." *Trends in Cognitive Sciences* (2003) 7:65–69.

Weinberger, N.M. "Effects of Conditioned Arousal on the Auditory System." In: *The Neural Basis of Behavior* (New York: Spectrum Publications, 1982), 63–91.

Weinberger, N.M. "Retuning the Brain by Fear Conditioning." In: *The Cognitive Neurosciences*, ed. M.S. Gazzaniga (Cambridge, MA: MIT Press, 1995), 1071–90.

Weinberger, N.M. "The Nucleus Basalis and Memory Codes: Auditory Cortical Plasticity and the Induction of Specific, Associative Behavioral Memory." *Neurobiology of Learning and Memory* (2003) 80:268–84.

Weinberger, N.M. "Auditory Associative Memory and Representational Plasticity in the Primary Auditory Cortex." *Hearing Research* (2007) 229:54–68.

Weiskrantz, L. "Behavioral Changes Associated with Ablation of the Amygdaloid Complex in Monkeys." *Journal of Comparative and Physiological Psychology* (1956) 49:381–91.

Weiskrantz, L. "Trying to Bridge Some Neuropsychological Gaps Between Monkey and Man." *British Journal of Psychology* (1977) 68:431–45.

Weiskrantz, L. "The Problem of Animal Consciousness in Relation to Neuropsychology." *Behavioural Brain Research* (1995) 71:171–75.

Weiskrantz, L. *Consciousness Lost and Found: A Neuropsychological Exploration* (New York: Oxford University Press, 1997).

Weisskopf, M.G., E.P. Bauer and J.E. LeDoux. "L-Type Voltage-Gated Calcium Channels Mediate NMDA-Independent Associative Long-Term Potentiation at Thalamic Input Synapses to the Amygdala." *Journal of Neuroscience* (1999) 19:10512–19.

Weisskopf, M.G., and J.E. LeDoux. "Distinct Populations of NMDA Receptors at Subcortical and Cortical Inputs to Principal Cells of the Lateral Amygdala." *Journal of Neurophysiology* (1999) 81:930–34.

Wemmie, J.A. "Neurobiology of Panic and pH Chemosensation in the Brain." *Dialogues in Clinical Neuroscience* (2011) 13:475–83.

Wemmie, J.A., M.P. Price, and M.J. Welsh. "Acid-Sensing Ion Channels: Advances, Questions and Therapeutic Opportunities." *Trends in Neurosciences* (2006) 29:578–86.

Wemmie, J.A., R.J. Taugher, and C.J. Kreple. "Acid-Sensing Ion Channels in Pain and Disease." *Nature Reviews Neuroscience* (2013) 14:461–71.

Wendler, E., et al. "The Roles of the Nucleus Accumbens Core, Dorsomedial Striatum, and Dorsolateral Striatum in Learning: Performance and Extinction of Pavlovian Fear-Conditioned Responses and Instrumental Avoidance Responses." *Neurobiology of Learning and Memory* (2014) 109:27–36.

Wenger, M.A., F.N. Jones, and M.H. Jones. *Physiological Psychology* (New York: Holt Rinehart Winston, 1956).

Whalen, P.J. "Fear, Vigilance, and Ambiguity: Initial Neuroimaging Studies of the Human Amygdala." *Current Directions in Psychological Science* (1998) 7:177–88.

Whalen, P.J., and E.A. Phelps. *The Human Amygdala* (New York: Guilford Press, 2009).

Whalen, P.J., et al. "Masked Presentations of Emotional Facial Expressions Modulate Amygdala Activity Without Explicit Knowledge." *Journal of Neuroscience* (1998) 18:411–18.

Wheeler, M.A., D.T. Stuss, and E. Tulving. "Toward a Theory of Episodic Memory: The Frontal Lobes and Autonoetic Consciousness." *Psychological Bulletin* (1997) 121:331–54.

Whissell, C.M. "The Role of the Face in Human Emotion: First System or One of Many?" *Perceptual and Motor Skills* (1985) 61:3–12.

Whitfield, C.L. "The 'False Memory' Defense Using Disinformation and Junk Science in and out of Court." *Journal of Child Sexual Abuse* (2000) 9:53–78.

Whiting, S.E., et al. "The Role of Intolerance of Uncertainty in Social Anxiety Subtypes." *Journal of Clinical Psychology* (2014) 70:260–72.

Whitmer, A.J., and I.H. Gotlib. "An Attentional Scope Model of Rumination." *Psychological Bulletin* (2013) 139:1036–61.

Whittle, N., et al. "Deep Brain Stimulation, Histone Deacetylase Inhibitors and Glutamatergic Drugs Rescue Resistance to Fear Extinction in a Genetic Mouse Model." *Neuropharmacology* (2013) 64:414–23.

Whorf, B.L. *Language, Thought, and Reality* (Cambridge, MA: Technology Press of MIT, 1956).

Wickens, J.R., et al. "Dopaminergic Mechanisms in Actions and Habits." *Journal of Neuroscience* (2007) 27:8181–83.

Wiers, R.W., and A.W. Stacy, eds. *Handbook of Implicit Cognition and Addiction* (Thousand Oaks, CA: Sage, 2006).

Wierzbicka, A. "Emotion, Language, and Cultural Scripts." In: *Emotion and Culture: Empirical Studies of Mutual Influence,* eds. S. Kitayama and H.R. Markus (Washington, DC: American Psychological Association, 1994), 133–96.

Wiggs, C.L., and A. Martin. "Properties and Mechanisms of Perceptual Priming." *Current Opinion in Neurobiology* (1998) 8:227–33.

Wilensky, A.E., et al. "Rethinking the Fear Circuit: The Central Nucleus of the Amygdala Is Required for the Acquisition, Consolidation, and Expression of Pavlovian Fear Conditioning." *Journal of Neuroscience* (2006) 26:12387–96.

Wilensky, A.E., G.E. Schafe, and J.E. LeDoux. "Functional Inactivation of the Amygdala Before but Not After Auditory Fear Conditioning Prevents Memory Formation." *Journal of Neuroscience* (1999) 19:RC48.

Wilensky, A.E., G.E. Schafe, and J.E. LeDoux. "The Amygdala Modulates Memory Consolidation of Fear-Motivated Inhibitory Avoidance Learning but Not Classical Fear Conditioning." *Journal of Neuroscience* (2000) 20:7059–66.

Wilhelm, F.H., and W.T. Roth. "The Somatic Symptom Paradox in DSM-IV Anxiety Disorders: Suggestions for a Clinical Focus in Psychophysiology." *Biological Psychology* (2001) 57: 105–40.

Williams, B.A. "Two-Factor Theory Has Strong Empirical Evidence of Validity." *Journal of the Experimental Analysis of Behavior* (2001) 75:362–65; discussion 367–78.

Williams, L.M., et al. "Mode of Functional Connectivity in Amygdala Pathways Dissociates Level of Awareness for Signals of Fear." *Journal of Neuroscience* (2006) 26:9264–71.

Willshaw, D.J., and J.T. Buckingham. "An Assessment of Marr's Theory of the Hippocampus as a Temporary Memory Store." *Philosophical Transactions of the Royal Society B: Biological Sciences* (1990) 329:205–15.

Wilson, M.A., and B.L. McNaughton. "Reactivation of Hippocampal Ensemble Memories during Sleep." *Science* (1994) 265:676–79.

Wilson, T.D. *Strangers to Ourselves: Self-Insight and the Adaptive Unconscious* (Cambridge, MA: Harvard University Press, 2002).

Wilson, T.D., and E.W. Dunn. "Self-Knowledge: Its Limits, Value, and Potential for Improvement." *Annual Review of Psychology* (2004) 55:493–518.

Wilson, T.D., et al. "Introspecting About Reasons Can Reduce Post-Choice Satisfaction." *Personality and Social Psychology Bulletin* (1993) 19:331–39.

Wilson-Mendenhall, C.D., L.F. Barrett, and L.W. Barsalou. "Situating Emotional Experience." *Frontiers in Human Neuroscience* (2013) 7:764.

Wilson-Mendenhall, C.D., et al. "Grounding Emotion in Situated Conceptualization." *Neuropsychologia* (2011) 49:1105–27.

Winkielman, P., and K.C. Berridge. "Unconscious Emotion." *Current Directions in Psychological Science* (2004) 13:120–23.

Winkielman, P., K.C. Berridge, and J.L. Wilbarger. "Emotion, Behavior, and Conscious Experience: Once More Without Feeling." In: *Emotion and Consciousness,* eds. L.F. Barrett et al. (New York: Guilford Press, 2005), 335–62.

Winkielman, P., K.C. Berridge, and J.L. Wilbarger. "Unconscious Affective Reactions to Masked Happy Versus Angry Faces Influence Consumption Behavior and Judgments of Value." *Personality and Social Psychology Bulletin* (2005) 31:121–35.

Wise, R.A. "Plasticity of Hypothalamic Motivational Systems." *Science* (1969) 165:929–30.

Wise, S.P. "Forward Frontal Fields: Phylogeny and Fundamental Function." *Trends in Neurosciences* (2008) 31:599–608.

Witten, I.B., et al. "Recombinase-Driver Rat Lines: Tools, Techniques, and Optogenetic Application to Dopamine-Mediated Reinforcement." *Neuron* (2011) 72:721–33.

Wittgenstein, L. *Philosophical Investigations* (Oxford, UK: Basil Blackwell, 1958).

Wolff, S.B., et al. "Amygdala Interneuron Subtypes Control Fear Learning Through Disinhibition." *Nature* (2014) 509:453–58.

Wolpe, J. *Psychotherapy by Reciprocal Inhibition* (Stanford, CA: Stanford University Press, 1958).

Wolpe, J. *The Practice of Behavior Therapy* (New York: Pergamon Press, 1969).

Wood, N.E., et al. "Pharmacological Blockade of Memory Reconsolidation in Posttraumatic Stress Disorder: Three Negative Psychophysiological Studies." *Psychiatry Research* (2015) 225:31–39.

Wood, S., et al. "Psychostimulants and Cognition: A Continuum of Behavioral and Cognitive Activation." *Pharmacological Reviews* (2014) 66:193–221.

Woodward, D.J., et al. "Modulatory Actions of Norepinephrine on Neural Circuits." *Advances in Experimental Medicine and Biology* (1991) 287:193–208.

Woody, C.D., ed. *Conditioning: Representation of Involved Neural Functions* (New York: Plenum Press, 1982).

Woody, S., and S. Rachman. "Generalized Anxiety Disorder (GAD) as an Unsuccessful Search for Safety." *Clinical Psychological Review* (1994) 14:743–53.

Wundt, W. *Principles of Physiological Psychology* (Leipzig: Engelmann, 1874).

Wynne, C.D. "The Perils of Anthropomorphism." *Nature* (2004) 428:606.

Xue, Y.X., et al. "A Memory Retrieval-Extinction Procedure to Prevent Drug Craving and Relapse." *Science* (2012) 336:241–45.

Yancey, S.W., and E.A. Phelps. "Functional Neuroimaging and Episodic Memory: A Perspective." *Journal of Clinical and Experimental Neuropsychology* (2001) 23:32–48.

Yang, E., et al. "On the Use of Continuous Flash Suppression for the Study of Visual Processing Outside of Awareness." *Frontiers in Psychology* (2014) 5:724.

Yang, F.C., and K.C. Liang. "Interactions of the Dorsal Hippocampus, Medial Prefrontal Cortex and Nucleus Accumbens in Formation of Fear Memory: Difference in Inhibitory Avoidance Learning and Contextual Fear Conditioning." *Neurobiology of Learning and Memory* (2013) 112:186–94.

Yang, Y.L., P.K. Chao, and K.T. Lu. "Systemic and Intra-Amygdala Administration of Glucocorticoid Agonist and Antagonist Modulate Extinction of Conditioned Fear." *Neuropsychopharmacology* (2006) 31:912–24.

Yates, A.J. *Behavior Therapy* (New York: Wiley, 1970).

Yehuda, R., and J. LeDoux. "Response Variation Following Trauma: A Translational Neuroscience Approach to Understanding PTSD." *Neuron* (2007) 56: 19–32.

Yerkes, R.M., and J.D. Dobson. "The Relation of Strength of Stimulus to Rapidity of Habit-Formation." *Journal of Comparative Neurology and Psychology* (1908) 18:458–82.

Yin, H., R.C. Barnet, and R.R. Miller. "Trial Spacing and Trial Distribution Effects in Pavlovian Conditioning: Contributions of a Comparator Mechanism." *Journal of Experimental Psychology: Animal Behavior Processes* (1994) 20:123–34.

Yin, J.C., and T. Tully. "CREB and the Formation of Long-Term Memory." *Current Opinion in Neurobiology* (1996) 6:264–68.

Yook, K., et al. "Intolerance of Uncertainty, Worry, and Rumination in Major Depressive Disorder and Generalized Anxiety Disorder." *Journal of Anxiety Disorders* (2010) 24:623–28.

Yook, K., et al. "Usefulness of Mindfulness-Based Cognitive Therapy for Treating Insomnia in Patients with Anxiety Disorders: A Pilot Study." *The Journal of Nervous and Mental Disease* (2008) 196:501–3.

Yoshida, W., et al. "Uncertainty Increases Pain: Evidence for a Novel Mechanism of Pain Modulation Involving the Periaqueductal Gray." *Journal of Neuroscience* (2013) 33:5638–46.

Young, L.J., Z. Wang, and T.R. Insel. "Neuroendocrine Bases of Monogamy." *Trends in Neurosciences* (1998) 21:71–75.

Zald, D.H. "The Human Amygdala and the Emotional Evaluation of Sensory Stimuli." *Brain Research. Brain Research Reviews* (Amsterdam) (2003) 41:88–123.

Zalla, T., and M. Sperduti. "The Amygdala and the Relevance Detection Theory of Autism: An Evolutionary Perspective." *Frontiers in Human Neuroscience* (2013) 7:894.

Zanchetti, A., et al. "Emotion and the Cardiovascular System in the Cat." *Ciba Foundation Symposium* (1972) 8:201–19.

Zanette, L.Y., et al. "Perceived Predation Risk Reduces the Number of Offspring Songbirds Produce Per Year." *Science* (2011) 334:1398–1401.

Zeidan, F., et al. "Neural Correlates of Mindfulness Meditation-Related Anxiety Relief." *Social Cognitive and Affective Neuroscience* (2014) 9:751–59.

Zeidner, M., and G. Matthews. *Anxiety 101* (New York: Springer, 2011).

Zeki, S., and O. Goodenough. "Law and the Brain: Introduction." *Philosophical Transactions of the Royal Society B: Biological Sciences* (2004) 359:1661–65.

Zeman, A. "The Problem of Unreportable Awareness." *Progress in Brain Research* (2009) 177:1–9.

Zhu, Y., et al. "Neural Basis of Cultural Influence on Self-Representation." *NeuroImage* (2007) 34:1310–16.

Zinbarg, R.E. "Concordance and Synchrony in Measures of Anxiety and Panic Reconsidered: A Hierarchical Model of Anxiety and Panic." *Behavior Therapy* (1998) 29:301–23.

Zoladz, P.R., and D.M. Diamond. "Linear and Non-Linear Dose-Response Functions Reveal a Hormetic Relationship Between Stress and Learning." *Dose-Response: A Publication of the International Hormesis Society* (2009) 7:132–48.

CREDITS

Figure 1.1: The Anguish of Laocoön and His Sons.
Reprinted with permission from © Marie-Lan Nguyen / Wikimedia Commons.

Figure 1.4: Some Variations on the Themes of Fear and Anxiety.
From Makari (2012). Reprinted with permission from Henning Wagenbreth.

Figure 3.3: Endocrine Support of Defense: Sympathoadrenal and Pituitary-Adrenal Systems.
From Rodrigues et al (2009), modified with permission from the *Annual Review of Neuroscience*, volume 32, © 2009 by Annual Reviews, http://www.annualreviews.org.

Figure 3.4: Fanselow's Predatory Imminence Theory.
From Fanselow and Lester (1988). Used with permission from Michael Fanselow.

Figure 4.1: The Brain in Brief.
Parts e. and i. from Bownds (1999). Used with permission of Deric Bownds.

Figure 4.2: How Cannon and Bard Produced Sham Rage.
From Purves et al: *Neuroscience*, Second Edition, Figure 29.1, modified with permission of Sinauer Associates.

Figure 4.3: Hypothalamically Elicited Rage and the Amygdala-Hypothalamic-PAG Rage Pathway (left).
From Flynn, © 1967 Rockefeller University Press and Russell Sage Foundation. In *Neurophysiology and Emotion*, pp. 40–59. Used with permission from Rockefeller University Press.

Figure 4.4: The Arousal System Then and Now.
(Top) Modified from Starzl, Taylor and Magoun (1951), *Journal of Neurophysiology*, 14, pp. 479-96). Used with permission. (Bottom) Modified with permission of American Academy of Sleep Medicine, from *Sleep*, España and Scammell, 34, 7, © 2011; permission conveyed through Copyright Clearance Center, Inc.

Figure 4.7: Hebbian Mechanisms Underlie Threat Conditioning.
(Bottom) Reprinted from *Cell*, 147/3, Johansen, Cain, Ostroff, LeDoux, "Molecular Mechanisms of Fear Learning and Memory," 509–24, © 2011, with permission from Elsevier.

Figure 4.8: Optogenetic Demonstration of Hebbian Learning in the Lateral Amygdala.
(Top) Buchen (2010), adapted by permission from Macmillan Publishers Ltd. *Nature News,* volume 465, pp. 26–28, © 2010.

Figure 5.1: Emotional Expressions: Duchenne and Messerschmidt.
Guillaume-Benjamin DUCHENNE de BOULOGNE Plate 7 in "Mécanisme de la physiognomie humaine (The mechanisms of human facial expression)." Paris: Chez Veuve Jules Renouard 1862 [Octavo edition], National Gallery of Australia, Canberra, purchased 2005. Lithograph by Matthias Rudolph Toma (1792–1869) depicting Franz Messerschmidt's "Character Heads" (1839). The Museum of Fine Arts, Budapest.

Figure 5.3: Ekman's Basic Emotions Expressed in the Face.
Used with permission of Paul Ekman (© 2011 by Paul Ekman, www.paulekman.com).

Figure 5.4: To Tell the Truth (or Not).
Used with permission of CBS News.

Figure 7.5: Summary of Explicit and Implicit Memory Circuits.
From Squire and Zola-Morgan (1991) *Science,* 253, pp 1380–86. Reprinted with permission from AAAS.

Figure 8.1: Innate (Prepared) and Conditioned Threat Stimuli in Human Studies.
Reprinted from Ewbank M.P., E. Fox, and A.J. Calder. (2010), *Frontiers in Human Neuroscience,* 7: July 2010. Copyright © 2010 Ewbank, Fox and Calder.

Figure 10.2: If Freud Did Exposure.
© 2013 Erik Rogers, EXPOSURE THERAPY, pen/pencil on paper, 8"x10."

Figure 11.3: Amygdala and Prefrontal Circuits Underlying the Conditioning and Extinction of Defense Responses.
Adapted from Lee et al (2013), http://creativecommons.org/licenses/by/3.0/.

INDEX

NOTE: *Italic page references* indicate figures and tables.